H. E. A. Schenk et al. (Eds.)

Eukaryotism and Symbiosis

Springer
Berlin
Heidelberg
New York
Barcelona
Budapest
Hong Kong
London
Milan
Paris
Santa Clara
Singapore
Tokyo

H. E. A. Schenk · R. G. Herrmann · K. W. Jeon
N. E. Müller · W. Schwemmler (Eds.)

Eukaryotism and Symbiosis

Intertaxonic Combination versus Symbiotic Adaptation

With 145 Figures

Springer

Prof. Dr. Hainfried E. A. Schenk
University of Tübingen
Botanical Institute
Auf der Morgenstelle
72076 Tübingen, Germany

Dipl. Biochem. Norbert E. Müller
University of Tübingen
Botanical Institute
Auf der Morgenstelle
72076 Tübingen, Germany

Prof. Dr. Reinhold G. Herrmann
University of Munich
Botanical Institute
Menzinger Str. 67
80638 München, Germany

Prof. Dr. Werner Schwemmler
Free University Berlin
Institute of Plant Physiology
Krahmerstr. 6
12207 Berlin, Germany

Prof. Dr. Kwang W. Jeon
University of Tennessee
at Knoxville
Dept. of Biochemistry
Knoxville, Tennessee 37996
USA

In collaboration with the International Society of Endocytobiology

Endocytobiology VI, updated edition of the Proceedings of the Sixth International Colloquium on Endocytobiology and Symbiosis, Tübingen, September 6–10, 1995

Sponsors:
Landesgirokasse, Stiftung Natur und Umwelt, Stuttgart
Prof. Dr. Reinhold G. Herrmann, München
Prof. Dr. Klaus Wolf, Aachen
Vereinigung der Freunde der Universität Tübingen (Universitätsbund e.V.)

ISBN-13: 978-3-642-64598-3 e-ISBN-13: 978-3-642-60885-8
DOI: 10.1007/978-3-642-60885-8

Die Deutsche Bibliothek – CIP-Einheitsaufnahme
Eukaryotism and symbiosis: intertaxonic combination versus symbiotic adaption; [proceedings of the Sixth International Colloquium on Endocytobiology and Symbiosis, Tübingen September 6–10, 1995] / H. E. A. Schenk ... [In collab. with the International Society of Endocytobiology]. – Berlin; Heidelberg; New York; Barcelona; Budapest; Hong Kong; London; Milan; Paris; Santa Clara; Singapore; Tokyo: Springer, 1997

Softcover reprint of the hardcover 1st edition 1997

Cover Design: Design & Production GmbH, Heidelberg

SPIN 10567054 31/3137-5 4 3 2 1 0 – Printed on acid-free paper

Preface

The rapidly growing interest in the fascinating field of Endocytobiology, the science of both organismic interactions leading to symbiotic adaptations at the cellular level and the integrative potential of cells (intertaxonic combination) originating in new taxonic entities, induced the establishment of a triennial cycle of meetings, the *International Colloquium on Endocytobiology and Symbiosis*, and the founding of the International Society of Endocytobiology (ISE) in 1983. Since 1980, the colloquium has been held in different places around the world, with an expanding range of topics especially in recent years, indicative of the growth, vitality and momentum of the field. This volume (with important reviews, updated until Spring 1997) contains the proceedings of the Sixth International Colloquium held at the University of Tübingen on September 6-10, 1995, under the auspices of the International Society of Endocytobiology.

Interdependence among species is a law of nature. The degree of interdependence may vary substantially, and it may range from entirely innocuous events or purely nutritional aspects to obligatorily symbiotic (mutualistic or parasitic) forms. Many organisms interact through cell surfaces or via some damaged tissues, but a significant fraction has evolved highly sophisticated mechanisms including those which allow for cellular endocytosis. Well-known examples are found in mycorrhizal fungi or among soil bacteria, Rhizobia, which in their symbiotic state behave as a quasi-organelle, and *Agrobacterium tumefaciens* which manipulates its host by direct gene transfer to provide nitrogenous and carbon compounds that only the bacterium can metabolize. Only very few of such interactions have made an enduring contribution to the development of Life, i.e. they led to new taxonic entities, most prominent among them being the eukaryotic cell(s). Mitochondria, kinetosomes, hydrogenosomes, and plastids, all organelles of fundamental importance in the living world, are proposed to have come from endocytobionts. Within this frame of diversity, the focussed topics for the Colloquium into two principal aspects (given by the subheadings) rather than giving an overview on the immense multiplicity of symbiotic associations and phenomena, that were subject of the earlier colloquia (see the previous proceedings: Endocytobiology I to V). The two topics were adaptation phenomena associated with extra- and intracellular symbioses, and the far-less understood processes and events at the physiological, structural and genomic levels, that lead to unequal intertaxonic combinations of cells from different taxons.

Advances in techniques of molecular and cell biology during the past decade have generated unprecedented opportunities for exploring new horizons in understanding the Life on our planet and its evolution. The molecular phylogenetic analysis to determine how complex organisms have developed, maintained genetic integrity over time, and interacted in biological associations is one of the most amazing and fascinating chapters in modern biology. However, our knowledge of the underlying fundamental processes is still limited, and related data are scattered among various disciplines, often with little relationship to each other. The Tübingen conference accounted for the interdisciplinary aspects and brought together researchers in various fields of biology, medicine, biochemistry, and biophysics. There it became obvious that an up-to-date account of the state of knowledge in the field was urgently needed, especially since there had been no recent treatment of individual aspects per se, or in combination. This book has been designed to fill a portion of the gap, primarily for the benefit of research workers (e.g., molecular evolutionists), university teachers and advanced students who wish to gain first-hand knowledge on the scope and direction. Written by prominent scientists, these 45 articles cover various topics in both fields. They present an overview of current accomplishments and potential future directions, argue against outdated notions from the traditional biology of the eukaryotic cell, and convey many of the remarkable, recently discovered aspects of the biology of complex cells, organisms, and of their interactions.

The organizers and editors thank all chairpersons for giving program proposals, especially P. Wrede, W.-H. Kunau and H.-D. Görtz. They thank the ISE members and other participants who came from Australia, Austria, France, Germany, Great Britain, Hungary, Italy, Japan, The Netherlands, Russia, Spain, Sweden, and USA, and worked out their presentations into stimulating reviews, quite frequently updating them until the last minute. They express their special gratitude to the sponsors of the Colloquium, the University of Tübingen, the Universitätsbund e.V., who provided support in financial, technical as well as other matters, and, regarding this book, to the Landesgirokasse Stuttgart, the Association of Friends of the Tübingen University, Prof. Dr. R.G. Herrmann, Munich, and Prof. Dr. K. Wolf, Aachen, who provided the printing cost. Without such financial support this book could not have been produced. The organizers are most grateful to the Dean of the Faculty of Biology, Prof. Hans-Ulrich Seitz, who kindly opened the conference, and also to our colleagues at the Institute of Botany, especially Prof. Hager and Prof. Hampp for their support.

An unforgettable event was the awarding of the Miescher-Ishida-Prize to Prof. Masahiro Sugiura for his outstanding work on plastid chromosomes by the President of the Society, Prof. K.W. Jeon. The ceremony included warm words of Dr. Hans Kössel to the prizewinner and a musical frame arranged by a young musical baroque trio of Noriko and Kazuhiro Fujiwara and Ilka Meyer-Schenk with pieces by Jacques Hoteterre.

We are grateful to many assistants and secretaries, workmen and graduate students at the Institute, particularly Albin Nickol, Birgit Blank and Fred Kippert. Finally the editors thank the Springer Verlag Company for their support producing this volume, and last not least Birgit Blank for her indefatigable effort in setting up the content of this volume camera ready.

The Editors

Introductory Remarks

During the past decade there has been a radical conceptual change in viewing at the living matter, owing to new biochemical techniques that have allowed a spectacular progress in the study of evolution at the molecular level. The concept of „molecular phylogeny", presently one of the most fertile fields in biology, has not only established unequivocally that all life forms fall into one of three lines of descent, Archaea, Eubacteria (both prokaryotes) and - relevant in the context - Eukarya, it has also allowed the formulation of a new concept of eukaryotism. It is now indisputable that the eukaryotic cell with its compartmentalized genome (e.g. nucleus/cytosol, mitochondria, chloroplasts), which alone possessed a potential to develop into advanced forms of life, is the result of processes that have changed (endo)cytobioses to the stage of this cell. It is also true that both its generation and its evolution have proceeded in a very complex manner. It can involve as many as five or six originally autonomous cells of different taxonomic origin (intertaxonic combination). Today, the cell may be considered as a „genomic laboratory" with an enormous potential of restructuring genetic material. This concept will ultimately lead to a natural, coherent and universal tree of life as well as to an understanding of (symbiotic, pathogenic and ecological) organismic interactions at the molecular genetic and/or physiological levels. The new topics include, for instance, exogenosomal organellar evolution in contrast to evolutionary adaptation of partners in symbiotic associations, aspects of gene transfer, RNA editing, organellar protein import, and even a related nomenclature.

The diversity of cases and the rapidly expanding knowledge, which inherently include changes in interpretation, render it difficult to formulate a comprehensive and unequivocal terminology that describes events and genetic make-up of eukaryotic cells and of processes related to their generation. This uncertainty can be observed in the use of distinct terms in the field of intertaxonic combination: for instance, it is a question of what the chloroplast is: a symbiont or a cell organelle. What is a eukaryotic cell? Is it a host with endosymbionts (like chloroplasts or mitochondria) or a host with endocytobiotic cell organelles or a cellular entity with distinct organelles some of which originate from earlier endocytobionts? These questions do not seem to be trivial as may be simply demonstrated with the terms cyanome and cyanelle. According to the original definition by Pascher, Glaucocystophyta (earlier called cyanomes) surely are not symbiotic consortia in comparison with real cyanomes, as the endocytobiotic consortia of *Geosiphon* with *Nostoc*, or *Rhizosolenia* with *Richelia intracellularis* are. It should also be obvious that a term like „symbiosis" can only be used for living systems and not for chemical (molecular) „complexes" at the level of inanimate matter. Clearly, there is a need for a generally agreed terminology that describes the outlined complex biology appropriately. The editors endeavored to standardize the text to some extent, taking care that introduced changes did not distort the intended meaning. The Glossary also provides some definitions which may contribute to resolve controversial points.

The Editors

Glossary

biont (Hawksworth 1988): biological system with the ability of genetically independent reproduction (genetically autonomous biological system).

cell organelle: cell compartment with distinct functions (normally surrounded by at least one envelope membrane) composed additionally by proteins of which at least some are encoded on nuclear DNA. Phylogenetically 2 forms are distinguished: exo- and endogenosomes.

complex plastid: an exogenosome of phyletic secondary or tertiary order.

cyanelle: genetically autonomous, symbiotic living cyanobacterium (like a zoochlorella, Pascher 1929: "Cyanelle bezeichnet eben eine endosymbiontisch lebende Blaualge gegenüber der freilebenden. Cyanelle soll so wenig ein systematischer Begriff sein wie Zoochlorelle und Zooxanthelle"). The term should not be used any more for photosynthetic glaucocystophytan chromatophores which are exogenosomes on the status of plastids (cyanoplasts).

cyanome (Pascher 1929): symbiotic consortium, consisting of host and cyanelle.

cyanoplast (Schenk 1990): eukaryotic, photo(hydro)trophic cell organelle, with regard to distinct cyanobacterial characteristics obviously originating from a former cyanelle by IITC.

cytobiont: a biont associated either endo- or epi-cytobio(n)tic with the host cell.

dibiont: a symbiotic consortium of two bionts, e.g. a dibiontic lichen (Hawksworth 1988).

endocytobiont: a facultative or obligate intracellular biont.

endogenosome (phylogenetic term, Schenk 1992): usually a membrane-bounded cell organelle with an endogenous (autogenous) origin.

endosymbiont: following Smith and Douglas (1987) a biont which „occurs within host cells“ (endocytobiont) „or“ (only in multicellular hosts also) „outside them (extracellular)“.

exogenosome (phylogenetic term, Schenk 1992a): *per definitionem* a cell organelle (containing DNA or not), descending from a former cytobiont, which finally and irreversibly has come under the genomic control of the former host cell by IITC (contrast: „endogenosome“, an organelle as autogenous product of a pro- or eukaryotic cell).

intertaxonic combination (ITC, Sitte 1991): intracellular processes (including gene transfer) changing a symbiotic consortium towards a new taxon by irreversible combination (mixing) of the original symbiotic genomes (Sitte 1991). The observed gene transfer is unequal (IITC), proceeding mainly from the cytobiont towards the former host nucleus ("endocytobiological rule", Sitte 1991). IITC changes a dibiont towards a monobiont system.

Serial Endosymbiosis Theory (SET, Taylor 1974): ITC is not only a unique event (Endosymbiosis Theory, EST, Margulis 1970), but happened several times in series towards complex exogenosomes (complex plastids).

symbiosis: "... Zusammenleben ungleichnamiger Organismen ..." (de Bary 1879), „a permanent or long-lasting (Smith and Douglas 1987) association of differently named bionts“. This broad definition includes both parasitic and mutualistic heterologous associations.

symbiosome (Roth, Jeon and Stacey 1988): cell compartment composed by one or more cells of the endocytobiont (symbiont) and by the symbiosome (outer envelope) membrane, also called perisymbiontic (e.g. perialgal, peribacteroid) membrane. In case of one symbiont per symbiosome its plasmalemma represents the inner envelope of the symbiosome. Symbiosomes can be newly formed, artificially or naturally, dependent on the ontogenetic stage of the partners, or can be transmitted from one to the next generations.

Contents

Part I

Intertaxonic Combination and the Origin and Differentiation of the Cell

1.1 Phylogeny of Exogenosomes

1.2 Intertaxonic Combination and Gene Transfer (Interspecific, Intracellular)

1.3 Protein Import into Cell Organelles - Exogenosomes and Endogenosomes

1.4 Metabolic Control and Ontogenetic Regulations Between Exogenosomes and Nucleus

1.5 Molecular evolution

Part II

Symbiotic Systems
Adaptation, Signal Transduction, Taxonomy and Evolution

2.1 Molecular Approach to Taxonomy of Endocytobionts

2.2 Endocytobionts in Protists and Invertebrates

2.3 Symbiotic Plant Microbe/Fungus Interactions

2.4 Intra- and Extracellular Interactions Between Phycobionts and Mycobionts

2.5 Vertebrate Evolution and Medical Significans

Part I

Intertaxonic Combination and the Origin and Differentiation of the Cell

1.1 Phylogeny of Exogenosomes

Origin and Evolution of Chloroplasts: Current Status and Future Perspectives

K. V. Kowallik
Department of Botany, Heinrich Heine University Düsseldorf, Universitätsstr. 1, 40225 Düsseldorf, Germany

Key words: Chloroplast genome, gene clusters, gene loss, ATPase, ribosomal proteins, introns, rDNA.

Summary: The evolution of plastids of all major algal lineages (Glaucocystophyta, Rhodophyta/Chromophyta, Chlorophyta) traces back to a single successful primary prokaryotic/eukaryotic endocytobiosis involving a cyanobacterium as endocytobiont and a heterotrophic flagellate as host. Therefore, all plastids possessing two envelope membranes, together with their eukaryotic hosts, reveal sister group relationships, whereas complex plastids surrounded by more than two membranes (Heterokontophyta, Cryptophyta, Prymnesiophyta, Dinophyta, Euglenophyta, Chlorarachniophyta) evolved independently of each other, and at different times, by secondary eukaryotic/eukaryotic intertaxonic combination. The evolutionary history of all extant plastid types, however, is ambiguous with respect to single gene phylogenies, mainly because basal branching orders separating the plastids into the major evolutionary lineages are poorly resolved. In contrast, the information contained in completely sequenced plastid genomes of the red alga *Porphyra purpurea*, the chromophyte *Odontella sinensis*, the glaucocystophyte *Cyanophora paradoxa*, the euglenophyte *Euglena gracilis*, and the ulvophyte *Bryopsis plumosa* unambiguously resolves the affiliations among the major algal lineages. Complex gene clusters that appear to have evolved only during chloroplast evolution, to the exclusion of cyanobacteria, reflect the strongest markers in chloroplast evolution. The overall gene complement which comprises 67% identical genes in any chloroplast lineage, cannot be explained as the result of convergent evolution. Without exception, both soluble and structural supramolecular complexes are differentially affected by gene relocation or gene loss/replacement, thereby disproving the idea of a selection mechanism responsible for a gene which is either retained within the chloroplast genome or transferred to the nuclear genome. Differences observed among extant chloroplast genomes may easily be explained by secondary (late) gene relocations/modifications and may therefore help to delineate evolutionary traits. Such lineage-specific modifications include the uptake and lateral propagation of group I and group II introns in green algal, but not in red algal and chromophyte plastids.

Introduction

In 1905 Mereschkowsky published his visionary and, at that time, revolutionary ideas about the origin of chloroplasts. Visionary, because there was no proof to verify the idea of free-living cyanobacteria that entered a heterotrophic host to eventually change into chloroplasts. Revolutionary, because the hypothesized endosymbiotic origin of chloroplasts implied that algae and plants are chimeras made up from animals and chloroplasts. Using diatoms as a model, Mereschkowsky's equation of 1910 effaced the traditional borders separating the plant and animal kingdom:

$$\text{ANIMALS PLUS CHLOROPLASTS = DIATOMS} \quad (1)$$

In addition, by simply transforming Eq. (1) Mereschkowsky anticipated organisms that through a loss of chloroplasts differentiate into secondary heterotrophs or animals, respectively:

$$\text{DIATOMS MINUS CHLOROPLASTS = ANIMALS} \quad (2)$$

Equation (1) was anticipated by a footnote in Schimper's treatise (1883) on the ontogeny of higher plant chloroplasts. It is now widely accepted as the endosymbiont theory.

Only recently, Eq. (2) gained support from molecular and biochemical data. Circular plasmids of 35 kb typical of parasitic protists of the phylum Apicomplexa resemble chloroplast genomes that have lost their photosynthetic capability. These plasmids are devoid of photosynthetic genes but contain most of the genes required for transcription/translation. In the malaria parasite *Plasmodium falciparum* (Wilson et al. 1996) conserved gene clusters and individual genes unknown from green algae and land plants suggest that the progenitor of the apicomplexan 35 kb plasmid derived from a red algal/chromophyte chloroplast genome. Obviously, photosynthetic genes that are no longer required in heterotrophic or parasitic habits (Wolfe et al. 1992) become inactivated and deleted in a relatively short period.

Based upon ultrastructure and 18S rRNA sequence data, sister-group relationships between photosynthetic and non-photosynthetic dinoflagellates also suggest a photosynthetic history of extant heterotrophs (Schnepf 1992). The well-known heterotroph euglenoid *Astasia longa* contains a chloroplast genome that has lost half of its genomic complexity compared with its photosynthetic counterpart *Euglena* (Siemeister et al. 1990). Most of the photosynthetic genes are missing in this euglenoid, except for a functional *rbcL* gene (Siemeister and Hachtel 1990). Therefore, loss of photosynthetic genes suggests protists that may have repeatedly acquired and lost chloroplasts, provided they experienced phagotrophy in addition to photosynthesis and did not become exclusively dependent on photosynthetic energy.

Following Mereschkowsky's hypothesis, which included six to nine independent primary prokaryotic/eukaryotic endosymbioses with at least three different types of prokaryotes involved (leading to the red algal, chromophyte, and chlorophyte lineages), the nature of the photosynthetic prokaryotes was heatedly discussed for some time (e.g., Whatley et al. 1979; Margulis and Obar 1985). More frequently considered were three independent primary endocytobioses (e.g., Raven 1970), one leading to red algal chloroplasts, another to chlorophyll *a+c*-containing plastids, and the third one giving rise to chloroplasts of green algae and land plants. Besides this scenario, cyanoplast containing protists were described as the result of independent endosymbioses including heterotrophic hosts and cyanobacterial endocytobionts (Schenk 1994).

In contrast, Cavalier-Smith (1982) counter-argued in favor of the supposition that only one successful primary prokaryotic/eukaryotic endocytobiosis led to all extant chloroplast lineages. Although the concept of the origin and evolution of plastid types in the Euglenozoa, Dinozoa, and Chromophyta, as discussed by Cavalier-Smith in 1982, has had to be modified since then, mainly on the basis of ultrastructural and molecular data, the idea of a monophyletic origin of all extant plastid types has now been widely accepted (Kowallik 1989, 1993, 1994; Gray 1989; Douglas 1994; Morden et al. 1992; Delwiche et al. 1995; Bhattacharya and Medlin 1995). Nucleotide and amino acid sequence comparisons using distance matrix and parsimony algorithms reveal that all extant plastid types including cyanoplasts cluster as a monophyletic group, to the exclusion of cyanobacteria and prochlorophytes. However, affiliations within the chloroplast/cyanoplast cluster are subject to the molecular marker used and, to a minor extent, to the kind and number of taxa involved as well as to the underlying algorithms and their sensitivity to different rates of nucleotide substitutions and changes in base compositional bias.

Evolutionary rates and the mode of nucleotide substitution are subject to various internal and external factors. For instance, nucleotide positions of ribosomal rRNA genes sequences, whose transcripts are capable of forming complex secondary structures, evolve at different rates (Rothschild et al. 1986). Chloroplast genes that are duplicates within the large inverted repeat sequence (IR segment) reveal evolutionary rates reduced by a factor of four when compared with those of single copy regions (Wolfe et al. 1987). Pairs of genes which result from a duplication of an ancestral gene and which fulfill the *bona fide* requirements for phylogenetic reconstructions (e.g., psaA,B; psbA,D; atpA,B; atpG,D) do not yield congruent phylograms, even if the same set of organisms is being used to calculate the tree (data not shown). In addition, phylogenetic trees may be difficult to interpret due to increased back mutations (homoplasy) affecting branches that have separated early in evolution (Chapman and Buchheim 1991). Sequences with a pronounced base compositional bias may also cause erroneous affiliations (Lockhart et al. 1992; Howe et al. 1993). As an example, the chloroplasts of *Euglena gracilis* have repeatedly been interpreted as close relatives of chromophyte plastids on the basis of rRNA gene phylogenies (Douglas and Turner 1991; Giovannoni et al. 1993). However, as already predicted by Gibbs (1978), the complete sequence of the *Euglena* plastid genome (Hallick et al. 1993) unambiguously reveals a typical chlorophytic origin, provided that the overall gene complement and rearranged gene clusters exclusively found in chlorophyll *a+b*-

containing algae are accorded a higher rank in evolutionary considerations than any given gene phylogeny.

In order to avoid inconsistencies in phylogenetic reconstructions based upon single gene phylogenies this article discusses recent data derived from completely sequenced chloroplast genomes and gene clusters of green and non-green algae that are considered as markers largely unaffected by different evolutionary rates and base compositional bias. There is good evidence that such characters shed light on the origin of chloroplasts and resolve the affiliations of extant algal lineages at basal branching orders more precisely than any evolutionary tree based upon single gene phylogenies.

Molecular Evolution of Chlorophyll *a+b* Containing Chloroplasts

In 16S rRNA sequence phylogenies, chloroplasts of green algae and metaphytes are rooted within prasinophyte-like scaly flagellates (Steinkötter et al. 1994; Friedl 1997). The inferred monophyly of green algal/metaphyte chloroplasts complements with nuclear-encoded 18S rRNA sequence phylogenies (M. Melkonian, pers. commun.). Also well supported is the sister-group relationship of green algae/metaphytes with red algae in various phylogenetic analyses, thus explaining the evolutionary history of plastids surrounded by two membranes that presumably reflect the outer and inner membrane of gram-positive eubacteria (reviewed in Bhattacharya and Medlin 1995). The apparent monophyly of cyanoplasts that constitute a sister clade to red algal and green chloroplasts in 16S rRNA phylogenies (Helmchen et al. 1995) is consistent with a monophyletic origin of all extant plastid types.

Despite the many diverged characters observed in ultrastructure, pigmentation, reserve carbohydrates, motility, and life history, the sister group relationship of the Chlorophyta and Rhodophyta implies that all differences on the molecular, biochemical, physiological, and morphological level became evident only after these two lineages had separated from a common ancestor, presumably more than one billion years ago.

The completely sequenced chloroplast genomes of the prasinophyte *Nephroselmis olivacea* (Turmel et al. 1996), the ulvophyte *Bryopsis plumosa* (Nitsch et al., unpublished results) and the trebouxiophyte *Chlorella pyrenoidosa* (Wakasugi et al. 1997), as well as gene cluster analysis in the prasinophyte *Tetraselmis carteriiformis* (Turmel et al. 1996) and the ulvophyte *Derbesia marina* (Behn et al., unpublished results), indicate that chloroplast genome evolution within the green algal/land plant lineage is mainly caused by gene cluster rearrangements, intron uptake and proliferation, and, to a much lesser extent, by gene relocations. Besides these evolutionarily important events, the uptake of sequences of other than cyanobacterial or chloroplast origin has to be taken into account as well. These characters reflect the major differences to the red algal/chromophyte lineage of descent (see next chapter). In *Bryopsis* a sequence of some 25 kb which does not

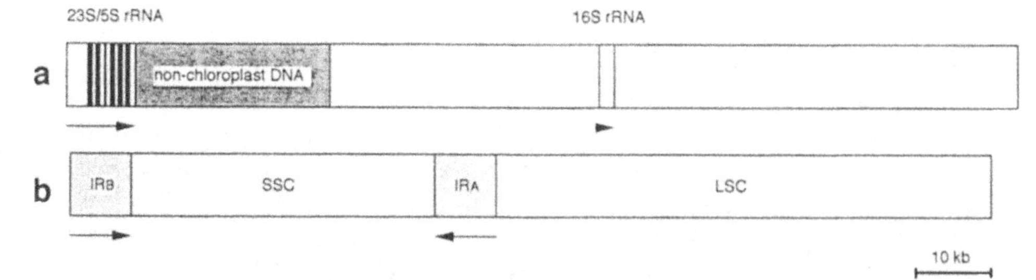

Fig. 1a, b. Chloroplast genomes of the siphonaceous green alga *Bryopsis plumosa* (a) and the centric diatom *Odontella sinensis* (b). The circular chromosomes are displayed in the linearized form obtained by a cut 5' to the single copy 23S rRNA gene in *Bryopsis* and a cut at the boundary between the IRB segment and the large single copy region (LSC) in *Odontella.* The IR segments enclose the small single copy region (SSC). Group I introns in the *Bryopsis* 23S rRNA gene are showns as black bars. Arrows define the direction of transcription of the rDNA in *Bryopsis* and the direction of the IR segements (lightly shaded regions) in *Odontella.* The 25 kb "non-chloroplast DNA" in the *Bryopsis* chloroplast genome follows downstream of the 23S/5S rRNA genes (dark-shaded region). The chloroplast genomes are drawn to scale

code for chloroplast proteins ("non-chloroplast DNA," Fig. 1a) is inserted into the genome downstream of the single-copy 23S/5S rRNA genes. Unusually large reading frames encoded in this sequence, together with a codon usage uncommon for chloroplast genes, reveal similarities with viral protein sequences (Nitsch 1995). Preliminary results indicate that in the siphonaceous green alga *Derbesia marina* also, reading frames, unknown from any other chloroplast lineage, are stable elements of this chloroplast genome. The insertion of "foreign" DNA into these genomes obviously preceded the separation of these two genera from a common ancestor.

The deep bifurcation of chloroplast history in green plants, which even splits the Prasinophyceae into two diverging groups (Melkonian 1996, see also Fig. 6), implies that the evolutionary history of chlorophyll *a+b*-containing chloroplasts may reveal traits specific for either the chlorophytic or the charophytic lineage. Unfortunately there are still pronounced gaps in our knowledge of chloroplast genome organization and evolution in different green algal lineages, permitting only a fragmentary and provisional picture of the events that have left their footprints in green algal chloroplast genomes with the course of time.

One of the major differences between green algal and land plant chloroplast genomes is found in the linear arrangement of genes and gene clusters. In land plant chloroplast genomes including *Marchantia* and seed plants the chloroplast genomes closely resemble each other. Differences are mainly due to the presence or absence of individual genes of which those involved in chlorophyll biosynthesis (present in *Marchantia* and *Pinus*, but absent in Angiospermes) or genes coding for NADH dehydrogenase subunits (present in Angiosperms and *Marchantia*, but partially lacking in black pine) may serve as two examples. The reason why land plant

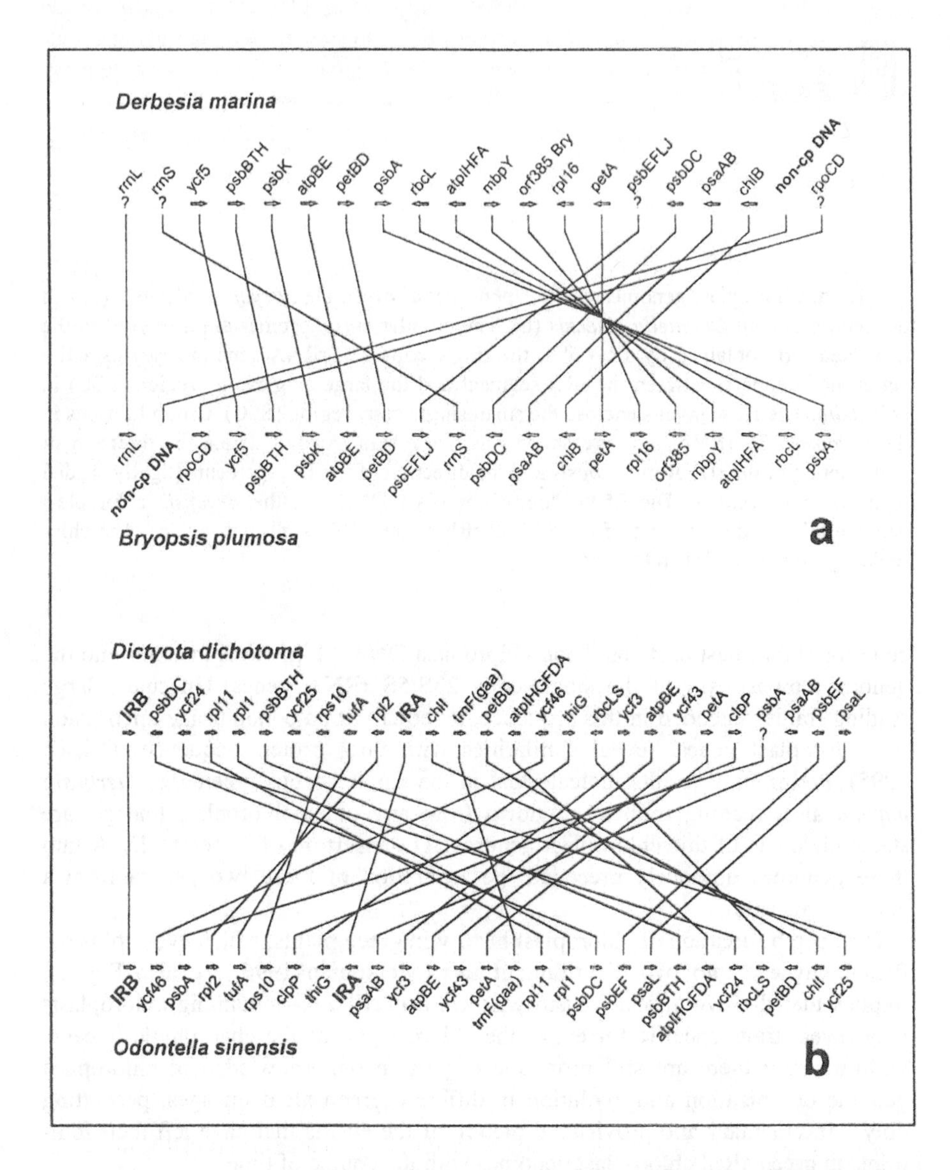

Fig. 2a, b. Rearranged chloroplast genomes of the siphonaceous green algae *Derbesia marina* and *Bryopsis plumosa* (a) and of the chromophytic algae *Dictyota dichotoma* and *Odontella sinensis* (b). The genomes are aligned using the 23S rRNA genes as start in (a) and one of the IR segments in (b). Homologous gene clusters are connected by lines to show the degree of sequence conversion. Arrows indicate the direction of transcription of gene clusters as predicted from sequencing. In the *Derbesia* and *Dictyota* chloroplast genome the transcription direction of a few gene clusters is unknown to date (indicated by question marks)

chloroplast genomes are so highly conserved was suspected to lie in their uniform architecture, which includes the presence of large inverted repeat (IR) sequences (Palmer and Thompson 1982). Nevertheless, in green algae (and also in chromophytes) chloroplast genomes appear to be highly rearranged, even though their gross architecture is comparable to that of metaphytes. The degree of sequence scrambling may exceed that of land plants by an order of magnitude, even among closely related species or among genera sharing a common history (Fig. 2a). As yet there is no plausible explanation for this apparent instability of algal chloroplast genomes.

Siphonaceous green algae, which are among the most ancient groups of algae, reveal characters not found in other chlorophyll *a+b*-containing chloroplasts. While the loss of the IR sequence occurred independently in the Charophyta (Manhart et al. 1989, 1990), Ulvophyta (Hedberg and Hommersand 1981; Nitsch 1995), and Metaphyta (Palmer and Thompson 1982), the disruption of the rRNA operon and the stable insertion of a DNA fragment not of chloroplast origin appear to be unique to members of the Bryopsidales. Also remarkable are numerous group I and group II introns that split rRNA and protein coding genes in different chlorophytic lineages. These introns may be used as evolutionary markers, as most of them are secondary introns, in accordance with the intron-late hypothesis. In particular, group I introns are known to contain reading frames for endonucleases (ENases) allowing them to be laterally transferred within a given genome (Yamada et al. 1994; Belfort and Perlman 1995) or even spread among different organelles (Turmel et al. 1995). On the other hand, group I introns may easily disappear through a molecular mechanism in which reverse transcriptases (RTases) are involved (Lambowitz and Belfort 1993).

While cyanobacteria, red algal, and chromophytic chloroplast genomes are usually devoid of intron sequences, green algal, euglenoid, and metaphyte chloroplast genomes exhibit introns to varying extents. In *Bryopsis plumosa*, nine group I and seven group II introns have been identified so far (Nitsch et al., in preparation), of which six group I introns contain reading frames for ENases and two group II introns code for RTases (Table 1).

With only one exception, all ENase-containing group I introns in the *Bryopsis plumosa* chloroplast genome are found in the single-copy 23S rRNA gene, which therefore encompasses 6.7% of the entire chloroplast genome (8295 bp out of 124 kb). In contrast, the 16S rRNA gene that maps almost opposite to the 23S rRNA gene on the circular chromosome, reveals an uninterrupted reading frame (Fig. 1a).

Most remarkably, the 23S rRNA introns 1,2,3,4, and 6 of *Bryopsis* are also found in 23S rRNA genes of different *Chlamydomonas* species at a few conserved sites (Turmel et al. 1993). It may therefore be concluded that these introns were already present in the ancestor common to both *Chlamydomonas* and *Bryopsis*. In addition, intron 2 is also found in the coccalean green alga *Trebouxia aggregata* (Trebouxiophyceae; M. Turmel, pers. commun.), thereby supporting the concept of monophyly of extant volvocalean (Chlorophyceae), coccalean (Pleurastrophyceae) and siphonaceous (Ulvophyceae) green algae. These sister group relationships are well documented in 18S rRNA phylogenies (Friedl 1997) and support

Table 1: Introns in the chloroplast genome of *Bryopsis plumosa*

Gene - Intron	Group	ORF	Function ORF
rpl 12	I A	-	-
atp A-1	I B	-	-
psb B	I B	300	Endonuclease
rrn L-1	I B	181	Endonuclease
rrn L-2	I B	224	Endonuclease
rrn L-3	I A	-	-
rrn L-4	I B	219	Endonuclease
rrn L-5	I B	276	Endonuclease
rrn L-6	I C	193	Endonuclease
rpl 5	II B	-	-
psa A	II A	483	RTase-Maturase
psb A	II A	527	RTase-Maturase
psb T	II B	-	-
atp A-2	II B	-	-
atp F	II B	-	-
ycf 3	II B	-	-

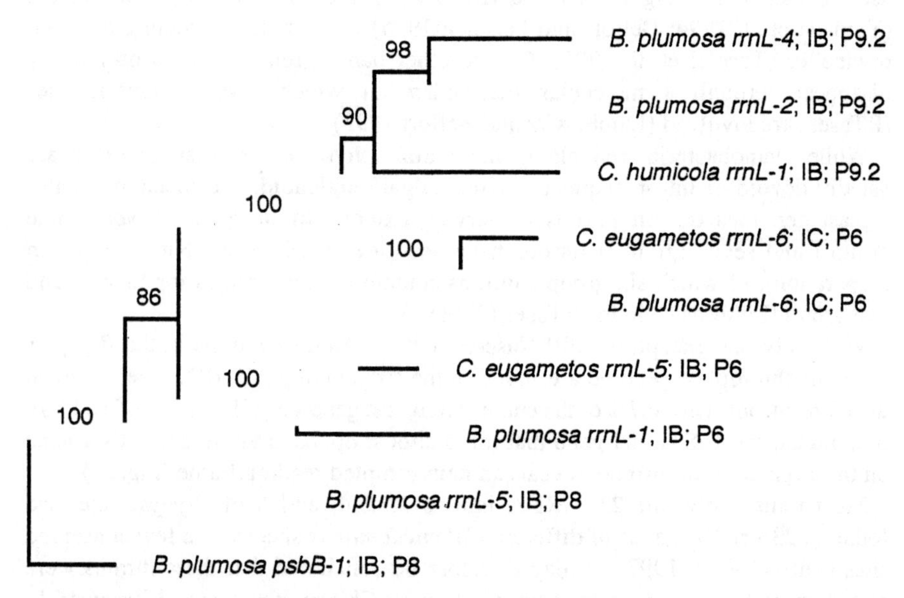

Fig. 3. Distance matrix analysis (TREECON; van de Peer and de Wachter 1993) of amino acid sequences predicted from endonucleases encoded within group I introns in the chloroplast genomes of *Chlamydomonas* species and of *Bryopsis plumosa.* Introns refer to those in Fig. 1a. Also included are the intron subgroups and the loops containing reading frames for site-specific endonucleases. Bootstrap values result from 100 replications. The endonuclease reading frame encoded in the *B. plumosa psbB* gene was used as outgroup

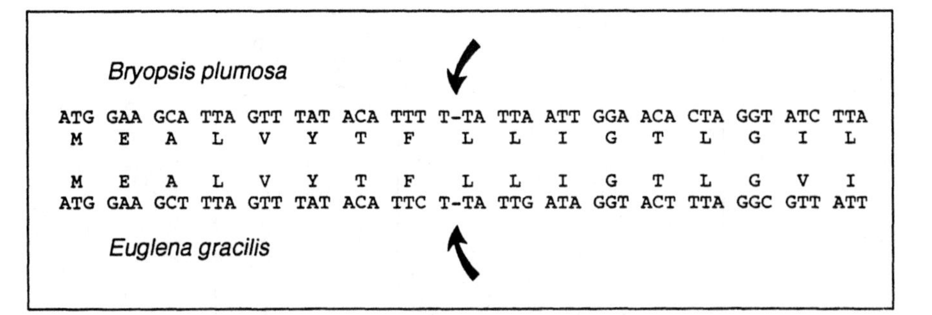

Fig. 4. Alignment of the 5' regions of the *psbT* genes of *Bryopsis plumosa* and of *Euglena gracilis*. The insertion sites of a homologous intron within the leucine codon TTA are arrowed. Amino acid residues are shown in the one-letter code

the modern concept of green algal systematics and evolution based upon ultrastructural evidence (Stewart and Mattox 1975).

The origin of introns 2 and 4 of the *Bryopsis* 23S rRNA gene deserves further attention, as these introns do not indicate independent origins, but may rather be interpreted by lateral transfer within the *Bryopsis* 23S rRNA gene. In a phylogenetic analysis using the aligned ENases of introns 1,2,4,5,6 together with their *Chlamydomonas* counterparts, introns 2 and 4 not only belong to the same subgroup B, but encode their ENases exclusively within loop P9.2 (Fig. 3; secondary structure models not shown). It is noteworthy that their counterpart in *Chlamydomonas humicola* encodes a homologous ENase within the same loop. From the phylogram it may be concluded that introns 2 and 4 of the *Bryopsis* 23S rRNA gene are sister introns that arose through a duplication of a pre-existing intron followed by lateral distribution and insertion of one of its offspring sequences (Nitsch 1995). Thus, one of these two introns displays the most recent event of intron insertion in the *Bryopsis* 23S rRNA gene.

Group II introns, which are characterized by their secondary structure and widely conserved 5' (GUGYG) and 3' (AY) intron/exon boundaries (Michel et al. 1989), may likewise delineate affiliations among genera, even in different evolutionary lineages. In *Bryopsis plumosa* a group II intron of 430 bp splits the gene *psbT* (Fig. 4), which in all chloroplast genomes is located at a conserved site downstream of the *psbB* gene. The insertion site is within the leucine codon T-TA. Most surprisingly, a group II intron of 1352 bp is found in the *psbT* gene of *Euglena gracilis* at the very same position. Several large insertions/deletions may account for the size differences of these two introns, which otherwise appear to be similar by sequence. Thus, these two homologous introns are strong evolutionary markers supporting the sister-group relationship between euglenoid and green algal chloroplasts. There is no doubt that this intron entered the *psbT* gene of a green algal chloroplast ancestral to both the ulvophyte and euglenoid chloroplasts.

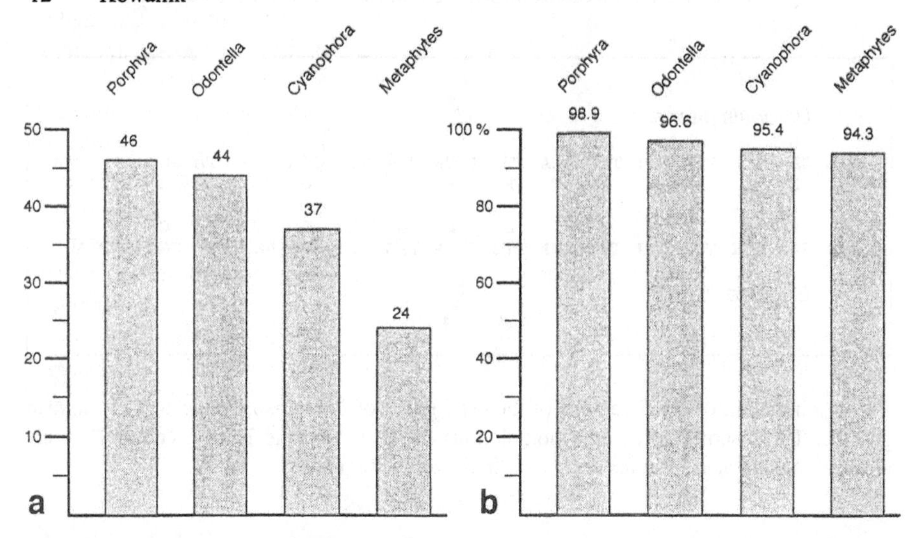

Fig. 5a. Number of ribosomal proteins (Rps, Rpl) encoded in the chloroplast genomes of *Porphyra purpurea*, *Odontella sinensis*, *Cyanophora paradoxa*, and metaphytes. **b.** Gene complement of the *Euglena gracilis* chloroplast genome (100%) contained in the completely sequenced chloroplast genomes of *Porphyra purpurea*, *Odontella sinensis*, *Cyanophora paradoxa*, and metaphytes including *Marchantia*, black pine, tobacco, rice, and maize

Two group II introns in the *Bryopsis* chloroplast genome that split the *psaA* and the *psbA* genes contain open reading frames with similarities to reverse transcriptases/maturases of a mitochondrial group II intron ORF of *Podospora anserina* (Faßbender et al. 1994). Such intron ORFs may play a role in the lateral propagation of the intron via an RNA intermediate by reverse splicing or cDNA integration (Lambowitz and Belfort 1993). Interestingly, the ORF-containing group II intron that splits the *psbA* gene in *Bryopsis plumosa* is found in the *rbcL* gene in *Bryopsis maxima* (Kono et al. 1991), which in *Bryopsis plumosa* reveals an uninterrupted reading frame. This indicates that the intron was present in the ancestor common to both *Bryopsis maxima* and *Bryopsis plumosa*, but evolved independently after these two species had separated from their common ancestor.

While gene cluster analysis along all major lineages is restricted by the fact that the gene content in chloroplast genomes of chlorophyll $a+b$-containing plants is reduced by about 50% compared with that of the chloroplast genome of *Porphyra*, ancestral cyanobacterial gene clusters, though truncated, may still be encountered. Even in the *Euglena* chloroplast genome, which contains the lowest number of genes of any photosynthetic organism known so far, and which therefore constitutes the basal gene complement present in all extant chloroplast genomes (Fig. 5b), the *atpA* gene cluster reflects evolutionary traits which are consistent with a green algal ancestry. The well-documented *atpA* gene cluster, which in cyanobacteria and non-green algae contains the genes *atpI,H,G,F,D,A* in the given transcription order (Pancic et al. 1992; Kostrzewa and Zetsche 1992; Leitsch and

Kowallik, unpublished results), experienced drastic modifications almost exclusively within the chlorophyll $a+b$ branch (Fig. 6). Whereas the gene *atpC* became nuclear prior to the basal split into the green and non-green lineages, presumably in the course of a massive shrinkage of the cyanobacterial genome by some 90%, the loss of the genes *atpD* and *atpG* is of phyletic implication for all chlorophyll $a+b$-containing organisms. The only additional loss of an ATPase gene, *atpI*, affected the cyanoplasts of *Cyanophora* (Stirewalt et al. 1995). Group I and group II introns split the remaining genes only in green plants. They clearly resemble late introns. However, nonhomologous insertion sites of a group II intron in the *atpF* gene, together with a nonhomologous primary sequence, is indicative of an independent event within the charophycean and chlorophycean line of descent. The *Euglena* introns splitting the ATPase genes also appear to be of independent origin, and most probably, the insertion of nine amino acid residues into the *atpH* gene product, including two cysteines, occurred only within the Metaphyta. Finally, the overlap between 3'*atpB* and 5'*atpE* that may have resulted from a 3' extension of the *atpB* gene into the *atpE* gene is found exclusively in land plants (Fig. 6, Kroth-Pancic et al. 1995).

Molecular Evolution of Chlorophyll a+c Containing Chloroplasts

Initial support for a single successful prokaryotic/eukaryotic endocytobiosis leading to all extant chloroplast lineages came from plastid genomes of four heterokont chromophytes that were chracterized by restriction site and gene mapping experiments (Kowallik 1989). As a result of these investigations, differences in ultrastructure and pigmentation that were traditionally accorded primary rank for discriminating between chloroplast types and served to delineate major traits in algal evolution, must be considered as modifications of primary prokaryotic characters. Consequently, the ultrastructure, biochemistry, and molecular biology that may be used to distinguish between individual plastid types actually reflect secondarily acquired character states.

Ultrastructural peculiarities of cryptomonads first led Greenwood et al. (1977) to suggest that chlorophyll $a+c$-containing algae whose plastids are surrounded by an ER cisterna may have evolved through secondary eukaryotic/eukaryotic intertaxonic combination. Complex plastids of chromophytic algae were interpreted as the result of multiple secondary endosymbioses including red algae as endocytobionts and various flagellated heterotrophs as hosts (Gibbs 1981; Whatley and Whatley 1981). This scenario, if correct, should have confounding effects on reconstructed phylogenies inferred from individual genes and must have left unequivocal traces in the physical character of chromophytic plastid genomes. Indeed, phylogenetic analyses based upon nucleotide as well as inferred amino acid sequences exhibit chromophytic sequences as sister groups of red algal sequences. Usually the bootstrap support does not even discriminate between red algal and chromophytic sequences which in "universal" trees frequently group together as a single clade (Melkonian 1996; Bhattacharya and Medlin 1995).

It seems plausible that the close relationship between red algal and chromophytic gene sequences may also be manifested at the structural level of their respective genomes. For this reason we cloned and completely sequenced the chloroplast genome of the marine centric diatom *Odontella sinensis* (Fig. 7; Kowallik et al. 1995; Freier 1995; Stöbe 1995) and compared the organization of its gene clusters with those of the previously sequenced chloroplast genome of the red alga *Porphyra purpurea* (Reith and Munholland 1995).

Like most other chloroplast genomes known to date the *Odontella* chloroplast genome reveals two single-copy regions separated by an IR sequence that contains the rRNA operon (Fig. 1b). Intron sequences are absent from this genome, as in *Porphyra*. As already stated (Kuhsel and Kowallik 1987; Kowallik 1989) the IR sequences of chromophytic plastid genomes are small in size. In *Odontella* the IR segment contains, in addition to the rRNA genes together with the spacer-encoded tRNAs trnI (gau) and trnA (ugc), an open reading frame ORF 355, which is so far unique to chloroplast/cyanobacterial genomes, together with the tRNA gene trnP (ugg) facing the boundary of the large single-copy region. Downstream of the 5S rRNA gene a conserved reading frame ycf32 contributes to the IR, which terminates within the ribosomal protein gene rpl32. One copy of this gene extends into the small single-copy region and is probably the functional (complete) gene, whereas the rpl32' reading frame (Fig. 7) is C-terminally truncated by three amino acid residues. It may be expected that through copy correction mechanisms, which result in an extension of IR sequences into both single-copy regions, the two copies of the rpl32 gene will eventually become colinear. Altogether, the total size of the IR of 7.714 bp measures only one-third to one-fourth of the size of an angiosperm IR. This typical IR segment does not, however, correspond to the structural organization of the rDNA in *Porphyra* (Reith and Munholland 1995), which lacks the large IR segment. Instead, the two rRNA operons, which are not fully identical by sequence, are decoded from the same strand. This character which is unusual in chloroplasts, may have been acquired within the red algae, provided it may also be found in species other than *Porphyra*.

The *Porphyra* chloroplast genome contains some 80 additional genes compared with that of *Odontella*, explaining its enhanced coding capacity and its ancestral condition. The products of these extra genes are involved in transcription/translation, phycobiliprotein complexes, and various biosynthetic pathways. Following secondary endosymbiosis these genes either became deleted (e.g., all genes involved in biliprotein synthesis and assembly) or replaced by nuclear homologs of the second host (probably most of the genes involved in various biosynthetic pathways) or relocated to the nuclear genome (mainly photosynthetic genes).

Differences in gene content between the red algal and diatom chloroplast genome give rise to new insights into the evolution of the Heterokontophyta that

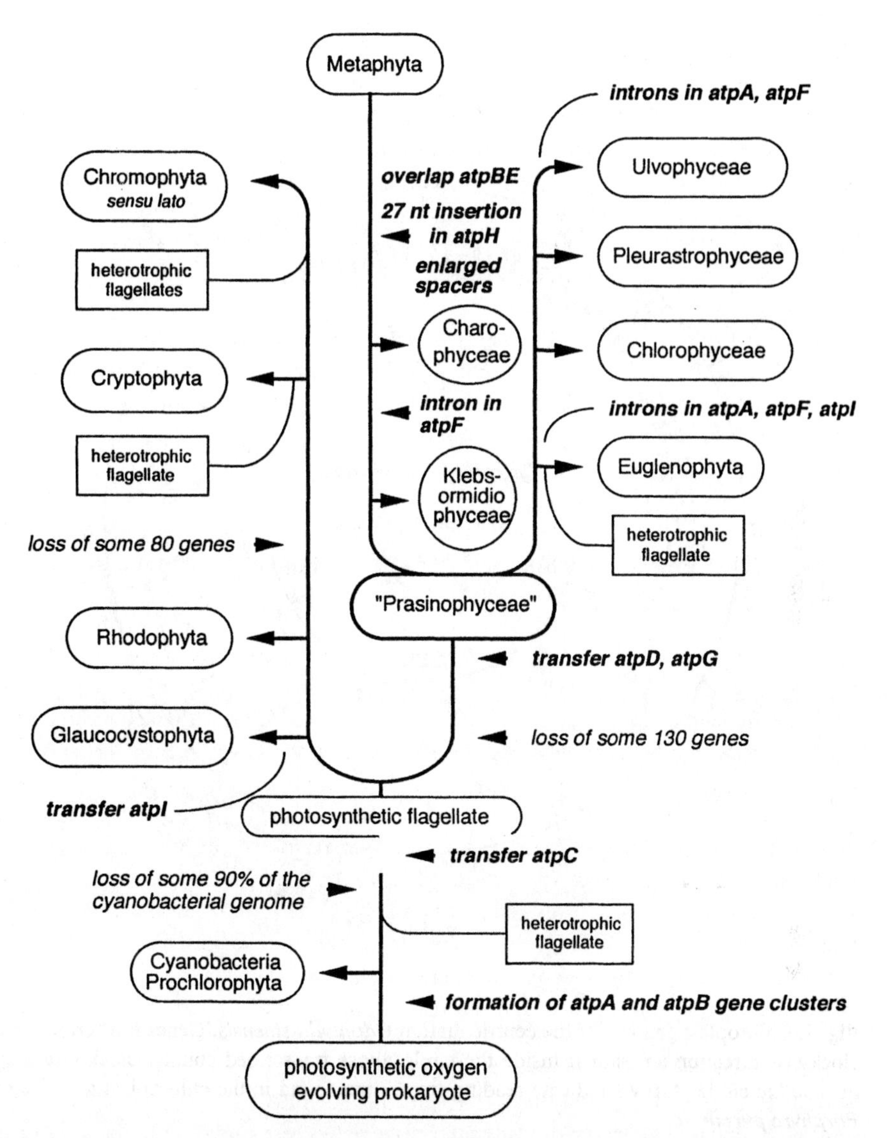

Fig. 6. Universal evolutionary tree of photosynthetic organisms combining the eukaryotic history and the origin and evolution of chloroplasts and cyanoplasts. Primary (prokaryotic/eukaryotic) as well as secondary (eukaryotic/eukaryotic) endosymbioses are shown together with some events that occurred during chloroplast genome evolution. This tree does not include the evolutionary history of dinoflagellates. For further description, see text

are believed to be monophyletic (van den Hoek et al. 1995). A preliminary survey of the chloroplast genome of the brown alga *Dictyota dichotoma*, however, may

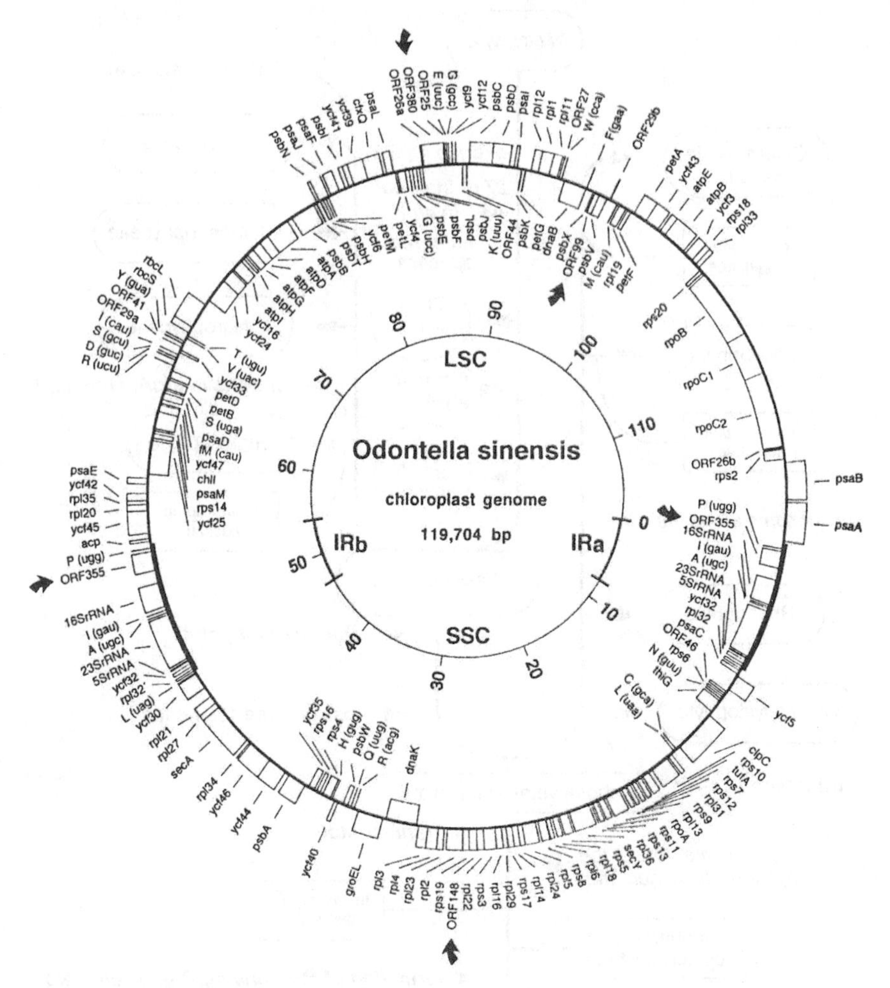

Fig. 7. Chloroplast genome of the centric diatom *Odontella sinensis*. Genes transcribed in a clockwise direction are shown inside the circle, those transcribed counter-clockwise map outside the circle. Arrows indicate reading frames not found in the chloroplast genome of *Porphyra purpurea*

indicate that the evolutionary histories of diatoms and brown algae are largely independent. This may be predicted from the individual gene content of the respective genomes as well as from the linear gene arrangement: with few exceptions *(rpl2/tufA/rps10,atpBE/ycf43/petA/clpP,psbEF/psaL)* the linear array of complex gene clusters is not preserved between diatoms and brown algae. If diatoms and brown algae do indeed trace back to a single secondary endosymbiosis,

Porphyra purpurea

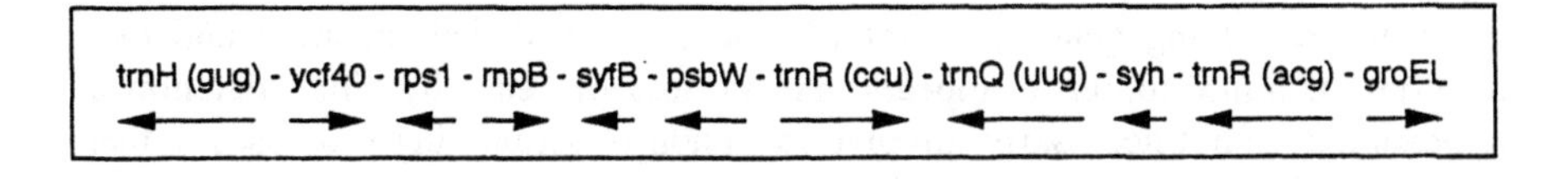

Odontella sinensis

Fig. 8. Segment of the *Porphyra purpurea* chloroplast genome, together with its counterpart in *Odontella sinensis.* Note that the direction of transcription of individual genes of the *Odontella* genome resembles that of *Porphyra,* though five genes at three different positions are missing in the *Odontella* chloroplast genome

gene cluster rearrangements must have occurred to a degree hitherto unknown among related taxa (see Fig. 2b).

Despite the loss of some 80 genes, the remaining gene complement in *Odontella* still resembles the ancestral red algal condition as far as complex gene clusters are concerned. Among the latter is the large ribosomal protein gene cluster which is colinear in both *Odontella* and *Porphyra* (Freier 1995) and which is unknown in *Synechocystis* PCC 6803 (Kaneko et al. 1996), *Cyanophora* (Stirewalt et al. 1995), and green plants. However, it is also found in the chloroplast genome of the cryptophyte *Guillardia theta* (S. Douglas, pers. commun.) and may therefore be considered a most confident evolutionary marker proving the descendence of chlorophyll *a+c*-containing plastids from red algal ancestors. In detail, the large red algal/chromophytic ribosomal protein gene cluster that contains 29 genes in *Porphyra* or 30 genes in *Odontella,* results from fusion processes of two previously independently transcribed operons, including the *E. coli* S10, spc, and α operons which in *Synechocystis* constitute a single gene cluster (*rpl3* through *rpl31*, cf. Fig. 7). As in *E. coli*, the str operon is distantly located in the *Synechocystis* genome. This arrangement was retained within the Chlorophyta although both the S10/spc/α and the str operon became reduced in size due to massive gene loss. In addition to fusion processes, splitting of the S10/spc/α gene cluster has also occurred during chloroplast evolution, as in *Cyanophora* and *Euglena* (Freier et al., in preparation).

Gene clusters that do not constitute transcriptional units, because their constituents are transcribed in different directions, may likewise serve as evolutionary markers. In Fig. 8 a series of eleven *Porphyra* genes that are transcribed from both strands are compared with the corresponding and remaining genes in *Odontella.*

Five genes altogether [*rps1, rnpB, syfB, trnR(ccu), syh*] were lost during the process of transformation of a red algal endocytobiont into the diatom plastid. Surprisingly, the remaining genes in *Odontella* have retained their position and their direction of transcription. It appears that the loss of genes by either deletion or relocation does not necessarily involve inversion or scrambling processes; rather, this stream-lining process resembles the excision of intron sequences to some extent, followed by the fusion of the 3' and 5' boundaries. Whether such molecular mechanisms may act during chloroplast genome evolution in chromophytes is hitherto unknown.

In comparison with metaphyte chloroplast genomes the *Odontella* chloroplast genome comprises some 50 additional genes, despite its reduced genomic complexity. Apart from the lack of introns, the reason for this is a tightly packed gene order, which only allows small spacer regions to separate the individual genes. This feature is also shared by cyanobacteria and the chloroplast genomes of other non-green algae and therefore reflects a primitive character. Similarly, but *vice versa*, the greatly reduced gene complement of land plant chloroplast genomes is to be considered a derived condition. In comparison with the green algal/land plant lineage the *Odontella* and *Porphyra* chloroplast genomes contain numerous genes that have obviously been relocated to the nucleus in the Chlorophyta. Among these are genes coding for polypeptides of photosynthetic complexes (e.g., *psaD,E,F,L; psbV,W,X; petF; atpD,G*), ribosomal proteins (*rpl1, 3, 4, 6, 11, 12, 13, 18, 19, 24, 25, 29, 31, 34, 35; rps5, 6, 9, 10, 13, 17, 20*), chaperonins (*groEL; secA,Y; dnaB; acpP*), as well as enzymes or subunits of enzymes involved in biosynthetic pathways (*rbcS, cfxQ; thiG*). In particular, the distribution of ribosomal protein genes and their retention within the chloroplast genomes of the major algal lineages has important phylogenetic implications (see Fig. 5a).

Conclusions and Perspectives

The molecular evolution of a genetic entity is always dependent on its counterparts in any given eukaryotic cell. Consequently, the co-evolution between the chloroplast and the nuclear genomes in at least those taxa that derive from a single prokaryotic/eukaryotic endosymbiosis (Rhodophyta, Chlorophyta, Glaucocystophyta) should reflect similar or identical traits among these genetic compartments. If mitochondria trace back to a single endosymbiotic event, similar phylogenies should be expected for mitochondrial genes as well. However, the highly differing rates of evolution in different mitochondrial lineages, and the fact that a mitochondrial genome represents a genetic mosaic rather than a mirror-image of an ancestral eubacterial genome (Gray 1989), renders this supposition difficult to assess.

Despite the differences in evolutionary rates in chloroplast and nuclear genomes, the anticipated co-evolution of these two genetic compartments is well substantiated in green algae and metaphytes by 18S rRNA-based phylograms and those inferred from various chloroplast-encoded rRNA and protein sequences. In contrast, algae that evolved through secondary (Chromophyta *sensu lato*, Chlorarachniophyta, Euglenophyta) or tertiary endosymbioses (some dinoflagellates) are expected to exhibit incongruent nuclear and organellar phylogenies, due to the different hosts involved and the different periods of time over which these algae evolved. Consequently, gene phylogenies inferred from nuclear genes of complex algae must be interpreted with a certain scepticism, since such genomes are chimeras that most probably retained genes of the disintegrated nucleus of the eukaryotic endocytobiont as well as organellar genes that became nuclear following primary endosymbiosis. The true origin of nuclear genes of complex algae may therefore be obscured or blurred by the superimposition of the ongoing process of the lineage-specific shrinkage of organellar genomes.

It becomes increasingly plausible that in certain algal lineages even repeated gain and loss of photosynthetic endocytobionts and/or interorganismal gene transfer has taken place, leaving their molecular traces behind. Even chloroplast genomes usually considered as a streamlined cyanobacterial genome may have stably incorporated genes or large DNA fragments from outside, some presumably of viral origin. The presence of intron-encoded reverse transcriptases/maturases and endonucleases in green algae renders it possible that the uptake of foreign DNA may have been accomplished by a reverse-transcribed RNA intermediate.

What we really need for a better understanding of algal chloroplast genome evolution is a closer view of those chloroplast genomes that may serve as representatives of all major evolutionary lineages. For example, PCR-based analyses both of conserved and of rearranged gene clusters in different evolutionary lineages may answer some questions that remain about rRNA of protein based phylogenetic tree constructions.

Apart from the necessary knowledge of a second red algal plastid genome that may reveal the degree of sequence divergence within this ancient group of algae, at least the chloroplast genomes of a brown alga, of chrysophytes, xanthophytes, and prymnesiophytes deserve more attention. In the green algal lineages those algae that evolved independently of their sister groups over a long period (e.g., the Oedogoniales, Dasycladales, Ulvales) should be candidates disclosing modes of chloroplast genome evolution different from those observed along the slowly evolving land plant chloroplast lineage. It may then eventually be possible even to reconstruct a hypothetical chloroplast genome of the last ancestor common to all extant plastid and cyanoplast genomes (see Löffelhardt et al., this volume p. 40).

An unprecedented source of new information provided by the complete sequence of the *Synechocystis* sp. PCC6803 genome opens new gates to a better understanding of the various chloroplast functions. While the structural components of chloroplasts have been largely unravelled within the past decade, our understanding of regulatory processes acting at the gene level and of components involved in the assembly of multisubunit complexes in chloroplasts is still fragmentary. Many new genes found as conserved reading frames (ycf designations) in the *Odontella* and *Porphyra* chloroplast genomes have counterparts in the

Synechocystis genome. This unicellular cyanobacterium, when used as a homologous transformation system, has already proven itself a powerful tool for elucidating the function of new genes. In the next few years much effort will be concentrated on attempts to gain a more sophisticated view of the regulatory processes in chloroplasts. In a similar way, most if not all of the additional genes in the chromophytic plastid genomes will be found as nuclear genes in green algae and metaphytes, thus allowing the importance of these reading frames to be analysed by well-established eukaryotic mutation and transformation systems.

Acknowledgements: I am grateful to my collaborators, Drs. Bettina Stöbe, Carola Leitsch, Wilfried Behn, Torsten Nitsch, and Ulrich Freier, for their permission to cite unpublished data. I thank Ms. Ina Schaffran and Mrs. Christel von der Lippe for their skilful and enthusiastic work. Supported by grants Ko 439/6-2 and 439/6-3 from the Deutsche Forschungsgemeinschaft.

References

Belfort M, Perlman S (1995) Mechanisms of intron mobility. J Biol Chem 270: 30237-30240

Bhattacharya D, Medlin L (1995) The phylogeny of plastids: A review based on comparisons of small-subunit ribosomal RNA coding regions. J Phycol 31: 489-498

Cavalier-Smith T (1982) The origin of plastids. Biol J Linn Soc 17: 289-306

Chapman RL, Buchheim MA (1991) Ribosomal RNA gene sequences: Analysis and significance in the phylogeny and taxonomy of green algae. Crit Rev Plant Sci 10: 343-368

Delwiche CF, Kuhsel M, Palmer JD (1995) Phylogenetic analysis of *tufA* sequences indicates a cyanobacterial origin of all plastids. Mol Phyl Evol 4: 110-128

Douglas SE (1994) Chloroplast origins and evolution. In: DA Bryant (ed) The Molecular Biology of Cyanobacteria. Kluwer Acad Publishers, Dordrecht, pp 91-118

Douglas SE, Turner S (1991) Molecular evidence for the origin of plastids from a cyanobacterium-like ancestor. J Mol Evol 33: 67-73

Faßbender S, Brühl K-H, Ciriacy M, Kück U (1994) Reverse transcriptase activity of an intron-encoded polypeptide. EMBO J 13: 2075-2083

Freier U (1995) Sequenz, Gene und phylogenetische Analyse der SSC- und IR-Region im Plastidengenom der Kieselalge *Odontella sinensis*. Thesis, University of Düsseldorf, 253 pp

Friedl T (1997) Evolution of green algae. In: D Bhattacharya (ed) The Origin of Algae and Plastids. Springer, Vienna, in press

Gibbs SP (1978) The chloroplasts of *Euglena* may have evolved from symbiotic green algae. Can J Bot 56: 2883-2889

Gibbs SP (1981) The chloroplasts of some algal groups may have evolved from eukaryotic algae. Ann NY Acad Sci 361: 193-208

Giovannoni SJ, Wood N, Huss VAR (1993) Molecular phylogeny of oxygenic cells and organelles based on small-subunit ribosomal rRNA sequences. In: RA Lewin (ed) Origins of Plastids. Chapman and Hall, New York, London, pp 159-70

Gray MW (1989) The evolutionary origins of organelles. TIG 5, 294-299

Greenwood AD, Griffiths HB, Santore UJ (1977) Chloroplasts and cell compartments in Cryptophyceae. Brit Phycol J 12: 119

Hallick RB, Hong L, Drager RG, Favreau MR, Monfort A, Orsat B, Spielmann A, Stutz E (1993) Complete sequence of *Euglena gracilis* chloroplast DNA. Nucleic Acids Res 21: 3537-3544

Hedberg MF, Huang YS, Hommersand MH (1981) Size of the chloroplast genome in *Codium fragile*. Science 213: 445-447

Helmchen TA, Bhattacharya D, Melkonian M (1995) Analyses of ribosomal RNA sequences from Glaucocystophyte cyanoplasts provide new insights into the evolutionary relationships of plastids. J Mol Evol 41: 203-210

Howe CJ, Beanland TJ, Larkum AWD, Lockhart PJ (1993) Plastid phylogeny and the problem of biased base composition. In: RA Lewin (ed) Origins of Plastids. Chapman and Hall, New York, London, pp 349-353

Kaneko T, Sato S, Kotani H, Tanaka A, Asamizu E, Nakamura Y, Miyajima N, Hirosawa M, Sugiura M, Sasamoto S, Kimura T, Hosouchi T, Matsuno A, Muraki A, Nakazaki N, Naruo K, Okumura S, Shimpo S, Takeuchi C, Wada T, Watanabe A, Yamada M, Yasuda M, Tabata S (1996) Sequence analysis of the genome of the unicellular cyanobacterium *Synechocystis* sp. strain PCC6803. II. Sequence determination of the entire genome and assignment of potential protein-coding regions. DNA Res 3: 109-136

Kono M, Satoh H, Okabe Y, Abe Y, Katsumi N, Okada M (1991) Nucleotide sequence of the large subunit of ribulose-1,5-bisphosphate carboxylase/oxygenase from the green alga *Bryopsis maxima*. Plant Mol Biol 17: 505-508

Kostrzewa M, Zetsche K (1992) Large ATP synthase operon of the red alga *Antithamnion* sp. resembles the corresponding operon in cyanobacteria. J Mol Biol 227: 961-970

Kowallik KV (1989) Molecular aspects and phylogenetic implications of plastid genomes of certain chromophytes. In: JC Green, BSC Leadbeater, WL Diver (eds) The Chromophyte Algae: Problems and Perspectives. The Systematics Association Special Vol 38: 101-124, Clarendon Press Oxford

Kowallik KV (1993) Origin and evolution of plastids from chlorophyll *a+c*-containing algae: Suggested ancestral relationship to red algal plastids. In: RA Lewin (ed) Origins of Plastids. Chapman and Hall, New York, London, pp 223-263

Kowallik KV (1994) From endosymbionts to chloroplasts: Evidence for a single prokaryotic / eukaryotic endocytobiosis. Endocytobiosis and Cell Res 10: 137-149

Kowallik KV, Stöbe B, Schaffran I, Freier U (1995) The chloroplast genome of a chlorophyll a+c-containing alga, *Odontella sinensis*. Plant Mol Biol Rep 13: 336-342

Kroth-Pancic PG, Freier U, Strotmann H, Kowallik KV (1995) Molecular structure and evolution of the chloroplast *atpB/E* gene cluster in the diatom *Odontella sinensis*. J Phycol 31: 962-969

Kuhsel M, Kowallik KV (1987) The plastome of a brown alga, *Dictyota dichotoma*. II. Location of structural genes coding for ribosomal RNAs, the large subunit of ribulose-1,5-bisphosphate carboxylase/oxygenase and for polypeptides of photosystems I and II. Mol Gen Genet 207: 361-368

Lambowitz AM, Belfort M (1993) Introns as mobile genetic elements. Annu Rev Biochem 62: 587-622

Lockhart PJ, Howe CJ, Bryant DA, Beanland TJ, Larkum AWD (1992) Substitutional bias confounds inference of cyanoplast origins from sequence data. J Mol Evol 34: 153-162

Manhart JR, Kelly K, Dudock BS, Palmer JD (1989) Unusual characteristics of *Codium fragile* chloroplast DNA revealed by physical and gene mapping. Mol Gen Genet 216: 417-421

Manhart JR, Hoshaw RW, Palmer JD (1990) Unique chloroplast genome in *Spirogyra maxima* (Chlorophyta) revealed by physical and gene mapping. J Phycol 26: 490-494

Margulis L, Obar R (1985) *Heliobacterium* and the origin of chloroplasts. BioSystems 17: 317-325

Melkonian M (1996) Systematics and evolution of the algae: Endocytobiosis and evolution of the major algal lineages. In: Progress in Botany 57: 281-311

Mereschkowsky C (1905) Über Natur und Ursprung der Chromatophoren im Pflanzenreiche. Biol Centralbl 25: 593-604

Mereschkowsky C (1910) Theorie der zwei Plasmaarten als Grundlage der Symbiogenesis, einer neuen Lehre von der Entstehung der Organismen. Biol Zentralbl 30: 278-367

Michel F, Umesono K, Ozeki H (1989) Comparative and functional anatomy of group II catalytic introns - a review. Gene 82: 5-30

Morden CW, Delwiche CF, Kuhsel M, Palmer JD (1992) Gene phylogenies and the endosymbiotic origin of plastids. BioSystems 28: 75-90

Nitsch T (1995) Das Chloroplastengenom der siphonalen Grünalge *Bryopsis plumosa*: Klonierung und Sequenzierung phylogenetisch relevanter Bereiche. Thesis, University of Düsseldorf, 260 pp

Palmer JD, Thompson WF (1982) Chloroplast DNA rearrangements are more frequent when a large inverted repeat sequence is lost. Cell 29: 537-550

Pancic PG, Strotmann H, Kowallik KV (1992) Chloroplast ATPase genes in the diatom *Odontella sinensis* reflect cyanobacterial characters in structure and arrangement. J Mol Biol 224. 529-536

Raven PH (1970) A multiple origin for plastids and mitochondria. Science 169: 641-646

Reith M, Munholland J (1995) Complete nucleotide sequence of the *Porphyra purpurea* chloroplast genome. Plant Mol Biol Rep 13: 333-335

Rothschild LJ, Ragan MA, Coleman AW, Heywood P, Gerbi SA (1986) Are rRNA sequence comparisons the Rosetta Stone of phylogenetics? Cell 47: 660

Schenk HEA (1994) *Cyanophora paradoxa*: Anagenetic model or missing link of plastid evolution? Endocytobiosis and Cell Res 10: 87-106

Schimper AFW (1883) Über die Entwickelung der Chlorophyllkörner und Farbkörper. Bot Zeitung 41: 105-114, 121-131, 137-146, 153-162

Schnepf E (1992) From prey via endosymbiont to plastid: Comparative studies in Dinoflagellates. In: RA Lewin (ed) Origins of Plastids. Chapman and Hall, New York, London, pp 53-72

Siemeister G, Hachtel W (1990) Structure and expression of a gene encoding the large subunit of ribulose-1,5-bisphosphate carboxylase (rbcL) in the colourless euglenoid flagellate *Astasia longa*. Plant Mol Biol 14: 825-833

Siemeister G, Buchholz C, Hachtel W (1990) Genes for plastid elongation factor Tu and ribosomal protein S7 and six tRNA genes on the 73 kb DNA from *Astasia longa* that resembles the chloroplast DNA of *Euglena*. Mol Gen Genet 220: 425-432

Steinkötter J, Bhattacharya D, Semmelroth I, Bibeau C, Melkonian M (1994) Prasinophytes form independent lineages within the Chlorophyta: evidence from ribosomal RNA sequence comparisons. J Phycol 30: 340-345

Stewart KD, Mattox KR (1975) Comparative cytology, evolution and classification of the green algae with some consideration of the origin of other organisms with chlorophylls *a* and *b*. Bot Rev 41: 104-135

Stirewalt VL, Michalowski CB, Löffelhardt W, Bohnert HJ, Bryant DA (1995) Nucleotide sequence of the cyanoplast genome from *Cyanophora paradoxa*. Plant Mol Biol Rep 13: 327-332

Stöbe B (1995) Ursprung und molekulare Evolution der Chloroplasten der centralen Kieselalge *Odontella sinensis*. Thesis, University of Düsseldorf, 222 pp

Turmel M, Otis C, Lemieux C (1996) Evolution of the chloroplast genome in green algae. Abstr SV-2.02 Annu Meeting German Bot Soc Düsseldorf

Turmel M, Gutell RR, Mercier J-P, Otis C, Lemieux C (1993) Analysis of the chloroplast large subunit RNA gene from 17 *Chlamydomonas* taxa. Three internal transcribed spacers and 12 group I intron insertion sites. J Mol Biol 232: 446-467

Turmel M, Coté C, Otis C, Mercier J-P, Gray MW, Lonergan KM, Lemieux C (1995) Evolutionary transfer of ORF-containing group I introns between different subcellular compartments (chloroplasts and mitochondrion). Mol Biol Evol 12: 533-545

van de Peer Y, de Wachter R (1993) TREECON: A software package for the construction and drawing of evolutionary trees. CABIOS 9: 177-182

van den Hoek C, Mann DG, Jahns HM (1995) Algae. An Introduction to Phycology. Cambridge University Press, 623 pp

Wakasugi T, Nagai T, Kapoor M, Sugita M, Ito M, Ito S, Tsudzuki J, Nakashima K, Tsudzuki T, Suzuki Y, Hamada A, Ohta T, Inamura A, Yoshinaga K, Sugiura M (1997) Complete nucleotide sequence of the chloroplast genome from the green alga *Chlorella vulgaris*: The existence of genes possibly involved in chloroplast division. Proc Natl Acad Sci USA 94: 5967-5972

Whatley JM, John P, Whatley FR (1979) From extracellular to intracellular: The establishment of mitochondria and chloroplasts. Proc Royal Soc Lond 204: 165-187

Whatley JM, Whatley FR (1981) Chloroplast evolution. New Phytol 87: 233-247

Wilson RJM, Denny PW, Preiser PR, Rangachari K, Roberst K, Roy A, Whyte A, Strath M, Moore DJ, Moore PW, Williamson DH (1996) Complete gene map of the plastid-like DNA of the malaria parasite *Plasmodium falciparum*. J Mol Biol 261: 155-172

Wolfe KH, Li W-H, Sharp PM (1987) Rates of nucleotide substitution vary greatly among plant mitochondrial, chloroplast and nuclear DNAs. Proc Natl Acad Sci USA 84: 9054-9058

Wolfe KH, Morden CW, Palmer JD (1992) Function and evolution of a minimal plastid genome from a non-photosynthetic parasitic plant. Proc Natl Acad Sci USA 89: 10648-10652

Yamada T, Tamura K, Tadanori A, Songsri P (1994) Self-splicing group I introns in eukaryotic viruses. Nucleic Acids Res 22: 2532-2537

What's Eating Eu? The Role of Eukaryote/Eukaryote Endosymbioses in Plastid Origins

G. I. McFadden and P. Gilson
Plant Cell Biology Research Centre, School of Botany, University of Melbourne, Parkville VIC 3052, Australia

Key words: Nucleomorph, endosymbiosis (endocytobiosis), endocytobiont, cryptomonads, *Chlorarachnion*, plastid evolution, signal peptides, plastid targeting.

Summary: The plastids of green algae, and their descendants the land plants, clearly arose from a cyanobacterium-like endocytobiont. An early eukaryote (thus far unidentified) is believed to have phagocytosed a photosynthetic prokaryote and retained it as an endocytobiont. Having relinquished its autonomy, the endocytobiont is now reduced to organelle status (exogenosome) within the former eukaryotic host. The two membranes surrounding the plastid probably represent the two membranes of the gram-negative endocytobiont. A third membrane (presumed to have been created from the host plasma membrane during the engulfment of the endocytobiont) is now apparently lost. This scenario for plastid origin in green algae is termed a *primary endosymbiosis*. Plastids of red algae probably also derive from a primary endosymbiosis, quite possibly the same one as led to green algal plastids. Origins of plastids in other algal groups are less clear. The three-membraned plastids of dinoflagellates and euglenoids could also have arisen from one or more primary endosymbioses but with the food vacuole membrane being retained. However, the paucity of molecular data, particularly from dinoflagellates, leaves the origin of their plastids open to several different interpretations. The origin of plastids surrounded by four membranes (as occur in heterokonts, haptophytes, cryptophytes, and chlorarachniophytes) is hypothesised to involve two sequential endocytobioses. First, a primary endosymbiosis between a eukaryote and a prokaryote created a photosynthetic eukaryote analogous to, or perhaps even homologous to, the archetypal algal cell. This primary endosymbiosis was then followed by secondary endosymbiosis between a phagotrophic eukaryote and the product of the primary endosymbiosis to create a eukaryote with a photosynthetic eukaryotic endocytobiont. Secondary endosymbiosis produces plastids with four membranes: two from the original gram-negative prokaryote, a third from the plasma membrane of the primary host, and a fourth from the food vacuole of the secondary host. Initially, the nucleus and cytoplasm of the endocytobiont probably persisted, but in the case of heterokonts and haptophytes these structures have now apparently vanished leaving an essentially empty space between the endocytobiont's plasma membrane and the two membranes of the plastid. In the case of cryptomonads and chlorarachniophytes, however, it has now been shown that vestiges of the endocytobiont's nucleus and cytoplasm are retained. Now much reduced, these nucleocytoplasmic remnants are an invaluable key to unraveling the history of plastid origins through eukaryote/eukaryote endocytobioses. The extra membranes surrounding the plastids pose novel challenges for understanding targeting of proteins into plastids and a hypothesis for targeting is presented. The extraordinarily compact nuclear genomes of the endocytobionts are now being sequenced and could prove to be useful models for the coming wave of eukaryotic genome research.

Introduction

A decade ago Pace et al. (1986) wrote that the origin of plastids from eubacterial ancestors is "beyond reasonable doubt." Indeed, the more details we acquire of plastid molecular biology, the more certain we can be that these organelles derive from a bacterial endocytobiont. The question has now been redirected to ask whether all plastids derive from a single endocytobiosis or whether plastid-producing partnerships between eukaryotic hosts and prokaryotic endocytobionts arose multiple times (Gray 1993; Howe et al. 1992; Cavalier-Smith 1992; Douglas 1992). An unequivocal answer is not yet forthcoming, and the debate swings each time a new piece of the puzzle is overturned (e.g. Douglas et al. 1990; Reith and Munholland 1993; Wolfe et al. 1994; Morse et al. 1995). In addition to the uncertainty regarding how many endocytobioses between a prokaryote and eukaryote are involved in plastid history, another layer of complexity has insinuated itself. In 1974, Max Taylor floated the idea that some plastids were acquired secondarily through eukaryote/eukaryote endocytobiosis. This process, wherein a phagotrophic eukaryote engulfs and retains at least part of a photosynthetic eukaryote results in lateral transfer of photosynthetic capacity in addition to vertical inheritance (Gibbs 1978,1990; Whatley et al. 1979). Since a eukaryote/eukaryote endosymbiosis (endocytobiosis) needs always to follow an endocytobiosis between a prokaryote and a eukaryote (primary endosymbiosis), it is sometimes termed secondary endosymbiosis.

Multiple Plastid Membranes

The key influence on early hypotheses for secondary endosymbioses was the number of membranes surrounding the plastid (Gibbs 1978,1990; Whatley et al. 1979). Primary endosymbiosis is believed to have produced a double-membrane-bound endocytobiont (the two membranes most probably represent the membranes of the gram-negative endocytobiont with the food vacuole having been lost (Cavalier-Smith 1992). Those algal plastids with three or four surrounding membranes could then be explained by invoking secondary endosymbioses, the extra membranes deriving from the plasma membrane of the engulfed eukaryote or the food vacuole of the secondary host, or both. While these hypotheses were an attractive way to explain the extra membranes, they were difficult to test. The heredity of the membranes was (and still is) technically inscrutable, there being no unequivocal way to determine a membrane's evolutionary origin. The focus therefore turned to biochemical and molecular comparisons of the plastid itself. These approaches have provided a wealth of data but have not satisfactorily tested hypotheses of secondary endosymbiosis. Indeed, given the entire nucleotide sequence of Euglena chloroplast DNA (Hallick et al. 1993), and considerable biochemical data about the plastid, we still cannot say whether its plastid has a secondary or primary origin (e.g. McFadden et al. 1995; Cavalier-Smith 1995).

In the earliest stages of a secondary endosymbiosis the engulfed cell presumably possessed all those structures we associate with a free-living cell (viz. nucleus, cytoplasm, mitochondrion, endomembrane system, cytoskeleton). Secondary endosymbiosis hypotheses posit that these structures were lost, leaving only the plastid and some supernumerary membranes (Gibbs 1978,1990; Whatley et al. 1979). In two groups of algae, however, there appeared to be traces of the endocytobiont's nucleus and cytoplasm lying between the second and third membranes, exactly where they would be expected to lie (Fig. 1). These algae, cryptomonads and chlorarachniophytes, were recognised to offer the best chance of testing secondary endosymbiosis hypotheses. If a vestige of the endocytobiont's nuclear genome was still present, then the rapidly emerging techniques of molecular genetics could be brought to bear on the problem.

When Greenwood (1974) discovered a small nucleus-like organelle riding pick-a-back on the chloroplast of cryptomonads, he named it the nucleomorph and suggested it might represent the vestigial remains of a eukaryotic nucleus (Greenwood 1974; Greenwood et al. 1977). A similar hypothesis was put forward to explain the presence of a plastid with four membranes and associated nucleus-like structure in amoeboid organisms known as chlorarachniophytes (Hibberd and Norris 1984).

Proof of Secondary Endosymbiosis

Early cytochemical tests confirmed that both cryptomonad and chlorarachniophyte nucleomorphs contained DNA, and also revealed a nucleolus-like region rich in eukaryotic rRNAs (reviewed in McFadden 1993). These data suggested that the nucleomorph DNA encodes rRNA components for ribosome-like particles thought to be vestiges of the endocytobiont cytoplasm. Proving this required demonstrating that there were rRNA genes in the nucleomorph that were different to those of the former host nucleus. Moreover, transcripts of these rRNAs should be localised in the surrounding ribosome-like particles. Three techniques were pivotal in satisfying these proofs of secondary endosymbiosis: (1) PCR (polymerase chain reaction), (2) nucleomorph isolation, and (3) in situ hybridization.

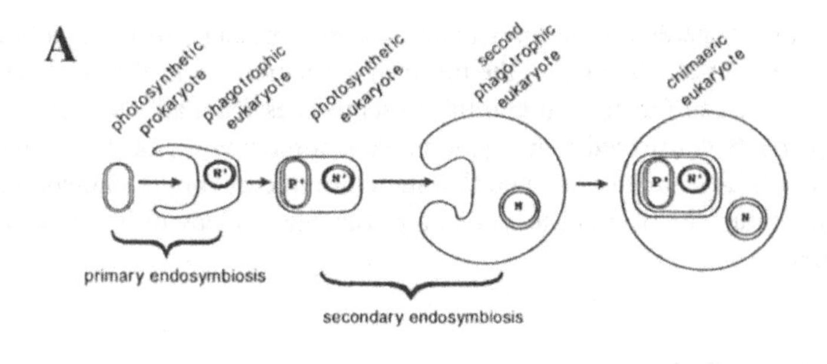

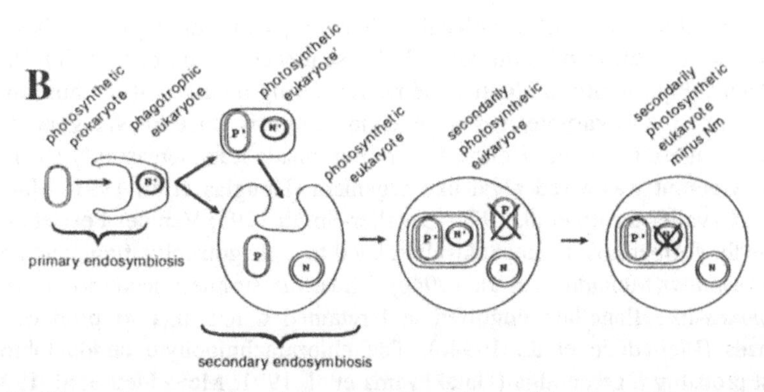

Fig. 1A, B. Schematic comparing origin of plastids **A** by traditional secondary endosymbiosis and **B** according to the modified hypothesis of Häuber et al. (1994) and Scherzinger et al. (1995). Common to both versions is that a primary endosymbiosis creates a plastid (P') in a eukaryote. In the original hypothesis this photosynthetic eukaryote was then engulfed and retained by a phagotroph (**A**). In the modified hypothesis (**B**), the photosynthetic eukaryote diverges into multiple lineages, one of which (photosynthetic eukaryote) is subsequently engulfed and retained by another (photosynthetic eukaryote) as a secondary endocytobiont. The primary endocytobiont (P) is then lost. The modified hypothesis predicts that genes transferred from plastid (P) to nucleus (N) are employed to supply proteins to the newly acquired plastid (P'). If nucleus (N) lacks the full complement of essential genes, then the endocytobiont nucleus (N') will persist (viz. chlorarachniophytes and cryptomonads). If however, in a separate secondary endosymbiosis, nucleus (N) contains all the genes essential for plastid (P'), then nucleus (N') could be redundant and disappear (viz. heterokont and haptophyte algae)

Douglas et al. (1991) employed PCR to amplify genes for the small subunit rRNA from total cryptomonad DNA and recovered two PCR products of different sizes, one approximately 200 bp longer than the other. The genes are divergent in sequence and considered likely to derive from phylogenetically remote eukaryotic lineages (Douglas et al. 1991), perhaps a host and an endocytobiont. The shorter gene was subsequently shown to occur in the host nucleus (Eschbach et al. 1991a) and encode transcripts for ribosomes in the host cytoplasm (McFadden et al.

1994a). The longer gene is found in isolated nucleomorphs (Hansmann and Eschbach 1991; Maier et al. 1991), and the transcripts accumulate in the small volume of cytoplasm between the second and third membranes (McFadden et al. 1994a). These experiments confirmed that cryptomonads contained a eukaryotic endocytobiont, and similar results confirmed that the chlorarachniophyte nucleomorph is also a vestigial nucleus of a photosynthetic eukaryotic endocytobiont (McFadden et al. 1994b).

Identity of the Partners

The quest to determine a molecular phylogeny of living organisms has resulted in the accumulation of numerous rRNA sequences. This burgeoning data base offers the opportunity to identify the nearest living relatives of the host and endocytobiont in a eukaryote/eukaryote endocytobiosis once rRNA gene for each component are available. Hence, for cryptomonads it is increasingly clear that the endocytobiont was a red algal-like organism (Douglas et al. 1991; Maier et al. 1991; Cavalier-Smith et al. 1994; Cavalier-Smith 1995; Van der Peer et al. 1996), while the former host is most closely related to a phagotrophic flagellate known as *Goniomonas* (McFadden et al. 1994c). Thus, the simplest scenario is that a *Goniomonas*-like flagellate engulfed and retained a red alga to produce cryptomonads (McFadden et al. 1994c). The chlorarachniophyte endocytobiont was most probably a green alga (Hatakeyama et al. 1991; McFadden et al. 1995), and newly developed phylogenetic methods suggest a close relationship to *Chlorella* and *Nannochlorum* (Van der Peer et al. 1996). The former host component of chlorarachniophytes is clearly related to a group comprising filose amoebae and sarcomonad flagellates (Bhattacharya et al. 1995a; Cavalier-Smith 1995), so chlorarachniophytes and cryptomonads almost certainly arose from two separate secondary endosymbioses and not from a single event as previously suggested (Cavalier-Smith 1992; Cavalier-Smith et al. 1994).

Role of the Nucleomorph

Once a secondary endosymbiotic origin for plastids of cryptomonads and chlorarachniophytes was proven unequivocally, focus again returned to the nucleomorph. Why should the nuclear genome and cytoplasmic ribosomes of the endocytobiont persist? With the knowledge that plastids are only semi-autonomous and dependent on the nucleocytoplasmic compartment to encode and synthesize the majority of their proteins, it seemed obvious that the nucleomorphs, being the relict nuclei of previously free-living photosynthetic organisms, were retained because they encode proteins essential to the acquired plastid (McFadden 1993; McFadden and Gilson 1995). Plastids probably contain in the order of 1000 proteins, but only about one-tenth of these are encoded and synthesized by the plastid itself. The remainder are believed, by default, to be produced from genes in the nucleus. Only a handful of these genes have been isolated. The nucleomorphs of cryptomonads and chlorarachniophytes were therefore anticipated to be the potential repository for the 900 or so still-to-be-identified nuclear-encoded plastid proteins (McFadden 1993). Since the nucleomorph genome was believed to have lost many non-essential functions, it was anticipated that the genes for plastid proteins would be readily identifiable.

Initial enthusiasm for the nucleomorphs as a wellspring of nuclear-encoded plastid protein genes was dampened with the establishment of nucleomorph genome size. Cryptomonad nucleomorphs contain only 550 - 660 kb of DNA (McFadden et al. 1994a; Eschbach et al. 1991b) and chlorarachniophyte nucleomorphs contain a mere 380 kb (Gilson and McFadden 1995). These genomes are clearly incapable of encoding anything approaching 900 plastid proteins, particularly as the demonstration that nucleomorphs encode rRNAs for their translation machinery suggested that they probably encoded components for their replication and expression (McFadden et al. 1994a, b). The great majority of chloroplast proteins (in cryptomonads at least) must therefore be encoded by the secondary host nucleus. (The chloroplast DNA of chlorarachniophytes has not been characterised so no estimate of how many plastid proteins it encodes can be made.)

Intracellular Gene Transfer and Secondary Endosymbiosis

How does the secondary host acquire the multitude of genes essential for plastid maintenance? Intracellular gene transfer (from the plastid to the secondary host, or from the endocytobiont nucleus to the secondary host) are both possible (McFadden and Gilson 1995). In addition, it has been postulated that the secondary host could originally have contained a primary endocytobiont (Fig. 1B) and the primary host nucleus (later to also become a secondary host) could have contained numerous plastid genes from the first (primary) endocytobiont which were then engaged for use with a secondarily acquired plastid (Häuber et al. 1994; Scherzinger et al. 1995). Thus, the secondary host could harbour a cocktail of genes for plastid proteins, some from the primary endocytobiont and some from the either the plastid or nucleus of the secondary endocytobiont, which, as we have seen above, may or may not contain the same primary endocytobiont.

The hypothesis that hosts of secondary endosymbioses were originally hosts for primary endosymbioses (Häuber et al. 1994; Scherzinger et al. 1995) has its basis in the belief that intracellular gene transfer is more likely between the multicopy plastid genome and host nucleus than between the secondary endocytobiont nucleus (usually single copy in cryptomonads and only several copies in some species of chlorarachniophytes) and secondary host nucleus. This is argued to permit more frequent gene transfer (Häuber et al. 1994; Scherzinger et al. 1995). Whether or not intracellular gene transfer will be more frequent from a prokaryote genome to a eukaryotic genome than it would from one eukaryotic to another is debatable. Certainly transfer between two eukaryotic genomes requires less changes in gene architecture for acquisition of expression than transfer between a prokaryote and a eukaryote would. Moreover, the argument that transfer from nucleomorph to secondary host nucleus is unlikely because the donor genome will be damaged (Scherzinger et al. 1995) is not entirely valid. Duplicated portions of parts or all of the nucleomorph chromosomes could be transferred by a variety of means, but still leave the donor genome intact. In any case, it is not necessary to infer that the endocytobiont remains intact throughout the development of the endocytobiosis. It is plausible, and in our opinion highly likely, that the endocytobiont was engulfed and lost multiple times before a stable integration was established, and gene transfer events could have occurred at various intervals during this process.

Advantages of Modified Secondary Endosymbiosis Hypothesis

The idea that one or several proteins required by the eukaryotic endocytobiont's plastid are relicts from a previous primary endosymbiosis (Häuber et al. 1994; Scherzinger et al. 1995) is attractive on several counts. Firstly, it has recently been shown that the sister group of the clade including cryptomonad hosts and *Goniomonas* is the glaucophytes, which contain descendants of primary endocytobionts in the form of cyanoplasts (Schenk 1994, Bhattacharya et al. 1995b; Van der Peer et al. 1995). In addition, the little-known cryptomonad *Peliaina* has a cyanelle-like (Margulis 1993) exogenosome. Moreover, at least one genus of the filose amoebae, which are the nearest relatives of the chlorarachniophyte hosts, also contains a cyanelle-like (Bhattacharya et al. 1995a; Van der Peer et al. 1995) exogenosome. Thus, the organisms from which the secondary hosts of cryptomonads and chlorarachniophytes derive are known to harbour primary endocytobionts, thereby satisfying one prediction (Häuber et al. 1994) of the modified secondary endosymbiont hypothesis.

In addition to predicting that the secondary hosts had primary endocytobionts, the modified hypothesis also has predictive power for the presence or absence of the nucleomorph in algae derived from secondary endosymbioses (Scherzinger et al. 1995). Nucleomorph presence would be dependent on the extent of gene transfer from plastid to nucleus in the two photosynthetic lineages (Fig. 1B). If the secondary host nucleus can supply all components for the secondarily acquired plastid, then the nucleomorph is redundant and can disappear. If, however, the secondary host nucleus is unable to supply all required components, then the nucleomorph may need to be retained. This means that if gene transfer between plastid and nucleus in the lineage that becomes the secondary endocytobiont is more extensive than gene transfer between plastid and nucleus in the lineage that becomes the secondary host, the resultant nucleomorph would contain genes not possessed by the secondary host nucleus and would need to be retained (Scherzinger et al. 1995). If, on the other hand, the gene transfer between plastid and nucleus in the lineage which becomes the secondary endocytobiont is less extensive than or equivalent to gene transfer between plastid and nucleus in the lineage that becomes the secondary host, then the secondary host nucleus would contain all the genes harboured by the resultant nucleomorph, making the latter redundant and perhaps dispensable. If (assuming the modified hypothesis is correct) we can generate complete gene catalogues for the plastid genomes of the primary and secondary endocytobionts for the various secondary endosymbioses, then we will be in better position to understand the existence of nucleomorphs.

Secondary Endosymbiosis and Selection

For any endocytobiosis to become permanently established there needs to be some form of interdependence. In its simplest form this interdependence could arise through loss of genetic autonomy of either or both partners (Cavalier-Smith and Lee 1985). If the endocytobiotic partnership is beneficial to both partners, one can envisage positive selection for any mechanism that stabilises the interaction. Hence, if a plastid gene were to become duplicated in the nucleus and acquire a mechanism for expression and transport of the product back into the plastid, then selection could favour mutations rendering the original plastid gene dysfunctional. If, on the other hand, the nucleus of a host already contained genes for plastid proteins (complete with targeting information) from a previous endocytobiont (the modified secondary endosymbiosis hypothesis), then the simple loss of function of one or more of these genes in the second endocytobiont could 'lock in' the new partner. Put another way, the host had the ability to complement a mutation in the endocytobiont genome, simultaneously rescuing the mutant and stabilising a mutually beneficial interaction. In this way one could imagine that hosts containing, or once having contained, primary endocytobionts may have a predisposition for obtaining novel photosynthetic endocytobionts (Häuber et al. 1994; Scherzinger et al. 1995).

Targeting Mechanisms

Plastid proteins encoded by the former host nuclei of cryptomonads and chlorarachniophytes and synthesized on cytoplasmic ribosomes must traverse the four enveloping membranes to reach the plastid stroma (Fig. 2). Those proteins destined for the thylakoid lumen must traverse five membranes (Fig. 2). No details of the targeting mechanisms are known, but some principles can be inferred by examining the topology of membranes surrounding the endocytobionts (e.g. Gibbs 1979; Rensing et al. 1994a; McFadden 1993; Hofmann et al. 1994).

In cryptomonads the endocytobiont lies within the lumen of the ER (endoplasmic reticulum), and the first stage of targeting could be achieved through a cotranslational insertion using a signal peptide (e.g. Rensing et al.

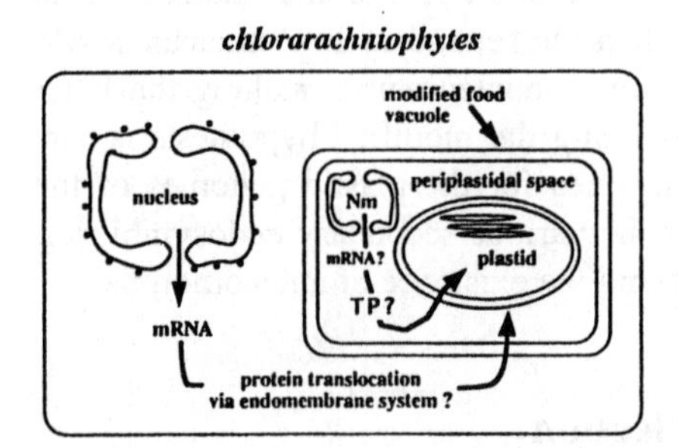

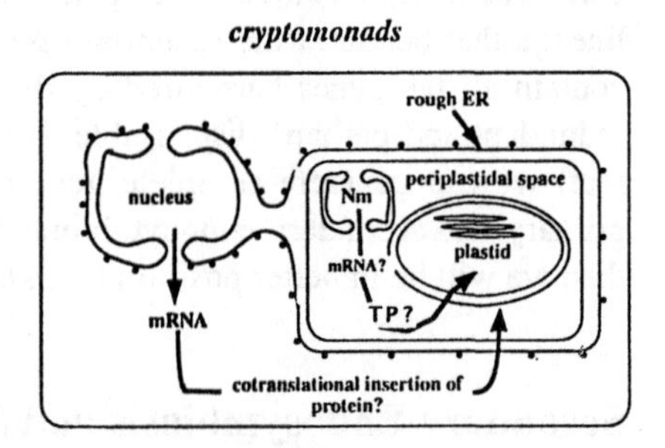

Fig. 2. Comparison of the subcellular location of endocytobionts in cryptomonads and chlorarachniophytes. Plastid proteins encoded by host nucleus could conceivably use a signal peptide to accomplish the initial targeting step across the outermost membrane, by cotranslational insertion into the ER lumen in cryptomonads and by vesicle transfer between the endomembrane system and the food vacuole in chlorarachniophytes. Mechanisms for further transport are discussed in the text. Plastid proteins encoded by the nucleomorph (Nm) in either organism could be delivered using a transit peptide (TP)

1994a; Hofmann et al. 1994), as demonstrated for nuclear-encoded plastid proteins of diatoms (whose plastids also lie in the ER lumen) (Grossman et al. 1990; Bhaya and Grossman 1991). Thus far only one plastid protein gene, the alpha subunit of phycoerythrin (*cpeA*), is known to be encoded by the secondary host nucleus in cryptomonads (Jenkins et al. 1990; G. I. McFadden, unpublished). This gene, which presumably derives from the cyanobacterial homologue present in the prokaryotic endocytobiont, could have transferred to the former host nucleus after the secondary endosymbiosis, or could have transferred to the primary host nucleus (which becomes the nucleomorph) and later to the secondary host nucleus. CpeA of cryptomonads has a 52-amino acid N-terminal extension that is removed *in vivo* (Jenkins et al. 1990). We speculate here that this leader is a multipartite targeting peptide involved in translocation of CpeA across the five membranes into its ultimate destination in the thylakoid lumen (Ludwig and Gibbs 1989).

Computer analyses of CpeA using an algorithm (Mantei 1992) and a neural network (Nielsen et al. 1995) both predict that the first 17 amino acids, which are part of a hydrophobic domain (Fig. 3A), could function as a eukaryotic signal peptide with cleavage predicted between alanine 17 and tyrosine 18. Comparison of the N-terminal leader peptide of CpeA with the signal peptide of diatom fucoxanthin/chlorophyll binding proteins (Grossman et al. 1990; Bhaya and Grossman 1991) reveals highly similar amino acid sequences in the first 17 amino acids (Fig. 3B) - exactly the region predicted to act as a eukaryotic signal peptide in CpeA (Fig. 3A). If the first 17 amino acids direct cotranslational insertion across the ER lumen, amino acids 18-52 (which are also removed from the mature protein)

(Jenkins et al. 1990) could play a role in transporting the protein across the remaining four membranes to the ultimate destination in the thylakoid lumen (Ludwig and Gibbs 1989). In this regard it is striking that neural network analysis (Nielsen et al. 1995) also predicts the existence of a second signal peptide at the C-terminal end of the leader sequence (Fig. 3A). Rather than being a eukaryotic signal peptide, the second signal peptide is prokaryote-like. Cleavage of the putative prokaryotic signal peptide is predicted to occur between alanine 52 and alanine 53 - exactly the site determined *in vivo* (Jenkins et al. 1990). The region of the CpeA leader ahead of the mature protein displays five features typical of prokaryotic signal peptides (cf. von Heijne 1983; 1985; Howe and Wallace 1990): compliance with the -1,-3 rule, an alanine residue at +1, a leucine residue at -8, a central hydrophobic region ("h-region"; see Fig. 3A), and positively charged amino acids (Arg-24 Arg-25, the "n-region") ahead of the hydrophobic subdomain. This indicates that transport of CpeA across the thylakoid membrane into the lumen most probably uses the Sec-dependent[1] prokaryotic signal peptide pathway (see Howe and Wallace 1990), which is sometimes described as conservative targeting. We postulate that targeting of CpeA in cryptomonads involves a multipartite leader sequence with at least three components: (1) a eukaryotic signal peptide directing cotranslational insertion into the ER lumen, (2) an undefined region (probably spanning amino acids 18 to about 25-30, which is hydrophilic in CpeA) that is involved in transit across the endocytobiont plasma membrane and which also contains a transit peptide[2] effecting transport across the two plastid membranes into the plastid stroma, and (3) a second, prokaryote-like signal peptide involved in Sec-dependent transport into the thylakoid lumen.[3]

The ability of the most N-terminal 17 amino acids to act as a first component for plastid targeting using the signal peptide pathway in cryptomonads could be tested in reconstituted in vitro systems using canine microsomes (Walter and Blobel 1983; Titus 1991), or isolated endocytobionts when available. The putative prokaryotic signal peptide, which we predict to be the final portion (part 3) of the multipartite targeting peptide, could be assayed for secretion competency in a variety of systems using transgenic bacteria. The possible role of other parts of the leader sequence in targeting could be tested in the recently developed in vitro

[1] Cryptomonad plastid DNA encodes a secY homologue that may have a role in transport into the thylakoid lumen (Douglas 1992; Rensing and Maier 1994).

[2] The recent demonstrations that transit peptides from a red algal nuclear-encoded plastid protein will effect targeting into plant plastids (Apt and Grossman 1993) while plant transit peptides effect import into cryptomonad plastids (Rensing and Maier 1995) suggest that the transit peptide principally works in a fundamentally similar way in divergent photosynthesizers and recycling of transit peptides for targeting to different plastids is thus plausible.

[3] The beta subunit of phycoerythrin in cryptomonads is encoded by the chloroplast genome but does not possess a bacterial-type signal peptide and must be transported into the lumen by another pathway (Reith and Douglas 1990).

import system for isolated cryptomonad plastids and thylakoids (Rensing and Maier 1995). Characterisation of genes for other plastid proteins would also help to define the roles of different parts of the targeting system.

In contrast to cryptomonads, the chlorarachniophyte endocytobiont lies in a vacuole rather than the ER lumen (Hibberd and Norris 1984) (see also Fig. 1, this paper). This vacuole, hypothesised to be the food vacuole, is potentially part of the endomembrane system and proteins could theoretically be targeted to the plastid through the endomembrane pathway (Häuber et al. 1994) using a signal sequence as the initial step. Alternatively, proteins could be targeted directly from the cytoplasm using targeting information located anywhere within the protein (e.g. Verner and Schatz 1988). Isolation of genes for nuclear-encoded plastid proteins in chlorarachniophytes will be necessary to begin testing these possibilities.

The Question of the Second Membrane

Transport across the ER membrane, the two chloroplast envelopes, and the thylakoid membrane can all be plausibly explained using pre-existing systems engaged in tandem. How the proteins in either cryptomonads or chlorarachniophytes cross the second membrane (counting from the outside) is not known. This membrane is hypothesised to represent the plasma membrane of the eukaryotic endocytobiont (see McFadden 1993), so a transporting mechanism would need to cross from the "outside" to the "inside" of the membrane with respect to its original topology. It has been suggested that the chaperonin hsp70, encoded by the nucleomorph and probably situated in the vestigial endocytobiont cytoplasm between the second and third membranes, could have an as yet undefined role in transport across the second membrane (Hofmann et al. 1994; Rensing et al. 1994b).

Targeting and the Modified Secondary Endosymbiosis Hypothesis

Targeting of nuclear-encoded proteins into a secondarily acquired plastid is obviously complex and has implications for the modified secondary endosymbiosis hypothesis. How do proteins originally destined for a primary endocytobiont become targeted to a secondary endocytobiont? We have already presented an hypothesis explaining a mechanism for this targeting and, with the exception of the second membrane, this system uses pre-existing components recycled into a multipartite tandem array. Whether the genes come from primary endocytobiont, the nucleomorph, or directly from the plastid of the secondary endocytobiont into the secondary host nucleus, the components used to assemble a targeting system could be the same.

What do Nucleomorphs Encode?

Characterisation of nucleomorph information content has been approached through random sequencing of nucleomorph DNA. In cryptomonads the heat shock chaperone hsp70 is the only nucleomorph protein gene characterised thus far (Hofmann et al. 1994; Rensing et al. 1994b). This protein is suggested to play a role in transport of nuclear-encoded plastid proteins across the cytoplasm separating the second and third membranes surrounding the plastid (Hofmann et al. 1994; Rensing et al. 1994b).

Fifteen genes have been characterised from the chlorarachniophyte nucleomorph (Gilson and McFadden 1996). These genes encode three rRNAs, one snRNA (U6), and several proteins (Gilson and McFadden 1996). The gene products have roles in mRNA processing and translation (Gilson and McFadden 1996). Interestingly, an hsp70 chaperonin is also encoded by the chlorarachniophyte nucleomorph (Rensing et al. 1994b; Gilson and McFadden 1997). The gene density in the chlorarachniophyte nucleomorph is very high, with more than 70% of the DNA being coding DNA, and some of the genes overlap and others are cotranscribed (Gilson and McFadden 1996). The DNA is relatively AT rich with 73% AT in coding regions and 83% in non-coding regions. The genetic code is assumed to be standard with ATG start codons and all three stop codons used (Gilson and McFadden 1996). No evidence of RNA editing has yet been encountered (Gilson and McFadden 1996). The protein gene transcripts are polyadenylated, probably using a standard eukaryotic polyadenylation motif (ATTAAA) (Gilson and McFadden 1996). The sequence data are consistent with the nucleomorph genome being a reduced eukaryotic nucleus with at least partial capacity to express its information content.

Only one chlorarachniophyte nucleomorph gene (*clpP*, the catalytic subunit of the ATP-dependent Clp or Ti protease) from the 15 thus far identified potentially encodes a protein for the plastid (Gilson and McFadden 1996). Clp proteases are common to plastids and bacteria, where they degrade incorrectly folded proteins (Gottesman et al. 1990), and the nucleomorph ClpP (which has an N-terminal extension relative to plastid-encoded ClpPs) could be targeted to the plastid (Gilson and McFadden 1996). Alternatively, the nucleomorph ClpP may function in the endocytobiont cytoplasm as animal and plant cells have recently been demonstrated to have nuclear genes that encode ClpP-like proteins (GenBank accession Z49073, T46236, H23955, H35529, T23272).

Why Has the Nucleomorph not Disappeared?

It now seems an inescapable conclusion that there has been extensive transfer of genes from the nucleomorph to the secondary host nucleus. Beyond the previously discussed selection for genetic interdependence, why so many genes should be transferred cannot be readily explained. However, given that so many genes have

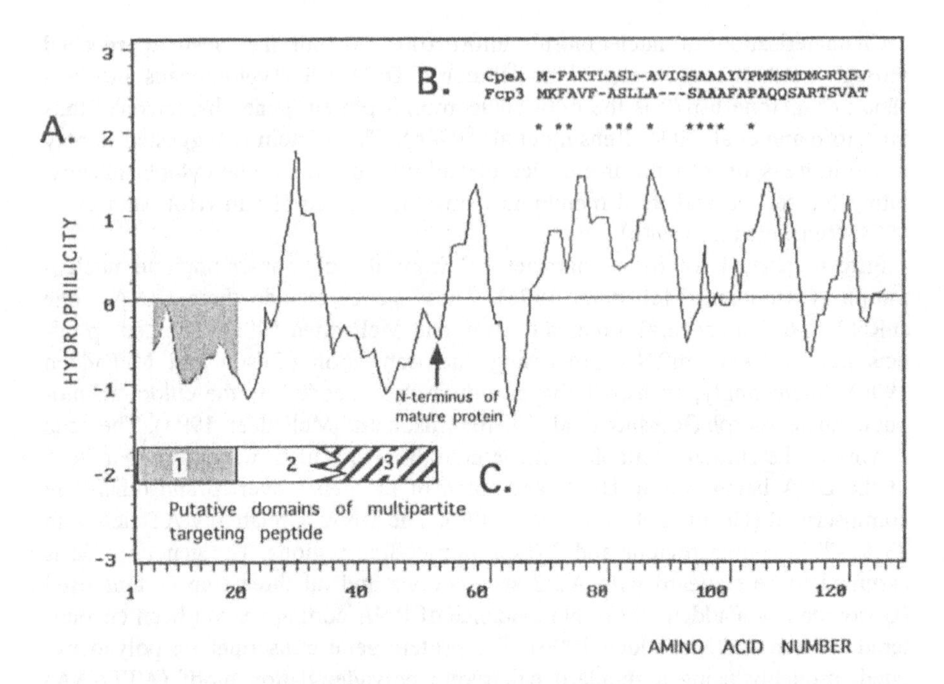

Fig. 3. Analysis of N-terminal leader for cryptomonad nuclear-encoded plastid protein. **A** Hydrophilicity plot of CpeA. The site where the first 52 amino acids are removed in vivo to produce the mature CpeA is indicated by an arrow. The first 24 amino acids form a hydrophobic domain, which is followed by a hydrophilic domain, and then a further hydrophobic domain ahead of the cleavage site. It is postulated that these domains represent minimally three portions of a multipartite targeting peptide involved in transport of CpeA across five membranes. A putative eukaryote-like signal sequence domain comprising 17 N-terminal amino acids is shaded. **B** Alignment of the N-terminal amino acid sequence of the diatom fucoxanthin/chlorophyll c-binding protein (Fcp3) and the cryptomonad phycoerythrin alpha subunit (CpeA). Both proteins are nuclear-encoded plastid proteins targeted into plastids surrounded by four membranes, the outermost of which is continuous with the rough ER. The N-terminal sequence of Fcp protein has been demonstrated to function as a signal peptide (Bhaya and Grossman 1991) and the sequence identity with CpeA suggests that a similar mechanism could operate in cryptomonads. **C** Model showing three putative domains for multipartite targeting peptide. Domain 1 would contain a eukaryotic signal peptide. Domain 2 would contain a transit peptide and perhaps also information involved in crossing the endocytobiont's[1] plasma membrane. Domain 3 would be a prokaryote-like signal peptide involved in transport into the thylakoid lumen. The boundary separating domains 2 and 3 cannot be defined yet

---

[1] At the editors request, the terms „endocytobiosis, endocytobiont, and exogenosome“ have been exchanged for the terms endosymbiosis, endosymbiont, and endosymbiotic organelle respectively, if they are used for mechanistic process desription.

moved or been lost one may ask: Why not all genes? Why hasn't the nucleomorph completely disappeared?

There are several possible explanations for the presence of the nucleomorph. Firstly, it is possible that the nucleomorphs are still in a state of reduction and that eventually they will disappear. However, given the similarity in karyotype between chlorarachniophytes and cryptomonad nucleomorphs (three small ca. 100- to 200-kb chromosomes each carrying rRNA genes) (McFadden and Gilson 1995), we believe they have converged at some similar end point of reduction and neither can now disappear altogether. A second explanation for nucleomorph persistence is that it is selfish DNA and is a mere passenger, but the extreme reduction combined with apparent functionality bespeak a contributory role for nucleomorphs. A third possibility is that the nucleomorph provides some as yet undetermined function for the former host that is not related to photosynthetic capacity. Such a gene product, or secondary product, could conceivably be exported to the host cytoplasm. Yet another possibility is that during reduction nucleomorphs have evolved some peculiar genetic process which precludes transfer of the last remaining plastid protein genes into the host nucleus, a molecular mismatch that somehow stemmed the flow of genes in the latter stage. One possibility for such a genetic peculiarity is the abundant small introns in chlorarachniophyte nucleomorph protein genes (Gilson and McFadden 1996). At 18-20 bases these are the smallest known spliceosomal introns (Gilson and McFadden 1996) and it may be that they have become so tiny, or so divergent in sequence, during the miniaturization of the nucleomorph genome that they would not be able to removed by the host nucleus' spliceosomal machinery should the genes become transferred to the host nucleus. It would be necessary to transform the host nucleus with a gene containing a nucleomorph intron to test this hypothesis.

The final possible explanation for the nucleomorph may not hinge on genes for plastid proteins at all. It is quite possible that the nucleomorph encodes no plastid proteins but simply exists to provide the pathway for proteins targeted to the plastid from the host cytoplasm. In this scenario we can imagine that the targeting mechanism has evolved to pass through the two outer membranes, across the endocytobiont cytoplasm, and then through the two inner membranes. Perhaps the nucleomorph's only role is to provide components of this pathway for the proteins to traverse in order to reach their correct destination.

Acknowledgements. We thank Uwe Maier and co-workers for discussion of ideas and access to preprints. G. McF. is a Senior Research Fellow of the Australian Research Council. P. G. is the recipient of an Australian Postgraduate Award. The project is supported by a grant from the Australian Research Council.

References

Apt RE, Hofmann WE, Grossman A (1993) J Biol Chem 268: 16208-16215

Bhattacharya D, Helmchen T, Melkonian M (1995a) J Euk Microbiol 42: 65-69
Bhattacharya D, Helmchen T, Bibeau C, Melkonian M (1995b) Mol Biol Evol (in press)
Bhaya D, Grossman AR (1991) Mol Gen Genet 229: 400-404
Cavalier-Smith T (1992) BioSystems 28: 91-106
Cavalier-Smith T (1995) In: Arai R (ed) Biodiversity and evolution. Tokyo: National Science Museum
Cavalier-Smith T, Lee JJ (1985) J Protozool 32: 376-379
Cavalier-Smith T Allsop MTEP, Chao EE, Mathews G (1994) Proc Natl Acad Sci USA 91: 11368-11372
Douglas SE (1992a) BioSystems 28: 57-68
Douglas SE (1992b) FEBS Lett 298: 93-96
Douglas SE, Durnford DG, Morden CW (1990) J Phycol 26: 500-508
Douglas SE, Murphy CA, Spencer DF, Gray MW (1991) Nature 350: 148-151
Eschbach S, Wolters J, Sitte P (1991a) J Mol Evol 32: 247-252
Eschbach SC, Hofmann CJB, Maier U-G, Sitte P, Hansmann P (1991b) Nucl Acids Res 19: 1779-1781
Gibbs SP (1978) Can J Bot 56: 2882-2889
Gibbs SP (1979) J Cell Sci 35: 253-266
Gibbs SP (1990) In: Wiessner W, Robinson DG, Starr RC (eds) Experimental Phycology, Vol 1. Cell walls and surfaces, reproduction and photosynthesis. Springer, Vienna, pp 145-157
Gilson PR, McFadden GI (1995) Chromosoma 103: 635-641
Gilson PR, McFadden GI (1996) Proc Natl Acad Sci USA 93: 7737-7742
Gilson PR, McFadden GI (1997) BioEssays 19: 167-173
Gottesman S, Squires C, Pichersky E, Carrington M, Hobbs M, Mattick JS, Dalrymple B, Kuramitsu H, Shoroza T, Foster T, Clark WP, Ross B, Squires CL, Maurizi MR (1990) Proc Natl Acad Sci USA 87: 3513-3517
Gray MW (1993) Curr Opin Genet Dev 3: 884-890
Gray MW (1994) Am Soc Microbiol News
Greenwood AD (1974) In: Sanders JV, Goodchild DJ (eds) Electron microscopy 1974, Australian Academy of Sciences, Canberra, pp 566-567
Greenwood AD, Griffiths HB, Santore UJ (1977) Br Phycol J 12: 119
Grossman AR, Manodori A, Snyder D (1990) Mol Gen Genet 224: 91-100
Hallick RB, Hong L, Drager RG, Favreau MR, Monfort A, Orsat B, Spielman A, Stutz E (1993) Nucl Acids Res 21: 3537-3544
Hansmann P, Eschbach S (1991) Eur J Cell Biol 52: 373-378
Hatakeyama N, Sasa T, Watanabe MM, Takaichi S (1991) J Phycol 27 (suppl), A154
Häuber MM, Müller SB, Speth V, Maier U-G (1994) Bot Act 107: 383-386
Hibberd DJ, Norris RE (1984) J Phycol 20: 310-330
Hofmann CJB, Rensing SA, Häuber MM, Martin WF, Müller SB, Couch J, McFadden GI, Sitte P, Igloi GL, Maier U-G (1994) Mol Gen Genet 243: 600-604
Howe CJ, Wallace TP (1990) Nucl Acids Res 18: 3417
Howe CJ, Beanland TJ, Larkum AWD, Lockhart PJ (1992) Trends Ecol Evol 7: 378-383
Jenkins J, Hiller RG, Speirs J, Godova-Zimmermann J (1990) FEBS Lett 273: 191-194
Ludwig M, Gibbs SP (1989) J Cell Biol 108: 875-884
Maier U-G, Hofmann CJB, Eschbach S, Wolters J, Igloi GL (1991) Mol Gen Genet 230: 155-160

Mantei N (1992): Signalase/Analyze is a Macintosh freeware program for applying the algorithm of von Heijne (Nucl Acids Res 14: 4683-4690, 1986) to the prediction and analysis of mammalian signal sequences
Margulis L (1993) Symbiosis in cell evolution. WH Freeman, New York
McFadden GI (1993) Adv Bot Res 19: 189-230
McFadden GI, Gilson PR (1995) Trends Ecol Evol 10: 12-17
McFadden GI, Gilson PR, Douglas SE (1994a) J Cell Sci 107: 649-657
McFadden GI, Gilson PR, Hofmann CJB, Adcock GA, Maier U-G (1994b) Proc Natl Acad Sci USA 91: 3690-3694
McFadden GI, Gilson PR, Hill DRA (1994c) Eur J Phycol 29: 29-32
McFadden GI, Gilson PR, Waller RF (1995) Arch Protistenk 145: 231-239
Morse D, Salois P, Markovic P, Hastings WJ (1995) Science 268: 1622-1624
Nielsen H, Engelbrecht J, Brunak S, von Heijne G (1995) Signalp World Wide Web Server. An improved method for prediction of signal peptides using neural networks. http://www.cbs.dtu.dk/signalp/cbssignalp.1.html
Pace NR, Olsen GJ, Woese CR (1986) Cell 45: 325-326
Reith M, Douglas SE (1990) Plant Mol Biol 15: 585-592
Reith M, Munholland J (1993) Plant Cell 5: 465-475
Rensing SA, Maier U-G (1994) Mol Phyl Evol 3: 187-191
Rensing SA, Maier U-G (1995) Curr Genet (in press)
Rensing SA, Müller SB, Hofmann CJB, Häuber MM, Maier U-G (1994a) Endocytobiosis and Cell Res 10: 259-260
Rensing SA, Goddemeier M, Hofmann CJB, Maier U-G (1994b) Curr Genet 26: 451-455
Schenk HEA (1994) Endocytobiosis and Cell Res 10: 87-106
Scherzinger MM, Hofmann CJB, Rensing SA, Maier U-G (1995) Bot Acta (in press)
Taylor FJR (1974) Taxon 23: 229-258
Titus DE (1991) Protocols and Applications Guide, Madison: Promega Corporation
Van der Peer Y, Rensing SA, Maier U-G, De Wachter R (1996) Proc Natl Acad Sci USA 93: 7732-7736
Verner K, Schatz G (1988) Science 241: 1307-1313
von Heijne G (1983) Eur J Biochem 133: 17-21
von Heijne G (1985) J Mol Biol 184: 99-105
Walter P, Blobel G (1983) Meth Enzymal 96: 50-84
Whatley JM, John P, Whatley FR (1979) Proc R Soc London B 204: 165-187
Wolfe GR, Cunningham FX, Durnford D, Green BR, Gantt E (1994) Nature 367: 566-568

The Complete Sequence of the Cyanelle Genome of *Cyanophora paradoxa*: The Genetic Complexity of a Primitive Plastid*

Löffelhardt[1], V.L. Stirewalt[2], C.B. Michalowski[3], M. Annarella[2], J.Y. Farley[2], W.M. Schluchter[2], S. Chung[2], C. Neumann-Spallart[1], J.M. Steiner[1], J. Jakowitsch[1], H.J. Bohnert[3] and D.A. Bryant[2]
[1]Institut für Biochemie und Molekulare Zellbiologie der Universität Wien und Ludwig-Boltzmann-Forschungsstelle für Biochemie, 1030 Vienna, Austria
[2]Department of Biochemistry and Molecular Biology, The Pennsylvania State University, University Park, Pennsylvania 16802, USA
[3]Department of Biochemistry, University of Arizona, Tucson, Arizona 85721, USA

Key words: *Cyanophora paradoxa,* Glaucocystophyta, cyanoplast, cyanelle (genome), complete sequence, protoplastid.

Summary: The completion of genome sequencing of the cyanoplasts (here still called cyanelles) of *Cyanophora paradoxa* and of other primitive plastids provides strong evidence in favor of a common origin for all plastid types from a semiautonomous ancestral photosynthetic organelle, the „protoplastid". The cyanoplasts representing an early off-branch are the only plastids that have retained the peptidoglycan wall inherited from the protoplastid (and, previously, from the endocytobiotic cyanobacterium, that participated in the singular primary endosymbiotic event). Cyanoplasts, as well as other plastids devoid of chlorophyll b, contain between 50 and 100 genes that are absent from the genomes of the more derived chloroplasts of higher plants.

Introduction

Present knowledge of the biochemistry, molecular biology, and cell biology of plastids leaves no room for doubt about their endocytobiotic origin. A monophyletic (cyanobacterial) origin and a singular event are assumed (Löffelhardt 1995; Reith and Munholland 1993; Löffelhardt and Bohnert 1994a,b). Support for these concepts will be presented during the discussion of the complete sequence of the cyanelle (cyanoplast) genome from *Cyanophora paradoxa* (Stirewalt et al. 1995) (Fig. 1). Considering that it was the primary endosymbiotic event that led to the establishment of the semiautonomous photosynthetic organelle while secondary endosymbioses resulted only in modifications (at the genomic and/or struc-

* Dedicated to Prof. Dr. Benno Parthier on occasion of his 65th birthday.

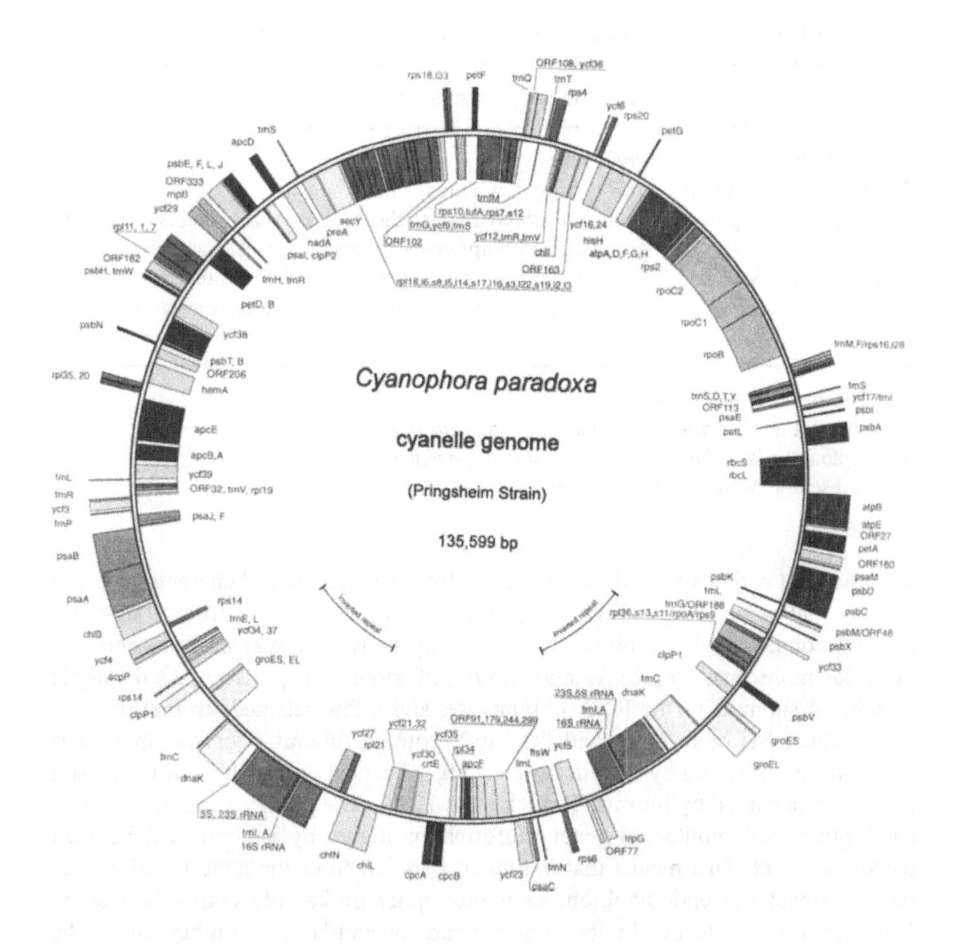

Fig. 1. Gene map of the *C. paradoxa* cyanelle genome. Transcription of the genes on the outer circle is clockwise. *Ycf*: open reading frame present in several plastid DNAs

tural level), it seems justified to say that all plastids are finally derived from a common precursor which we may call the "protoplastid".

Plastid genomic data sets

The data set available on completely sequenced plastid genomes comprises at present five higher plants, one green alga, one euglenoid alga, one red alga (Reith and Munholland, 1995), one diatom (Kowallik et al. 1995), and one glaucocystophyte (*C. paradoxa*). From a comparative analysis, a standard set of approximately 80 genes emerged that are shared by almost all plastid types regardless of

Table 1. Additions to the standard set of plastid genes

Plastids from	Additional genes present
Higher plants	*ndh* genes[a]
Chlorophyta, Euglenoids	*tufA*, chlorophyll biosynthesis[b]
Chromophyta, Cryptomonads, *Porphyra*, *Cyanophora*	*tufA*, chlorophyll biosynthesis, ribosomal proteins, chaperonins, phycobiliproteins[c], preprotein translocase subunits, metabolite and ion transport[d] fatty acid synthetase, amino acid biosynthesis, isoprenoid biosynthesis, transcription factors

[a]pseudogenes in *Pinus thunbergii.*
[b]Also present in *Marchantia polymorpha* and *Pinus thunbergii.*
[c]In Cryptomonads, red algae, and *Cyanophora paradoxa.*
[d]Also present in *Marchantia polymorpha.*

their pigment composition, their more primitive or more derived characteristics, or the fact that some (*Euglena gracilis*, *Odontella sinensis*) arose from a superimposed secondary endosymbiotic event. Among these common plastid genes are those for more than 30 rRNAs and tRNAs, 20 ribosomal proteins, 27 thylakoid proteins, 4 subunits of the RNA polymerase, and a few selected stromal proteins such as the LSU of Rubisco and the ClpP protease subunit. This minimal set is most closely displayed by *E. gracilis* (143 kb, 97 genes), where 40% of the plastid genome is occupied by introns (Hallick et al. 1993). The protoplastid, on the other hand, presumably contained genetic information higher by a factor of 2.5-3 than the standard set. This means that the major part (90%) of the gene transfer from the cyanobacterial endocytobiont (a phototrophic prokaryote comprising about 2500 genes) to the host cell nucleus had already taken place. A combination of the features of the extant cyanelles of *C. paradoxa* (genome size 135.6 kb, 193 genes) and the rhodoplasts of *Porphyra purpurea* would most closely describe the protoplastid, the large rhodoplast genome (191 kb) harboring the highest number of genes (250) thus far reported for a plastid (Reith and Munholland 1993, 1995) and the peptidoglycan wall of the cyanelles (Fig. 2).
The gene contents of all plastid types fall between these extremes. The gene transfer to the nucleus has continued but at a diminished rate and with quantitative and qualitative differences between the various evolutionary lines. Green algae and higher plants show a similar gene content: the *ndh* genes have been retained in the latter only, indicating a rather late transfer of these genes in all algal phyla. Cyanelles, red algae, and the chromophytic and cryptophytic algae originating from a secondary endosymbiotic event show a considerable surplus of genes, especially in respect of nonphotosynthetic functions of plastids (Table 1).

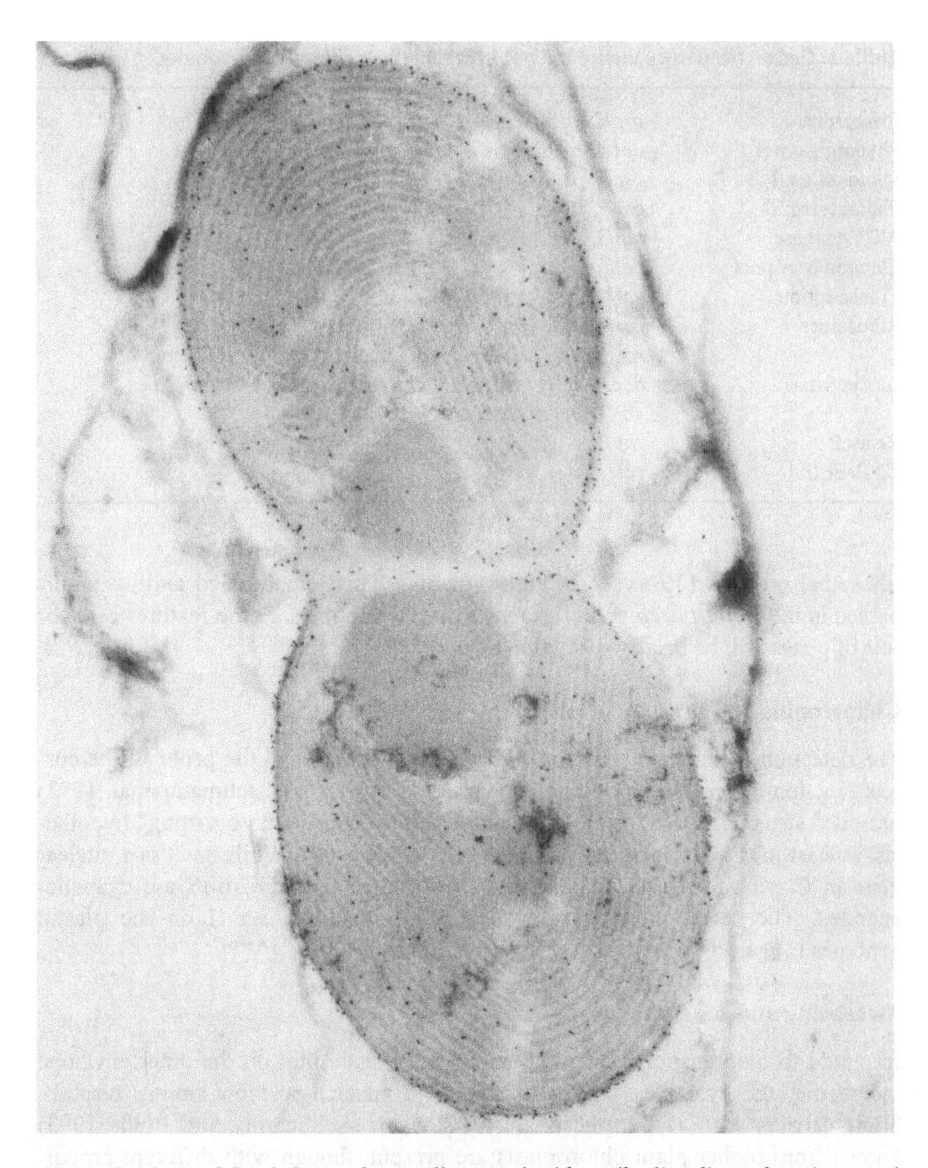

Fig. 2: Sections of fixed *Cyanophora* cells treated with antibodies directed against peptidoglycan (*E. coli*) and subsequently with Protein A-Gold. Deposition of gold particles occurs mainly in the envelope and the newly formed septum of the dividing cyanelle

The Surplus Genes of the Cyanelle Genome

The surplus (as compared to higher plant chloroplasts) of more than 50 genes found on the cyanelle genome (Fig. 1, Table 2) is most pronounced in respect of

Table 2. Genes found in cyanelles but not in higher plant chloroplast genomes

Chaperones	*groES, groEL, dnaK*
Phycobilisomes	*apcA, apcB, apcD, apcE, apcF, cpcA, cpcB*
Photosystem I	*psaE, psaF, (psaM)*
Photosystem II	*psbV, psbX*
ATP synthase	*atpG, atpD*
Electron transport	*petF*
Transcription	*ycf27, ycf29, ycf30*
Ribosomes	*rps5, rps9, rps10, rps13, rps17*
	rpl1, rpl3, rpl5, rpl6, rpl18, rpl19, rpl28, rpl34, rpl35
Biosynthesis	*acpP, chlB, chlI, chlL, chlN, crtE, ftsW, hemA, hisH*
	nadA, preA, rbcS, trpG
RNaseP	*rnpB*
CAB/ELIP	*ycf17*

ribosomal proteins (36 vs 21 in higher plants). This is paralleled and even surpassed in the *P. purpurea* and *O. sinensis* plastid genomes. Some features of other sets of genes will be briefly discussed below.

Chaperonins and Protein Transport

The detection of the *secY* gene on the cyanelle genome and the proof of the correct function of the corresponding gene product in *E. coli* (Flachmann et al. 1993) provided strong support for the correctness of the "conservative sorting" hypothesis, at least in the case of plastids (Smeekens et al. 1990). While *secA* is a nuclear gene in *C. paradoxa*, the chaperonins DnaK, GroEL, and GroES are cyanelle-encoded. The retention of these genes (even including *secA*) on the plastid genomes is again typical of algae without chlorophyll b.

Metabolite and Ion Transport

In regard to the transport systems for ions and metabolites on the inner envelope membrane, the cyanelles may also occupy a special position among plastids. Some carriers such as the phosphate translocator (Schlichting and Bothe 1993) known from higher plant chloroplasts are present, though with different properties. Others are absent, such as the malate/oxaloacetate shuttle and the ADP/ATP translocator (Schlichting et al. 1990). Four cyanelle ORFs (open reading frames) are likely to encode subunits of two typical prokaryotic transporters comprising a periplasmic binding protein, one or two integral membrane proteins and a peripheral membrane protein (cytoplasmic side) containing the ATP-binding site. The only other documented case with respect to higher plant plastid genes is *mbpX* and *mbpY* from *Marchantia polymorpha*, which may be involved in sulfate transport (Laudenbach and Grossman 1991).

Recently three clustered genes for components of an ABC transporter for manganese ions were identified in the cyanobacterium *Synechocystis* 6803 (Bartsevich and Pakrasi 1995). The ATP-binding protein, encoded by *mntA,* shows 40% identity to the cyanelle ORF244, and the intrinsic membrane protein (*mntB*) 33% identity to the cyanelle ORF299 that is adjacent. The conservation in sequence and gene arrangement is significant and exceeds the values found for the ATP-binding component of various other prokaryotic transporters (<30%), where often only the regions surrounding the Walker boxes contribute. The second pair comprises cyanelles *ycf16* (ORF 259) and *ycf24* (ORF486). Homologs to both have been detected on red algal and diatom plastid genomes, with identity scores as high as 60% to 68% (putative ATP-binding protein, ORF259) and 78% to 81% (putative membrane protein, ORF486). While it is obvious that these clustered genes also encode ABC transporter components, a likely function cannot be assigned at present.

Protein Degradation

The chloroplast gene for the proteolytic subunit of the Clp protease, *clpP*, is highly conserved (Sugiura 1992; Maurizi et al. 1990). Interestingly, two genes, *clpP1* and *clpP2*, are present on the cyanelle genome, showing only moderate sequence similarity (Table 3), in contrast to the two counterparts found in the cyanobacterium *Synechocystis* sp. strain 6803 (Kaneko et al. 1995). Furthermore, the conserved serine and histidine residues found in cyanelle *clpP1* (positions 98 and 123, respectively) and all other *clpP* genes are replaced in cyanelle *clpP2* by glycine and glutamine, respectively. Since ClpP is a serine protease (Maurizi et al. 1990) this might result in an alteration of the cleavage specificity. Genes for the ATP-binding large subunit of the Clp protease (*clpC*), which is nucleus-encoded in higher plants and in *C. paradoxa,* have been detected on the plastid genomes of *P. purpurea* and *O. sinensis*. Surprisingly, in the latter case the *clpP* gene contained within the standard set of plastid genes is absent.

Genes for Biosynthetic Enzymes

The numerous biosynthetic processes performed by the plastid are of great importance for the plant cell as a whole and in many cases are confined to this organelle. Often their prokaryotic nature is evident, but in higher plants these processes are carried out by the products of eukaryotic, i. e., nuclear genes (Löffelhardt 1995). The case is different for algae without chlorophyll b: former endocytobiont genes for various functions (Table 1) have been retained in the organelle. Some genes have only been found on the cyanelle genome (Table 2), e. g., *nadA*, encoding quinolinate synthase, but in general they are also present in the *O. sinensis* plastome (Kowallik et al. 1995) and in even higher numbers in the *P. purpurea* plastome (Reith and Munholland 1995). *crtE*, encoding geranyl geranyl pyrophosphate synthase, shows 51% sequence identity to the nucleus-encoded counterpart from *Capsicum annuum* (Kuntz et al. 1992). *preA*, originally named *crtE*,

Table 3. Amino acid identity scores (%) of *clpP* genes from prokaryotes and from plastids

	C. paradoxa-1	*E. coli*	Syn. 6803-1	*N. tabacum*	*C. paradoxa*-2	Syn. 6803-2
C. paradoxa-1	--	45.9	54.3	42	36.4	52
E. coli		--	64.8	42.4	43.4	67
Syn. 6803-1			--	50	41.7	74.1
N. tabacum				--	33.3	48.3
C. paradoxa-2					--	41.1

resembles the octaprenylpyrophosphate synthase from *E. coli* (Asai et al. 1994) and thus may catalyze the biosynthesis of the plastoquinone side chain. Alternatively, its function could be that of an undecaprenyl pyrophosphate synthase. This is the only gene of peptidoglycan biosynthesis that still awaits identification. *E. coli* transformed with cyanelle *preA* showed some resistance towards bacitracin, an antibiotic interfering with the recycling of undecaprenyl pyrophosphate (C. Hink and D. Mengin-Lecreulx, unpublished data). The other genes listed in Table 2 specify acyl carrier protein (*acpA*), enzymes involved in the biosynthesis of amino acids (*trpG, hisH*) and of chlorophyll (*chlB, chlI, chlL, chlN*) and glutamyl-tRNA reductase (*hemA*).

Peptidoglycan Biosynthesis

It is now clear that the cyanelle wall is not a rudimentary one but even exceeds that of *E. coli* with respect to the number of layers. The observed substitution of the D-isoglutamyl residue in the peptide side chain with N-acetylputrescine might be a key feature of cyanelle walls (Pfanzagl et al. 1996a,b). As a possible reason why the peptidoglycan has been maintained in cyanelles only, impeded transfer of the respective genes to the nuclear genome (finally leading to their loss in all other plastid types) was assumed. Much to our surprise, however, this is not the case. More than 20 eukaryotic genes specifying enzymes some of which have been detected in cyanelles and are responsible for building up the prokaryotic organelle wall must reside in the nuclear genome of *C. paradoxa*. Just one gene of that kind, a homolog to the *E. coli* cell division gene *ftsW*, was found on the cyanelle genome. *FtsW* is not directly involved in murein biosynthesis but was shown to be essential for the activity of distinct PBPs (penicillin binding proteins) (Donachie 1993). Upon sequencing of the plastid genome of the green alga *Chlorella vulgaris* a homolog of *minD* was found, mutations of which lead to minicell formation in *E. coli* (M. Sugiura, personal communication). Finally, a nucleus-encoded homolog of *ftsZ*, the gene product of which is crucial for septation in bacteria, has been detected in *Arabidopsis thaliana*. Higher plant FtsZ is synthesized as a precursor containing a typical stroma-targeting peptide and is imported *in vitro* into isolated chloroplasts (Osteryoung and Vierling 1995). These data point to some residual prokaryotic features in organelle division although the peptidoglycan wall has been abandoned in all plastid types with the exception of the cyanelles.

Conclusion

In summary, the peculiarities in the organization of some plastid genomes e. g. the vast number of introns present in *Euglena gracilis* (Hallick et al. 1993) should not be overemphasized compared to the numerous parallels revealed by phylogenetic analysis and, especially, by the increasing amount of data on plastid genome organization in different algal phyla. Differences in pigment composition are now considered to be less crucial. It is clear that chlorophyll b could be acquired at quite different evolutionary levels: at the prokaryotic level (Palenik and Haselkorn 1991), after the primary endosymbiotic event, and after the secondary endosymbiotic event. Alternatively, the cyanobacteria-like ancestral photosynthetic prokaryote could be imagined to possess both types of light-harvesting systems: chlorophyll a/b antennae as well as phycobilisomes (Bryant 1992). In this scenario loss of one or the other function could have occurred in different evolutionary lines. Remnants of chlorophyll a/b protein genes on cyanobacterial genomes and in biliprotein-containing algae or nonfunctional genes for phycobilisome components in chlorophytes would lend support to this hypothesis. The cyanelle ycf17 shows significant sequence similarity to ORFs specifying small proteins in red algal plastid and cyanobacterial genomes and to LHCII and ELIP genes from higher plants.

Acknowledgements. This work has primarily been supported by a grant from the US Department of Agriculture (National Research Initiative, Plant Genome Program) to H.J.B. and D.A.B. Additional support was provided by the National Science Foundation (DMB-8818997; D.A.B.), the Arizona Agricultural Experimental Station (H.J.B.), and the Austrian National Bank (Jubiläumsfondsprojekt 5436; W.L.)

References

Asai K, Fujisaki S, Nishimura Y, Nishino T, Okada K, Nakagawa T, Kawamukai M, Matsuda H (1994) The identification of *Escherichia coli ispB* (*cel*) gene encoding the octaprenyl diphosphate synthase. Biochem Biophys Res Commun 202: 340-354

Bartsevich VV, Pakrasi H (1995) Molecular identification of an ABC transporter complex for mangenese. Analysis of a cyanobacterial mutant strain impaired in the photosynthetic oxygen evolution process. EMBO J 14: 1845-1853

Bryant DA (1992) Puzzles of chloroplast ancestry. Curr Biol 2: 240-242

Donachie WD (1993) The cell cycle of *Escherichia coli*. Annu Rev Microbiol 47: 199-230

Flachmann R, Michalowski CB, Löffelhardt W, Bohnert HJ (1993) SecY, an integral subunit of the bacterial preprotein translocase is encoded by a plastid genome. J Biol Chem 268: 7514-7519

Hallick RB, Hong L, Drager RG, Favreau MR, Montfort A, Orsat B, Spielmann A, Stutz E (1993) Complete sequence of *Euglena gracilis* chloroplast DNA. Nucleic Acids Res 21: 3537-3544

Kaneko T, Tanaka A, Sato S, Kotani H, Sazuka T, Miyajima N, Sugiura M, Tabata S (1995) Sequence analysis of the genome of the unicellular cyanobacterium *Synechocystis* sp. strain PCC6803. I. Sequence features in the 1 Mb region from map positions 64% to 92% of the genome. DNA Res 2: 153-166

Kowallik KV, Stoebe B, Schaffran I, Freier U (1995) The chloroplast genome of a chlorophyll a+c containing alga, *Odontella sinensis*. Plant Mol Biol Rptr 13: 336-342

Kuntz M, Römer S, Suire C, Hugueney P, Weil J-H, Schantz R, Camara B (1992) Identification of a cDNA for the plastid-located geranylgeranyl pyrophosphate synthase from *Capsicum annuum*: correlative increase in enzyme activity and transcript level during ripening. Plant J 2: 25-34

Laudenbach D, Grossman A (1991) Characterization and mutagenesis of sulfur-regulated genes in a cyanobacterium: evidence for a function in sulfate transport. J Bacteriol 173: 2739-2750

Löffelhardt W (1995) Molecular Analysis of Plastid Evolution. In: Joint I (ed) Molecular Ecology of Aquatic Microbes. Springer-Verlag, Berlin-Heidelberg, pp 265-278 (NATO ASI Series G, vol 138)

Löffelhardt W, Bohnert HJ (1994a) Molecular Biology of Cyanelles. In: Bryant DA (ed) The Molecular Biology of the Cyanobacteria. Kluwer Academic Publishers, Dordrecht, pp 65-89

Löffelhardt W, Bohnert HJ (1994b) Structure and Function of the Cyanelle Genome. In: Jeon KW, Jarvik J (eds) International Review of Cytology. Academic Press, Orlando, vol 151: 29-65

Maurizi MR, Clark WP, Kim S-H, Gottesman S (1990) ClpP represents a unique family of serine proteinases. J Biol Chem 265: 12546-12552

Osteryoung CM and Vierling E (1995) Conserved cell and organelle division. Nature 376: 473-474

Palenik B, Haselkorn R (1991) Multiple evolutionary origins of prochlorophytes, the chlorophyllb-containing prokaryotes. Nature 355: 265-267

Pfanzagl B, Zenker A, Pittenauer E, Allmaier G, Martinez-Torrecuadrada J, Schmid ER, de Pedro MA, Löffelhardt W (1996a) Primary structure of cyanelle peptidoglycan of *Cyanophora paradoxa*: a prokaryotic cell wall as part of an organelle envelope. J Bacteriol 178: 332-339

Pfanzagl B, Allmaier G, Schmid ER, de Pedro MA, Löffelhardt W (1996b) N-Acetylputrescine as a characteristic constituent of cyanelle peptidoglycan in glaucocystophyte algae. J Bacteriol 178: 6994-6997

Reith M, Munholland J (1993) A high-resolution gene map of the chloroplast genome of the red alga *Porphyra purpurea*. Plant Cell 5: 465-475

Reith M, Munholland J (1995) Complete nucleotide sequence of the *Porphyra purpurea* chloroplast genome. Plant Mol Biol Rptr 13: 332-335

Schlichting R, Bothe H (1993) The cyanelles (organelles of a low evolutionary scale) possess a phosphate translocator and a glucose carrier in *Cyanophora paradoxa*. Bot Acta 106: 428-434

Schlichting R, Zimmer W, Bothe H (1990) Exchange of metabolites in *Cyanophora paradoxa* and its cyanelles. Bot Acta 103: 392-398

Smeekens S, Weisbeek P, Robinson C (1990) Protein transport into and within chloroplasts. Trends Biochem Sci 15: 73-76

Stirewalt VL, Michalowski CB, Löffelhardt WL, Bohnert HJ, Bryant DA (1995) Plant Mol Biol Rptr 13: 327-332

Sugiura M (1992) The chloroplast genome. Plant Mol Biol 19: 149-169

Plastid-Like Organelles in Anaerobic Mastigotes and Parasitic Apicomplexans

J.H.P. Hackstein[1], H. Schubert[2], J. Rosenberg[3]; U. Mackenstedt[4], M. van den Berg[5], S. Brul[1], J. Derksen[6], and H.C.P. Matthijs[2]
[1]Department of Microbiology and Evolutionary Biology, Faculty of Science, Catholic University of Nijmegen, 6525 ED Nijmegen, The Netherlands
[2]Laboratorium voor Microbiologie, University of Amsterdam, Nieuwe Achtergracht 127, 1018 WS Amsterdam, The Netherlands
[3]Department of Animal Physiology, Ruhr-Universität, 44780 Bochum, Germany
[4]Department of Parasitology, Ruhr-Universität, 44780 Bochum, Germany
[5]Department of Biochemistry, Faculty of Medicine, E.C.Slater Institute, University of Amsterdam, Meibergdreef 15, 1105 AZ Amsterdam, The Netherlands
[6]Department of Experimental Botany, Faculty of Science, Catholic University of Nijmegen, 6525 ED Nijmegen, The Netherlands

Psalteriomonas lanterna is a bacterivorous, free-living, microaerophilic amoebomastigote. It lacks dictyosomes, microbodies, and mitochondria (Broers et al. 1990). However, in both mastigote and amoeba stages of this organism, we discovered a novel organelle that we named *thylakosome* (Hackstein et al. 1994). Electron microscopy consistently revealed the presence of a more or less constant number of "vacuoles" that contained conspicuous membraneous structures. After the use of improved anaerobic fixation procedures it became evident that these membranes resembled thylakoids in non-photosynthetic plastids or photosynthetic prokaryotes (Fig. 1). Certain stages of thylakosome development were morphologically similar to etiolated plastids of green algae or the chromoplasts of higher plants (Kowallik and Herrmann 1972). These observations suggested that the thylakosomes might be highly specialized plastids with essential functions.

In order to provide more evidence for potential photosynthetic functions of the thylakosomes we analyzed the pigments of *Psalteriomonas* mastigotes. *Psalteriomonas* is nearly colourless, but high pressure liquid chromatography (HPLC) and mass spectrometry revealed the presence of traces of chlorophyll a in association with a carotenoid (Hackstein et al. 1995b). Neither the carotenoid nor the chlorophyll was detectable in an extract made from a 1000 times concentrated mastigote-free culture medium containing the enriched food bacteria. Since also the co-elution of chlorophyll and carotenoid argued against a contamination, the presence of the carotenoid might be due to the thylakosomes and indicative for a plastidic metabolism. DAPI and ethidium bromide stained the organelles, indicating the presence of DNA. Polymerase chain reaction (PCR) amplifications on total DNA extracted from *Psalteriomonas* mastigotes with cyanelle-specific primers for the small subunit of the ribosomal genes suggested the presence of ribosomal genes as inverted repeats - an orientation that is characteristic of plastids (J. H. P. Hackstein, D. Bhattacharya, and M. Melkonian, unpublished). Low-

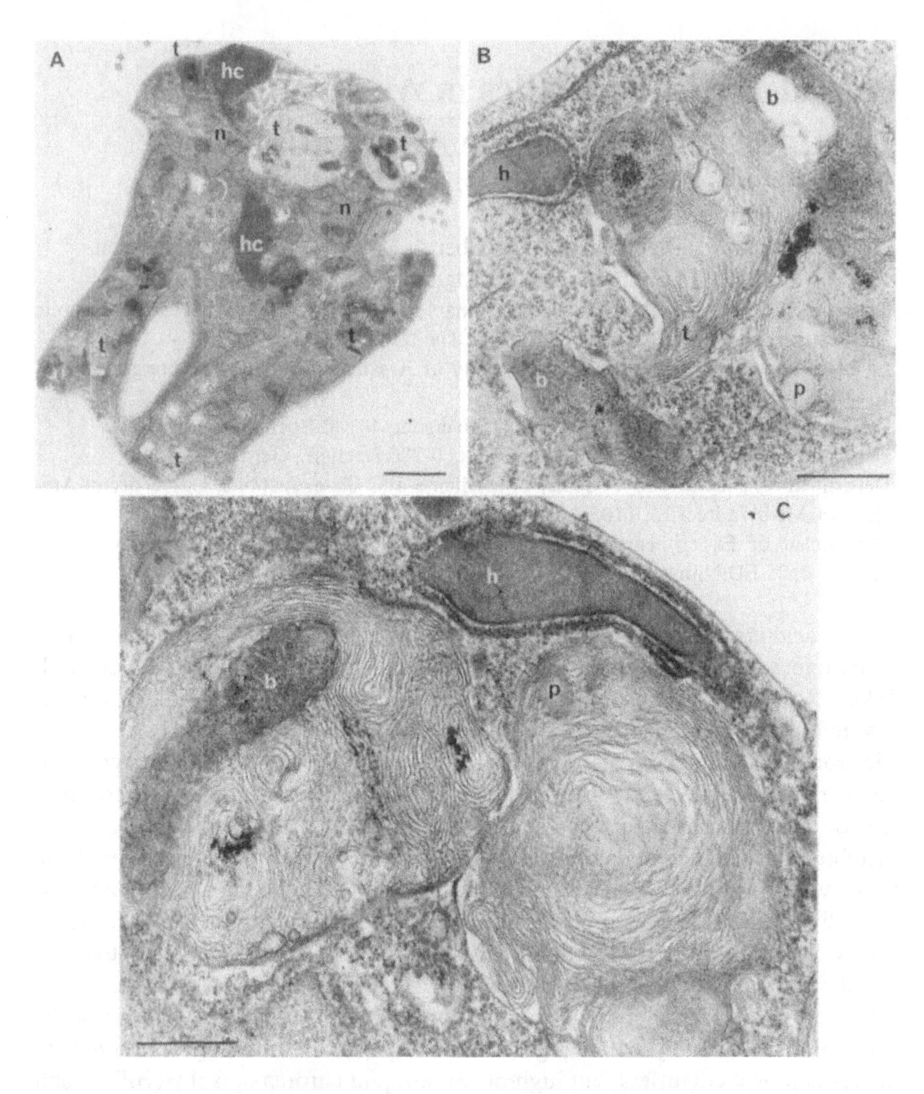

Fig. 1A. A mastigote of *Psalteriomonas lanterna*. Many thylakosomes (t) and 2 hydrogenosomal complexes (hc) are seen. n: 2 of the 4 nuclei of the mastigote. Bar: 2 µm **B** An extension of a mature thylakosome (t) can be seen close to a bacterium (b) that is surrounded by a phagosomic membrane. Remnants of ingested bacteria (b) can be seen inside the thylakosome. Bar 0.5 µm **C** Two mature thylakosomes, the left one with an ingested bacterium. The right one contains a pyrenoid-like structure (p). The immature hydrogenosome (h) is surrounded by an ER cysterna, covered with ribosomes. Such structures had been misinterpreted as mitochondria by Broers et al. (1990). Bar 0.5 µm

temperature fluorometry at 77K and polarography in Mehler-type experiments suggested the presence of photosystems PS I and PS II. Gold-labelled antisera

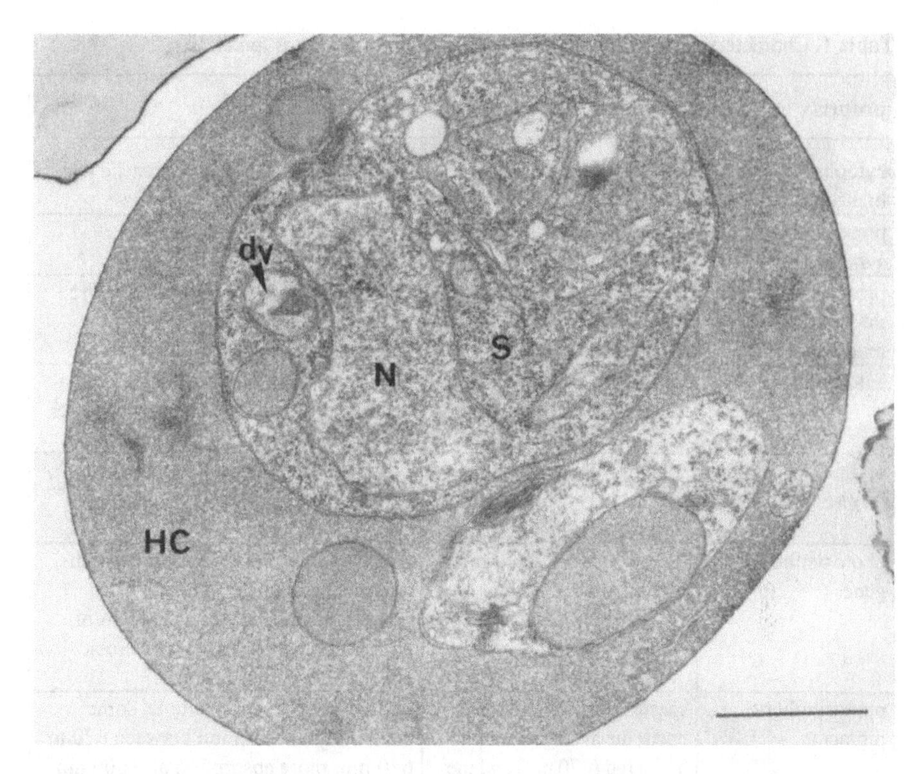

Fig. 2. *Plasmodium flaciparum,* schizont (S) in its host cell (HC). Note the granules (ribosomes?) and the stroma-like, dark structure inside the double-walled vesicle (dv). N nucleus. Bar 1 µm

against cytochrome-oxidase (COX) and ribulose bisphosphate carboxylase (Rubisco) labelled the thylakosomes, indicating the presence of both enzymes in the organelles. Table 1 summarizes the evidence that characterizes the thylakosome as a highly specialized plastid. Its photosynthetic properties are rudimentary, and it is likely that the organelle has both anabolic and catabolic functions in the metabolism of the anaerobic amoebomastigote *Psalteriomonas*.

Many, though not all, apicomplexan parasites harbour elusive organelles that have been described as "*Hohlzylinder*" or "double-walled vesicles" (van der Zypen and Piekarski 1967; Mehlhorn 1988; Siddal 1992). Electron microscopy revealed a number of morphological traits (e.g. membraneous - and stroma-like structures, putative 70S ribosomes) similar to the thylakosomes of *Psalteriomonas* (Fig. 2, 3). These organelles divide in a plastid-like manner (Hackstein et al. 1997).

The concentration of chlorophyll a in apicomplexan parasites was too low to be detected by HPLC and spectrophotometry. However, with the aid of 77K fluorometry we were also able to record emissions with maxima at 685 and 716 nm in *Toxoplasma gondii* trophozoites (Hackstein et al. 1995a). In photosyn-

Table 1. Characterization of the thylakosomes of *Psalteriomonas lanterna*

property	method of detection	conclusion
cytoplasmic membrane. structures	electron microscopy	"*thylakosome*", unknown organelle with potential photosynthetic activities
presence of photosynth. enzymes	immune-gold-labelling of RuBisCo	CO_2 fixation occurs in the thylakosomes
presence of respiratory terminal oxidase	imm.-gold-labelling of cytochrome-oxidase aa_3, Western-blotting[2]	cytochrome-c-oxidase present in the thylakosomes
autoreplication	DAPI and EtBR staining, electron microscopy	thylakosomes contain DNA; 10-20 organelles/host cell have been maintained over more than 6 years of laboratory culture.
organelle protein synthesis	inhibition of COX biosynthesis by doxycycline, Western-blotting[3]	organelle harbours functional protein synthesis machinery
photosynthesis genes	PCR with conserved primers for *psaB* and *psbA*	DNA fragments of the expected length amplified:800 bp (*psaB*) and 570 bp (*psbA*); no amplification with DNA of food bacteria or *Trimyema* (anaerobic Ciliate without thylakosomes)
photosynthetic pigments[1]	spectroscopy of acetone extracts; the average absorption in the red (670 nm) and the carotenoid Soret band at 418 nm were used for quantification of pigments. Excitation of carotenoids yields fluorescence in the red	carotenoid peaks in the blue, some chlorophyll absorption between 670 to 680 nm, more absorption at ± 600 nm (consequence of possible covalent linkage between carotenoids and chlorophyll)
	HPLC separation and mass spectrometry (MS)[4]	major HPLC peak eluted at the retention time of chl b; absorption spectra reveal presence of carotenoid + chl a, and a peak at 600 nm. Following electron impact MS of the HPLC peak, fragments of mass 892 (chl a), 550 and 540 (carotenoid), and a weak set of parent peaks of mass 1432 and 1442 were found
	molar concentrations of pigments; assuming a molar extinction coefficient of 1800 at 440 nm for carotenoids and of 460 for chl a at 675 nm	$\pm 5 \times 10^9$ molecules carotene /host cell , $\pm 2.5 \times 10^8$ molecules chl a /host cell (c.f. chloroplast: 1.5×10^8 molecules carotenoid and 6.7×10^8 molecules of chl)
photosystems[1]	77K fluorescence spectroscopy[5] excitation 380-420 nm	fluorescence emission at 685 and 722 nm indicates presence of PS II and PS I reaction centers

Table 1: Continuation

property	method of detection	conclusion
electron transfer via PS I[1]	ascorbate/DCIP to MV, in presence of DCMU and KCN, as light dependent O_2 uptake	PS I catalyzed electron transfer occurs: 3.5 mol O_2 /mol chl a/ min
electron transfer via PS II[1]	DCBQ/K_3 $Fe(CN)_6$ as light dependent O_2 release	PS II catalyzed electron transfer occurs: 1.3 mol O_2 /mol chl a /min
dark respiration[1]	O_2 uptake in the dark	substantial electron transfer, Knallgas-reaction? : 4.7 nmol O_2 / ml cell suspension/min
lowered H_2 release in the light	co-cultivation of *Psalteriomonas* and methanogenic bacteria, analysis of gas composition in the head-space of the culture vessels	light-enhanced H_2 uptake

[1] done with 1000-fold enriched mastigotes that were harvested by electromigration following Broers et al. (1992);
[2] c.f. Brul et al. (1993);
[3] doxycycline is a potent inhibitor of mitochondrial and prokaryotic protein synthesis, c.f. van den Bogert et al. (1988);
[4] HPLC was performed as described in Mantoura and Llewellyn (1983) and van der Staay et al. 1992. The 20.2 min retention time peak fraction of a HPLC run of acetone extracts of *P. lanterna* cells was collected, concentrated, and subjected to mass spectrometry with a JMS-SX/SX 102A (JEOL Inc., Japan) 4 sector mass spectrometer operated in the fast atom bombardment mode;
[5] The 77K fluorescence emission spectra of whole cells of *P. lanterna* were recorded with a Hitachi F4010 fluorescence spectrophotometer (Hitachi, Japan), and corrected for detector sensitivity. The PMT tube was protected by a 620 nm cut-on filter.

thetic organisms, such emissions indicate the presence of photosystems PS I and PS II (Rijgersberg and Amesz 1980; Rijgersberg et al. 1979). A faint red fluorescence (> 650 nm) from living *Toxoplasma gondii* trophozoites was recorded with a residual light CCD camera after excitation in the blue that is likely to be derived from the emission of chlorophyll a. Therefore, the presence of photosynthetic organelles in these parasites is likely. Using PCR techniques, a highly conserved *psbA* gene of *Sarcocystis muris* was identified and partially sequenced. Phylogenetic analysis located this gene in the neighbourhood of the primitive "green" plastids (Hackstein et al. 1995a). Because a 35-kb plastidic genome has been identified in many apicomplexans (Feagin 1994, McFadden et al. 1996, Wilson et al. 1996, Köhler et al. 1997), it is likely that the "double-walled vesicles" are highly specialized plastids that fulfil important functions in the parasite's metabolism. The herbicide-derivative Toltrazuril (Baycox) was effective in killing both *Psalteriomo-*

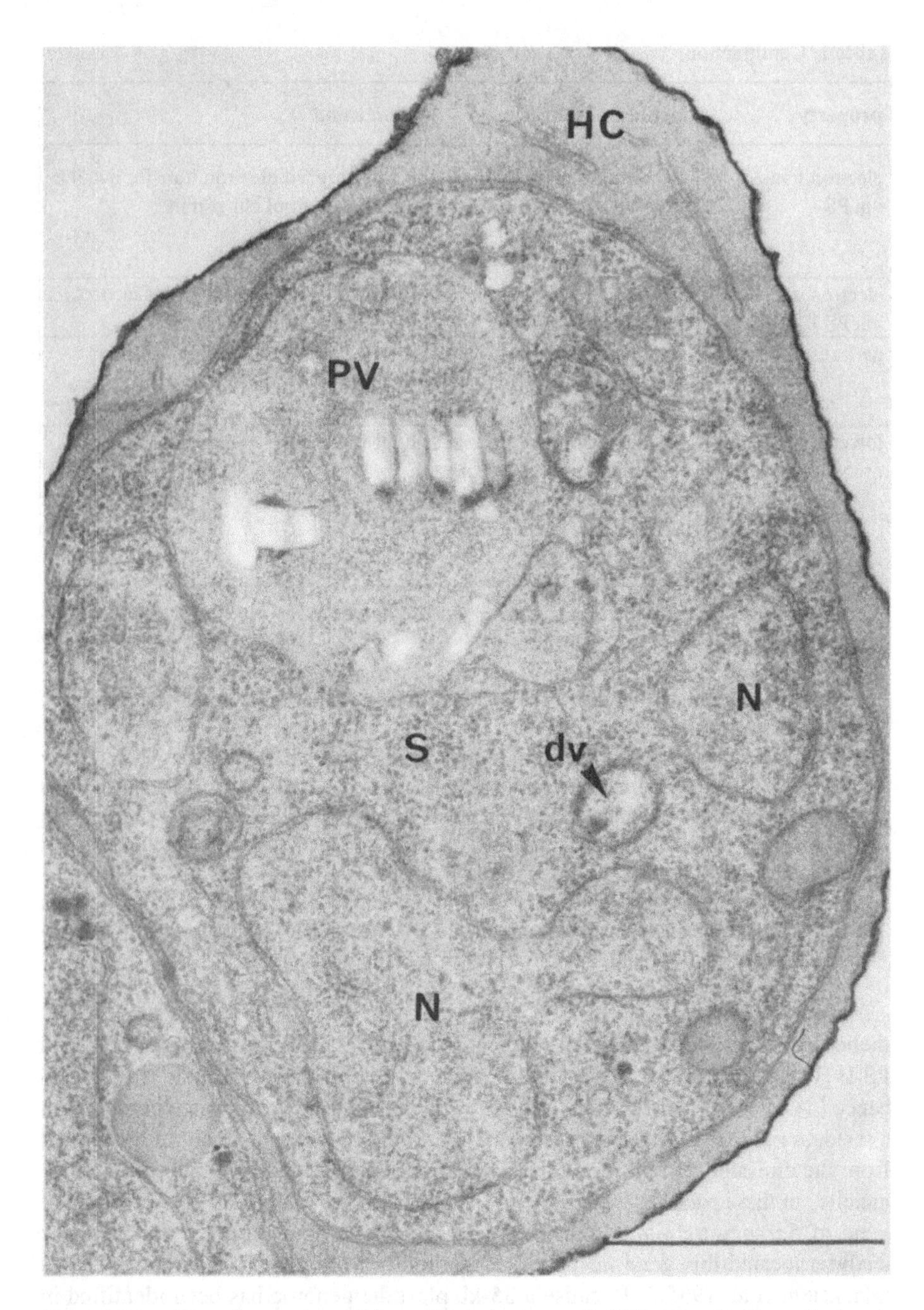

Fig. 3. *Plasmodium flaciparum,* schizont (S) in its host cell (HC). The double-walled vesicle (dv) contains membraneous structures and granular particles that might be ribosomes. N nucleus, PV parasitophorous vacuole. Bar 1 µm

nas and many apicomplexan parasites (Mehlhorn et al. 1984; Hackstein et al. 1997). Therefore, the presence and function of PS II seems to be crucial for both organisms. One may speculate about the role of plastids in the metabolism of organisms that live heterotrophically or parasitically in environments with low oxygen tensions. It seems reasonable to speak of "early" acquisition of plastids by the eukaryotic progenote, since the evolution of eukaryotes preceded the appearance of the higher atmospheric oxygen concentrations necessary for the function of mitochondria (Hackstein and Mackenstedt 1995).

References

Bogert C van den, Muus P, Haanen C, Pennings A, Melis TE, Kroon AM (1988). Mitochondrial biogenesis and mitochondrial activity during the progression of the cell cycle of human leukemic cells. Exp Cell Res 178: 143-153

Broers CAM, Stumm CK, Vogels GD, Burgerolle G (1990) *Psalteriomonas lanterna* gen. nov., sp. nov., a free-living amoeboflagellate isolated from freshwater anaerobic sediments. Eur J Protistol 25: 369-380

Broers CAM, Molhuizen HOF, Stumm CK, Vogels GD (1992) An electromigration technique, applied for the concentration of anaerobic protozoa from mass cultures. J Microbial Meth 14:217-220

Brul S, Kleijn J, Lombardo MCP, Berg M van den, Stumm CK, Vogels GD (1993) Coexistence of hydrogenosomes and mitochondrial structures in the amoeboflagellate *Psalteriomonas lanterna*. J Euk Microbiol 40:17A

Feagin JE (1994) The extrachromosomal DNAs of apicomplexan parasites. Annu Rev Microbiol 48:81-104

Hackstein JHP, Mackenstedt U (1995) A "photosynthetic" ancestry for all eukaryotes? Trends Ecol Evol 10: 247

Hackstein JHP, Mackenstedt U, Mehlhorn H, Meijerink JPP, Schubert H, Leunissen JAM (1995a) Parasitic apicomplexans harbour a chlorophyll a - D1 complex, the potential target for therapeutic triazines. Parasitol Res 81: 207-216

Hackstein JHP, Schubert H, Berg M van den, Brul S, Derksen J, Matthijs HCP (1994) A novel photosynthetic organelle in anaerobic mastigotes. Endocytobiosis and Cell Res 10: 261

Hackstein JHP, Schubert H, Berg M van den, Brul S, Derksen JWM, Matthijs HCP (1995b) A photosynthetic organelle in anaerobic flagellates. J Euk Microbiol 42: p 4A

Hackstein JHP, Schubert H, Rosenberg J, Berg M van den, Brul S, Matthijs HCP, Derksen JMW (1997) A novel photosynthetic organelle in anaerobic mastigotes. (submitted)

Köhler S, Delwiche CF, Denny PW, Tilney LG, Webster P, Wilson RJM, Palmer JD, Roose DS (1997) A plastid of probable green algal origin in apicomplexan parasites. Science 275: 1485-1489

Kowallik KV, Herrmann RG (1972) Do chromoplasts contain DNA? I. Electron-microscopic investigation of *Narcissus* chromoplasts. Protoplasma 74: 1-6

McFadden GI, Reith ME, Munholland J, Lang-Unnasch N (1996) Plastid in human parasites. Nature 381: 482

Mantoura RFC, Lewellyn CA (1983) The rapid determination of algal chlorophyll and carotenoid pigment and their breakdown products in natural waters. Anal Chim Acta 151: 297-314

Mehlhorn H, Ortmann-Falkenstein G, Haberkorn A (1984) The effects of sym. triazinones on developmental stages of *Eimeria tenella*, *E. maxima* and *E. acervulina*: a light and electronmicroscopic study. Z Parasitenkd 70: 173-182

Mehlhorn H (ed) (1988) Parasitology in focus. Springer, Berlin Heidelberg New York

Rijgersberg CP, Amesz J, Thielen PGM, Swager JA (1979) Fluorescence emission spectra of chloroplast and subchloroplast preparation at low temperature. Biochim Biophys Acta 545: 473-482

Rijgersberg CP, Amesz J (1980) Fluorescence and energy transfer in phycobiliprotein-containing algae at low temperature. Biochim Biophys Acta 593: 261-271

Siddal ME (1992) Hohlzylinder. Parasitol Today 8: 90-91

Staay GWM van der, Brouwer A, Baard RL, Mourik F van, Matthijs HCP (1992) Separation of photosystems I and II from the oxychlorobacterium (prochlorophyte) Prochlorothrix hollandica and association of chlorophyll b binding antenna with photosystem II. Biochim Biophys Acta 1102: 220-228

Wilson RJM, Denny PW, Preiser PR, Rangachari K, Roberts K, Roy A, Whyte A, Strath M, Moore SJ, Moore PW, Williamson DH (1996) Complete gene map of the plastid-like DNA of the malaria parasite *Plasmodium falciparum*. J Mol Biology 261: 155-172

Zypen E van der, Piekarski G (1967) Ultrastrukturelle Unterschiede von *Toxoplasma* gondii. Z Bakt Parasit Infekt Hygiene 203: 495-517

Complete Mitochondrial DNA Sequence of Budding yeast *Hansenula wingei* Indicates its Intermediary Characteristics Between Those of Yeasts and Filamentous Fungi

T. Sekito, K. Okamoto, H. Kitano, and K. Yoshida
Department of Biological Science, Faculty of Science, Hiroshima University, Higashi-Hiroshima, 739, Japan

Key words: Complete mitochondrial DNA sequence, *Hansenula wingei*, NADH dehydrogenase subunit gene, phylogenetic analysis, mitochondrial genome reorganization.

Summary: We have determined the complete DNA sequence of budding yeast *Hansenula wingei* mitochondrial genome. The sequence data clearly reveal characteristics uniquely intermediary between those of yeasts and those of filamentous fungi. Nevertheless, phylogenetic analysis indicates that *H. wingei* is closely related to the yeast *Saccharomyces cerevisiae*. These results suggest that drastic reorganization occurred in yeast mitochondrial genomes during evolution.

Introduction

The genes for NADH dehydrogenase (ND) subunits 1-6 and 4L are encoded by the mitochondrial genome of almost all organisms. However, yeast mitochondrial genomes so far investigated, such as *Saccharomyces*, *Kluyveromyces*, and *Candida*, do not encode them. Hoeben and Clark-Walker (1986) indicated the possibility that *Brettanomyces* yeasts encode an *ND1* gene on that mitochondrial genome, by Southern analysis using *Aspergillus nidulans ND1* probe. Recently, we have completely sequenced the mitochondrial genome of budding yeast *Hansenula wingei* (Sekito et al. 1993, Sekito et al. 1994, Sekito et al. 1995a, Sekito et al. 1995b). This is the first published report of a complete sequence of mitochondrial genome for budding yeasts. In addition, we identified *ND1-6* and *ND4L* genes on the *H. wingei* mitochondrial genome. The *H. wingei* mitochondrial genome has characteristics uniquely intermediary between those of yeasts and filamentous fungi.

Materials and Methods

Yeast *H. wingei* strain 21 (*21 ade, his*) was used. Isolation and cloning of mitochondrial DNA have been described previously (Sekito et al. 1995b, Okamoto et al. 1991). The nucleotide sequences were determined according to procedures adapted from Sanger et al. (1977), using an automatic sequencer model 373A (ABI). The sequences of subcloned plasmids were analyzed with GENETYX software (Software Development Co., Tokyo, Japan). The sequence homology search was carried out using a DNA database in DDBJ (DNA Database Japan). Phylogenetic analyses were performed using the neighbor-joining method program constructed by N. Saito of the National Institute of Genetics, Japan.

Results and Discussion

The size of the *H. wingei* mitochondrial genome is 27 694 bp, which is relatively small for yeasts. It encodes the genes for 15 proteins, large and small rRNAs, and 25 tRNAs as shown in Table 1 and Fig. 1. The gene organization reflects combined characteristics of yeast and filamentous fungi, because both *ND* genes and *VAR1* gene could be found on the same genome. *ND* genes, which almost all organisms including filamentous fungi encode on their mitochondrial genomes, had not previously been found on yeast mitochondrial genomes. *H. wingei* is the first yeast of which the mitochondrial genome is found to encode *ND* genes (Okamoto et al. 1993, Okamoto et al. 1994). *VAR1* gene is a specific gene for yeast mitochondrial genome. Therefore, despite its encoding *ND* genes, we can regard the *H. wingei* mitochondrial genome as a typical yeast mitochondrial genome.

The structure of the *H. wingei* mitochondrial genome is quite different from that of the yeast *S. cerevisiae* mitochondrial genome which has been fully investigated (de Zamaroczy and Bernardi 1986). The most striking difference is the length of the intergenic region: the intergenic region of the *H. wingei* mitochondrial genome is much smaller than that of *S. cerevisiae*. In this respect, the *H. wingei* mitochondrial genome is similar to the mitochondrial genome of the filamentous fungus *A. nidulans*, which has a small intergenic region and few introns (Brown 1993).

Upstream some genes on the *H. wingei* mitochondrial genome, the nonanucleotide motif (A/T)TATAAG(T/A)(A/T), which is found to function as a promoter in *S. cerevisiae* mitochondria, was identified (Fig. 1). On the other hand, the genetic code of the *H. wingei* mitochondrial genome is essentially identical with that of filamentous fungi mitochondria, which has only one change from UGA-stop to UGA-Trp, whereas that of *S. cerevisiae* mitochondria has two more changes, AUA-Ile→Met and CUN-Leu→Thr.

The nucleotide sequences of *H. wingei* mitochondrial genes are most homologous to those of *S. cerevisiae* (Table 2). Phylogenetic analysis based on rRNA sequence data also indicates that *H. wingei* and *S. cerevisiae* mitochondria diverged from the same lineage as shown in Fig. 2 (K. Okamoto et al., unpublished observations). These results are consistent with the relationships based on morphological characteristics, electrophoretic karyotype (Yoshida et al. 1989), and restriction fragment length polymorphism (RFLP) analysis of mitochondrial DNA (Okamoto et al. 1991).

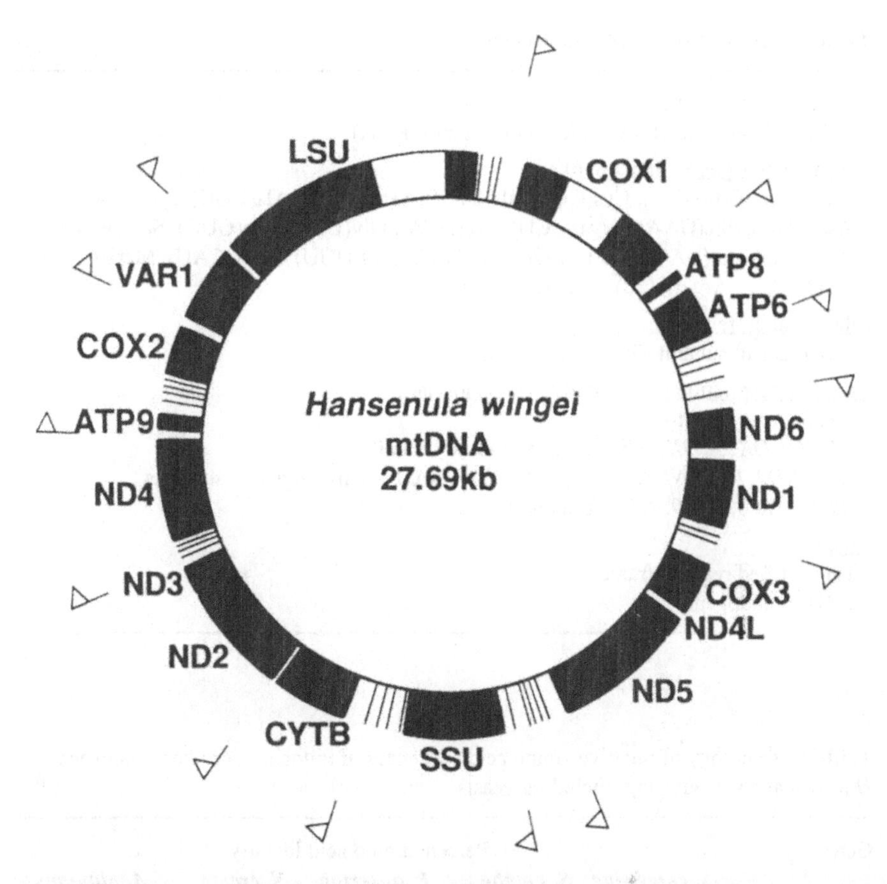

Fig. 1. Map of the mitochondrial genome of the budding yeast *H. wingei*. Genes and exons are shown as black blocks, introns are white. tRNA genes are indicated as thin bars with the one letter code for the cognate amino acid. Flags indicate the position of promoter sequence (A/T)TATAAG(T/A)(A/T). All genes are transcribed clockwise

Although the *H. wingei* mitochondrial genome has uniquely intermediary characteristics between those of yeast and filamentous fungi mitochondrial genomes, *H. wingei* itself is phylogenetically much closer to *S. cerevisiae* as described above. This suggests that there was drastic reorganization of the mitochondrial genome during the divergence between *H. wingei* and *S .cerevisiae*. Recently, H. Kitano et al. (Kitano et al. 1995) and Nozef and Fukuhara (Nozef and Fukuhara 1994) found the *ND* genes on yeast *Pichia*, *Yallowia*, and *Candida* mitochondrial genomes. Therefore, it is possible that the majority of yeasts encode *ND* genes on their mitochondrial genomes. In other words, yeasts so far investigated, including *S. cerevisiae*, *Candida glabrata*, and *S. pombe*, etc., are rather exceptional cases,

Table 1. Genes on *H. wingei* mitochondrial genome

Ribosomal RNA genes
SSU(small subunit rRNA), *LSU*(large subunit rRNA)

Transfer RNA genes (anticodon)
Thr(UGU), Glu(UUC), Cys(GCA), Ile(GAU), Ser1(UGA), Ala(UGC) Tyr(GUA), Asn(GUU), Phe(GAA), Arg1(UCU), Trp(UCA), Gly(UCC) Asp(GUC), Ser2(ACU), Pro(UGG), Val(UAC), Leu1(UAA), Gln(UUG) Lys(UUU), Met1(CAU), Met2(CAU), Leu2(UAG), Arg2(AGC), His(GUC) Met3(CAU)

Ribosomal protein gene
VAR1: Small subunit ribosomal protein

Genes for respiration and oxidative phosphorylation
CYTB: Apocytochrome b
COX1, *COX2*, *COX3*: Cytochrome oxidase subunits
ND1, *ND2*, *ND3*, *ND4*, *ND4L*, *ND5*, *ND6*: NADH dehydrogenase subunits
ATP6, *ATP8*, *ATP9*: ATPase subunits

Others
Unidentified reading frame
tRNA pseudogene

Table 2. Homology of putative amino acid sequences of mitochondrial genes between *H.wingei* and various fungi including yeasts

Gene	Percent amino acid identity				
	S. cerevisiae	***S .pombe***	***P .anserina***	***N. crassa***	***A. nidulans***
COX1	79.8	60.9	63.2	67.4	61.5
COX2	77.3	50.8	59.2	60.9	54.0
COX3	73.9	49.2	54.6	52.8	54.3
CYTB	76.7	56.9	62.2	58.7	62.8
ATP6	64.1	41.9	53.6	53.0	51.2
ATP8	83.3	56.3	56.3	52.1	54.0
ATP9	86.8	65.8	-	65.3	61.1
ND1	-	-	49.7	48.7	49.7
ND2	-	-	30.2	30.3	?
ND3	-	-	34.3	?	39.4
ND4	-	-	38.5	?	38.9
ND4L	-	-	45.7	45.7	?
ND5	-	-	42.5	39.6	39.5
ND6	-	-	38.1	?	34.0
VAR1	34.7	-	-	-	-

S. pombe, *Schizosaccharomyces pombe*; *P. anserina*, *Podospora anserina*; *N. crassa*, *Neurospora crassa*

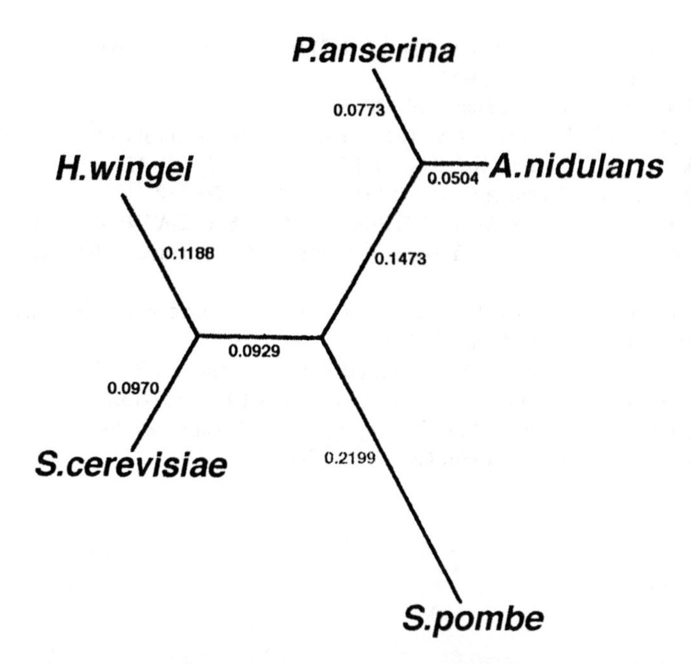

Fig. 2. Phylogenetic relationship of fungi based on mitochondrial SSU rRNA sequences. The most parsimonious relationship among various fungal mitochondria is shown. This is based on analysis of highly conserved regions of the corresponding mitochondrial SSU rRNAs. The number near each branch indicates the weighted mutational distance

and in fact the *S. cerevisiae* mitochondrial genome has unusual characteristics, such as the large intergenic region, some unidentified reading frames, and many intergenic GC-rich sequences. To clarify the aberrant evolutional changes in yeast mitochondrial genome, further investigation including sequence analysis of yeast mitochondrial genomes is indispensable.

Acknowledgements. This study is supported in part by grants from the Ministry of Science, Culture and Education, Japan to K. O. and K. Y., and by a grant from the Electric Foundation of Chugoku to K. Y., K. O. holds Fellowships of the Japan Society for Promotion of Science for Japanese Junior Scientists. We thank Dr. N. Saito for providing computer programs.

References

Brown TA (1993) In: O'Brien SJ (ed) Genetic maps. Cold Spring Harbor Laboratory Press, Cold Spring Harbor, NY, pp 3.85-3.86

Hoeben P, Clark-Walker GD (1986) Curr Genet 10: 371-379

Kitano H, Sekito T, Ishitomi H, Yoshida K (1995) In: Nucleic Acids Symposium Series 34. Oxford University Press, pp 23-24
Nozef N, Fukuhara H (1994) J Bacteriol 176: 5622-5630
Okamoto K, Sekito T, Yoshida K (1993) XV International Botanical Congress, 504
Okamoto K, Sekito T, Yoshida K (1994) Mol Gen Genet 243: 473-476
Okamoto K, Suzuki K, Yoshida K (1991) Jpn J Genet 66: 709-718
Sanger F, Nicklen S, Coulson AK (1977) Proc Natl Acad Sci USA 74: 5463-5467
Sekito T, Okamoto K, Kitano H, Yoshida K (1993) XV International Botanical Congress, 503
Sekito T, Okamoto K, Kitano H, Yoshida K (1994) In: Nucleic Acids Symposium Series 31. Oxford University Press, pp 233-234
Sekito T, Okamoto K, Kitano H, Yoshida K (1995) Curr Genet 28: 39-53
Sekito T, Okamoto K, Kitano H, Yoshida K (1995) Yeast 11: 1317-1321
Yoshida K, Hisatomi T, Yanagishima N (1989) J Basic Microbiol 29: 99-128
Zamaroczy M de, Bernardi G (1986) Gene 47: 155-177

Biogenesis of Hydrogenosomes in *Psalteriomonas lanterna* : No Evidence for an Exogenosomal Ancestry

J.H.P. Hackstein[1], J. Rosenberg[2], C.A.M. Broers[1], F.G.J. Voncken[1], H.C.P. Matthijs[3], C.K. Stumm[1], and G.D. Vogels[1]

[1]Department of Microbiology and Evolutionary Biology, Faculty of Science, Catholic University of Nijmegen, 6525 ED Nijmegen, The Netherlands

[2]Department of Animal Physiology, Ruhr-Universität, 44780 Bochum, Germany

[3]Laboratorium voor Microbiologie, University of Amsterdam, Nieuwe Achtergracht 127, 1018 WS Amsterdam, The Netherlands

Introduction

Hydrogenosomes are cellular organelles that occur exclusively in certain anaerobic, unicellular eukaryotes and some anaerobic fungi (reviewed in Müller 1993). The key enzymes of this organelle are a hydrogenase and a pyruvate:ferredoxin oxidoreductase (Lindmark and Müller 1973; Müller 1988). Hydrogenosomes are engaged with the anaerobic cellular energy metabolism, and they generate hydrogen and acetate if supplied with their characteristic substrates pyruvate or malate (Müller 1993). Hydrogenosomes in trichomonads have been studied extensively, but knowledge about such organelles in most of the other anaerobes is fragmentary. Frequently, identification as a hydrogenosome depends solely on a cytochemical hydrogenase assay and on ultrastructural similarities. Their occurrence in distantly related organisms, however, and substantial differences in shape and fine structure strongly suggest that the hydrogenosomes of the various protists and fungi are not the same (Fig. 1; cf. Coombs and Hackstein 1995).

The elucidation of the structure, function, and evolutionary origin of the hydrogenosomes present in various protists and fungi is a challenging project that demands multidisciplinary efforts. Since it has not been possible yet to demonstrate the presence of DNA, RNA, or ribosomes in the archetypal hydrogenosomes of *Trichomonas* (Müller 1993), basic problems arise for the evaluation of their ancestry (Margulis 1993, Müller 1997). The lack of even a rudimentary genome and an organellar protein-synthesizing machinery hampers the DNA sequence analysis and phylogenetic analysis that could reveal descent from a formerly endocytobiotic, autonomously replicating organism. Only the analysis of the nuclear encoded proteins and their putative import and targeting signals can potentially provide suggestive evidence in favor of an exogenosomal or a cellular origin. Further, the study of the mode of multiplication and maturation of the hydrogenosomes might help to decide between an exogenosomal or an endogenosomal origin.

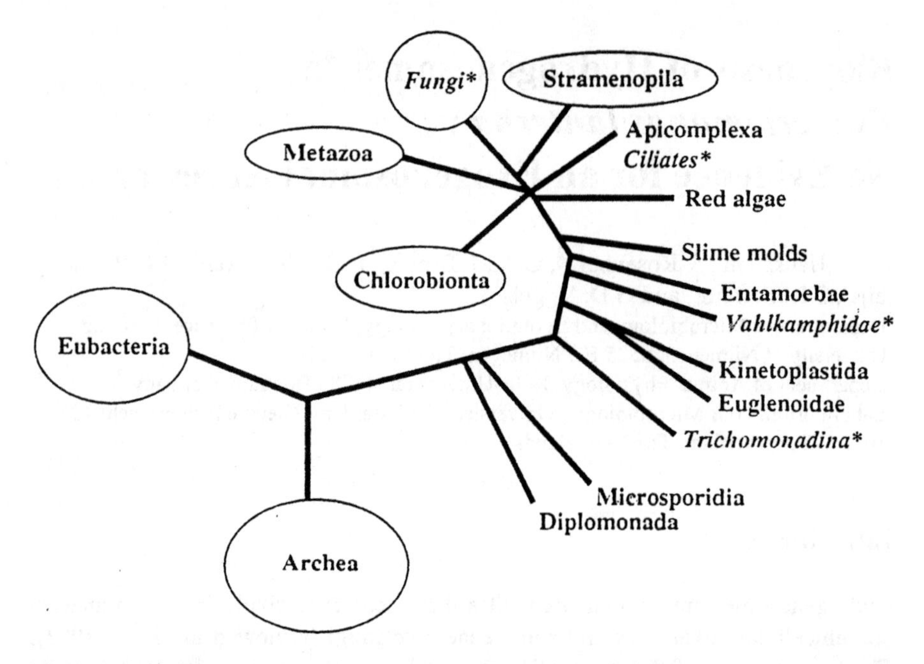

Fig. 1. An unrooted phylogenetic tree based on SSrDNA sequence analysis. Hydrogenosomes are found in the taxa indicated by * and italics. A polyphyletic origin of the hydrogenosomes is likely - even among the ciliates (see Brul and Stumm 1994, Embley et al. 1995, Schlegel 1994)

The Hydrogenosomes of *Psalteriomonas lanterna*

The free-living, microaerophilic amoebomastigote *Psalteriomonas lanterna* harbors two populations of hydrogenosomes (Hackstein et al. 1995a). The mature, active organelles form a large complex in the center of the mastigote cell that exhibits a characteristic autofluorescence at 590 nm (Hackstein et al. 1995a). Cytochemical staining with BSTP [2-(2'-benzothiazolyl)-5-styryl-3-(4' phthalhydrazidyl)-tetrazolium chloride] as electron acceptor and hydrogen as a substrate suggests that this complex exhibits hydrogenase activity (cf. Zwart et al. 1988; Broers 1992). In addition, the complex can be stained with an FITC-labeled antiserum against the hydrogenase of *Trichomonas vaginalis*. Electron microscopic studies reveal that the complex consists of a stack of individual organelles that resemble the hydrogenosomes of trichomonads (Broers et al. 1990; Hackstein et al. 1995a). Endosymbiotic methanogenic bacteria known from earlier studies are absent. Consequently, the formation of the hydrogenosomal complex does not depend on the presence of methanogenic bacteria (cf. Broers et al. 1990; Broers 1992). The individual hydrogenosomes are surrounded by a double membrane; their matrix is very homogeneous. There are no indications for the presence of nucleoids and ribosomes, nor for the existence of tubuli or cristae. Since both DAPI and ethidium bromide fail to stain the hydrogenosomal complex, it may be

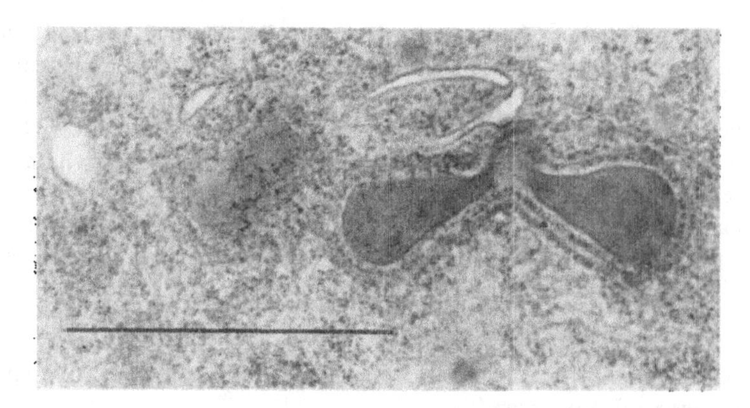

Fig. 2. Electron micrograph of a budding hydrogenosome of *Psalteriomonas lanterna*. The hydrogenosome is surrounded by rough ER cisterns. Bar = 1.0 μm

concluded that the mature hydrogenosomes of *Psalteriomonas* are also devoid of nucleic acids, like those of *Trichomonas* (Müller 1993).

Ontogeny of the Hydrogenosomes of *Psalteriomonas lanterna*

In exponentially growing cultures of mastigotes of *Psalteriomonas lanterna* electron microscopy reveals the presence of sausage- and dumbbell-shaped organelles in the periphery of the cells. These organelles closely resemble the hydrogenosomes in the hydrogenosomal complex (Hackstein et al. 1995a). Since the peripheral organelles are surrounded by one or two cisterns of rough endoplasmic reticulum (ER) (Fig. 2), they have also been interpreted as modified mitochondria (Broers et al. 1990; Broers 1992). However, immuno-gold labeling with an antiserum against mitochondrial cytochrome oxidase (COX) showed that cross-reacting proteins are present exclusively in the thylakosomes and not in the putative hydrogenosomes (Hackstein et al. 1994; Hackstein et al. 1995b). The BSTP reaction fails to indicate hydrogenase activity in these organelles, but the antiserum directed against hydrogenase also stains the cytoplasmic organelles (Hackstein et al. 1995a). Because Nielsen and Diemer (1976) interpreted dumbbell- and sausage-shaped hydrogenosomes of *Trichomonas* as propagation stages of the organelles, it may be speculated whether the cytoplasmic organelles of *Psalteriomonas* represent immature propagation stages of hydrogenosomes. Electron micrographs reveal that virtually all organelles are surrounded by rough ER, and, in addition, many of them are attached to a central vesicle (Fig. 3; Hackstein et al. 1995a). Their shape and a morphometric analysis of randomly chosen sections of peripheral hydrogenosomes suggest that two lobes of developing hydrogenosomes are budding off the central vesicle rather than dividing by fission. Electron microscopy fails to detect ribosomes or nucleoids in these developing hydrogenosomes either, and the stereotyped presence of surrounding cisterns of rough ER suggests a cotranslational protein import into the organelle. Since DAPI and ethidium bromide also fail to label these stages of hydrogenosomes, it may be concluded that both the mature and developing hydrogenosomes of *Psalteriomonas lanterna* lack nucleic acids and depend exclusively on protein import. Moreover, the mode of their propagation is unlike that of all other cellular organelles with an established endosymbiotic ancestry, i.e., mitochondria and chloroplasts. Rather, the development of the hydrogenosomes of *Psalteriomonas* resembles certain developmental stages of peroxisomes and other microbodies the ancestry of which is uncertain (Hruban and Rechcigl 1969).

Can Analysis of Proteins and Their Import Signals Show an Exogenosomal Ancestry?

The evidence presented so far characterizes the hydrogenosomes of *Psalteriomonas lanterna*, and also those of *Trichomonas*, as nonautonomously propagating cellular compartments without a genome and without a protein-synthesizing machinery. Apparently, their enzymes are encoded by nuclear genes. In both protists the enzymes seem to be synthesized in the cytoplasm: in *Trichomonas* on polysomes (Lahti and Johnson 1991; Johnson et al. 1993), in *Psalteriomonas* on the rough ER surrounding the developing organelles (Figs. 2, 3). This situation allows two alternative interpretations: (1) The genome of the putative ancestral prokaryotic endosymbiont has been lost completely, with the exception of those genes encoding the hydrogenosomal enzymes. These genes have been transferred to the host's nucleus, thereby acquiring a eukaryotic codon usage and GC/AT ratio. (2) Alternatively, the genes encoding the hydrogenosomal enzymes have been recruited from other - perhaps various - sources. In both cases the enzymes had to acquire suitable targeting signals that direct the proteins into the organelle. These targeting signals may provide some clues to the ancestry of the hydrogenosomes, but only if they are sufficiently conserved, like the peroxisomal targeting signals (Gould et al. 1990; Keller et al. 1991).

The enzyme content of a hydrogenosome per se cannot be informative about its ancestry, since a match with other enzyme sets of other organelles or free-living prokaryotes may be the result of convergent evolution or may even be purely accidental. The amino acid sequence of hydrogenosomal proteins is basically informative, but DNA sequence analysis of a number of genes encoding enzymes characteristic of hydrogenosomes reveals that the enzymes in hydrogenosomes of *Trichomonas* are related neither to a putative eubacterial ancestor nor to mitochondrial enzymes (Länge et al. 1994; Hrdy and Müller 1995a,b). Neither can the DNA sequence information available from *Psalteriomonas* (Brul et al. 1994) reveal a significant relationship to a prokaryotic or mitochondrial ancestry. The presence of import signals has so far been inferred from the DNA sequences; it has yet to be proven experimentally. However, even the putative import signals of hydrogenosomal isozymes in *Trichomonas* are not conserved. An alignment re-

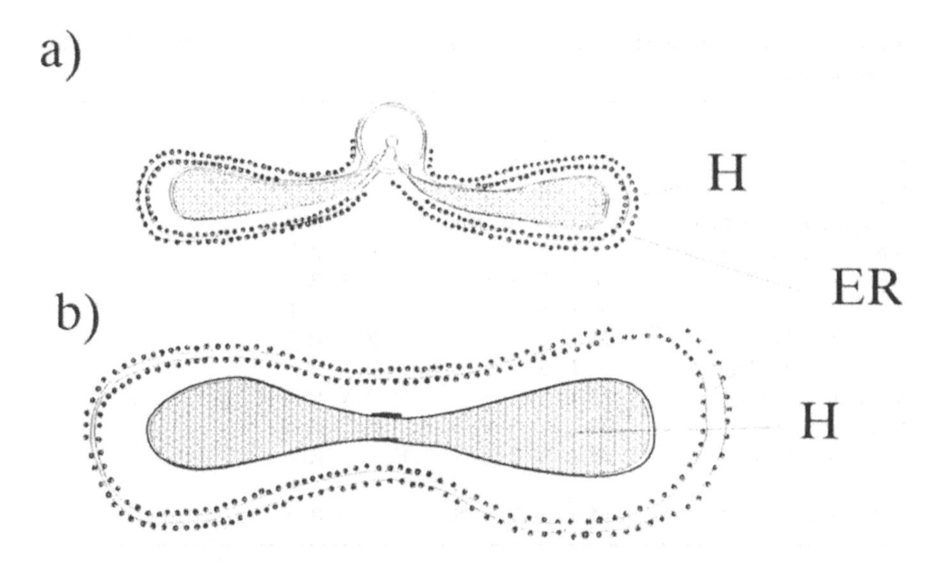

Fig. 3a, b. Cartoon of the budding process of a hydrogenosome of *Psalteriomonas lanterna*. **a** The ER cistern is in close contact with the central vesicle from which the two lobes of the hydrogenosomes (H) are budding off. **b** More advanced stage of hydrogenosomal development. The central vesicle has become resorbed or shed. The matrix of the hydrogenosome appears nearly crystalline. (After Hackstein et al. 1995a)

veals the presence of a motif also found in the *Psalteriomonas* ferredoxin (Table 1). This motif is different from the import signals known from mitochondria, plastids, and peroxisomes, and points, like the coding sequences, to independent evolution of the hydrogenosomes. Since this information cannot provide evidence of any relationship to free-living prokaryotes or semiautonomously replicating cellular organelles either, it is futile to dispute about an exogenosomal ancestry of the hydrogenosomes, for the lack of sequence conservation makes it unlikely that this hypothesis can ever be proved.

A Polyphyletic Endogenosomal Ancestry of Hydrogenosomes?

Hydrogenosomes are found in widely divergent taxa of anaerobic protists and fungi, such as amoebomastigotes, flagellates, trichomonads, ciliates, and chytridiomycetes (Fig. 1). However, as mentioned earlier, their characterization is mainly based on a cytochemical stain and their ultrastructure (Coombs and Hackstein 1995). This is especially true of the hydrogenosomes of anaerobic ciliates. In these organisms ultrastructural studies suggest similarities to mitochondria. Since it is likely that anaerobic ciliates evolved from mitochondria-bearing aerobic species (Cavalier-Smith 1993), it is tempting to speculate whether the hydrogenosomes of anaerobic ciliates are highly specialized mitochondria (Finlay and Fenchel

Table 1. Alignment of the putative amino-terminal targeting sequences of hydrogenosomal proteins

PFO-A	M	L	R	S								F
PFO-B	M	L	R	N								F
Ferredoxin	M	L	S	Q	V	C					R	F
α-Succinate-thiokinase - 1	M	L	A	G	D				F	S	R	N
α-Succinate-thiokinase - 2	M	L	S	S	S				F	E	R	N
α-Succinate-thiokinase - 3	M	L	S	S	S				F	E	R	N
β-Succinate-thiokinase	M	L	S	S	S				F	A	R	N
β-Succinate-thiokinase	M	L	S	A	S	S	N		F	A	R	N
Adenylate kinase	M	L	S	T	L	A	K	R	F			
Malic enzyme A	M	L	T	S	S	V	S	V	P	V	R	N
Malic enzyme B	M	L	T	S	V	S	Y		P	V	R	N
Malic enzyme C	M	L	T	S	S	V	B	F	P	A	R	E
PSALT-Fd	M	V	S	G	V	S					R	N

Modified from Hrdy and Müller 1995a,b

1989, Palmer 1997). However, in as much as no additional molecular, biochemical, and ultrastructural studies have supported this hypothesis, the suggestion of a mitochondrial descent of the hydrogenosomes of ciliates remains speculation.

The putative hydrogenosomes of anaerobic fungi have been studied in more detail, but even here the analysis of their physiological and molecular properties is still fragmentary. As in anaerobic ciliates, rumen fungi are likely to be derived from aerobic ancestors (Schlegel 1994). However, in contrast to the hydrogenosomes of ciliates, electron microscopy does not relate them to mitochondria. Rather, their ultrastructure strongly resembles that of peroxisomes (Yarlett et al. 1986; Marvin-Sikkema et al. 1992). A cross-reaction of antisera directed against the conserved carboxy-terminal ("SKL") targeting signal of peroxisomes has also been taken as evidence of the presence of peroxisomal import signals in the fungal hydrogenosomes (Marvin-Sikkema et al. 1993). However, direct experimental and molecular genetic evidence is still lacking. In contrast, first DNA sequencing data on putative hydrogenosomal enzymes of anaerobic fungi suggest a mitochondrial ancestry of these proteins and the presence of potential import signals at both the amino- and carboxy-terminal positions of the coding sequence (F. G. J. Voncken et al. pers. comm.). Therefore, a chimeric origin of the hydrogenosomes of chytridiomycete fungi is likely.

In conclusion, the available data strongly suggest a polyphyletic or even chimeric origin of hydrogenosomes, but they can support neither an exogenosomal nor an endogenosomal ancestry of the hydrogenosome. The data, including DNA sequencing information, reveal a divergence and heterogeneity that make it futile to speculate about an endocytobiotic origin of these organelles. Since there are no indications for the presence of a rudimentary genome or an organellar protein-synthesizing machinery either, it will be impossible to establish a descent from an endocytobiotic prokaryotic ancestor. Moreover, all available evidence favors the interpretation of the hydrogenosomes as an authentic cellular compartment of the eukaryotic cell.

Acknowledgements. We thank Theo van Alen and Holger Schlierenkamp for excellent technical assistance.

References

Broers CAM (1992) Anaerobic Psalteriomonad amoeboflagellates.PhD Thesis, Catholic University of Nijmegen

Broers CAM, Stumm CK, Vogels GD, Burgerolle G (1990) *Psalteriomonas lanterna* gen. nov., sp. nov., a free-living amoeboflagellate isolated from freshwater anaerobic sediments. Eur J Protistol 25: 369-380

Brul S, Stumm CK (1994) Symbionts and organelles in anaerobic protozoa and fungi. Trends Ecol Evolution 9: 319-324

Brul S, Veltman RH, Lombardo MCP, Vogels GD (1994) Molecular cloning of hydrogenosomal ferredoxin cDNA from the anaerobic amoeboflagellate *Psalteriomonas lanterna*. Biochim Biophys Acta 1183: 544-546

Cavalier-Smith T (1993) Kingdom protozoa and its 18 phyla. Microb Rev 57: 953-994

Coombs GH, Hackstein JHP (1995) Anaerobic protists and anaerobic ecosystems. In: Brugerolle G, Mignot J-P (eds) Protistological actualities. Proceedings of the Second European Congress of Protistology, Clermont-Ferrand, France, pp 90-101

Embley TM, Finlay BJ, Dyal PL, Hirt RP, Wilkinson M, Williams AG (1995) Multiple origins of anaerobic ciliates with hydrogenosomes within the radiation of aerobic ciliates. Proc R Soc Lond B 262: 87-93

Finlay BJ, Fenchel T (1989) Hydrogenosomes in some anaerobic protozoa resemble mitochondria. FEMS Microbiol Lett 65: 311-314

Gould SJ, Keller G-A, Schneider M, Howell SH, Garrard LJ, Goodman JM, Distel B, Tabak H, Subramani S (1990) Peroxisomal protein import is conserved between yeast, plants, insects and mammals. EMBO J 9: 85-90

Hackstein JHP, Schubert H, Berg M van den, Brul S, Derksen JWM, Matthijs HCP (1994) A photosynthetic organelle in anaerobic flagellates. Jahrestagung der Deutschen Sektion der International Society of Endocytobiology, Blaubeuren/Germany, April 7-9, 1994. Endocytobiol Cell Res 10: 261

Hackstein JHP, Rosenberg J, Broers CAM, Matthijs HCP, Stumm CK, Vogels GD (1995a) Biogenesis of hydrogenosomes in *Psalteriomonas lanterna.* Second European Congress of Protistology/Eighth European Conference on Ciliate Biology, July 21-26, Clermont-Ferrand, France 1995, abstract 281, p 97

Hackstein JHP, Schubert H, Berg M van den, Brul S, Derksen JWM, Matthijs HCP (1995b) A photosynthetic organelle in anaerobic flagellates. J Eukaryot Microbiol 42: 4A

Hrdy I, Müller M (1995a) Primary structure and eubacterial relationships of the pyruvate:ferredoxin oxidoreductase of the amitochondriate eukaryote *Trichomonas vaginalis.* J Mol Evolution 41: 388-396

Hrdy I, Müller M (1995b) Primary structure of the hydrogenosomal malic enzyme of *Trichomonas vaginalis* and its relationship to homologous enzymes. J Eukaryot Microbiol 42: 593-603

Hruban Z, Rechcigl M Jr (1969) Microbodies and related particles. Morphology, biochemistry, and physiology. Academic Press, New York

Johnson PJ, Lahti CJ, Bradley PJ (1993) Biogenesis of the hydrogenosome in the anaerobic protist *Trichomonas vaginalis*. J Parasitol 79: 664-670

Keller G-A, Krisans S, Gould SJ, Sommer JM, Wang CC, Schliebs W, Kunau W, Brody S, Subramani S (1991) Evolutionary conservation of a microbody targeting signal that targets proteins to peroxisomes, glyoxysomes, and glycosomes. J Cell Biol 114: 893-904

Lahti CJ, Johnson PJ (1991) *Trichomonas vaginalis* hydrogenosomal proteins are synthesized on free polyribosomes and may undergo processing upon maturation. Mol Biochem Parasitol 46: 307-310

Länge S, Rozario C, Müller M (1994) Primary structure of the hydrogenosomal adenylate kinase of *Trichomonas vaginalis* and its phylogenetic relationships. Mol Biochem Parasitol 66: 297-308

Lindmark DG, Müller M (1973) Hydrogenosome, a cytoplasmic organelle of the anaerobic flagellate *Tritrichomonas foetus*, and its role in pyruvate metabolism. J Biol Chem 248: 7724-7728

Margulis M (1993) Symbiosis in cell evolution. Freeman, New York

Marvin-Sikkema FD, Lahpor GA, Kraak MN, Gottschal JC, Prins RA (1992) Characterization of an anaerobic fungus from llama faeces. J Gen Microbiol 138: 2235-2241

Marvin-Sikkema FD, Kraak MN, Veenhuis M, Gottschal JC, Prins RA (1993) The hydrogenosomal enzyme hydrogenase from the anaerobic fungus *Neocallimastix sp. L2* is recognized by antibodies directed against the C-terminal microbody protein targeting signal SKL. Eur J Cell Biol 61: 86-91

Müller M (1988) Energy metabolism of protozoa without mitochondria. Annu Rev Microbiol 134: 465-488

Müller M(1993) The hydrogenosome. J Gen Microbiol 139: 2879-2889

Müller M(1997) Evolutionary origins of trichomonad hydrogenosomes. Parasitol Today 13: 166-167

Nielsen MH, Diemer NH (1976) The size, density, amd relative area of chromatic granules ("hydrogenosomes") in *Trichomonas vaginalis* Donne from cultures in logarithmic and stationary growth. Cell Tissue Res 167: 461-465

Palmer JD (1997) Organelle genomes: going, going, gone. Science 275: 790-791

Schlegel M (1994) Molecular phylogeny of eukaryotes. Trends Ecol Evolution 9: 330-335

Yarlett N, Orpin CG, Munn EA, Yarlett NC, Greenwood CA (1986) Hydrogenosomes in the rumen fungus *Neocallimastix patriciarum*. Biochem J 236: 729-739

Zwart KB, Goosen NK, van Schijndel MW, Broers CAM, Stumm CK, Vogels GD (1988) Cytochemical localization of hydrogenase activity in the anaerobic protozoa *Trichomonas vaginalis*, *Plagiopyla nasuta*, and *Trimyema compressum*. J Gen Microbiol 134: 2165-2170

1.2 Intertaxonic Combination and Gene Transfer (Interspecific, Intracellular)

Eukaryotism, Towards a New Interpretation*

R. G. Herrmann
Botanisches Institut der Ludwig-Maximilians-Universität, Menzinger Str. 67, 80638 München, Germany

Key words: Multiple endosymbioses, cell conglomerates, genome compartmentation, cell organelles (exogenosomes), intertaxonic combination, rearrangements of genetic material, loss and gain of function, integrated genetic machinery, two-step evolution.

Introduction

The phylogenetic-comparative molecular approach has caused a fundamental conceptual change in the way we view living matter. The new approach is based on the consequent utilization of genotypic comparisons, thus bypassing the limitations of phenotype. It differs in this respect fundamentally from the previous attempts of classification of life by Haeckel (1866), Chatton (1938), Copeland (1938), or Whittacker (1969). The approach rests on the decipherment of the genetic code, the development of potent strategies for nucleic acid sequencing, and on the assumption that each change *per se*, and its perpetuation, alteration, or relationship to other changes in an information-storing molecule, as in genomes in general, represents a phylogenetic event and is thus a direct record of evolutionary history. Consequently, sequence analysis may be expected to provide the means by which phylogeny may be deduced and - refined and complemented by analysis of the evolution of functions and structures - allow a global genealogical system to be established, i.e. to permit *all* lifeforms to be classified and related to each other, since any organism can be related to any other, unambiguously and quantitatively, and the properties of organisms can be understood in the context of their relatives. This has now been generally accepted, replaces the previous view in textbooks, and has resulted in a fundamental conceptual change in the thinking of biologists. Although even central aspects are still the subject of critical debate (see, e.g. Doolittle 1996; Martin 1996), it is highly probable that the approach will ultimately lead to a natural, coherent, and universal tree of life.

Two decades of work in the field of "molecular phylogeny" have revealed that

*Dedicated to Prof. Dr. Eberhard Schnepf on occasion of his 65th birthday.

present understanding of the living diversity on our planet is superficial. Sequence comparison, initially of rRNA sequences, has prompted the beginnings to restructuring of all major groups of organisms, and has substantially changed our view of the biological world by revealing that the earlier prevailing opinion which placed all organisms in one of two primary categories (prokaryotes and eukaryotes) was incorrect, but rather that all life falls into one of *three* principal lines of descent. These lines of descent, designated phylogenetic domains (Woese 1987; Woese et al. 1990; Olsen et al. 1994), comprise two categories of prokaryotes (Archaea and Eubacteria or Bacteria) and Eukarya. Until this approach became available, the problem of a microbial phylogeny was regarded as intractable because of the lack of morphological diversity and other characteristics with a natural, phylogenetic foundation, and the absence of adequate techniques. Now correct description of the microbial world becomes possible through the sequence analysis of the coding potentials of organisms in their entirety, and we are beginning to understand in outline how biological structure and pathways evolved.

In a comparable way, present understanding of eukaryotism is still superficial, and taxonomic classification of eukaryotes with their basic and (five) crown taxa, which reflect an early evolution of protistean lineages and a relatively simultaneous appearance of several advanced groups (plants, animals, fungi, alveolates, stromenopiles), often incorrect (e.g. Sogin 1994, Doolittle 1996). Furthermore, various forms such as lower photoautotrophs had until recently received astonishingly little taxonomic definition. This has now begun to change as well. However, the elaboration of an accurate genealogy for eukaryotes has required and requires a different set of chronometer molecules and modified strategies. For plants, these were initially organelle DNAs, especially the analysis of operon structures and genes involved in photosynthesis or in the flux of genetic information in organelles; nowadays nuclear genes are more and more employed, particularly those involved in plastid biology (Cerff 1995; Herrmann 1996; Martin and Schnarrenberger 1997). The latter represent an increasingly important source of information for a better understanding of eukaryotism. They have significantly extended our knowledge by uncovering additional, fundamental aspects of the biology of eukaryotes and have allowed delineation of the principles and formulation of a new concept of the eukaryotic cell and of eukaryotism. Both appear in a new light, phylogenetically and functionally.

Strictly speaking, the eukaryotic cell is a phylogenetic curiosity (Herrmann 1996) which originated in a conglomerate of cells, two in animals, three, four or even five in plants which arose through the incorporation and integration of unicellular prokaryotic or eukaryotic organisms respectively into a protoeukaryotic or eukaryotic cell (intertaxonic combination; Fig. 1). These conglomerates had to acquire and to develop both a common metabolism and appropriate ways of inheritance. Molecular biology, complemented by ultrastructural and biochemical work, has not only amply demonstrated that mitochondria and plastids originated as endocytobionts, both probably monophyletically, it also verified plastids of primary, secondary, or even tertiary phylogenetic origin. It has also begun to deduce those organisms to which the organelles are affiliated, although the nature of the early host cells as well as the progenitors of the cytoskeleton, membrane flux, and nucleus still remain uncertain (Sogin 1994; Langer et al. 1995; Doolittle 1996), and has demonstrated that the evolution of eukaryotic cells and organisms must have been complex and diverse, especially in the case of their autotrophic forms.

This chapter intends to approach eukaryotism from a *functional* point of view, and to provide a synopsis and a framework which attempts to conceptually organize the diverse findings and scattered literature into a coherent view, in part with a new interpretation. The origin and "entrance" of the eukaryotic "Urwirt" is not considered here. The focus will be on plant organisms. Only a limited fraction of the information available in the literature and in data bases is described; the examples are chosen to illustrate basic points. This concept has been outlined in part in a recent review in another context (Herrmann 1996).

Outline of a General Phylogeny of Eukaryotic Photoautotrophs

Organelle Chromosomes

The genetic information of plastids, the plastome, is contained in a single circular DNA molecule generally ranging in size between 100 and 200 kb. The evidence of complete sequences for more than a dozen plastid chromosomes, including information previously lacking for anything other than chlorophyll *a/b*-containing organisms, together with appreciable sequence data from other plastid DNAs, support a common history of *all* photosynthetic plastids, regardless of differences in antenna design, pigmentation, envelope structure, and storage products (e.g. Kowallik 1994). This had already been supposed for quite some time due to a remarkable similarity of the basic machinery for oxygenic photosynthesis in pro- and eukaryotes (Blankenship 1992; Bryant 1992). The plastid chromosomes sequenced, representing seven algal phyla of widely separated chlorophytic, rhodophytic, and chromophytic thallophytes, bryophytes, and vascular plants, display striking similarities with regard to coding capacity and overall structural and functional organization, notwithstanding differences in detail. Notably gene clusters, e.g. those encoding components for the photosynthetic or the organelles' translational machineries such as the *atp* or *rrn* operons (see, e.g. Bohnert et al. 1979; Cozens and Walker 1987; Nelson 1992, also Herrmann et al. 1992), bear striking similarities to each other and to the corresponding gene clusters of prokaryotes. They are therefore more diagnostic for a deduction of ancestry than the mere individual sequences that were frequently selected initially for study and were sometimes misinterpreted. The primary differences between plastid chromosomes appear to be caused primarily by occasional rearrangements due to intra- or intermolecular recombination, fusion and fragmentation of operons, and losses of genes from operons or of operons to varying extents (see, e.g. Palmer 1992; Whittier and Sugiura 1992). These all are considered to represent derived traits, as are antennae, their pigments, or other previously used features.

All plastid lineages including cyanoplasts (formerly designated cyanelles), the photosynthetic organelles with a residual murein containing cell wall of the freshwater protozoon *Cyanophora paradoxa* (Schenk 1970) and of related members of Glaucocystophyceae (cf. Löffelhardt and Bohnert 1994), share approximately half, nearly 100 genes, of the gene complement found in plastid chromosomes studied to date. The chromosomes of rhodoplasts, cyanoplasts, and of plastids from chromophytic algae house additional genes when compared to those of the chlorophyll *a/b* lineage, frequently members of discrete operons. In the chlorophyll *a/b* lineage these are either nuclear-coded [such as *SecA* and *SecY* (Nakai et al. 1994; Yuan et al. 1994; Berghöfer et al. 1995; Laidler et al. 1995)], or unknown as yet. *Porphyra purpurea* plastid chromosomes contain an additional 120 (Reith and Munholland 1995), that of *Odontella sinensis* an additional 40 (Kowallik et al. 1995), and cyanoplast DNA an additional 60 genes (Stirewalt et al. 1995). Apparently, gene loss, intracellular gene translocation, and gene management by the nucleus (see below, pp. 91, 101f) have reached different stages in the chlorophyll *a/c*, *a/d* and *a/b* lineages of plants. The last mentioned lineage is the most advanced in terms of intracellular gene transfer, disregarding relatively minor, but nevertheless telling differences within this group (see, e.g. Baldauf et al. 1990). An exception may be *Acetabularia* (Tymms and Schweiger 1989). Apart from the limited extent of the overall changes compared to the source genome(s), the differences between plastid chromosomes of the major lineages are not surprising. For instance, the complexity of envelopes in secondary organelles with their phylogenetically different origin (Fig. 1) may impose structural and functional constraints both with regard to the lateral transfer of genes from the organelle into the nucleus and to the product return/translocation into the organelle (see below, pp. 85, 96). Collectively, the coincidence between corresponding prokaryotic and plastid gene and operon structures, substantial sequence conservation, and an analysis of appropriate nuclear "marker" genes leave no doubt about both their common ancestry and the adoption of genes lacking in plastid transcription units by the nucleus (see below, pp. 86). They suggest that the basic gene arrangement of plastid chromosomes was established *prior to* the divergence of the principal lineages, and consequently supports, but does not prove, a *monophyletic* origin of the organelle.

Complete sequences of mitochondrial genomes (chondriomes) are available from protists including fungi, animals (Gray 1989; Wolstenholme 1992), the liverwort *Marchantia polymorpha* (Oda et al. 1992), and the higher plant *Arabidopsis thaliana* (Unseld et al. 1996). Unlike plastomes, the gross phylogenetic delineation of plant chondriomes and its relation to the evolution of plastomes have not been settled, primarily due to the lack of an equivalent body of sequence information including that for the genome of an ancestral representative. The DNA molecules constituting higher plant chondriomes display three unique features. They can greatly vary in sizes, between 200 and >2000 kb, and may thus be one order of magnitude or more larger than those of animals, fungi, and liverworts which share generally densely packed genomes in the range of 16,000 to 100,000 nucleotides and, as plastids, a basic set of genes (less than 100, encoding components of the respiratory chain assemblies of the inner mitochondrial membrane and of the organelle's own translational machinery) (Gray 1989; Wolstenholme 1992). Although they contain additional genes (see, e.g. Gonzalez et al. 1993, Schuster et al. 1993), the substantial size differences are caused primarily by an appreciable fraction of sequence duplications, "noncoding" and integrated plastid and nuclear DNA (see below. pp. 86f). Furthermore, higher plant chondriomes are greatly rearranged and may occur in single, isomeric molecules, but also in different subgenomic molecules that may be circular or linear (Bendich 1996). Subgenomic molecules result from intramolecular recombination at widely separated sequence repetitions, and may exist in a membrane-attached form, some also in an episomal form. Depending on the number of such sites, structural chondriome complexity can be substantial. Nevertheless, the multipartite chondriomes appear to be genetically stable although individual DNA molecules can vary and change appreciably in quantity (see, e.g. Grayburn and Bendich 1987; Small et al. 1989). Finally, a relatively large number of the organelle tRNAs appear to originate in translocated plastid DNA or even have to be imported from the nucleo/cytosolic compartment, although plant mitochondria basically possess sufficient coding potential to encode a full tRNA complement (Maréchal-Drouard et al. 1993). Some features, which differ from those of plastid chromosomes (see, e.g. Palmer 1992), cannot be properly explained at present, but the evidence as it stands suggests that basic aspects of evolution are similar for both organelle genomes (see below, pp. 99f).

Primary, Secondary and Tertiary Plastids (Exogenosomes)

The photosynthetic organelles of Rhodophyta, Chlorophyta (red and green algae, bryophyta, vascular plants), and cyanoplasts are surrounded by two envelopes and should thus have originated from the acquisition of a prokaryotic, most likely cyanobacteria-like endocytobiont by a nonphotosynthetic eukaryotic or protoeukaryotic cell (prokaryote/eukaryote or proto-eukaryote symbiosis; Fig. 1b). They are therefore defined as descendants of (a) primary endocytobiont(s). The most appealing candidates for plastid ancestors, the chlorophyll *a*- and *b*-containing prochlorophyta, are not only phylogenetically quite heterogeneous, but they also do not appear to be the ancestors of Chlorophyta (Palenik and Haselkorn 1992; Urbach et al. 1992). This has generated the problem of the origin of chlorophyll *b* in seemingly unrelated groups of organisms. Present data favor the idea that the ancestor contained antenna systems based on both phycobilin and chlorophyll *b* pigments that split into the basic lineages, rather than that chlorophyll *b* was invented twice in evolution, although the relatively simple biochemical metabolic path from chlorophyll *a* and the existence of chlorophyll *b* variants (Goericke and Repeta 1996) would not preclude this *a priori*. A potential representative of the "missing link" has recently been found in *Prochlorococcus marinus*, a highly abundant autotroph at the bottom of the euphotic layer (Chisholm et al. 1992) that appears to possess both antenna systems (Hess et al. 1996).

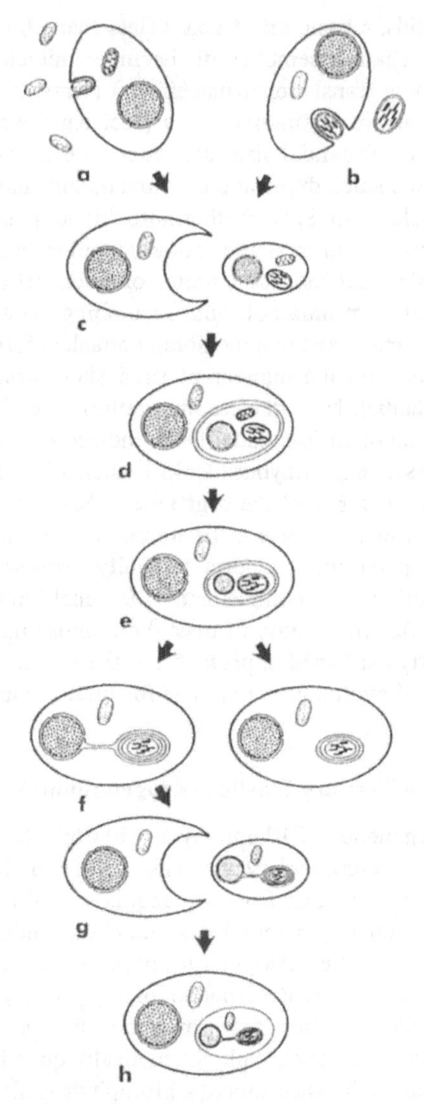

Fig. 1. Principal steps in the evolution of photosynthetic organelles and of eukaryotism. Mitochondria and primary plastids resulting from phagocytosis of (a) a heterotrophic or (b) a photoautotrophic prokaryote into a protoeukaryotic cell. (c) - (h) Complex plastids generated by secondary and tertiary endosymbioses. (c) A phototrophic eukaryotic microbe is incorporated by an autotrophic cell. Alteration of the endocytobiont includes the loss of mitochondria and the reduction of the nucleus to a nucleomorph (e), or its loss (f). This results in an organism with photosynthetic organelles surrounded by three (right) or four (left) envelope membranes which, in turn, may be incorporated by a tertiary flagellated host to form a tertiary organelle, such as in fucoxanthin-containing Dinophyta. Cryptomonads and Chloroarachniophyceae (c) are intermediates or links between the hypothetical stage (d) and stage (f). (Modified from Herrmann 1996, see also text)

All other major algal groups house plastids that are surrounded by *three* (Euglenophyceae, most Dinophyceae) or *four* (Chrysophyceae or Heterokontophyta, Haptophyceae, Chlorarachniophyceae, Cryptophyceae) membranes and probably represent chimaeras of two or even three distinct eukaryotic cells (Fig. 1e - h), although detail is still subject to some controversy. The latter group includes the plastids of apicomplexan protozoan parasites (Köhler et al. 1997). The formation of such plastids, complex or "secondhand" plastids (Sitte 1987; McFadden 1993), is apparently the result of a eukaryote/eukaryote endosymbiosis that should therefore have been progressed through two successive stages, a primary event involving a phototrophic prokaryote and a eukaryotic or protoeukaryotic host (Fig. 1b), and a second endosymbiosis involving the resulting photosynthetic primary eukaryotic protist (microalga) and a second eukaryote as host (Fig. 1c). Circumstantial evidence suggests that some nuclear genes may have originated in a transiently present endocytobiont that may have been replaced by a different one (Häuber et al. 1994). If correct, "direct" and "replaced" complex endocytobionts/organelles may have to be distinguished, but such cases may be tedious to untangle taking processes such as lateral gene transfer into account. Tertiary endocytobionts, in turn, would have resulted from serial endosymbiosis based on secondary eukaryotic cellular entities engulfed by eukaryotic, generally flagellated hosts (Fig. 1g - h). They are found for example in fucoxanthin-containing Dinophyceae (Jeffrey and Vesk 1976; Chesnick et al. 1996).

In the three-layered plastids of euglenoids and dinoflagellates, the second membrane is assumed to have been lost and is frequently considered to represent the phagosome membrane (phagocytic vacuole) of the host (Fig. 1f, Gibbs 1979; 1993). It is usually quite fragile. However, it has been argued that the triple-layered plastid envelopes of euglenoids could reflect a primary endosymbiosis as well (see, e.g. Cavalier-Smith 1995; McFadden et al. 1995), although the biology, structures, and sequence data of this group make this an unlikely possibility. The key to the answer to the crucial question of membrane identities resides in the phylogenetic origin of plastid envelopes, which has not been settled in most instances, not even for primary endocytobionts. Apart from the fact that the sequence data available on this are sparse, lipid and protein compositions, often chosen for diagnosis, may be phylogenetically adaptable and, like other biochemical data, do not provide decisive criteria *per se*. In primary endocytobionts, the envelope membranes could theoretically have been derived from the double-membrane of the endocytobiont (Cavalier-Smith 1992), or from one of these and the host plasmalemma or food vacuole membrane; in euglenoids all three might have been retained. In secondary organelles the additional membranes may derive from the engulfed eukaryote or from the plasmalemma/food vacuole membrane and ER of the secondary host.

Although details of the history of plastid membranes and hosts from secondary and tertiary endocytobionts are currently under debate, it is undisputed that these complex organisms have arisen *polyphyletically* both with regard to symbiont and host, and that not all phylogenetic intermediates have been vanished (see, e.g.

Lewin 1992; Sitte 1993; McFadden and Gilson 1995). Cryptophyceae, Chlorarachniophyceae and fucoxanthin-containing Dinophyceae appear to represent "intermediate stages" in the development of complex plastids. The periplastid compartment between the two pairs of plastid membranes in the former cases can house a DNA-containing organelle, the nucleomorph (Greenwood 1974; Greenwood et al. 1977; Hibbert and Norris 1984), and an additional category of 80S ribosomes that appear to be functional (McFadden et al. 1994a, b). They are considered to represent the vestigial nucleus and cytosol, respectively. Fucoxanthin-containing Dinophyceae, in turn, are binucleate and can contain reduced chlorophyll *c*-containing endocytobionts with a mesokaryotic nucleus (Fig. 1 h, Jeffrey and Vesk 1976; Chesnick et al. 1996) that apparently are not viable *per se*. Sequence evaluation has revealed two phylogenetically distinct 18S rRNA genes in the nucleus and nucleomorph of Cryptophyceae suggesting that they derived from different lineages, the latter in this instance probably from a rhodophyte-like eukaryotic endocytobiont (Douglas et al. 1991; Maier et al. 1991; Van de Peer et al. 1995, see also Liaud et al. 1996). In Chlorarachniophyceae, amoeboid algae, and probably in apicomplexan protozoan parasites the secondary plastid appears to originate in an ancient chlorophytic microalga (McFadden et al. 1995; Van de Peer et al. 1996; Köhler et al. 1997). Fucoxanthin-containing Dinophyceae, in turn, appear to have a secondary endocytobiont antecedent of plastid related to diatoms (Chesnick et al. 1996). The host component of Chlorarachniophyceae seems to stem from a barely studied lineage of phagotrophic amoeboid flagellates (Bhattacharya et al. 1995). However, the phototrophic members of that group appear to form independent evolutionary branches including as well chlorophyll *a/c*-containing, Haptophyceae-related organelles that lack nucleomorphs (Cavalier-Smith et al. 1996). Some of these cases provide unequivocal examples of chimaeras with different evolutionary histories even within (present) taxonomic classes, reinforcing the polyphyletic origin of this category of eukaryotes. More than half-a-dozen independent events have been amply demonstrated for anucleomorphic and nucleomorph-containing plastids from secondary endocytobioses (cf. Lewin 1992), and a comparable figure can be expected for tertiary events.

The Nucleomorph

All presently available data are commensurate with the interpretation that the nucleomorph of Cryptophyceae and Chlorarachniophyceae represents the relict nucleus of previously free-living phototrophs, although some of the features of its DNA resemble strikingly those known from the reduced genomes of organelles. In those instances studied, nucleomorphs house a densely packed subgenome comprised of three small chromosomes of a total of about 600 kb in the former case and less than 400 kb in the latter (Eschbach et al. 1991; McFadden et al. 1994a; Rensing et al. 1994; McFadden and Gilson 1995) with discrete telomers (Gilson and McFadden 1995). This corresponds to an estimated number of 200 - 300 genes, a figure that has to be related to some 20.000 - 40.000 genes per provenance nucleus. Available information on coding capacity, gene organization and expression predominantly from a Chlorarachniophyta nucleomorph has uncovered genes for nucleomorph maintenance, components involved in the flux of genetic information (replication, transcription, processing, translation, chaperones, protein degradation), signal transduction, and an abundance of (unusually short) spliceosomal introns, generally eukaryote-type components. On the other hand, the "miniaturized genome" displays an unexpected structural and functional compaction, reminiscent of organelle genomes, such as only short intercistronic regions, overlapping transcription and at least one instance of bicistronic transcription, i.e. operon-like DNA segments (Gilson and McFadden 1996).

The estimated figure of nucleomorph genes is too low to manage the integration of plastid and residual cytosol into the cellular genetic context. Remarkably, only a few genes have been found in nucleomorph DNA so far that are involved in plastid management (see, e.g. Hofmann et al. 1994; Gilson and McFadden 1996). This implies that the majority of the plastid proteins must be encoded in the genome of the secondary host and that the nucleomorph (*plus* surrounding cytosol) is of dual genetic origin in a double sense: It still influences plastid biogenesis and function, but is itself influenced by the host's nuclear genome.

Mitochondria, Hydrogenosomes (Exogenosomes)

Mitochondria appear to originate monophyletically in the α section of proteobacteria (see, e.g. Whatley et al. 1979; Falah and Gupta 1994), and hence are descendants of a primary endosymbiosis. Secondary forms as known from plastids have not been detected. However, it is relevant to emphasize that hydrogenosomes, (the only) double membrane-bounded, functionally specialized organelles found in amitochondriate eukaryotes, share a common host with mitochondria, but lack DNA (Müller 1993; Bui et al. 1996; Horner et al. 1996; see below, pp. 102).

Functional Aspects of Endosymbiont Integration - The Concept of a *Functional* Molecular Phylogeny

Evolution is based on selection of function. Although mere sequence analysis has been and is useful for the deduction of phylogeny, without a knowledge of the underlying functional aspects it does not unravel how and why a particular development has occurred, and how the eukaryotic entity operates at present. That plastids and mitochondria each derived from a free-living prokaryote or, in the former case, may be degenerate descendants of engulfed eukaryotes had profound consequences not only for the evolution, but also for the biogenesis and physiology of that cell type. Actually, the complexity of the evolution of the eukaryotic cell as deduced from mere sequence studies is based on an analogous complexity in the evolution of its function and structure. The contours of these changes are emerging, and an understanding of them is essential for an understanding of crucial aspects of that cell. In other words, the current concept of "molecular phylogeny" has to progress to a "functionally defined molecular phylogeny". *In my opinion, the eukaryotic cell can ultimately be understood only on the basis of its history.*

The prokaryotic predecessors of the organelle genomes must have possessed coding sequences for all components necessary for life. The roughly 3.6 Mb genome of the free-living cyanobacterium *Synechocystis* sp. PCC6803, of which the entire sequence has recently been determined, houses ca. 3,100 genes (Kaneko et al. 1996; equivalent data are not yet available for proteobacterial and nucleomorph-related nuclear genomes). On the other hand, current plastid chromosomes encode only a fraction, in the order of 5% or less, of the components needed for the function and maintenance of a cell. This mere comparison illustrates the substantial modification, probably rapid at first, that the cell association has undergone. The changes that occurred during the establishment of endosymbiosis and stabilization of the cell conglomerate, up until the point when all that remains is the eukaryotic cell with its photosynthetic and/or respiratory organelles, can be summarized in six main points that in my opinion characterize that cell type and, implicitly, eukaryotism. *They cannot be reconciled with the currently prevailing semiautonomy concept* (see below, pp. 104f). The framework is provided first of all by (a) a massive restructuring of genetic information in the entire unit that has been accompanied by (b) the establishment of nuclear regulatory dominance and (c) a reductive evolution of organelle genomes. The alterations include both loss and gain of information from/for the entire system. They occur at the subgenic as well as the genic level as judged from *three* phylogenetic categories of nuclear genes/information involved in the biology of organelles (d - f) that have been deduced from functional analyses (cf. Herrmann 1996).

Loss of Genes

A first category of nuclear genes in regard to organelle management is related to information that has been lost from the entity. This loss of genes or suites of genes occurred progressively and predominantly from the endocytobionts, protoorganelles or organelles, including the nucleomorph, since today much of the basic metabolism is cytosolic and consequently encoded by the nucleus. Substantial gene losses have also occurred from the host compartment as will be outlined below. Examples are found in genes that were initially redundant in the endocytobiotic conglomerate, and one was removed, e.g. to avoid metabolic competition. The loss was intercompartmentally complemented, in a complex manner since genes derived from the eubacterial endocytobionts replaced equivalent host genes quite frequently (see, e.g. Brinkmann and Martin 1996; Martin and Schnarrenberger 1997). Other genes, such as those for the endocytobiont cell walls, may have been dispensable for different reasons. Cell wall remnants are found around cyanoplasts (Schenk 1970; Löffelhardt and Bohnert 1994). In secondary endocytobionts, mitochondria and related genes, and, remarkably also genes of the endomembrane system, cytoskeleton, etc., are or appear to be lacking (Gibbs 1978; 1990; Whatley et al. 1979). An instance in which an eukaryotic component has replaced a prokaryotic form might be found in acetyl-CoA carboxylase (Sasaki et al. 1995). This multimeric enzyme, as fatty acid synthase assemblies, exists in prokaryotic and eukaryotic forms, of which the latter is found in yeasts and animals. Both forms are found in dicotyledonous plants, the prokaryotic variant (700 kDa) in plastids, the eukaryotic one (500 kDa) in the cytosol. Monocots appear to possess only the latter with isoforms in both compartments. Available data suggest an enormous selective pressure and a rapid drift especially of duplicated or nonessential information (Ainsworth et al. 1987; Mardsen et al. 1987; Chao et al. 1989). They also indicate the existence of appropriate mechanisms in the nucleus, e.g. in allopolyploid nuclei (Chao et al. 1989), to tolerate gene redundancy only if they are of functional use, e.g. to adapt or facilitate the synthesis of a component in response to different environmental or developmental stimuli, or to generate an appropriate structural ambience, e.g. for fine-tuning of antenna systems (see below), in heterodimeric photosynthetic reaction centers, or F-/V-ATPases (Nelson 1992; Herrmann 1996). The novel element was (and is) that such mechanisms monitor *inter*compartmental duplications for dispensability of information, possibly by epigenetic processes (see below, pp. 89).

Transfer of Genes/DNA Segments

The genes lost from the endocytobionts fall basically into two classes. A second class of now nuclear genes was endocytobiont-specific, and a substantial fraction of them was not present in the host cell. Such information may have been relinquished from the engulfed organisms as well. However, unlike the former class, it has been preserved within the entire system since it was symbiose-relevant, or otherwise of advantage for the entire unit. Examples are found in components that are essential for extant plastid function or in processes conferred by plastids, such as autotrophy by the photosynthetic machinery, which was one of the most useful features of the entrapped unit. It basically solved the energy demand of the novel cell. A similar scenario could apply to the respiratory machinery in mitochondria which may have conferred on the entity bacterial "achievements" that were advantageous, for instance, for adaptation during the global changes from reducing to oxidizing environment (see, e.g. Fenchel and Bernard 1993; Doolittle 1996).

The maintenance of such genetic information was accompanied by a transfer of genes from the endocytobiont into the nucleus and product return during the phylogenetic transformation to the organelle (Weeden 1981), either of part of it in the case of supramolecular structures, or perhaps complete, in that for single- or simple-chain enzymes. A typical example of the latter case is represented by one of the plastid-located GAPDHs, which is the reimported product of a translocated gene (Martin and Schnarrenberger 1997). The transfer process is obviously more advanced in mitochondria, and probably terminated in hydrogenosomes (see below, pp. 101). For complex organelle structure the result was a dual genetic origin (Kawashima and Wildman 1972), such as for the photosynthetic and respiratory membranes (see, e.g. Herrmann et al. 1991), multimeric soluble enzymes of the chloroplast stroma, e.g. ribulose bisphosphate carboxylase/oxygenase, acetyl-CoA carboxylase (Sasaki et al. 1995), or Clp proteases (Adam 1996), organelle ribosomes, or for other parts of the translation machinery. Remarkably, only part of the tRNA complement appears to be derived from the original chondriome in some lower and in higher plants, probably also in the plastome of the parasite *Epifagus* (Wolfe et al. 1992). An appreciable number of tRNA genes utilized in such mitochondria are of plastid origin, and a fraction of organelle tRNAs have to be imported from the nucleo/cytosolic compartment (Joyce and Gray 1989; Maréchal-Drouard et al. 1993). This, the recent surprising finding of nuclear-coded genes for the catalytic subunit of ribulose-1,5-bisphosphate carboxylase/oxygenase in dinoflagellates that resembles the L2-type enzyme found in purple bacteria (Morse et al. 1995; Rowan et al. 1996), of possible gene transfer from "replaced" secondary endocytobionts, and other diverse origins of nuclear genes including those that clearly indicate „replacement of replaced genes“ cannot be reconciled with the idyllic picture of a simple gene transfer and product return („product specificity corollary“; Weeden 1981). Rather DNA rearrangements in the eukaryotic cell must have been appreciably more complex than currently supposed, and include "compartmental detour" of genes and/or gene products.

Particularly instructive in this respect is the work on the genes for Calvin cycle enzymes and their cytosolic homologues of the glycolytic, gluconeogenetic and oxidative pentose phosphate pathways (Martin et al. 1993; 1996; Martin and Schnarrenberger 1997) which illustrates impressively the complexity of the rearrangements that the genetic material of the entity has undergone. Even if lateral gene transfer between bacteria, gene transfer from engulfed (or fused) cells, recurrent (early) gene duplications, differential gene losses, and subsequent functional specialization still obfuscate the picture, a highly complex mosaic pattern of nuclear gene evolution emerges including three potential sources (archaebacteria, proteobacteria, cyanobacteria). In most instances only one or two of the genes has/have survived to the present. Most of the now nuclear genes of the Calvin cycle in higher plants appear to have been obtained from eubacterial (proteobacterial and cyanobacterial) sources with the possible exception of transketolase. This implies that the pristine nuclear genes and even subsequent proteobacterial copies may have been replaced, apparently quite frequently. Further complicating the matter, this is not uniform in the eukaryotic "crown" taxa. A few examples may illustrate this (Martin and Schnarrenberger 1997). Genes for the small Rubisco subunit and GAPDH genes appear as a family that can be traced to a series of ancient, preendosymbiotic duplications. An example for duplications that occurred after gene transfer and led to functionally distinct isozymes in different compartments has been found in phosphoglycerate kinase that exhibits a high affinity to the cyanobacterial genes and has apparently replaced the previous, phylogenetically probably intermediate proteabacterial component still found in heterotrophs. Yet another example of this kind though illustrating a proteobacterial (mitochondrial) origin for a plastid enzyme is represented by triosephosphate isomerase. The plastid-located NAD^{+}-dependent GAPDH, in turn, unlike the above mentioned plastid isozym, appears to be derived from the duplication of a

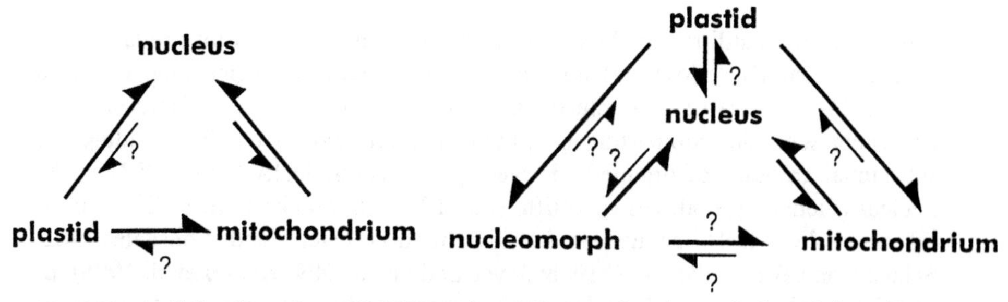

Fig. 2. Intracellular flux of genetic information accompanying the evolution of the eukaryotic cell, in a primary and terminal secondary system (left), and an intermediate secondary endocytobiotic system (right)

proteobacterial antecedent that encodes the cytosolic isoform. *Obviously, each enzyme displays a particluar history.*

Intracellular gene transfer in secondary endocytobionts reflects this complexity as well since translocations between host nucleus, plastid, and nucleomorph, and a two-step serial process involving the latter compartments are likely (Fig. 2). Suggestive for this are, for instance, the nuclear-coded phycoerythrin α-subunit of cryptophytes and sequence data on nucleomorph DNA (Jenkins et al. 1990; McFadden and Gilson 1995; Gilson and McFadden 1996). The above mentioned coding capacity and compaction of the nucleomorph genome, which appears to include a decline of genes involved in plastid management reflects a massive gene flux from the nucleomorph to the host nucleus, and implicitly a two-step gene transfer, early from the (primary) plastid to the nucleomorph predecessor, and, after secondary endocytobiosis, from this compartment to the host genome. It is conceivable that the presence or absence of a nucleomorph may simply depend on taxonomic relationships of the symbiotic partners, and on the "kinetics"/degree of gene transfer that may not be unrelated. A relict nucleus may have to be maintained if the gene transfer from the plastid in the free-living microalga that became engulfed were quite advanced and hence that from the endocytobiont nucleus/nucleomorph to the host system lasting. In Cryptophyceae, the eukaryotic remnant may have lasted for the functional compensation of the phylogenetically different plastid chromosome (see above, pp. 80f). Furthermore, it has been postulated that the host genomes of such organisms may even possess genes from plastids of an extinct primary endocytobiont that has been replaced by a secondary endocytobiont (Häuber et al. 1994). The finding of glyceraldehyde phosphate dehydrogenase genes possibly acquired from eubacterial donors in early-branching protists lacking mitochondria indicates such an existence of ephemeral cell associations and losses (Henze et al. 1995) that may have preceded stable integration. Assuming this supposition to be correct, this could have required fewer changes for the functional integration of the multienveloped system. Nothing is known from tertiary endocytobionts in this respect.

Information is still relatively scarce as to how and how often intracellular phylogenetic gene/DNA loss and transfer have occurred, and what happened to a gene/DNA segment or gene cluster after its transfer, but all available data are consistent with the compartmental phylogeny outlined so far. The findings of substantial amounts of organelle-derived, promiscuous DNA (Ellis 1982) in the nucleus (Pichersky et al. 1991; Ayliffe et al. 1992; Brennicke et al. 1993), of plastid or nuclear DNA sequences in mitochondria (Stern and Lonsdale 1982; Schuster and Brennicke 1987; 1988; Joyce and Gray 1989; Knoop et al. 1996), or of initially plastid-derived nucleomorph sequences in nuclear DNA (see above, pp. 81f) suggest that such processes are relatively frequent, specifically if the possibility is considered that some of the rearranged sequences are lost, that only a fraction of the transferred sequences survive in evolution, and that translocation is not unidirectional. It is important to emphasize that it probably occurs (or has occurred) between all genetic compartments, even if it has not yet been demonstrated in all instances (Fig. 2).

Circumstantial evidence has been obtained for reverse transcription of ancestral edited mitochondrial RNA intermediates (Schuster and Brennicke 1987; Nugent and Palmer 1991; Covello and Gray 1992; Grohmann et al. 1992; Brennicke et al. 1993), and direct DNA transfer between compartments (Ayliffe et al. 1992). The former is probably relatively rare, but would provide the preselection for functional sequences and hence obviate obstacles to gene transfer such as the excision of compartment-specific introns or posttranscriptional editing events. The latter could involve leaky membranes, membrane vesiculation or fusion. As to the organelle RNA polymerases (see below, pp. 91, 102), genetic changes could result from horizontal gene transfer as well, such as viruses, mobile elements (transposons and retrotransposons; Knoop et al. 1996), or result from mechanisms known from Ti plasmids of agrobacteria. It is also imaginable that gene transfer during digestion of an engulfed cell contributed to genome evolution, or even genes relinquished from transitory endocytobionts that were lost before they were stably integrated.

Available information also indicates that there is no gross preselection of organelle sequences for transfer, since no correlation with gene sizes has been noted. However, an inspection of the nature of the translocated genes reveals that in plastids only certain categories, predominantly those involved in regulatory processes or those encoding single- or simple-chain catalytic polypeptides, appear to have managed successfully (see below, pp. 101). Gene transfer from mitochondria appears to be more advanced in general, and has included other components in addition. The comparison of *atp* operons from different prokaryotes and plastid chromosomes, in turn, in which the first translocated gene (*AtpC*) (Pancic et al. 1992; Pancic and Strotmann 1993) can form a separate transcription unit in the cyanobacterium *Synechococcus* 6716 (Falk and Walker 1988), has suggested that the excision of a gene-bearing segment from organelle DNA may not occur entirely at random (Herrmann et al. 1993), nor may the functional integration into a genome (see, e.g. Koncz et al. 1992).

Minimal prerequisites for the success of a nuclear gene copy are that a translocated gene is integrated into a favorable sequence context or that an appropriate context is recruited by rearrangement after integration, i.e. that it acquires DNA segments which possess the potential to develop into promoters and protein sorting signals (transit peptides or uncleaved sorting signals), and that regulatory constraints do not exclude this. The probability of picking up an appropriate DNA segment is not low (Baker and Schatz 1987; Koncz et al. 1992), in the order of 1 - 3%. Relatively little is known about whether the promoter *plus* the segment encoding the transit peptide in a given gene were acquired simultaneously or by different phylogenetic events. Evidence for both alternatives has been obtained, the latter *via* exon shuffling (Wolter et al. 1988; Nugent and Palmer 1991; Covello and Gray 1992; Oelmüller et al. 1993; Long et al. 1996) and in-phase fusions (Grohmann et al., 1992). Evidence exists also for disappearance of intervening sequences at domain junctions (Wolter et al. 1988, Kadowaki et al. 1996), that the same gene may be equipped with different transit peptides and *vice versa* (e.g. Kadowaki et al. 1996), and that exon shuffling can lead to compartmental redirection of a gene product (Long et al. 1996). The findings that promoters and segments for transit peptides between different nuclear genes encoding stromal or thylakoid proteins (see below, pp. 89) are unlike, that those of the same gene in different organisms resemble each other more or less, that a significant fraction of nuclear-encoded proteins for the photosynthetic machinery originates in single- or low-copy genes, and that nuclear genes for components of the Calvin cycle (Chao et al. 1989) or of the thylakoid membrane (Pillen et al. 1996) are scattered in the genome have suggested that gene losses and functional translocations generally occur individually.

Individual gene translocations can also be valid for gene families, which may originate in a single transfer event followed by subsequent duplication and distribution within the genome. However, repeated translocations have also been deduced on the basis of the same criteria (Wedel et al. 1992). For instance, the *Lhc* genes and relatives that encode multiple biologically active peptides [chlorophyll *a/b* antenna apoproteins (LHCPs) associated with photosystems I and II, CP29, CP24, ELIPs, and the product of the *PsbS* gene] with functions in light collection, energy transfer or stress protection (Green et al. 1991) differ both in their functional complexity and phylogenetically from the *RbcS* gene family that encodes the small subunits of ribulose bisphosphate carboxylase/oxygenase or (for instance) from the fructose-1,6-bisphosphatase and sedoheptulose-1,7-bisphosphatase genes (Martin et al. 1996). The marked differences in transit peptides (and promoters, where known) between, but much less within the *Lhc* subgroups (Kim et al. 1992; Wedel et al. 1992; Cai et al. 1993), and positional studies by genetic mapping indicating that the *Lhc* single-copy genes and the genes of subfamilies may each occupy different chromosomal positions within a genome, are commensurate with the idea that the latter constitute a *super*family comprised of several subfamilies (Wedel et al. 1992). These would have originated from multiple translocation events, independently for each subfamily before the ancestral organelle locus was eliminated. The alternative of spreading of a single translocated gene and subsequent divergence is less likely in this instance. On the other hand, the *RbcS* gene family and the plastid and cytosolic isozymes of fructose-1,6-bisphosphatase as well as of sedulose-1,7-biphsophatase apparently arose through gene duplication after gene transfer, the latter group from an eubacterial bifunctional fructose/sedoheptulose bisphosphatase antecedent. The plastid enzyme then underwent functional specialization into the fructose and sedoheptulose phosphatases by acquiring substrate specificity (Martin et al. 1996). *PsbS* was suggested as an intermediate progenitor in the *Lhc* family (Wedel et al. 1992; see also Dolganov et al. 1995). That it has not been found in the *Synechococcus* 6803 genome (Kaneko et al. 1996) does not preclude that it fulfils this role, unless it is absent in the genomes of chlorophyll *a/b* antenna- *plus* phycobilisome-containing cyanobacteria (Hess et al. 1996) or prochlorophyta.

Once operating, the subsequent functional integration and streamlining processes, e.g. to account for differences in compartmental codon usage or regulatory sequences, as well as the competition with and the elimination of the plastome-encoded copies, probably proceeded relatively fast, since transitory states of functional genes for the same component in two compartments have generally not been observed. The functional nuclear genes and seemingly silent corresponding copies in mitochondria and plastids that may be lacking in close relatives (Sebald and Hoppe 1981; Baldauf et al. 1990; Covello and Gray 1992; Nugent and Palmer 1991; Grohmann et al. 1992; Sánchez et al. 1996) suggest that gene removal from organelles is a multistep process that includes gene inactivation, e.g. by sequence drift or partial deletion, and gradual loss of the obsolete copies by segregation of the altered chromosome within the organelle, of organelles with altered information from those with unaltered information, and, finally, of cells housing only altered organelles. The loss of a sequence from an individual organelle chromosome should usually be without risk to the functionality of the entire system (Wedel et al. 1992), due to the high degree of plastome (and chondriome) reiteration per cell (see, e.g. Herrmann and Possingham 1980). However, the conditions necessary for the accomplishment of nucleo/cytosolic dominance, that is, how a single nuclear gene is activated, competes with, and manages against the thousands of organelle gene copies per cell, the mechanisms that eliminate the locus from one of the subgenomes, and the functional consequences with regard to coordinated compartmental expression for the stoichiometry of protein production or intercompartmental signalling processes have remained enigmatic, as has the cause of the enormous plastome and chondriome reiteration. Ideas and observations are often virtually contradictory. One of several conceivable possibilities to acquire nuclear dominance could be that the expression of a functionally translocated gene may not underlie regulation limits initially, e.g. by light. Another may be an enhanced stability of mRNA levels conferred by intron containing mRNAs (Rose and Last 1997).

In promoters of translocated genes, qualitatively and quantitatively operating *cis*-elements frequently reside within the first 200 - 300 nucleotides upstream of the respective transcription start sites, but they may be located far upstream (>1000 nucleotides), downstream of the transcription start site within the 5'-untranslated leaders, at the initiative ATG, and even downstream of the initiator ATG (Dickey et al. 1992; 1994; Flieger et al. 1993; Bolle et al. 1996 a, b). They may be experimentally separable or intermingled (see, e.g. Flieger et al. 1993, Kusnetzov et al. 1996). Notwithstanding the lack of common *cis*-elements that would resemble heat shock or some hormone-responsive elements, or similarities among promoters in *cis*-element arrangement, many nuclear genes encoding constituents of a particular organelle structure, such as the thylakoid membrane or ribosomes, are seemingly regulated in a comparable way, e.g. coordinately expressed by light or expressed in an organ-specific way (Lübberstedt et al. 1994; Harrak et al. 1995). It is therefore not surprising that comparative studies on nuclear genes have uncovered an unexpected complexity of transcriptional control. The distinctively different sequences of the promoters of genes for thylakoid proteins, their different architectures and strengths, qualitatively different expression-relevant *cis*-elements (e.g. for light responsiveness) and distinctively different distributions of such elements and the corresponding *trans*-acting factors, even between genes that code for different subunits of the same membrane assembly (Lübberstedt et al. 1994) or of organelle ribosomes (Villain et al. 1994; Li et al. 1995), highlight two fundamental aspects of eukaryotic cell biology (Herrmann 1996): How is the transcription of nuclear genes with such different regulatory sequences coordinated both *per se* and with that of organelle genes, and how have these different promoter sequences resulting from phylogenetically independent and single transfer events of originally organelle-encoded genes to the nucleus, been integrated phylogenetically into the network of signal transduction chains and metabolic pathways to ensure the correct biogenesis of organelle structures? No superior regulatory elements coordinating the activities of subsets of nuclear genes involved in the biology of organelles have yet been deciphered, and as yet no reasonable concept has emerged for the observed largely synchronous expression and transcript accumulation in a particular substructure.

An equivalent situation exists for the transit peptides of organelle proteins of nuclear origin. For instance, with the exception of some components of the outer plastid envelope (Salomon et al. 1990; Li et al. 1991) or of the mitochondrial envelopes (see, e.g. Stuart and Neupert 1996) which operate with uncleaved sorting epitopes, all known organelle proteins of nuclear origin are synthesized with transient signals. The targeting signals fall into two principal, functionally and structurally distinct classes with stroma- and stroma-/thylakoid-, or matrix- and matrix/inner mitochondrial membrane-targeting quality (von Heijne 1986; von Heijne et al. 1989). The former specify translocation across the envelope membranes, in plastids apparently at a common entry site and by a common mechanism (Cline et al. 1993; import of NADPH:chlorophyllide oxidoreductase may be a possible exception; Reinbothe et al 1995). The latter represent bipartite structures with two signals in tandem that operate in a two-phase pathway successively with two distinct translocation systems, for protein delivery to the stromal phase and thylakoid transfer (Fig. 3), or transfer to the matrix and inner mitochondrial membrane/intermembrane space, respectively. Such transit peptides, as promoters, exhibit high variability in terms of sequence and length, although some sectoring into functional regions has been noted (Chaddock et al. 1995 and see below, pp. 96). Work with authentic precursor proteins or engineered translocation substrates has shown that the distribution of topogenic signals within the different precursors, as *cis*-elements in promoters, is quite diverse. For instance, a critical epitope for the integration into and translocation across the thylakoid membrane is a hydrophobic domain (Clausmeyer et al. 1993) which may be positioned transitorily or internally. The appealing view that hydrophobic epitopes have to be transitory and consequently part of the transit peptide for hydrophilic, lumenal proteins, but may operate as uncleaved signals in integral proteins such as the LHCP or the CP24 apoproteins associated with photosystem II (Lamppa 1988; Bartling et al. 1990; Reith and Douglas 1990; Cai et al. 1993), turned out not to be correct. The association with a given type of transit peptide is not strictly correlated with the polarity nor with the intraorganelle location of a protein. For instance, functional equivalents in chloroplasts and mitochondria may contain different categories of sorting signals. A precedent is found in the Rieske FeS protein of cytochrome complexes, which in plastids operates with a mere import, in mitochondria with a bipartite transit peptide (publications quoted in Bartling et al. 1990 and Pfanner 1992, respectively). Furthermore, various integral, often bitopic thylakoid proteins such as CFo-II, PsaF, PsbW, or PsbX, subunits of the ATP synthase, photosystems I and II respectively, depend on bipartite transit peptides (see below, pp. 97f). On the other hand, the plastid-encoded ß-phycoerythrin of Cryptophyceae, located in the thylakoid lumen, is apparently made without an N- or C-terminal extension (Reith and Douglas 1990). An equivalent situation exists in mitochondria (Hartl and Neupert 1990; Pfanner 1992), a more complex one in "secondary" plastids (see below, pp. 96). The enormous diversity in the arrangement of basic targeting epitopes in precursor proteins and their relation to topological locations of thylakoid proteins are not yet fully understood. They may at least in part be explained phylogenetically, as is evident from the similarity between different categories of sorting signals and from the conservation of part of the translocation machineries. This will be outlined by a few examples below.

Gain or Generation of Genes/Information

The functional and controlled genetic and metabolic integration of the endocytobiont into the novel biological entity required the development of a comprehensive set of devices and general intercompartmental regulatory controls to coordinate inheritance, gene expression and product fluxes *between* genetic compartments, and also between tissues. These include information for protein, RNA, or metabolite translocation across organelle envelopes, and an extensive machinery for the regulation of gene expression as well as of structure and composition in the organelles. At the multicellular level cells differentiate into a variety of specialized types with distinct structures and functions. In higher plants this includes architecturally different and physiologically specialized modifications of the plastid that are involved in various fundamental biological processes. Higher organisms may therefore be considered as complex cellular societies, organized in tissues each assigned a limited range of functions (Herrmann et al. 1992). The long distance interactions including partitioning and allocation of hormones, C and N assimilates, that is, the production and export of surplus photosynthetic energy from "source tissues", the import/storage processes of photosynthates into/in "sink tissues", and the regulatory relationship between photosynthetic carbon metabolism and nitrogen fixation illustrate this further complexity.

Intra- and intercellular controls appear to operate at almost all levels of regulation. Their establishment was again accompanied by massive genetic changes, but often of different kind. They have occurred at the subgenic as well as the genic level, and depend largely, though not exclusively (see, e.g. Oelmüller 1989; Taylor 1989; Hess et al., this volume), on nuclear-encoded components. These genes may be phylogeneti novel, or represent restructured/redirected information of preexisting genes or of gene modules. Four sets of examples may highlight changes at the subgenic stage (example 1), and this (third) category of (nuclear) genes (examples 2 - 4) which, like the intracellular rearrangement of genetic material, is often underestimated currently with regard to quantity and functional consequences.

Example 1: The finding of spatial (organ- and tissue-specific) transcriptional regulation in nuclear genes which were usually already translocated at the unicellular stage, the (little studied) adaptation of the promoter of such a gene for expression in different ecological habitats, similar promoter changes in the establishment of C4 biochemistry from orthologous genes in C3 plants (Stockhaus et al. 1994), the consequences of mitochondrial sequence duplication or of integrating a mitochondrial intron sequence into the upstream region of one of two largely conserved lectin genes (Knoop and Brennicke 1991; Wissinger et al. 1991, see also Sánchez et al. 1996), or the adjustment of some of the uncleaved, often hydrophobic domains of integral proteins to serve targeting functions besides membrane anchoring or membrane attachment (Michl et al. 1994) indicate a substantial adaptive sequence drift and reshuffling of elements that represents a gain of regulatory information after gene translocation. This information can be quantitative or qualitative in character. The feature is not restricted to nuclear genes. When compared to their prokaryotic equivalents, a similar restructuring of sequences must have occurred in plastid and mitochondrial chromosomes including their promoters. The demonstration that two distinct polymerases, encoded in different compartments and probably of different phlyogenetic origin (see next paragraph and below, pp. 102), cooperate to promote accurate transcription initiation in plastids (Lerbs-Mache 1993; Pfannschmidt and Link 1994; Suck et al. 1996) suggests that transcription in the organelle is substantially more complex than in its prokaryotic counterpart. Apparently, plastid chromosomes house three classes of genes with regard to promoters, two functional with only one each of the polymerases, and a third group operating with both (unpublished observations). In various instances this must necessarily have been accompanied by an increase of promoter complexity when compared to the prokaryotic ancestor, since both RNA polymerases operate on different promoters, often with multiple initiation sites, and only on subsets of genes in the organelle. In mitochondria, only one polymerase appears to operate, conceivably not the one of the bacterial progenitor, implying promoter changes in the chondriome as well (see below, pp. 103). Promotor changes can also be inferred from the compaction and gene organisation for nucleomorph DNA (see above, pp. 80). Another illustrative example has recently been noted in a functional comparison of prokaryotic and eukaryotic plastocyanins and their docking partner in photosystem I, PsaF. Eukaryotic PsaF subunits contain an additional approximately 20 amino acid residue epitope at their N-terminus capable of forming an amphipathic helix, which causes an increase in the electron transfer route to photosystem I by more than one order of magnitude (Haehnel et al. 1996). Remarkably, this segment exhibits notable homology to the epitope in the α subunit of distinct G-proteins involved in the interaction with the ß subunit (H. Fulgosi, personal communication).

Example 2: A variety of regulatory mechanisms and devices control gene expression and the flux of genetic information in the organelle. They first became apparent from differences in the transcript patterns of corresponding prokaryotic and plastid operons, which at the eukaryotic level generally exhibit an extensive posttranscriptional modification (see Herrmann et al. 1985). The diversity of components and processes involved includes genes in different compartments for two polymerases, genes for factors operating in the transcription of plastid operons (see, e.g. Mullet et al. 1990; Kim and Mullet 1995), for σ factors for the plastid-encoded *E. coli*-like, multisubunit RNA polymerase (genes: *rpo*) (Liu and Troxler 1996; Tanaka et al. 1996) that operate on σ^{70}-type promoters in plastid chromosomes resembling equivalent *E. coli* promoters with -10 and -35 consensus elements (reviewed in Igloi and Kössel 1992; Gruissem and Tonkyn 1993; Link 1996), or even an additional, nuclear-encoded, plastid-located RNA polymerase that appears to control the genes encoding this polymerase, as the expression of the plastid chromosome in general (Hess et al. 1993; Lerbs-Mache 1993; Pfannschmidt and Link 1994; Suck et al. 1996). Selection of σ^{70}-type promoters and modulation of their activity depends on σ-like factors and factors interacting with upstream elements (Sun et al. 1989; Iratni et al. 1994; Allison and Maliga 1995; Kim and Mullet 1995). Although the existence of a second transcription system based on a nuclear-coded polymerase is now unequivocal, the identity of the enzyme is not clear. A single-chain-type RNA polymerase, as found in T3 or T7 phages, and in mitochondria (Lerbs-Mache 1993; see below, pp. 102), or a complex enzyme (Pfannschmidt and Link 1994) have been described but the activities have not been obtained from axenically grown material or have not been checked on authentic templates, respectively. The greatly reduced, yet functional plastome of the Orobrancheacean parasite *Epifagus virginiana*, which has not only lost all genes for photosynthesis and chlororespiration but also those for the plastid-encoded RNA polymerase, presumably operates solely with the imported enzyme and a subset of organelle promoters (Wolfe et al. 1992). This would resemble the situation in mitochondria. Comparable results have been obtained from transcript analysis of *rpoB-*, *rpoA-*, and *rpoC1*-deficient tobacco mutants (Allison et al. 1996, and own unpublished work). The latter finding, as findings with ribosome-lacking plastids (Hess et al. 1993), suggests also that DNA replication and organelle multiplication are largely under nuclear control. Remarkably, the gene for such an enzyme is lacking in the sequenced *Synechococcus* 6803 genome (Kaneko et al. 1996), consistent with the notion that all bacterial organisms rely apparently on only one type of RNA polymerase (see below, pp. 102).

Other examples are provided by the multitude of factors involved in the subsequent steps of transcript modification or transcript stability, such as endonucleolytic activities, truncation of 3' ends, stabilization of 3' ends by polypeptides, *cis-* and *trans*-splicing, transcript editing (Westhoff and Herrmann 1988; Rochaix 1992; Freyer et al. 1993; Gray and Covello 1993; Kössel et al. 1993; Abrahamson and Gruissem 1996), or mRNA decoding in plastids and in mitochondria. They are evident from biochemical work and from nuclear mutant phenotypes affecting RNA processing (see, e.g. Rochaix 1992; Barkan et al. 1994). It is relevant to note that various of them appear to be derived rather than retained traits. For instance, RNA editing appears to exist predominantly in higher eukaryotes (Malek et al. 1996). The exact point of appearance of this process is still not entirely clear but available data favor the idea that RNA editing is a process acquired during the later development of eukaryotes, rather than an ancient one. In plant organelles, it can be quite substantial and generally involves C to U transitions, or occasionally the reverse (Gualberto et al. 1990; Schuster et al. 1990). Estimates range in the order of 1000 sites or more in mitochondria (Pring et al. 1994; Malek et al. 1996), and 300 for plastids. Editing can change the coding context, but in non-protein coding segments it may modulate expression (Pring et al. 1993; Araya et al. 1994). The underlying mechanism(s) and the origin of the biochemical machinery have largely escaped analysis. The reciprocal activities and other data suggest the operation of deamination/transamination or, probably less likely, transglycosylation reactions. Small guide RNAs have been postulated to confer the specificity of the reaction. However, the latter would have to be reconciled with the high number of edited sites, their relatively rapid phylogenetic change, and with the variability of surrounding sequences both within and between organisms in both organelles (Araya et al 1994; Bock et al. 1994; Yu and Schuster 1995). The majority of the compounds mentioned are encoded in the nucleus, disregarding a few exceptions such as the putative plastid-encoded maturase in the *trnK* intron.

Example 3: The lateral transfer of genes to the nucleus, and consequently the return of the corresponding gene products that are generally synthesized on free cytosolic ribosomes and translocated posttranslationally to the respective organelle, require not only the generation of organelle-specific targeting and sorting signals but also the concomitant establishment of efficient polypeptide (and metabolite) translocation machineries. These ensure accurate delivery of a protein to the respective organelle as well as to its correct destination in the organelle interior. The same is valid for metabolite exchanges across organelle membranes. Translocation systems in envelopes and internal membrane systems recognize transit peptides or uncleaved internal signals, and metabolites. The recent dissection of protein and metabolite translocation machineries in organelle envelopes has demonstrated that their constituents are nuclear-encoded. While each of the protein import machineries of the organelles appears to represent a unique translocation system, since no progenitors for some of the central components have yet been deduced (Pfanner 1992; Hirsch et al. 1994; Schnell et al. 1994; Seedorf et al. 1995; Lübeck et al. 1996; Muckel and Soll 1996), some translocation machineries involved in the intraorganelle sorting processes resemble each other and those of bacteria to a considerable extent. Also, transit peptides and translocation machineries of different origin can often be mutually substituted in these cases (see, e.g. Apt et al. 1993; Seidler and Michel 1990; Haehnel et al. 1994). Thus, despite of differences in detail there are clear-cut phylogenetic links for the latter processes (see also Schatz and Dobberstein 1996), but not for the former. It is relevant to note that protein transport, and implicitly receptor integration, would be reverse, i.e. inside out, if the outermost envelope were a derivative of the host plasmalemma. Therefore, the question of whether an import or an export machinery (Schatz and Dobberstein 1996) had to be inserted or generated depends critically on the nature and phylogenetic origin of a membrane, a point that is not finally settled as outlined above (see above, pp. 79).

The import, stepwise transport and assembly of nuclear-encoded proteins through and into various locations within the organelles, as transcriptional and posttranscriptional processes, are highly regulated (Fig. 3). Signals and sorting mechanisms are essential at every step for proper traffic regulation as well as for subcellular differentiation. The import of macromolecules is not restricted to proteins. In mitochondria (Maréchal-Drouard et al. 1993; Kumar et al. 1996) and possibly *Epifagus* plastids (Wolfe et al. 1992) it can include structural RNAs (tRNAs or tRNA/protein complexes) (Tarassov et al. 1995), which are terminal gene products, as proteins. It is striking that the pattern of tRNA gene dispersal differs significantly between plant species (Kumar et al. 1996). It is not known whether import includes mRNAs, genetically intermediate macromolecules, which would have to account for differences in compartmental codon usage.

Various steps of protein translocation and sorting are linked to the availability of energy (ATP, membrane potential, proton gradient) and can include the interaction with specific factors, such as chaperones or cofactors, which may also keep a component import- and routing-competent. The basic aspects of these events in plastids, including the nucleomorph in descendants of secondary endocytobionts, and mitochondria resemble each other, apart from some distinctive differences in detail, e.g. in the design of transit peptides which ensures organelle specificity, in the energy requirement at individual steps, or that signals are required in plastids that discriminate between protein transfer to the photosynthetic membrane and inner envelope (von Heijne 1986; von Heijne et al. 1989; von Heijne and Nishikawa 1991; Keegstra and von Heijne 1992; Pfanner 1992). The nucleomorph-encoded Hsp70 of Cryptophyceae may mediate transport of polypeptides from the eukaryotic endocytobiont remnant cytosol into the plastid matrix (Hofmann et al. 1994).

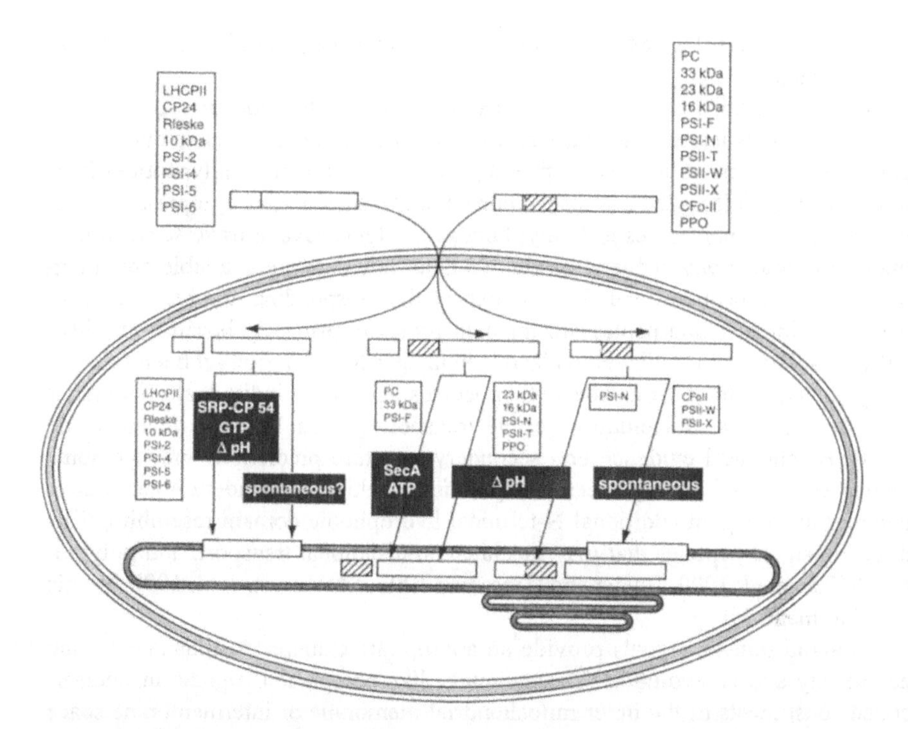

Fig. 3. Complexity of protein sorting and assembly processes in higher plant chloroplasts (for explanation see text). Plastocyanin, the 33 kDa protein of the oxygen-evolving system, and subunit Pja-F of photosystem I, follow the SEC route, while the 16- and 23- kDa proteins of the oxygen-evolving system, subunit N of photosystem I, subunit T of photosystem II, and lumenal polyphenol oxidase follow the pH-dependent route. The ATP synthase subunit CFo-II, PsbW, and PsbX of photosystem II, in turn, integrate spontaneously, and a LHCII apoprotein utilizes an SRP-like path (see text)

Import sequences are removed during or following translocation across the target membrane by a distinct metallo endopeptidase located in the plastid stroma or mitochondrial matrix (Hawlitschek et al. 1988; Yang et al. 1988; Oblong and Lamppa 1992). Bipartite presequences are often, but not always (see below, pp. 98), processed in two steps, with an additional intrinsic enzyme active at the lumenal face of stroma-exposed thylakoids (Kirwin et al. 1988) or in the inner mitochondrial membrane (Hartl and Neupert 1990). A third, ancient endopeptidase, CtpA, located in the thylakoid lumen, removes the C-terminus of the precursor of the D1 polypeptide located in the innermost core of the photosystem II reaction center (Anbudurai et al. 1994). Again, for both organelles the endopeptidases and other known "auxiliary" components originate in nuclear genes. Some of the genes for these components have been detected at the prokaryotic level and hence were probably translocated, the origin of others that may include the metalloprotease is unknown.

Additional complexity becomes apparent with the translocation of nuclear-encoded proteins into plastids derived from secondary endocytobionts due to their particular envelope structures. Depending on their destination, polypeptides have to be transported from the host or the nucleomorph-surrounding cytoplasms across two to four membranes, respectively; lumenal proteins have to traverse five membranes of various phylogenetic origin. To date, few data are available about signals and transit peptides, and about an expected corresponding number of translocation machineries and peptidases for their removal, primarily because the difficulty of isolation of multilayered plastids in integer form has not yet been satisfactorily solved. Different translocation processes across the individual membranes have been postulated including vesicle transport (Gibbs 1979; Hofmann et al. 1994). Biochemical evidence and secondary structure predictions indicate some unique features of presequences and proteins, multipartite tandem signal/transit peptides including an additional N-terminal hydrophobic domain resembling ER-directing signal peptides that may mediate cotranslational transport, and polyproteins (Chan et al. 1990; Bhaya and Grossman 1991; Shashidara et al. 1992; Pancic and Strotmann 1993).

Thylakoid transfer signals provide an appropriate example to illustrate sorting complexity and its evolution. These signals, like comparable signals in nuclear-coded constituents of the inner mitochondrial membrane or intermembrane space (Hartl and Neupert 1990), share key features with prokaryotic signal peptides. The finding that the second domain of bipartite transit peptides exists in corresponding cyanobacterial polypeptides (Kuwabara et al. 1987; Briggs et al. 1990; Chitnis et al. 1991), the presumed progenitors of chloroplast proteins, and the observation that this domain is capable of functionally replacing the prokaryotic equivalents (Seidler and Michel 1990; Haehnel et al. 1994) has originally suggested that only the import domains were acquired phylogenetically, and that all these proteins, in analogy to the import process into plastids, are transported across thylakoid membranes by a common, SEC-(secretory)-type mechanism inherited from the prokaryotic progenitor of plastids. The differences in primary sequence and the predicted general structural similarity of these targeting epitopes have also suggested that the basic signals of transit peptides are provided by topogenic elements rather than by primary sequence. Although various polypeptides, including those of plastid origin (Zak et al. 1997), utilize the ancestral route, as evidenced by the requirement of stromal factors and nucleoside triphosphates, the inhibition by sodium azide, known to be an antimetabolite of the SEC route in prokaryotes, and by the isolation of cDNAs for the plastid-located SecA and SecY proteins (Nakai et al. 1994; Yuan et al. 1994; Berghöfer et al. 1995; Laidler et al. 1995) which are still plastid-encoded in formerly secondary endocytobionts, Rhodophyta, and cyanoplasts (Douglas et al. 1991; Kowallik et al. 1995; Reith and Munholland 1995; Stirewalt et al. 1995), the recent comparative study of transit peptides and sorting processes for thylakoid precursor proteins has uncovered an unexpected diversity and subtlety of pathway modifications for targeting of nuclear-encoded proteins into and across the thylakoid membrane. Four distinct pathways have been delineated to date (Fig. 3). In three of these "bipartite" presequences are involved. The findings suggest that the phylogeny of the eukaryotic cell was not only accompanied by an intracellular rearrangement of genes that necessitated the acquisition of promoters and interacting *trans*-factors, segments for transit peptides or sequence modifications leading to protein-internal sorting information, and the complementary envelope import machinery for components of nucleo/cytosolic origin, it has also demonstrated marked effects of phylogenetic gene translocation in the late stages of biogenesis. These included the generation of new (regulatory?) luminal components that originate in nuclear genes (e.g. the 16- and 23-kDa polypeptides of the oxygen-evolving complex of photosystem II), and led probably to new variants for membrane translocation in plastids, to bipartite presequences of different origins, and also to different modes of integration. Comparable phylogenetically traceable sorting diversity has been noted for mitochondria (Stuart and Neupert 1990; Pfanner 1992).

In plastids, an integration route distinct from the SEC route involves a signal recognition particle (SRP), and resembles the SRP-mediated pathway of protein translocation in bacteria and across the ER membranes in eukaryotes (see, e.g. Schatz and Dobberstein 1996). For instance, LHC apo- or preapoproteins interact with the (azide-insensitive) stromal homologue of SRP54. This integration is dependent on GTP and stimulated by a ΔpH (Yuan et al. 1994; Nakai et al. 1994; Li et al. 1995 and references in these publications). However, a significant number of plastid components are translocated by a different route independently of soluble stromal factors, ATP or GTP, and require a transthylakoidal ΔpH gradient only (Fig. 3) (Cline et al. 1993; Henry et al. 1994; Robinson and Klösgen 1994). A fourth discrete pathway specific for the ATP synthase subunit CFo-II, the PsbW (Lorkovic et al. 1995), and PsbX subunits in photosystem II (the latter are lacking in the cyanobacterial genome; Kaneko et al. 1996) appears to operate *via* spontaneous integration without involvement of a proteinaceous receptor (Michl et al. 1994), with a mechanism reminiscent of that of M13 coat protein export (Kuhn et al. 1986). It is relevant to note in this context that the principal determinants for routing specificity reside generally, but not always completely, in the transit peptides (see, e.g. Robinson and Klösgen 1994). Neither a translocation mechanism that depends exclusively on a protonmotive force, nor the latter integration class have yet been detected in cyanobacteria. The ATP synthase subunits CFo-I (b)/CFo-II (b') (genes: *AtpF* and *AtpG*, respectively), which are characteristic of thylakoid-located ATP synthases (Herrmann et al. 1993), have provided an appealing (and the only) model to illustrate this. Their genes exist at the prokaryotic stage, still reside in the plastid chromosomes of originally secondary endosymbionts, Rhodophyta and cyanoplasts (Pancic et al. 1992; Kowallik et al. 1995; Reith and Munholland 1995; Stirewalt et al. 1995), but are found in different subcellular compartments in the chlorophyll *a/b*-lineage of organisms (Herrmann et al. 1993). In this case, both proteins integrate with different epitopes and mechanisms into the thylakoid membrane. CFo-I operates with a single hydrophobic domain for both targeting to and anchoring within the membrane, while integration of CFo-II requires an additional hydrophobic domain in its transit peptide.

The CFo-II transit peptide bears all the characteristics of bipartite presequences, but its second domain is lacking in the corresponding plastome-encoded and prokaryotic subunits. This suggests that its phylogenetic origin, as that of PsbW and PsbX that are not found in prokaryotes, differs from that of others, in that it seems to represent a modified import signal, comparable to the similarly structured transit peptide of CP24. The latter, however, functions as a mere import signal (Cai et al. 1993). Remarkably, the CFo-II transit peptide also lacks an intermediate cleavage site (Michl et al. 1994). The phylogenetic quality of that characteristic is not clear, however. Although the genes of other proteins, like PsaN, that lack such a cleavage site are not found in the sequenced *Synechocystis* genome (Kaneko et al. 1996), the 23- and 16-kDa precursor proteins, PsbW, and PsbX that are also not present possess cleavable bipartite transit peptides.

Example 4: As "open systems", organisms not only add matter, they also exchange it with their environment by constantly modifying structures to ensure efficiency under ever-changing environmental conditions. The molecular processes controlling this efflux of matter for maintenance, acclimatization, stress protection, and repair, or its resorption during senescence may be different from those that operate in the generation of structure, but their contribution to the function of supramolecular structure must be of comparable quality to that of those generating it. In fact, circumstantial evidence suggests that these processes are as rigorously regulated as the influx of matter, with comparable complexity and subtlety, and with a different set of compounds (see, e.g. Herrmann 1996). These include, for instance, protein kinases and associated phosphatases, other modifying enzymes, at least two dozen protease activities involved in protein turnover which have recently been detected in chloroplasts (unpublished data), or stress-induced proteins such as light stress proteins (ELIPs) and heat shock proteins (chaperones). Again, although some components such as Clp or FtsH proteases possess bacterial antecedents (see, e.g. Adam 1996), the majority of them are encoded in nuclear genes, and several of them may lack prokaryotic counterparts.

Collectively, it appears that regulation occurs at all levels, that the expression of *individual* organelle genes and the control of their products involves an *unexpectedly large number* of nuclear genes and factors, and that a very significant fraction of the nuclear genome, possibly 20% or more of its gene complement, operates for both organelles or is involved in the control of their genes. The sources, progenitors, and evolution of this catogory of nuclear genes are still largely enigmatic, and it is often not yet possible to discriminate between transfer of genes, modification of preexisting information, or generation of information. Knowledge of this information may become accessible as more data accumulate, despite of modular DNA rearrangements and sequence drift that may obfuscate and aggravate the delineation. The important task is therefore not looking only for homologues, but tracing events of gene duplication, gene/modul rearrangements, and changes in function.

Synopsis and Conclusions

Cellular cohabitation that led to eukaryotism (intertaxonic combination) must have had an enormous selective advantage, since it arose and was perpetuated polyphyletically (mitochondria including derivatives and plastids, plastids derived from primary, "direct" and "replaced" secondary, and probably even tertiary endosymbioses; Fig. 1). The integration of the photosynthetic and respiratory machineries was probably forced by environmental changes (see above, pp. 83). Cellular cohabitation permitted the development of advanced (true multicellular) life, in combination with oxygenic photosynthesis also of terrestrial life (Herrmann 1996), with an enormous morphogenetic potential that in the plant kingdom is evident, spatially and temporarily, from unicellular (protophytic) through thallophytic to kormophytic organisms with their nuclear phase changes and complex generation cycles. So far the reasons for that advantage are not entirely clear. They may reside in an intracellular division of labor which offered the opportunity of combining and streamlining the biogenetic potential in this entity by both complementing and divergent compartmental evolution. Although elements considered to be specific for eukaryotes, such as plasmodesmata, the functional differentiation of cells, or even more complex morphogenetic potential that contribute to division of labor within cells and among tissues, exist at the prokaryotic level (e.g. heterocysts, hormogonia, spore formation, or fruiting bodies of myxobacteria), pro- and eukaryotes obviously followed and developed different strategies for management of life.

The *fundamental feature* of the eukaryotic cell resides in an enormous capacity of restructuring genetic material during evolution which includes both loss and gain of function. These are equivalent in quality. The process is not unidirectional, from the organelle(s) including the nucleomorph, where existing, to the nucleus (Fig. 2). It appears to operate in (nearly) all directions though in different quantities, is complex, still ongoing, and not restricted to DNA and its product "protein". It can include multiple gene transfer/replacement, gene transfer that resulted in product detour (Martin and Schnarrenberger 1997), possibly gene transfer from an initial endocytobiont in a "replaced" secondary endosymbiont, or the translocation of RNA molecules or RNA/protein complexes across organelle envelopes (Maréchal-Drouard et al. 1993; Tarrasov et al. 1995). Clearly, this design diminishes the identity (Ellis 1982) and the independence of all genetic compartments, and its functional and phylogenetic consequences cannot be reconciled with the currently favored concept of a semiautonomy of organelles. To obtain a coherent view of the result, *the mosaic genome structure and an integrated genome of eukaryotes* (see below, pp. 103), it will be of considerable interest to evaluate in detail which part of the pathways that regulate energy, basic or secondary metabolism in plants originates in the former endosymbionts, whether and which genes of current pathways are derived from the host pathways or were replaced by corresponding ones from the endosymbionts, or were even integrated by lateral gene transfer, and whether, how, and to what extent the combined genetic potential of the cells has been rearranged to form novel pathways or to contribute to the generation of morphogenetic potential, e.g. by progressing from intra- to intercellular communication, or in the development of multicellularity and morphological diversity. The cause and underlying mechanisms of gene transfer are not entirely clear. Teleologically seen, it seems that the entity checks by trial and error where a given piece of DNA within the cell operates best to streamline or generate biogenetic potential. The play with gene redundancy from different phylogenetic sources offers the opportunity of subtle selection of genes/enzymes for the respective cellular context. The fact that the crucial steps of both metabolic and genetic energy conversion reside in the organelles has suggested that intracellular division of labor exists not only at the physiological but also at the genetic level, at least to some extent.

While the actual events of sequence changes are becoming less of a mystery, their role in evolution or their selective advantage is only rarely being studied, if at all. How can our current information on the set-up of genetic information in eukaryotes be reconciled and put into a coherent picture without overargumentation? In general terms, the basic tenet that appears is that *no eukaryote exists without an endosymbiotic event.* This opinion differs from traditional concepts. Membrane flux and cytoskeletal systems, characteristics of eukaryotic cells, were prerequisites for endocytosis, a process that was legion, but included that of two energy-producing eubacterial cells which escaped digestion and thus, with the generation of a new entity, made an enduring contribution to Life. The basic set-up to allow endocytosis could well have once existed in an ancient true symbiont-free cell. This cell may already have been a fusion chimera (or may have gained substantial genetic information from an engulfed organism) including a basic archaebacterial component (see, e.g. Langer et al. 1995; Doolittle 1996), but would have been clearly of different nature than the eukaryotic cell. Such a presymbiotic cell *did not possess the principal attribute of genetic compartimentation characteristic of eukaryotism*, and should therefore designated in a different way, e.g. protoeukaryotic. (Archae)bacterial homologues of components of the eukaryotic cytoskeleton and of elements of membrane flux should be discovered to elucidate these aspects (see also Erickson 1995). The phylogenetic position of amitochondriate eukaryotes is briefly considered below.

Recent evidence suggests that the primary endosymbiotic events (considered to have occurred approximately 1.5 x 10^9 years ago, concomitantly or relatively shortly after development of the oxygenic environment) and (less) the secondary ones were followed by massive DNA rearrangements. The rates of the major DNA losses (>80% of the engulfed genomes) were probably fast and losses repeatedly, in some 400 million years (K.V. Kowallik and W. Martin, personal communication). If this was the case, *eukaryote evolution must have progressed in two principal steps* consistent with the basic and crown taxa type phylogeny (see, e.g. Sogin 1994; Doolittle 1996). Much of the eukaryotic evolution was "cell evolution" (step 1), since multicellularity may cause at least "spatial" constraints to gene translocation. Only after the majority of genome was restructured and "silenced", multicellularity with its entirely different morphogenetic potential developed (about 800 million years ago), probably polyphyletically. Only relatively few genes have been translocated at the multicellular level.

The restructuring of genetic material was accompanied by the establishment and maintenance of a comprehensive nuclear regulatory dominance in the novel biological entity. This dominance, like intercompartmental controls in general, appears to operate at many levels. This is evident in the mere fact of intracellular gene translocation (Herrmann 1996), and in the predominant loss of genes from the endocytobiont, which resulted in a reductive evolution of plastome and chondriome. Reductive evolution of the nucleomorph (see above, pp. 80) highligths again the unique molecular biology that has accompanied the generation of that cell type with an integrated genetic system. These processes include the establishment of *inter*compartmental controls that are most conspicuous in the elimination of redundant genes. They are indicative of genetic integration as well. Nuclear dominance is reinforced by the trend that in plastids only a discrete type of sequence has been translocated, preferentially individual genes or products that are involved in regulatory processes, and genes that encode functionally comparable structural, or simple-chain catalytic components. For mitochondria and hydrogenosomes, gene translocation is more advanced and consequently includes other components as well. The different extent of gene transfer between organelles probably reflects both a *hierarchy of gene transfer* (genes encoding regulatory and single chain polypeptides initially) and a time scale of the endosymbioses (the proteobacterial earlier than the cyanobacterial). Transferring genes for a complex structure may not be trivial. The mitochondrial genome stage suggests also that translocation constraints discussed for hydrophobic (membrane) proteins have to be re-considered, last not least since such proteins can be imported into isolated organelles (Zak et al. 1997). However, some products, because of their hydrophobicity, may well interact unspecifically with any membrane and cause adverse effects (R.-B. Klösgen, personal communication).

The different degree of gene transfer from the organelles which in chlorophyll *a/b*-containing plastids, animal mitochondria and hydrogenosomes is most advanced, the complementation of plastome and chondriome mutants by artificial gene relocations *via* compartment-*alien* transformation, that is, the insertion of an organelle gene, modified for expression in the nucleus and for the import of the resulting product into the organelle (Banroques et al. 1986; Nagley and Devenisch 1989; Kanevski and Maliga 1994; Gray et al. 1996), the findings that the gene encoding the ATP synthase subunit CFo-II, but not that of the related subunit CFo-I, and the gene for the mitochondrial cytochrome c_1 but not that for the equivalent cytochrome f in plastids have been transferred, suggest all that the process of gene transfer from plastid chromosomes to the nucleus, or between other compartments, has not yet reached an "equilibrium". The smallest plastome known, that of *Euglena*, still houses a few more than 50 polypeptide genes. Comparison of the translocated mitochondrial genes encoding proteins shows that only those for two components, cytochrome *b* and subunit I of cytochrome oxidase, have apparently never been transferred functionally (U. Kück, personal communication). This striking finding again illustrates that gene rearrangements in mitochondria, as in plastids, are not terminated. The possible terminal situation is more difficult to envisage at present for plastids than for mitochondria, since it is not known whether constraints preclude the functional transfer of all the organelle genes (see Herrmann 1996). Mitochondria may approach the situation of hydrogenosomes, the DNA lacking organelle of various amitochondriate, functionally specialized eukaryotic protists. Both organelles (and kinetoplasts) appear to possess a common eubacterial ancestor (Müller 1993; Bui et al. 1996). Due to the functional specialization of hydrogenosomes (anaerobiosis), most products of mitochondria, notably for complex structure, appear to be unnecessary and therefore probably absent in hydrogenosomes. For that reason the perpetuation of a genome may have been dispensible.

The distribution of hydrogenosomes in independent, quite diverse phylogenetic branches, including the eukaryotic crown taxa, where they nest among species housing mitochondria, is commensurate with the idea that these organisms arose polyphyletically, that their appearance followed ecological and physiological, rather than taxonomic criteria, and that at least some of them lost their mitochondria secondarily (Germot et al. 1996, Roger et al. 1996). It is probably too premature to postulate that all amitochondriate eukaryotes are the result of secondary losses, but recent data indicate that early branching forms may be taxonomically misplaced.

Nuclear dominance, finally, is also evident from the establishment of a multitude of regulatory devices that control the maintenance and expression of genetic information, and the product transport into or product stability in the organelles. This includes the nuclear-coded, single-chain RNA polymerase in plastids and mitochondria. The situation of plastids is unique among "prokaryotes" which according to present knowledge rely on only one basic enzyme. The emerging picture in plastids is group-specific gene control by two enzymes. During ontogenesis, the activity of the second, nuclear-coded RNA polymerase probably operates on only a subset of genes, and appears to precede that of the plastid-encoded, *E. coli*-like RNA polymerase. If correct, the nuclear-derived enzyme may function as a "plastome-controlling transcriptase"[1], if it initiates transcription of organelle genes, among them the *rpo* genes. While the details of plastid promoter/polymerase interaction remain imprecisely known that work has raised a number of tantalizing questions, highly relevant for eukaryotism. What is the source of the gene(s) for this enzyme, in particular the phylogenetic relation between the plastid, mitochondrial, and possibly phage pendants, or the stage at

---

[1]We prefer this term since, in our opinion, it describes the function of the enzyme in an appropriate way.

which it was integrated and how it was streamlined during evolution? The fact that this enzyme finds analogues in phage genomes suggests a crucial contribution of horizontal gene transfer not only to prokaryotic, but also to eukaryotic genome evolution. Remarkably, *rpo* gene disruption in *Chlamydomonas*, different from that in tobacco, has not led to homoplastomic plants (Rochaix 1995), perhaps because the enzyme of nuclear origin is not present in that alga. If this were the case, a qualitative change in intercompartmental regulation would have to be postulated since perpetuation of the plastid in the alga would require the permanent presence of the organelle-encoded enzyme whereas in higher plants this would not be necessary. It will be interesting to see whether such changes contributed to the expansion of plastid chromosomes that has been noted in higher plants and concerns predominantly intergenic regions. The expansion of higher plant chondriomes has certainly different, yet unknown reasons. It will also be interesting to explore whether the presence of this enzyme is linked to multicellularity or land life, and whether plastids again approach the mitochondrial situation, in other words that the plastid-encoded polymerase is on its way to be phylogenetically eliminated.

Whatever the *status quo*, each genetic change in this entity has to occur in the context of stabilizing the alliance of cells and their derivatives at each stage of its evolution. The findings outlined and a multitude of others, such as related autotrophic/heterotrophic forms like *Euglena* and *Astasia* with different plastomes (Siemeister et al. 1990), the greatly reduced, yet functional plastome of *Epifagus virginiana* (Wolfe et al. 1992), possibly even the integration of endobacteria, like mycoplasms with their mere 482 genes that depend on the host cell (Fraser et al. 1995), the unexpected multitude of mechanisms and components operating for individual genes, or the multiple routes of protein translocation/integration machineries into/across thylakoid membranes, some of which seem to appear at the eukaryotic stage, show that the integrated genetic system has an enormous plasticity enabling it to buffer and balance newly arisen situations in the organism as essential genes are depleted or transferred.

The restructuring of genetic material in the original conglomerate of cells caused an entirely novel biological entity, the *eukaryotic cell*. This *cell is characterized by an integrated genetic machinery* in which the extent of compartmental genetic interdependence is greater than coexistence alone suggests, implying that information in *all* compartments is essential for its full physiological and biochemical competence. In this cellular entity compartmental *control and dependence are mutual*, and *none* of the genetic organelles (exogenosomes)/compartments including the nucleo/cytosolic space can now exist as a separate living entity. In "intermediate" secondary and tertiary endocytobionts the functional integration includes the nucleomorph genome (and surrounding cytosol) which is too small to encode all necessary components for the photosynthetic organelle). Both interdependence (Cavalier-Smith and Lee 1985) and the establishment of regulatory dominance of one of the compartments are characteristics of an integrated system and prerequisites for the stable integration of endocytobionts. This includes the development of both a common metabolism and inheritance, and relates to both primary (ontogenetic programme) and secondary (processes after reaching competence) differentiation. Modern concepts of eukaryotism have to account for this fundamental feature in the genetic structure, and avoid the use of reigning customary terms like "semiautonomy". These are in fact misleading.

The basic features of the unique genetic design of the eukaryotic cell and its implications were unequivocally described in the early data of formal genetics (Renner 1934; Stubbe 1959; see also Harte 1994). This work has received surprisingly little attention, probably since the underlying genetics was complex and publication of the data was scattered, sometimes in remote places, and in German, which was an additional deterrent. Then, *specificity* of the genetic interplay between plastid and nucleus became clear after the transfer (or exchange) of the plastids from one species into the nuclear background of another, possible due to specific assets in the genetic structure of materials like *Oenothera*. Interspecific or intergeneric genome/plastome (plastid/nuclear) hybrids or cybrids can exhibit appreciable developmental disturbances, even between related, still interbreedable species. This phenomenon is widespread (Kirk and Tilney-Bassett 1978), reversible, and may be valid for the animal kingdom, as reciprocal crosses between horse and donkey indicate. Interspecific grafting with *Acetabularia* and later work on *Solanaceae* by means of somatic genetics (Kushnir et al. 1987; Nguyen and Medgyesy 1989) have demonstrated that the extent of compartmental specialization can differ between taxonomic groups. The molecular basis of intercompartmental genetic specificity is unknown, but the implications are *twofold*. From a functional point of view, it is the *entire* integrated genetic system, the *genetic compartmentation itself*, that has become and is inherited, and underlies regulation in time, in space (in multicellular organisms), and in quantity (Herrmann et al. 1996). Stoichiometric regulation must account for the high reiteration of chondriomes and plastomes per cell. The cause of and the relationships between genome, plastome, and chondriome dosage with regard to function are not known (Herrmann and Possingham 1980). From a phylogenetic point of view, the fact that the genetic compartments even of related species may not be arbitrarily exchangeable without the risk of disturbing a harmonious intracellular genetic balance reflects both compartmental integration and coevolution. *This coevolution is an integral factor in the speciation of eukaryotes*. Findings such as that in interspecific or intergeneric plastome/genome hybrids housing the plastids of two species, a rapidly multiplying but incompatible (off-green to white) plastid will outgrow a compatible (green) but slower proliferating one (and lead to the death of the hybrid), suggest that compartmental genetic interaction places constraints with regard to progressive alteration. Obviously, a mutation in any one of the compartments has to satisfy the requirement that the entire system remains fully functional (Herrmann and Possingham 1980).

The *functional consequences* of the genetic alterations, i.e. the integration of the expression of two, three, or more genomes (intergenomic integration), are complex and difficult to unravel or judge, since changes that occurred at all regulatory levels, in the subgenic range and at the level of genes, are interwoven. Many aspects are only partly understood functionally and phylogenetically. For instance, the above mentioned changes in the epitope distribution and in the way in which subunits CFo-I and CFo-II are integrated were not a necessary consequence of the gene translocation by converting a possible cotranslational (CFo-I) into a post-translational (CFo-II) integration process, for CFo-I, like other plastid-encoded proteins, is correctly routed to the organelle and integrated into ATP synthase when fused to a stroma-targeting transit peptide and imported (Michl et al. 1994; Zak et al. 1997). Similarly, although a wealth of sequence data including information for entire plastid and mitochondrial chromosomes from various sources is now available, important details, the respective basic transcriptional plans and the economy of their expression remain enigmatic, in particular how the circular DNA molecules both have been and are functionally integrated into the overall cellular context. The complexity of the functional organization of plastid chromosomes and of individual operons that recent work has shown and included the detection of multiple promoters and transcription initiation sites within gene clusters (see, e.g. Christopher and Mullet 1994; Kapoor et al. 1994; Vera and Sugiura 1994; Vera et al. 1996), the enormous differences and changes in transcrption rates (Mullet 1993) as well as the involvement of two polymerases suggest that transcriptional control represents a major level of expression control in the organelles. Clearly, previous opinion on that point has to be revised. However, primary polycistronic transcripts in plastids usually undergo extensive processing as well (Barkan 1988; Westhoff and Herrmann 1988). This implies that a significant proportion of generally non-protein coding organelle DNA or RNA is involved in the correct expression of genes. The underlying signals and their cooperation and interaction with factors are largely unknown, as are those in mitochondrial and nucleomorph DNA. Why and how such a complex regulation has evolved is not known, but it is conceivable that the answer lies in the specific genetic compartmentalization of the eukaryotic cell. Under these conditions a convenient way to meet the increased demand for regulation might be to increase the sophistication of transcription and processing, or other RNA modification controls (Herrmann et al. 1992; 1993) requiring signals in or segments of often nonconserved RNA that are largely lacking in prokaryotes. Since the nucleus itself appears to rely heavily on RNA processing to regulate the expression of its genes, and the majority of components required for processing plastid RNA are likely to be coded in nuclear DNA, the nucleus will opt for more elaborate processes if it has to provide more complex regulation of gene expression in the organelles. *These features clearly show that the regulation of organelle gene expression does not follow mere bacterial paradigms as frequently stated, but that crucial features have been changed to eukaryotic characters to account for the situation in that cell.* Finally, multistep processes and advanced developmental patterns are subject to elaborate control mechanisms, embedded in a complex compartmental interplay of regulatory signals that can be modulated by both environmental and internal factors. All physiological, biochemical, and genetic work has uncovered a highly sophisticated regulation of the biogenesis and maintenance of the eukaryotic cell, at all expression levels studied, and with unexpected subtlety. Although individual regulatory mechanisms have been defined to some extent, there is not yet much information as to their relative contribution in a given expression pattern, or their *interaction*, and hence no coherent picture of the overall regulation. Nevertheless, the basic mechanisms in intercompartmental regulation are probably similar in outline throughout nature. Since the gross structure of plastid chromosomes is basically the same, it will be interesting to see whether and how positional rearrangements and sequence changes are related to function, reflect only their evolutionary history, or both.

To understand the evolution of the eukaryotic cell, it is not only desirable to establish catalogues of its subgenomes and their antecedents in a comparative fashion. The final goal of all biology is a description of the *functional* evolution of complete biogenetic potentials, their operation and maintenance. It becomes more and more obvious that the eukaryotic cell cannot be understood without its history. The insight into the phylogeny of eukaryotism represents one of the most fascinating chapters of recent research in the field, as of biology in general.

Acknowledgements: The author thanks Dr. W. Martin (Braunschweig) for very valuable suggestions. This work was supported by the German Research Foundation (DFG, SFB 184), the Fonds der Chemischen Industrie, and the Human Frontier Science Program (HFSP).

References

Abrahamson S, Gruissem W (1996) Transcriptional and post-transcriptional gene regulation in plastids. In: Frontiers in Molecular Biology, Molecular Genetics of Photosynthesis. Andersson B, Salter AH, Barber J (ed) Oxford, New York, Tokyo: IRL press, pp 104-122

Adam Z (1996). Protein stability and degradation in chloroplasts. Plant Mol Biol 32: 773-783

Ainsworth CC, Miller TE, Gale MD (1987) α-amylase and ß-amylase homoeoloci in species related to wheat. Genet Res 49: 93-103

Allison LA, Simon LD, Maliga P (1996) Deletion of *rpoB* reveals a second distinct transcription system in plastids of higher plants. EMBO J 15: 2802-2809

Allison LA, Maliga P (1995) Light-responsive and transcription-enhancing elements regulate the plastid *psbD* promoter. EMBO J 14: 3721-3730

Anbudurai PR, Mor TS, Ohad I, Shestakov SV, Pakrasi HB (1994). The *ctpA* gene encodes the carboxyl-terminal processing protease for the D1 protein of the photosystem II reaction center complex. Proc Natl Acad Sci USA 91: 8082-8086

Apt KE, Hoffmann NE, Grossman AR (1993). The c subunit of R-phycoerythrin and its possible mode of transport into the plastid of red algae. J Biol Chem 268: 16208-16215

Araya A, Bégu D, Litvak S (1994). RNA editing in plants. Physiol Plant 91: 543-550

Ayliffe MA, Timmis JN (1992) Plastid DNA sequence homologies in the tobacco nuclar genome. Mol Gen Genet 236: 105-112

Baker A, Schatz G (1987) Sequences from a prokaryotic genome or the mouse dihydrofolate reductase gene can restore the import of a truncated precursor protein into yeast mitochondria. Proc Natl Acad Sci USA 84: 3117-3121

Baldauf SL, Manhart JR, Palmer JD (1990) Different fates of the chloroplast *tuf*A gene following its transfer to the nucleus in green algae. Proc Natl Acad Sci USA 87: 5317-5321

Banroques J, Delahodde A, Jacq C (1986) A mitochondrial RNA maturase gene transferred to the yeast nucleus can control mitochondrial mRNA splicing. Cell 46: 837-844

Barkan A (1988) Proteins encoded by a complex chloroplast transcription unit are each translated from both monocistronic and polycistronic mRNAs. EMBO J 7: 2637-2644

Barkan A, Walker M, Nolasco M, Johnson D (1994) A nuclear mutation in maize blocks the processing and translation of several chloroplast mRNAs and provides evidence for the differential translation of alternative mRNA forms. EMBO J 13: 3170-3181

Bartling D, Clausmeyer S, Oelmüller R, Herrmann RG (1990) Towards epitope models for chloroplast transit sequences. Bot Mag (Tokyo) special issue 2: 119-144

Bendich AJ (1996) Structural analysis of mitochondrial DNA molecules from fungi and plants using moving pictures and pulsed-field gel electrophoresis. J Mol Biol 255: 564-588

Berghöfer J, Karnauchov I, Herrmann RG, Klösgen RB (1995) Isolation and characterization of a cDNA encoding the SecA protein from spinach chloroplasts: evidence for azide-resistance of Sec-dependent protein translocation across thylakoid membranes in spinach. J Biol Chem 270: 18341-18346

Bhattacharya D, Helmchen T, Melkonian M (1995) Molecular evolutionary analyses of nuclear-encoded small subunit ribosomal RNA identify an independent rhizopod lineage containing the Euglyphina and the Chlorarachniophyta. J Euk Microbiol 42: 65-69

Bhaya D, Grossman A (1991) Targeting proteins to diatom plastids involves transport through an endoplasmic reticulum. Mol Gen Genet 229: 400-404

Blankenship RE (1992) Origin and early evolution of photosynthesis. Photosynthesis Res 33: 91-111

Bock R, Kössel H, Maliga P (1994) Introduction of heterologous editing site into the tobacco plastid genome: the lack of RNA-editing leads to a mutant phenotype. EMBO J 13: 4623-4628

Bohnert HJ, Driesel AJ, Crouse EJ, Gordon K, Herrmann RG, Steinmetz A, Mubumbila M, Keller M, Burkard G, Weil JH (1979) Presence of a transfer RNA gene in the spacer sequence between the 16S and 23S rRNA genes of spinach chloroplast DNA. FEBS Lett 103: 52-56

Bolle C, Herrmann RG, Oelmüller R (1996a) Intron sequences are involved in the plastid- and light-dependent expression of the spinach *psaD* gene. Plant J 10: 919-924

Bolle C, Kusnetsov VV, Herrmann RG, Oelmüller R (1996b) The spinach *AtpC* and *AtpD* genes contain elements for light-regulated, plastid-dependent and organ-specific expression in the vicinity of the transcription start sites. Plant J 9: 21-30

Brennicke A, Grohmann L, Hiesel R, Knoop V, Schuster W (1993) The mitochondrial genome on its way to the nucleus: different stages of gene transfer in higher plants. FEBS Lett 325: 140-145

Brinkmann H, Martin W (1996) Higher-plant chloroplast and cytosolic 3-phosphoglycerate kinases: a case of endosymbiotic gene replacement. Plant Mol Biol 30: 65-75

Bryant DA (1992) Molecular biology of photosystem I. In: Current topics in photosynthesis. Barber J (ed) Amsterdam: Elsevier, pp 501-549

Bui ETN, Bradley PJ, Johnson PJ (1996) A common evolutionary origin for mitochondria and hydrogenosomes. Proc Natl Acad Sci USA 93: 9651-9656

Cai D, Herrmann RG, Klösgen RB (1993) The 20 kDa apoprotein of the CP24 complex of photosystem II: an alternative model to study import and intra-organellar routing of nuclear-encoded thylakoid proteins. Plant J 3: 383-392

Cavalier-Smith T (1995) Membrane heredity, symbiogenesis, and the multiple origins of algae. In: Biodiversity and Evolution. Arai R, Kato M, Doi Y (eds) Natl. Sci. Museum, Tokyo, pp 75-114

Cavalier-Smith T (1992). The number of symbiotic origins of organelles. BioSystems 28: 91-106

Cavalier-Smith T, Lee JJ (1985) Protozoa as hosts for endosymbioses and the conversion of symbionts into organelles. Protozool 32: 376-379

Cavalier-Smith T, Allsopp MTEP, Chao EE (1994) Chimeric conundra: Are nucleomorphs and chromists monophyletic or polyphyletic? Proc Natl Acad Sci USA 91: 11368-11372

Cavalier-Smith T, Allsopp MTEP, Häuber MM, Gothe G, Chao EE, Couch JA, Maier U-G (1996) Chromobiote phylogeny: the enigmatic alga *Reticuloshaera japonensis* is an aberrant haptophyte, not a heterokont. Eur J Phycol 32: 255-263

Cerff R (1995) The chimaeric nature of nuclear genomes and the antiquity of introns as demonstrated by the GAPDH gene system. In:Tracing biological evolution in protein and gene structures. Proceedings of the 20th Taniguchi International Symposium, Division of Biophysics. Go M, Schimmel P (eds) Amsterdam: Elsevier Science, pp 205-227

Chaddock AM, Mant A, Karnauchov I, Brink S, Herrmann RG, Klösgen RB, Robinson C (1995) A new type of signal peptide: central role of a twin-arginine motif in transfer signals for the DpH-dependent thylakoid protein translocase. EMBO J 14: 2715-2722

Chan RL, Keller M, Canaday J, Weil JH, Imbault P (1990) Eight small subunits of *Euglena* ribulose 1-5 bisphosphate carboxylase/oxygenase are translated from a large mRNA as a polyprotein. EMBO J 9: 333-338

Chao S, Raines CA, Longstaff M, Sharp PJ, Gale MD, Dyer TA (1989) Chromosomal location and copy number in wheat and some of its close relatives of genes for enzymes involved in photosynthesis. Mol Gen Genet 218: 423-430

Chatton E (1938) Titres et ravoux scientifiques (1906-1937) de Edouard Chatton. Sottano E (ed) Sète, France

Chesnick JM, Morden CW, Schmieg AM (1996) Identity of the endosymbiont of *Peridinium foliaceum* (Pyrrophyta): analysis of the *rbcLS* operon. J Phycol 32: 850-857

Chisholm SW, Frankel SL, Goericke R, Olson RJ, Palenik B, Waterbury JB, West-Johnsrud L, Zettler ER (1992) *Prochlorococcus marinus*, nov. gen. nov. sp., an oxyphototrophic marine prokaryote containing divinyl chlorophyll *a* and *b*. Arch. Microbiol 157: 297-300

Christopher DA, Mullet JE (1994) Separate photosensory pathways co-regulate blue light/ultraviolet-a-activated *psbD-psbC* transcription and light-induced D2 and CP43 degradation in barley (*Hordeum vulgare*) chloroplasts. Plant Physiol 104: 1119-1129

Clausmeyer S, Klösgen RB, Herrmann RG (1993) Protein import into chloroplasts: The hydrophilic lumenal proteins exhibit unexpected import and sorting specifities in spite of structurally conserved transit peptides. J Biol Chem 268: 13869-13876

Cline K, Henry R, Li C, Yuan J (1993) Multiple pathways for protein transport into or across the thylakoid membrane. EMBO J 12: 4105-4114

Copeland HF (1938) The kingdoms of organisms. Quart Rev. Biol. 13: 383-420

Covello PS, Gray MW (1992) Silent mitochondrial and active nuclear genes for subunit 2 of cytochrome *c* oxidase (*cox2*) in soybean: Evidence for RNA-mediated gene transfer. EMBO J 11: 3815-3820

Cozens AL, Walker JE (1987) The organization and sequence of the genes for ATP synthase subunits in the caynobacterium Synechococcus 6301. Support for an endosymbiotic origin of chloroplasts. J Mol Biol 194: 359-383

Dickey LF, Gallo-Meagher M, Thompson WF (1992) Light regulatory sequences are located within the 5' portion of the Fed-1 message sequence. EMBO J 11: 2311-2317

Dickey LF, Nguyen T-T, Allen GC, Thompson WF (1994) Light modulation of ferredoxin mRNA abundance requires an open reading frame. Plant Cell 6: 1171-1176
Dolganov NAM, Bhaya D, Grossman AR (1995) Cyanobacterial protein with similarity to the chlorophyll *a/b* binding proteins of higher plants: Evolution and regulation. Proc Natl Acad Sci USA 92: 636-640
Doolittle WF (1996) Some aspects of the biology of cells and their possible evolutionary significance. In: Evolution of microbial life. Roberts DMcL, Sharp P, Alderson G, Collins M (eds) Cambridge Univ. Press, pp 1-21
Douglas SE, Murphy CA, Spencer DF, Gray MW (1991) Cryptomonad algae are evolutionary chimaeras of two phylogenetically distinct unicellular eukaryotes. Nature 350: 148-151
Ellis J (1982) Promiscuous DNA - chloroplast genes inside plant mitochondria. Nature 299: 678-679
Erickson HP (1995) FtsZ, a prokaryotic homolog of tubulin? Cell 80: 367-370
Eschbach SC, Hofmann CJB, Maier U-G, Sitte P, Hansmann P (1991) A eukaryotic genome of 660 kb: electrophoretic karyotype of nucleomorph and cell nucleus of the cryptomonad alga *Pyrenomonas salina*. Nucleic Acids Res 19: 1779-1781
Falah M, Gupta RS (1994) Cloning of the *hsp70* (*dnaK*) genes from *Rhizobium meliloti* and *Pseudomonas cepacia*: phylogenetic ananlyses of mitochondrial origin based on highly conserved protein sequence. J Bact 176: 7748-7753
Falk G, Walker JE (1988) DNA sequence of a gene cluster coding for subunits of the Fo membrane sector of ATP synthase in *Rhodospirillum rubrum*. Biochem J 254: 109-122
Fenchel T, Bernard C (1993) A purple protist. Nature 362: 300
Flieger K, Tyagi A, Sopory S, Cséplö A, Herrmann RG, Oelmüller R (1993) A 42 bp promotor fragment of the gene for subunit III of photosystem I (*psaF*) is crucial for its activity. Plant J 4: 9-17
Fraser CM, Gocayne JD, White O, Adams MD, Clayton RA, Fleischmann RD, Bult CJ, Kerlavage AR, Sutton G, Kelley JM (1995) The minimal gene complement of *Mycoplasma genitalium*. Science 270: 397-403
Freyer R, Hoch B, Neckermann K, Maier RM, Kössel H (1993) RNA editing in maize chloroplasts is a processing step independent of splicing and cleavage to monocistronic mRNAs. Plant J 4: 621-629
Germot A, Philippe H. and Guyader (1996) Presence of a mitochondrial-type 70-kDa heat shock protein in *Trichomonas vaginalis* suggests a very early mitochondrial endosymbiosis in eukaryotes. Proc Natl Acad Sci USA 93: 14614-14617
Gibbs SP (1978) The chloroplasts of *Euglena* may have evolved from symbiotic green algae. Canad J Bot 56: 2883-2889
Gibbs SP (1979) The route of entry of cytoplasmatically synthesized proteins into chloroplasts of algae possessing chloroplast. ER. J Cell Sci 35: 253-266
Gibbs SP (1990) The evolution of algal chloroplasts. In: Experimental phycology, vol 1, Cell walls and surfaces, reproduction and photosynthesis. Wiessner W, Robinson DG, Starr RC (eds) Wien: Springer, pp 145-157
Gibbs SP (1993) The evolution of algal chloroplasts. In: The origins of plastids: symbiogenesis, prochlorophytes and evolution. Lewin R (ed). Chapman and Hall, New York, p 145-157
Gilson P, McFadden GI (1995) The chlorarachniophyte: a cell with two different nuclei and two different telomeres. Chromosoma 103: 635-641
Gilson PR, McFadden GI (1996) The miniaturized nuclear genome of a eukaryotic endosymbiont contains genes that overlap, genes that are cotranscribed, and the smallest known spliceosomal introns. Proc Natl Acad Sci USA 93: 7737-7742

Goericke R, Repeta D (1996) The pigment of *Prochlorococcus marinus*: The presence of divinyl-chlorophyll *a* and *b* in a marine prokaryote. Limnol Oceanogr 37: 425-433

Gonzalez DH, Bonnard G, Grienenberger J-M (1993) A gene involved in the biogenesis of cytochromes is co-transcribed with a ribosomal protein gene in wheat mitochondria. Curr Genet 24: 248-255

Gray MW (1989) Origin and evolution of mitochondrial DNA. Annu Rev Cell Biol 5: 25-50

Gray MW, Covello PS (1993) RNA editing in plant mitochondria and chloroplasts. FASEB 7: 64-71

Gray RE, Law RHP, Devenish RJ, Nagley P (1996) Allotopic expression of mitochondrial ATP synthase genes in nucleus of *Saccharomyces cerevisiae*. Methods Enzymol 264: 369-389

Grayburn WS, Bendich AJ (1987) Variable abundance of a mitochondrial DNA fragment in cultured tobacco cells. Curr Genet 12: 257-261

Green BR, Pichersky E, Kloppstech K (1991) Chlorophyll *a/b*-binding proteins: an extended family. TIBS 16: 181-186

Greenwood AD (1974) The Cryptophyta in relation to phylogeny and photosynthesis. In: Electron microscopy 1974. Sanders JV, Goodchild DJ (eds). Canberra: Australian Academy of Sciences, pp 566-567 (Abstr.)

Greenwood AD, Griffiths HB, Santore UJ (1977) Chloroplasts and cell compartments in Cryptophyceae. Br Phycol J 12: 119 (Abstr.)

Grohmann L, Brennicke A, Schuster W (1992) The mitochondrial gene encoding ribosomal protein S12 has been translocated to the nuclear genome in *Oenothera*. Nucl Acids Res 20: 5641-5646

Grossman A, Manadori A, Snyder D (1990) Light-harvesting proteins of diatoms: Their relationship to the chlorophyll *a/b* binding proteins of higher plants and their mode of transport into plastids. Mol Gen Genet 224: 91-100

Gruissem W, Tonkyn JC (1993) Control mechanisms of. plastid gene expression. Crit Rev Plant Sci 12: 19-55

Gualberto JM, Weil J-H, Grienenberger J-M (1990) Editing of the wheat *cox*III transcript: evidence for twelve C to U and one U to C conversions and for sequence similarities around editing sites. Nucleic Acids Res 18: 3771-3776

Haeckel E (1886) Generelle Morphologie der Organismen Georg Reimer, Berlin

Haehnel W, Jansen T, Gause K, Klösgen RB, Stahl B, Michl D, Huvermann B, Karas M, Herrmann RG (1994) Electron transfer from plastocyanin to photosystem I. EMBO J 13: 1028-1038

Harrak H, Lagrange T, Bisanz-Seyer C, Lerbs-Mache S, Mache R (1995) The expression of nuclear genes encoding plastid ribosomal proteins precedes the expression of chloroplast genes during early phases of chloroplast development. Plant Physiol 108: 685-692

Harte C (1994) Oenothera. Berlin, Heidelberg, New York: Springer

Hartl F-U, Neupert W (1990) Protein sorting to mitochondria: evolutionary conservations of folding and assembly. Science 247: 930-938

Häuber MM, Müller SB, Speth V, Maier U-G (1994) How to evolve a complex plastid? - A hypothesis. Bot Acta 107: 383-386

Hawlitschek G, Schneider H, Schmidt B, Tropschug M, Hartl F-U, Neupert W (1988) Mitochondrial protein import: Identification of processing peptidase and of PEP, a processing enhancing protein. Cell 53: 795-806

Henry R, Kapazoglou A, McCaffery M, Cline K (1994) Differences between lumen targeting domains of chloroplast transit peptides determine pathway specificity for thylakoid transport. J Biol Chem 269: 10189-10192

Henze K, Badr A, Wettern M, Cerff R, Martin W (1995) A nuclear gene of eubacterial origin in *Euglena gracilis* reflects cryptic endosymbioses during protist evolution. Proc Natl Acad Sci USA 92: 9122-9126

Herrmann RG (1996) Photosynthesis research - aspects and perspectives. In: Frontiers in Molecular Biology; Molecular Genetics in Photosynthesis. Andersson B, Salter AH, Barber J (eds) Oxford Univ. Press, pp 1-44

Herrmann RG, Possingham JV (1980) Plastid DNA - the plastome. In: Results and problems in cell differentiation. Reinert J (ed) Berlin, Heidelberg, New York: Springer. vol 10, pp 45-96

Herrmann RG, Oelmüller R, Bichler J, Schneiderbauer A, Steppuhn J, Wedel N, Tyagi AK, Westhoff P (1991) The thylakoid membrane of higher plants: genes, their expression and interaction. In: Plant molecular biology 2. Herrmann RG, Larkins B (eds) New York: Plenum, pp 411-427

Herrmann, RG, Westhoff P, Link G (1992) Biogenesis of plastids in higher plants. In: Plant gene research, vol 6. Herrmann RG (ed) Wien, New York: Springer, pp 275-349

Herrmann RG, Steppuhn J, Herrmann GS, Nelson N (1993) The nuclear-encoded polypeptide CFo-II is a real ninth subunit of chloroplast ATP synthase. FEBS Lett. 326: 192-198

Hess WR, Partensky F, van der Staay GWM, Garci-Fernandez JM, Börner T, Vaulot D (1996) Coexistence of phycoerythrin and a chlorophyll *a/b* antenna in a marine prokaryote. Proc Natl Acad Sci USA 93: 11126-11130

Hess WR, Prombona A, Fieder B, Subramanian AR, Börner T (1993) Chloroplast *rps15* and the *rpoB/C1/C2* gene cluster are strongly transcribed in ribosome-deficient plastids: evidence for a functioning non-chloroplast-encoded RNA polymerase. EMBO J 12: 563-571

Hibbert DJ, Norris RE (1984) Cytology and ultrastructure of *Chlorarachnion reptans* (Chlorarachniophyta divisio nova, Chlorarachniophyceae classis nova). J Phycol 20: 310-330

Hirsch S, Muckel E, Heemeyer F, von Heijne G, Soll J (1994) A receptor component of the chloroplast protein translocation machinery. Science 266: 1989-1992

Hofmann CJB, Rensing SA, Häuber MM, Martin WF, Müller SB, Couch J, McFadden GI, Igloi GL, Maier U-G (1994) The smallest known eukaryotic genomes encode a protein gene: towards an understanding of nucleomorph functions. Mol Gen Genet 243: 600-604

Horner DS, Hirt RP, Kilvington S, Lloyd D, Embley, TM (1996) Molecular data suggest an early acquisition of the mitochondrion endosymbiont. Proc. R Soc Lond B 263: 1053-1059

Igloi GL, Kössel H (1993) The transcriptional apparatus of chloroplasts. Crit Rev Plant Sci 10: 525-558

Iratni R, Baeza L, Andreeva A, Mache R, Lerbs-Mache S (1994) Regulation of rDNA transcription in chloroplasts: promoter exclusion by constitutive repression. Genes Develop 8: 2928-2938

Jeffrey SW, Vesk M (1976) Further evidence for a membrane-bound endosymbiont within the dinoflagellate *Peridinium foliaceum*. J Phycol 12: 450-455

Jenkins J, Hiller RG, Speirs J, Godovac-Zimmermann J (1990) A genomic clone encoding a cryptophyte phycoerythrin Â-subunit. FEBS Lett 273: 191-194

Joyce PBM, Gray MW (1989) Chloroplast-like transfer RNA genes expressed in wheat mitochondria. Nucleic Acids Res 17: 5461-5477

Kadowaki K, Kubo N, Ozawa K and Hirai A (1996) Targeting presequence acquisition after mitochondrial gene transfer to the nucleus occurs by duplication of existing targeting signals. EMBO J 15: 6652-6661

Kaneko T, Sato S, Kotani H, Tanaka A, Asamizu E, Nakamura Y, Miyajima N, Hirosawa M, Sugiura M, Sasamoto S, Kimura T, Hosouchi T, Matsuno A, Muraki A, Nakazaki N, Naruo K, Okumura S, Shimpo S, Takeuchi C, Wada T, Watanabe A, Yamada M, Yasuda

M, Tabata S (1996) Sequence analysis of the genome of the unicellular cyanobacterium *Synechocystis* sp. strain PCC6803. II. Sequence determination of the entire genome and assignment of potential protein-coding regions. DNA Res 3: 109-136

Kanevski I, Maliga P (1994) Relocation of the plastid *rbcL* gene to the nucleus yields functional ribulose-1,5-bisphosphate carboxylase in tobacco chloroplasts. Proc Natl Acad Sci USA 91: 1969-1973

Kapoor S, Wakasugi T, Deno H, Sugiura M (1994) An *atpE*-specific promoter within the coding region of the *atpB* gene in tobacco chloroplast DNA. Curr Genet 26: 263-268

Kawashima N, Wildman SG (1972). Studies on fraction I protein. IV. Mode of inheritance of primary structure in relation to whether chloroplast or nuclear DNA contains the code for a chloroplast protein. Biochim Biophys Acta 262: 42-49

Keegstra K, von Heijne G (1992) Transport of proteins into chloroplasts. In Plant Gene Research, vol 6. Herrmann RG (ed) Wien, New York: Springer, pp 353-370

Kim M, Mullet JE (1995) Identification of a sequence-specific DNA binding factor required for transcription of the barley chloroplast blue light-responsive *psbD-psbC* promoter. Plant Cell 7:1445-1457

Kim S, Sandusky P, Bowlby NR, Aebersold R, Green BR, Vlahakis S, Yocum CF, Pichersky E (1992) Characterization of a spinach *psbS* cDNA encoding the 22 kDa protein of photosystem II. FEBS Lett 314: 67-71

Kirk JTO, Tilney-Bassett RAE (1978) The plastids. Amsterdam, New York, Oxford: Elsevier/North Holland

Kirwin PM, Elderfield PD, Williams RS, Robinson C (1988) Transport of proteins into chloroplasts. J Biol Chem 263: 18128-18132

Knoop V, Brennicke A (1991) A mitochondrial intron sequence in the 5'-flanking region of a plant nuclear lectin gene. Curr Genet. 20: 423-425

Knoop V, Unseld M, Marienfeld J, Brandt P, Sünkel S, Ullrich H, Brennicke A (1996) *copia*-, *gypsy*- and LINE-like retrotransposon fragments in the mitochondrial genome of *Arabidopsis thaliana*. Genetics 142: 579-585

Köhler S, Delwiche CF, Denny PW, Tilney LG, Webster P, Wilson RJM, Palmer JD, Roos DS (1997) A plastid of probably green algal origin in apicomplexan parasites. Science 275: 1485-1489

Koncz C, Németh K, Rédei GP, Schell J (1992) T-DNA insertional mutagenesis in *Arabidopsis*. Plant Mol Biol 20: 963-976

Kössel H, Hoch B, Maier RM, Igloi GL, Kudla J, Zeltz P, Freyer R, Neckermann K, Ruf S (1993) RNA editing in chloroplasts of higher plants. In: Plant mitochondria. Brennicke A, Kück U (eds) Weinheim: VCH, pp 93-102

Kowallik KV (1994) From endosymbionts to chloroplasts: evidence for a single prokaryotic/eukaryotic endocytobiosis. Endocytobiosis and Cell Res 10: 137-149

Kowallik KV, Stoebe B, Schaffran I, Freier U (1995) The chloroplast genome of a chlorophyll *a* + *c* containing algae, *Odontella sinensis*. Plant Mol Biol Rep 13: 336-342

Kumar R, Maréchal-Drouard L, Akama K, Small I (1996) Striking differences in mitochondrial tRNA import between different plant species. Mol Gen Genet 252: 404-411

Kushnir SG, Shlumukov LR, Pogrebnyak NJ, Berger S, Gleba Y (1987) Functional cybrid plants possessing a Nicotiana genome and an Atropa plastome. Mol Gen Genet 209: 159-163

Kusnetsov V, Bolle C, Lübberstedt T, Sopory S, Herrmann RG, Oelmüller R (1996) Evidence that the plastid signal and light operate *via* the same *cis*-acting elements in the promoters of nuclear genes for plastid proteins. Mol Gen Genet 252: 631-639

Kuwabara T, Reddy KJ, Sherman LA (1987) Nucleotide sequence of the gene from the cyanobacterium *Anacystis nidulans* R2 encoding the Mn-stabilizing protein involved in photosystem II water oxidation. Proc Natl Acad Sci USA 84: 8230-8234

Laidler V, Chaddock AM, Knott TG, Walker D, Robinson C (1995) A SecY homolog in *Arabidopsis thaliana*. J Biol Chem 270: 17664-17667

Lamppa GK (1988) The chlorophyll *a/b*-binding protein inserts into the thylakoids independent of its cognate transit peptide. J Biol Chem 263: 14996-14999

Langer D, Hain J, Thuriaux P, Zillig W (1995) Transcription in Archaea: Similarity to that in Eucarya. Proc Natl Acad Sci USA 92: 5768-5772

Lerbs-Mache S (1993) The 110-kDa polypeptide of spinach plastid DNA-dependent RNA polymerase: Single-subunit enzyme or catalytic core of multimeric enzyme complexes? Proc Natl Acad Sci USA 90: 5509-5513

Lewin RA (ed) (1992) Origin of plastids. New York, London: Chapman and Hall

Li H-M, Moore T, Keegstra K (1991) Targeting of proteins to the outer envelope membrane uses a different pathway than transport into chloroplasts. Plant Cell 3: 709-717

Li X, Henry R, Yuan J, Cline K, Hoffman NE (1995) A chloroplast homologue of the signal recognition particle subunit SRP54 is involved in the posttranslational integration of a protein into thylakoid membranes. Proc Natl Acad Sci USA 92: 3789-3793

Li YF, Zhou DX, Claboult G, Bisanz-Seyer C, Mache R (1995) *Cis*-acting elements and expression pattern of the spinach *rps22* gene coding for a plastid-specific protein. Plant Mol Biol 28: 595-604

Liaud M-F, Brandt U, Scherzinger M, Cerff R (1996) Evolutionary origin of cryptomonad microalgae: Two novel chloroplast/cytosol-specific GAPDH genes as potential markers of ancestral endosymbiont and host cell components. J Mol Evol (Special Issue) 44: 28-37

Link G (1996) Green life: control of chloroplast gene transcription. BioEssays 18: 465-471

Liu B, Troxler RF (1996) Molecular characterization of a positively photoregulated nuclear gene for a chloroplast RNA polymerase sigma factor in *Cyanidium caldarium*. Proc Natl Acad Sci USA 93: 3313-3318

Löffelhardt W, Bohnert HJ (1994) Structure and function of the cyanelle genome. Int Rev Cytol 151: 29-65

Long M, de Souza SJ, Rosenberg C, Gilbert W (1996) Exon shuffling and the origin of the mitochondrial targeting function in plant cytochrome c1 precursor. Proc Natl Acad Sci USA 93: 7727-7731

Lorkovic ZJ, Schröder WP, Pakrasi HB, Irrgang K-D, Herrmann RG, Oelmüller R (1995) Molecular characterization of PsbW, a novel and the only nuclear-encoded component of the photosystem II reaction center complex in spinach. Proc Natl Acad Sci USA 92: 8930-8934

Lübberstedt T, Bolle C, Sopory S, Flieger K, Herrmann RG, Oelmüller R (1994) Promoters from genes for plastid proteins possess regions with different sensitivies towards red and blue light. Plant Physiol 104: 997-1004

Lübeck J, Soll J, Akita M, Nielsen E, Keegstra K (1996) Topology of IEP110, a component of the chloroplastic protein import machinery present in the inner envelope membrane. EMBO J 15: 4230-4238

Ludwig M, Gibbs SP (1989) Localization of phycoerythrin at the lumenal surface of the thylakoid membrane in *Rhodomonas lens*. J Cell Biol 108: 875-884

Maier U-G, Hofmann CJB, Eschbach S, Wolters J, Igloi GL (1991) Demonstration of nucleomorph-encoded eukaryotic small subunit ribosomal RNA in cryptomonads. Mol Gen Genet 230: 155-160

Malek O, Lättig K, Hiesel R, Brennicke A, Knoop V (1996) RNA editing in bryophytes and a molecular phylogeny of land plants. EMBO J 15: 1403-1411

Maréchal-Drouard L, Weil J-H, Dietrich A (1993) Transfer RNAs and transfer RNA genes in plants. Annu Rev Plant Physiol Plant Mol Biol 44: 13-32

Mardsen JE, Schawager SJ, May B (1987) Single-locus inheritance in the tetraploid tree frog *Hyla versicolor* with an analysis of expressed progeny ratios in tetraploid organisms. Genetics 116: 299-311

Martin W, Schnarrenberger C (1997) Evolutionary transfer of a mosaic pathway from eubacterial to nuclear chromosomes: The Calvin cycle enzymes of higher plants. Curr Genet (in press)

Martin WF (1996) Is something wrong with the tree of life? BioEssays 18: 523-527

Martin W, Lydiate D, Brinkmann H, Forkmann G, Saedler H, Cerff R (1993) Molecular phylogenesis in angiosperm evolution. Mol Biol Evolution 10: 140-162

Martin W, Mustafa A-Z, Henze K, Schnarrenberger C (1996) Higher plant chloroplast and cytosolic fructose-1,6-bisphosphatase isoenzymes: origins *via* duplication rather than prokaryote-eukaryote divergence. Plant Mol Biol (in press)

McFadden GI (1993) Second-hand chloroplasts: evolution of cryptomonad algae. Adv Bot Res 19: 189-230

McFadden GI, Gilson P (1995) Something borrowed, something green: lateral transfer of chloroplasts by secondary endosymbiosis. Trends Ecol Evol 10: 12-17

McFadden GI, Gilson PR, Douglas SE (1994) The photosynthetic endosymbiont in cryptomonad cells produces both chloroplast and cytoplasmic-type ribosomes. J Cell Sci 107: 649-657

McFadden GI, Gilson PR, Waller RF (1995) Molecular phylogeny of Chlorarachniophyta based on plastid rRNA and *rbcL* sequences. Arch Protistenkd 145: 231-239

McFadden GI, Gilson PR, Hofmann CJB, Adcock GJ (1994) Evidence that an amoeba acquired a chloroplast by retaining part of an engulfed eukaryotic alga. Proc Natl Acad Sci USA 91: 3690-3694

Michl D, Robinson C, Shackleton JB, Herrmann RG, Klösgen RB (1994) Targeting of proteins to the thylakoids by bipartite presequences: CFo-II is imported by a novel, third pathway. EMBO J 13: 1310-1317

Morse D, Salois P, Markovic P, Hastings JW (1995) A nuclear-encoded form II RubisCO in dinoflagellates. Science 265: 1622-1624

Muckel E, Soll J (1996) A protein import receptor of chloroplasts is inserted into the outer envelope membrane by a novel pathway. J Biol Chem 271: 23846-23852

Müller M (1993) The hydrogenosome. J Gen Microbiol 139: 2879-2889

Mullet JE (1993) Dynamic regulation of chloroplast transcription. Plant Physiol 103: 309-313

Mullet JE, Klein PG, Klein RR (1990) Chlorophyll regulates accumulation of the plastid-encoded chlorophyll apoproteins CP43 and D1 by increased apoprotein stability. Proc Natl Acad Sci USA 87: 4038-4042

Nagley P, Devenish RJ (1989) Leading organellar proteins along new pathways: the relocation of mitochondrial and chloroplast genes to the nucleus. TIBS 14: 31-35

Nakai M, Goto A, Nohara T, Sugita D, Endo T (1994) Identification of the SecA protein homolog in pea chloroplasts and its possible involvement in thylakoidal protein transport. J Biol Chem 266: 31338-31341

Nelson N (1992) Evolution of organellar proton-ATPases. Biochim Biophys Acta 1100: 109-124

Nguyen DT, Medgyesy P (1989) Limited chloroplast gene transfer *via* recombination overcomes plastome-genome incompatibility between *Nicotiana tabacum* and *Solanum tuberosum*. Plant Mol Biol 12: 87-93

Nugent JM, Palmer JD (1991) RNA-mediated transfer of the gene *coxII* from the mitochondrion to the nucleus during flowering plant evolution. Cell 66: 473-481

Oblong JE, Lamppa GK (1992) Identification of two structurally related proteins involved in proteolytic processing of precursors targeted to the chloroplast. EMBO J 11: 4401-4409

Oda K, Yamato K, Ohta E, Nakamura Y, Takemura M, Nozato N, Akashi K, Kanegae T, Ogura Y, Kohchi T, Ohyama K (1992) Gene organization deduced from the complete sequence of liverwort *Marchantia polymorpha* mitochondrial DNA. A primitive form of plant mitochondrial genome. J Mol Biol 223: 1-7

Oelmüller R (1989) Photooxidative destruction of chloroplasts and its effect on nuclear gene expression and extraplastidic enzyme levels. Photobiol 49: 229-239

Oelmüller R, Bolle C, Tyagi A, Niekrawietz N, Breit S, Herrmann RG (1993) Characterization of the promoter from the single-copy ferredoxin $NADP^+$-oxidoreductase gene from spinach. Mol Gen Genet 237: 261-272

Olsen GJ, Woese CR, Overbeek R (1994) The winds of (evolutionary) change: Breathing new life into microbiology. J Bacteriol 176: 1-6

Palenik B, Haselkorn R (1992) Multiple evolutionary origins of prochlorophytes, the chlorophyll *b*-containing prokaryotes. Nature 355: 265-267

Palmer JD (1992) Comparison of chloroplast and mitochondrial genome evolution in plants. In: Cell organelles. Herrmann RG (ed) Wien, New York: Springer, pp 99-133

Pancic PG, Strotmann H (1993) Structure of the nuclear encoded gamma subunit of CFoCF1 of the diatom *Odontella sinensis*, including its presequence. FEBS Lett 320: 61-66

Pancic PG, Strotmann H, Kowallik KV (1992) Chloroplast ATPase genes in the diatom *Odontella sinensis* reflect cyanobacterial characters in structure and arrangement. J Mol Biol 224: 529-536

Pfanner N (1992) Components and mechanisms in mitochondrial protein import. In Plant Gene Res., vol. Cell organelles. Herrmann RG (ed) Wien, New York: Springer, pp 371-400

Pfannschmidt T, Link G (1994) Separation of two classes of plastid DNA-dependent RNA polymerases that are differentially expressed in mustard (*Sinapis alba* L.) seedlings. Plant Mol Biol 25: 69-81

Pichersky E, Logsdon JMJr, McGrath JM, Stasys RA (1991) Fragments of plastid DNA in the nuclear genome of tomato: prevalence, chromosomal location, and possible mechanism of integration. Mol Gen Genet 225: 453-458

Pillen K, Jung C, Herrmann RG (1996) Genetic mapping of loci for thirteen nuclear-encoded polypeptides associated with the thylakoid membrane in *Beta vulgaris* L. FEBS Lett 395: 58-62

Pring D, Brennicke A, Schuster W (1993) RNA editing gives a new meaning to the genetic information in mitochondria and chloroplasts. Plant Mol Biol 21: 1163-1170

Reinbothe S, Runge C, Reinbothe C, van Cleve B, Apel K (1995) Substrate-dependent transport of the NADPH:protochlorophyllide oxidoreductase into isolated plastids. Plant Cell 7: 161-172

Reith M, Douglas SE (1990) Localization of ß-phycoerythrin to the thylakoid lumen of *Cryptomonas* Phi does not involve a signal peptide. Plant Mol Biol 15: 585-592

Reith M, Munholland J (1995) Complete nucleotide sequence of the *Porphyra purpurea* chloroplast genome. Plant Mol Biol Rep 13: 333-345

Renner O (1934) Die pflanzlichen Plastiden als selbständige Elemente der genetischen Konstitution. Ber Verh Sächs Akad Wiss Leipzig, Math-Phys Kl 86: 214-266

Rensing SA, Goddemeier M, Hofmann CJB, Maier U-G (1994) The presence of a nucleomorph *hsp70* gene is a common feature of Cryptophyta and Chlorarachniophyta. Curr Genet 26: 451-455

Robinson C, Klösgen RB (1994) Targeting of proteins into and across the thylakoid membrane - a multitude of mechanisms. Plant Mol Biol 26: 15-24

Rochaix J-D (1992) Post-transcriptional steps in the expression of chloroplast genes. Annu Rev Cell Biol 8: 1-28

Rochaix J-D (1995) *Chlamydomonas reinhardtii* as the photosynthetic yeast. Annu Rev Genet 29: 209-230

Roger AJ, Clark CG and Doolittle WF (1996) A possible mitochondrial gene in the early branching amitochondriate protist Trichomonas vaginalis. Proc Natl Acad Sci USA 93: 14618-14622

Rose AB and Last RL (1997) Introns act post-transcriptionally to increase expression of the *Arabidopsis thaliana* tryptophan pathway gene *PAT1*. Plant J 11: 455-464

Rowan R, Whitney SM, Fowler A, Yellowlees D (1996) Rubisco in marine symbiotic dinoflagellates: form II enzymes in eukaryotic oxygenic phototrophs encoded by a nuclear multigene family. Plant Cell 8: 539-553

Salomon M, Fischer K, Flügge U-I, Soll J (1990) Sequence analysis and protein import studies of an outer chloroplast envelope polypeptide. Proc Natl Acad Sci USA 87: 5778-5782

Sánchez H, Fester T, Kloska S, Schröder W, Schuster W (1996) Transfer of *rps19* to the nucleus involves the gain of an RNP-binding motif which may functionally replace RPS13 in *Arabidopsis* mitochondria. EMBO J 15: 2138-2149

Sasaki Y, Konishi T, Nagano Y (1995) The compartmentation of acetyl-coenzyme A carboxylase in plants. Plant Physiol 108: 445-449

Schatz G, Dobberstein B (1996) Common principles of protein translocation across membranes. Science 271: 1519-1526

Schenk HEA (1970) Nachweis einer lysozmyempfindlichen Stützmembran der Endocyanellen von *Cyanophora Paradoxa* Korschikoff. Z Naturf 25b: 640

Schnell DJ, Kessler F, Blobel G (1994) Isolation of components of the chloroplast protein import machinery. Science 266: 1007-1012

Schuster W, Brennicke A (1987) Plastid, nuclear and reverse transcriptase sequences in the mitochondrial genome of *Oenothera*: is genetic information transferred between organelles via RNA? EMBO J 6: 2857-2863

Schuster W, Brennicke A (1988) Interorganellar sequence transfer: plant mitochondrial DNA is nuclear, is plastid, is mitochondrial. Plant Sci 54: 1-10

Schuster W, Combettes B, Flieger K, Brennicke A (1993) A plant mitochondrial gene encodes a protein involved in cytochrome *c* biogenesis. Mol Gen Genet 239: 49-57

Schuster W, Hiesel R, Wissinger B, Brennicke A (1990) RNA editing in the cytochrome *b* locus of the higher plant *Oenothera berteriana* includes a U-to-C transition. Mol Cell Biol 10: 2428-2431

Sebald W, Hoppe J (1981) On the structure and genetics of the proteolipid subunit of the ATP synthase complex. Curr Top Bioenerg 12: 1-64

Seedorf M, Waegemann K, Soll J (1995) A constituent of the chloroplast import complex represents a new type of GTP-binding protein. Plant J 7: 401-411

Seidler A, Michel H (1990) Expression in *Escherichia coli* of the *psbO* gene encoding the 33 kd protein of the oxygen-evolving complex from spinach. EMBO J 9: 1743-1748

Shashidhara LS, Lim SH, Shackleton JB, Robinson C, Smith AG (1992) Protein targeting across the three membranes of the *Euglena* chloroplast envelope. J Biol Chem 267: 12885-12891

Siemeister G, Buchholz C, Hachtel W (1990) Genes for the plastid elongation factor Tu and ribosomal protein S7 and six tRNA genes on the 73 kb DNA from *Astasia longa* that resembles the chloroplast DNA of *Euglena*. Mol Gen Genet 220: 425-432

Sitte P (1987) Zellen in Zellen: Endocytobiose und die Folgen. In: Experiment und Theorie in Naturwissenschaft und Medizin. Lüst, R. (ed) Stuttgart: Wissenschaftliche Verlagsgesellschaft, pp 431-446

Sitte P (1993) Symbiogenic evolution of complex cells and complex plastids. Eur J Protist 29: 131-143

Small I, Suffolk R, Leaver CJ (1989) Evolution of plant mitochondrial genomes via substoichiometric intermediates. Cell 58: 69-75

Sogin ML (1994) The origin of eukaryotes and evolution into major kingdoms. In: Early life on earth. Bengtson S (ed) New York: Columbia UP, pp 181-192

Stern DB, Lonsdale DM (1982) Mitochondrial and chloroplast genomes of maize have a 12-kilobase DNA sequence in common. Nature 299: 698-702

Stirewalt VL, Michalowsky CB, Löffelhardt W, Bohnert HJ, Bryant DA. (1995) Nucleotide sequence of the cyanelle genome from *Cyanophora paradoxa*. Plant Mol Biol Rep 13: 327-332

Stockhaus J, Poetsch W, Steinmüller K, Westhoff P (1994) Evolution of the phosphoenolpyruvate carboxylase promoter of the C4 dicot *Flaveria trinervia*: an expression analysis in the C3 plant tobacco. Mol Gen Genet 245: 286-293

Stuart RA, Neupert W (1990) Apocytochrome *c*: an exceptional mitochondrial precursor protein using an exceptional import pathway. Biochimie 72: 115-121

Stuart RA, Neupert W (1996) Topogenesis of inner membrane proteins in mitochondria. TIBS 21: 261-266

Stubbe W (1959) Genetische Analyse des Zusammenwirkens von Genom und Plastom bei *Oenothera*. Z Vererbungslehre 90: 288-298

Suck R, Zeltz P, Falk J, Acker A, Kössel H, Krupinska K (1996) Transcriptionally active chromosomes (TACs) of barley chloroplasts contain the Â-subunit of plastome-encoded RNA polymerase. Curr Genet 30: 515-521

Sun E, Wu B-W, Tewari K (1989) In vitro analysis of the pea chloroplast 16S rRNA gene promoter. Molec. Cell Biol 9: 5650-5659

Tanaka K, Oikawa K, Ohta N, Kuroiwa H, Kuroiwa T, Takahashi H (1996) Nuclear encoding of a chloroplast RNA polymerase sigma subunit in a red alga. Science 272: 1932-1934

Tarassov I, Entelis N, Martin RP (1995) An intact protein translocating machinery is required for mitochondrial import of a yeast cytoplasmic tRNA. J Mol Biol 245: 315-323

Taylor WC (1989) Regulatory interactions between nuclear and plastid genomes. Annu. Rev. Plant Physiol. Plant Mol Biol 40: 211-233

Tymms MJ, Schweiger H-G (1989) Significant differences between the chloroplast genomes of two *Acetabularia mediterranea* strains. Mol Gen Genet 219: 199-203

Unseld M, Marienfeld JR, Brandt P, Brennicke A (1996) The mitochondrial genome of *Arabidopsis thaliana* contains 57 genes in 366,924 nucleotides. Nature Genet 15: 57-61

Urbach E, Robertson DL, Chisholm SW (1992) Multiple evolutionary origins of prochlorophytes within the cyanobacterial radiation. Nature 355: 267-270

Van de Peer Y, Rensing SA, Maier U-G, De Wachter R (1996) Substitution rate calibration of small subunit ribosomal RNA identifies chlorarachniophyte endosymbionts as remnants of green algae. Proc Natl Acad Sci USA 93: 7732-7736

Vera A, Sugiura M (1995) Chloroplast rRNA transcription from structurally different tandem promoters: an additional novel-type promoter. Curr Genet 27: 280-284

Vera A, Hirose T, Sugiura M (1996) A ribosomal protein gene (*rpl32*) from tobacco chloroplast DNA is transcribed from alternative promoters: similarities in promoter region organization in plastid housekeeping genes. Mol Gen Genet 251: 518-525

Villain P, Clabault G, Mache R, Zhou D-X (1994) S1F binding site is related to but different from the light-responsive GT-1 binding site and differentially represses the spinach *rps*1 promoter in transgenic tobacco. J Biol Chem 269: 16626-16630

von Heijne G (1986) Mitochondrial targeting sequences may form amphiphilic helices. EMBO J 5: 1335-1342

von Heijne G, Nishikawa K (1991) Chloroplast transit peptides. The perfect random coil? FEBS Lett 278: 1-3

von Heijne G, Steppuhn J, Herrmann RG (1989) Domain structure of mitochondrial and chloroplast targeting peptides. Eur J Biochem 180: 535-545

Wedel N, Klein R, Ljungberg U, Andersson B, Herrmann RG (1992) The single-copy gene *psbS* codes for a phylogenetically intriguing 22 kDa polypeptide of photosystem II. FEBS Lett 314: 61-66

Weeden NW (1981) Genetic and biochemical implications of the endosymbiotic origin of the chloroplast. J Mol Evol 17: 133-139

Westhoff P, Herrmann RG (1988) Complex RNA maturation in chloroplasts: The *psb*B operon from spinach. Eur J Biochem 171: 551-564

Whatley JM, John P, Whatley FR (1979) From extracellular to intracellular: the establishment of mitochondria and chloroplasts. Proc Roy Soc Lond B 204: 165-187

Whittacker RH (1969) New concepts of kingdoms of organisms. Science 163: 150-160

Whittier RF, Sugiura M (1992) Plastid chromosomes from vascular plants - genes. In: Plant gene research. Herrmann RG (ed), Wien, New York: Springer, vol 6, pp 164-182

Wissinger B, Hiesel R, Schobel W, Unseld M, Brennicke A, Schuster W (1991) Duplicated sequence elements and their function in plant mitochondria. Z Naturforsch 46: 709-716

Woese CR (1987) Bacterial evolution. Microbiol Rev 51: 221-271

Woese CR, Kandler O, Wheelis ML (1990) Towards a natural system of organisms: Proposal for the domains Archaea, Bacteria, and Eucarya. Proc Natl Acad Sci USA 87: 4576-4579

Wolfe KH, Morden CW, Palmer JD (1992) Function and evolution of a minimal genome from a nonphotosynthetic parasitic plant. Proc Natl Acad Sci USA 89: 10648-10652

Wolstenholme DR (1992) Genetic novelties in mitochondrial genomes of multicellular animals. Curr Opinion Genet Devel 2: 918-925

Wolter F, Fritz C, Willmitzer L, Schell J, Schreier P (1988) *RbcS* genes in *Solanum tuberosum*: Conservation of transit peptide and exon shuffling during evolution. Proc Natl Acad Sci USA 85: 846-850

Yang M, Jensen RE, Yaffe MP, Oppliger W, Schatz G (1988) Import of proteins into yeast mitochondria: The purified matrix processing protease contains two subunits which are encoded by the nuclear *MAS1* and *MAS2* genes. EMBO J 7: 3857-3862

Yu W, Schuster W (1995) Evidence for a site-specific cytidine deamination reaction involved in C to U RNA editing of plant mitochondria. J Biol Chem 270: 18227-18233

Yuan J, Henry R, McCaffery M, Cline K (1994) SecA homolog in protein transport within chloroplasts: evidence for endosymbiont-derived sorting. Science 266: 796-798

Zak E, Sokolenko A, Unterholzner G, Altschmied L, Herrmann RG (1997) On the mode of integration of plastid-encoded components of the cytochrome *bf* complex into thylakoid membranes. Planta 201: 334-341

Obituary Hans Kössel (1934-1995)

On December 24, 1995, Professor Hans Kössel passed away after a heart attack. For the last 20 years of his life he was Professor of Molecular Biology and Genetics in the Institute of Biology III at the University of Freiburg im Breisgau.

Hans Kössel grew up in Landsberg am Lech, where he was born on December 20, 1934. A scholarship from the state of Bayern for especially gifted high school graduates made it possible for him to study chemistry at the Ludwig Maximilian University in Munich. After receiving his diploma, he started his doctoral work with Adolf Butenandt at the Max Planck Institute for Biochemistry and received his degree in 1962. He stayed on for two more years at this renowned institute, where his activities at the time were strongly influenced by the Nobel laureates A. Butenandt and F. Lynen. In the laboratory of Wolfram Zillig, Hans Kössel investigated the chemical modification of nucleotides, especially their reactions with diazonium salts (Kössel 1964).

During this time, he married Ingeborg Heydkamp. In 1964, with her and two (later three) children, Hans Kössel went to the USA, where he was employed as Project Associate in the group of Professor Har Gobind Khorana, who later received a Nobel Prize. There, at the Institute for Enzyme Research of the University of Wisconsin-Madison, he worked on the then extremely popular topic of the elucidation of the genetic code (Khorana et al. 1966) and the enzymatic synthesis of polynucleotides (Kössel at al. 1967). Because of his contributions in this area and in the area of translation termination, he was brought back to Germany by

Carsten Bresch at the newly founded Institute of Biology III (Genetics and Molecular Biology) of the University of Freiburg. It should also be mentioned that Hans Kössel wrote a short textbook, Molecular Biology (Kössel 1966) during this time; the book, with a foreword by Adolf Butenandt and published by Klett Verlag in Stuttgart, contributed greatly to the establishment of this new field in Germany.

Various studies on the enzymatic hydrolysis of peptidyl-tRNAs and on the synthesis of oligonucleotides and polynucleotides (Kössel and Rajbhandary 1968) that Hans Kössel carried out partly in collaboration with Professor J. Lapidot (Lapidot et al. 1969) at the Hebrew University of Jerusalem, led finally to the sequencing and targeted mutagenesis of phage DNA. Kössel was also one who suggested the enzymatic synthesis of DNA as a method for sequencing DNA (Schott et al. 1973). At the invitation of Frederick Sanger to work in these areas, Hans Kössel went twice, in 1971 and in 1973, to the Medical Research Council Laboratory for Molecular Biology in Cambridge, England (Sanger et al. 1973, Sanger et al. 1974), where DNA sequencing by the dideoxy method was later fully developed for application. (F. Sanger received his second Nobel Prize for this achievement.)

In the last 15 years of his life, Hans Kössel, then leader of his own extremely active laboratory in Freiburg, devoted himself completely to the investigation of the chloroplast genome. In 1980, he published, together with Z. Schwarz, the sequence of the 16S rRNA from the chloroplasts of *Z. mays*, and proved without doubt the homology of the sequence with that of *E. coli* (Schwarz and Kössel 1980). With this evidence, the corresponding postulation of the endosymbiotic theory was verified beyond the shadow of a doubt. Investigations of the rRNA operon from *Z. mays* chloroplasts were extended further (Koch et al. 1981, Schwarz et al. 1981, Strittmatter et al. 1985) and culminated in the publication of the entire nucleotide sequence of this plastid genome in 1995 (Maier et al. 1995). Already before this, the Kössel group had discovered the editing of plastid mRNAs (Hoch et al. 1991) and continued to investigate this area successfully until the end (summarized in the following review).

Hans Kössel was an extremely successful researcher, whose perception and understanding substantially and strongly influenced the development of molecular genetics, gene technology, and endocytobiosis. More than 120 publications, many published in prestigious journals, persuasively bear witness to this. From 1981 he was an elected member of the European Molecular Biology Organization. Through all of this, Hans Kössel remained modest and unpretentious; he was never one to boast about his success. As a colleague, he was always cooperative and willing to help; a warm human nature was his. His fine, deeply moving words in honor of Professor Sugiura on the occasion of the award ceremony of the Miescher-Ishida Prize at the Sixth International Colloquium on Endocytobiology and Symbiosis in Tübingen will never be forgotten. On such occasions, one could perceive that Hans Kössel was not only a scientist - he was also a musician, a pianist and organist of high standard, who in his youth had even considered following a musical career.

Hans Kössel passed away too soon, much too soon for everything that he himself was still resolved to do and by which he could have furthered the progress of research. Through his numerous successes in his studies, he established a lasting, honorable monument. For us, who are allowed to go on his consistent and patient work and his refined human nature will linger as a wonderful memory and a challenging example.

Peter Sitte (Freiburg im Breisgau)

References

Hoch B, Maier RM, Appel K, Igloi HL, Kössel H (1991) Editing of a chloroplast mRNA by creation of an initiation codon. Nature 353: 178-180

Khorana HG, Büchi H, Ghosh H, Gupta N, Jacob TM, Kössel H, Morgan AR, Narang SA, Ohtsuka E, Wells RD (1966) Polynucleotide synthesis and the genetic code. Cold Spring Harbor Symp Quant Biol 31: 39-49

Koch W, Edwards K, Kössel H (1981) Sequencing of the 16S-23S spacer in a ribosomal RNA operon of *Zea mays* chloroplast DNA reveals two split tRNA genes. Cell 25: 203-213

Kössel H (1964) Zur Reaktion von Nucleotiden und Nucleinsäuren mit Diazoniumsalzen. Zeitschrift für analytische Chemie 205: 445-453

Kössel H (1966) Molekulare Biologie. Ein Studienbuch. Klett Verlag, Stuttgart

Kössel H, Morgan AR, Khorana HG (1967= Studies on polynucleotides. LXXIII. The synthesis *in vivo* of polypeptides containing repeating tetranucleotide sequences: direction of reading of messenger RNA. J Mol Biol 26: 449-475

Kössel H, Rajbhandary UL (1968) Studies on polynucleotides. LXXVI. Enzymatic hydrolysis of N-acylaminoacyl transfer RNA. J Mol Biol 35: 539-560

Lapidot Y, Inbar D, de Groot D, Kössel H (1969) Enzymatic hydrolysis of N-substituted Met-$tRNA_M$ and Met-$tRNA_F$. FEBS Lett 3: 253-256

Maier RM, Neckermann K, Igloi GL, Kössel H (1995) Complete sequence of the maize chloroplast genome: gene content, hotspots of divergence and fine tuning of genetic information by transcript editing. J Mol Biol 251: 614-621

Sanger F, Donelson J, Coulson AR, Kössel H, Fischer D (1973) The use of DNA polymerase I primed by a synthetic oligonucleotide to determine a nucleotide sequence in phage fl DNA. Proc Natl Acad Sci USA 70: 1209-1213

Sanger F, Donelson J, Coulson AR, Kössel H, Fischer D (1974) Determination of a nucleotide sequence in bacteriophage fl DNA by primed synthesis with DNA polymerase. J Mol Biol 90: 315-333

Schott H, Fischer D, Kössel H (1973) Synthesis of phage specific DNA fragments. II. Synthesis of four undecanucleotides complementary to a region of the coat protein cistron of phage fd. Biochemistry 12: 3447-3453

Schwarz Zs, Kössel H (1980) The primary structure of 16S rDNA from *Zea mays* chloroplast is homologous *to E. coli* 16S rRNA. Nature 283: 739-742

Schwarz, Zs, Kössel H, Schwarz E, Bogorad L (1981) A gene coding for $tRNA^{Val}$ is located near the 5' terminus of the 16S rRNA gene in the *Zea mays* chloroplast genome. Proc Natl Acad Sci USA 78: 4748-4752

Strittmatter G, Gozdzicka-Jozefiak A, Kössel H (1985) Identification of an rRNA operon promoter from *Zea mays* chloroplasts which excludes the proximal $tRNA^{Val}{}_{GAC}$ from the primary transcript. EMBO J 4: 599-604

Transcript Editing in Chloroplasts of Higher Plants

R. Bock, F. Albertazzi, R. Freyer, M. Fuchs, S. Ruf, P. Zeltz, and R. M. Maier
Institut für Biologie III, Universität Freiburg, Schänzlestraße 1, 79104 Freiburg, Germany

Correspondence to R.M. Maier

This review is dedicated to the memory of our late teacher and fatherly friend Hans Kössel. Having made outstanding contributions to the elucidation of the genetic code at the beginning of his career, he strongly believed in the code throughout his life, a belief that led him to discover plastid RNA editing in 1991.

Introduction

Posttranscriptional alterations of single nucleotides within an mRNA were first described for transcripts encoded by the kinetoplast DNA of trypanosomes (Benne et al. 1986). In this case, insertions and deletions of U residues directed by small guide RNAs can result in a substantial deviation of the mature mRNA sequence from the corresponding DNA template. This phenomenon was termed "RNA editing" by analogy to an editor's job modifying a manuscript and delivering a ready-for-press version. Soon after their initial discovery in kinetoplastid mitochondria, editing processes were detected in a variety of other genetic systems, including the nuclei of mammalian cells (Chen et al. 1987; Powell et al. 1987; Sommer et al. 1991) and the mitochondria, and chloroplasts of higher plants (Covello and Gray 1989; Gualberto et al. 1989; Hiesel et al. 1989; Hoch et al. 1991). A summary of the RNA editing systems known to date is given in Table 1. It appears useful to formally distinguish between two major types of RNA editing: insertional/deletional and substitutional editing. The above-mentioned kinetoplastid editing is of the insertional/deletional type, whereas editing in mammalian nuclei, plant mitochondria, and chloroplasts is of the substitutional type. Editing in mitochondria of the slime mold *Physarum polycephalum* is exceptional since in this system a mixture of base conversions and nucleotide insertions occurs (Gott et al. 1993). The case of paramyxoviral editing belongs to the gray zone around RNA editing in the strict sense since the guanosine nucleotides are inserted co-transcriptionally owing to stuttering of the RNA polymerase (Thomas et al. 1988).

The scope of this review is to summarize our present knowledge about RNA editing in higher plant chloroplasts, to discuss mechanistic aspects, and to present current evolutionary concepts of plastid RNA editing.

Table 1. Summary of the RNA editing systems described to date

Taxon	Type of editing	Cellular compartment	Reference
Trypanosomes	U insertion/deletion	mitochondrion	Benne (1994)
Physarum polycephalum	C-to-U conversion nucleotide insertions	mitochondrion	Mahendran et al. (1991); Gott et al. (1993)
Paramyxoviruses	G insertion	host cytoplasm	Thomas et al. (1988)
Higher plants	C-to-U conversion U-to-C conversion	mitochondrion	Bonnard et al. (1992); Schuster and Brennicke, (1994)
Higher plants	C-to-U conversion	chloroplast	Kössel et al. (1993)
Mammals	C-to-U conversion A-to-I conversion	nucleus	Chen et al. (1987); Powell et al. (1987); Sommer et al. (1991)
Snails, Monotremata, Marsupialia	various nucleotide conversions	mitochondrion	Yokobori and Pääbo (1995a,b); Janke and Pääbo (1993)
Drosophila melanogaster	A-to-I conversion	nucleus	Petschek et al. (1996)

Alteration of Chloroplast mRNAs by Transcript Editing

The discovery of RNA editing in chloroplasts was provoked by the curious case of the maize *rpl*2 gene, which was found to lack an ATG initiator codon at the DNA level (Kavousi et al. 1990). The conventional ATG codon was known to be present in *rpl2* genes of numerous other plant species, such as tobacco (Shinozaki et al. 1986) and the liverwort *Marchantia polymorpha* (Ohyama et al. 1986). The ACG codon found in the homologous position in maize was hypothesized to serve as an exceptional initiator codon (Hiratsuka et al. 1989). Given the prokaryotic nature of higher plant organelles this would have been a rather surprising feature of the chloroplast translational apparatus: protein synthesis in prokaryotes usually initiates at a standard AUG codon or at a GUG, never at an ACG triplet. AUG and GUG were known also to be used in chloroplasts (Shimada and Sugiura 1991), the use of ACG, however, would add a chloroplast-specific codon to the known prokaryotic initiator codons. Sequencing of maize *rpl2*-derived cDNAs, however, revealed the presence of a correct AUG initiator codon at the transcript level (Hoch et al. 1991) implying that the genomic ACG is converted either co- or post-transcriptionally into an AUG codon. Thus, the *rpl2* mRNA sequence deviates in a single position from the sequence of its DNA template, a phenomenon known as RNA editing. A similar C-to-U conversion was demonstrated for the initiator codons of the *psb*L genes from tobacco, bell pepper and spinach (Kudla et al.

1992; Kuntz et al. 1992; Bock et al. 1993) as well as of the *ndhD* gene of several species (Neckermann et al. 1994). It seems reasonable to assume that creation of the consensus AUG codon by RNA editing converts these transcripts from a non-translatable form into a translatable message. Recently, the ATG start codon of the *Chlamydomonas reinhardtii petA* gene was replaced with an ACG triplet by chloroplast transformation (Chen et al. 1995). Translation was shown to initiate at the mutant start codon with only very low efficiency, thus providing direct evidence for the functional importance of the start codon conversion by RNA editing.

The occurrence of ACG codons at potential initiator codons is rather rare. To test whether editing is a more general RNA processing mechanism in higher plant chloroplasts, several protein-coding regions were screened for possible internal RNA editing sites. Such a screen involves (1) an alignment of amino acid sequences derived from known DNA sequences of a chloroplast gene from several plant species, (2) selection of deviating positions for which amino acid identity can be restored by C-to-U transitions at the mRNA level, and (3) experimental verification of the candidate editing site by cDNA sequencing. As an example of this strategy, the identification of editing sites in the maize *ndhA*-encoded transcript by amino acid sequence comparison with homologous *ndh*A proteins from other species and with the evolutionary related mitochondrial *nad*1-encoded polypeptides (Maier et al. 1992a) is illustrated in Fig. 1. At site I, for which homologous sequences are absent from the mitochondrial proteins, editing in maize chloroplasts creates a UUG leucine codon that corresponds to the leucine residue encoded in the plastid DNAs of the other three plant species. Thus, this editing event results in the conservation of a presumably functionally important amino acid residue. Similarly, the UUA leucine codons produced by editing at sites II and III restore highly conserved leucine residues observed in other *ndh*A genes and in most of the homologous mitochondrial *nad*1 genes. Finally, the UCC serine codon at site IV, which is present in maize, rice, and tobacco *ndh*A, is converted into a UUC phenylalanine codon. In conclusion, the codon transitions introduced by the four editing events restore evolutionary conserved amino acid residues which are presumably essential for the structure and/or function of the *ndh*A-encoded polypeptide (a subunit of a putative plastid-localized NAD(P)H-dehydrogenase; Guedeney et al. 1996). This conclusion also holds true for nearly all the other editing sites identified to date in plastid mRNAs.

The availability of technology for higher plant chloroplast transformation (Svab et al. 1990; Svab and Maliga 1993) allowed us to address these functional aspects of RNA editing directly *in vivo*. Replacement of the tobacco plastid *psbF* gene with a non-editable version resulted in a phenylalanine-to-serine substitution owing to the lack of editing in a UCU codon. The affected amino acid residue is located within a transmembrane helix of the *psbF*-encoded peptide (the β-subunit of cytochrome b559). The incorporation of serine in the absence of RNA editing is associated with a phenotype (slower growth, lowered chlorophyll content, high chlorophyll fluorescence) characteristic of photosynthetic mutants. This finding

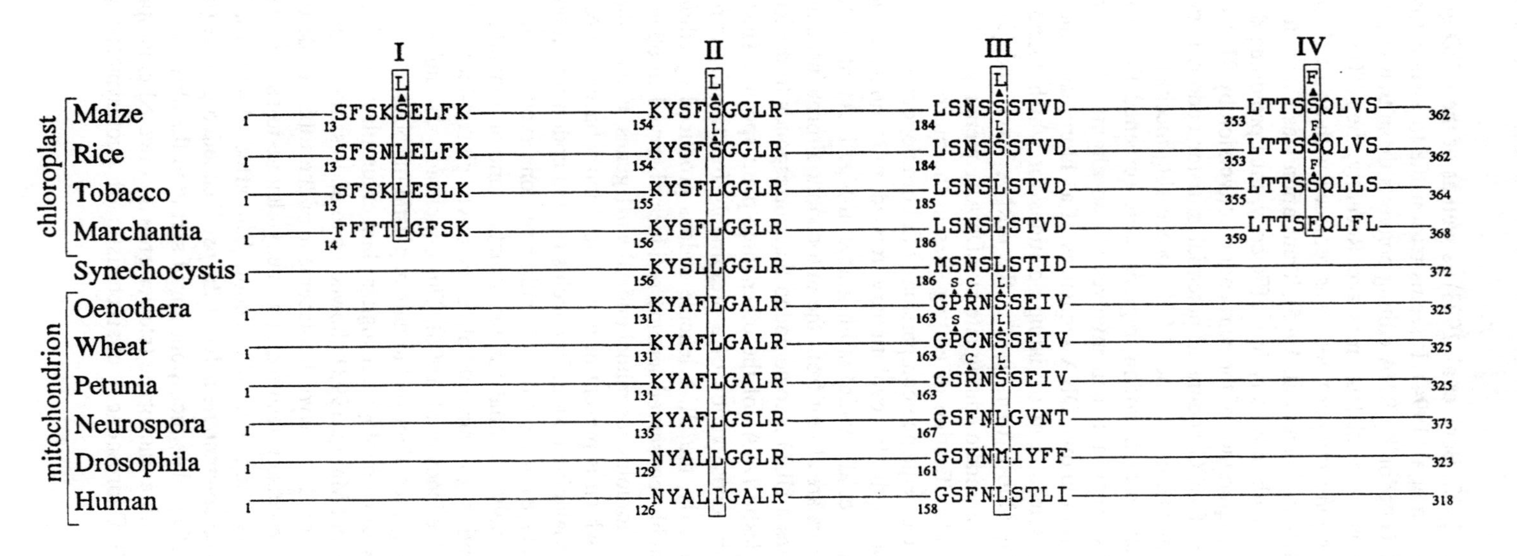

Fig. 1. Partial amino acid sequence alignment of *ndh*A-encoded polypeptides from selected species with the corresponding regions from the mitochondrial homologue (*ndh1*-encoded proteins). Four plastid RNA editing positions (I to IV) were identified by a computer search and subsequently verified experimentally (Maier et al. 1992a). Amino acid substitutions resulting from editing are marked by arrowheads. Mitochondrial editing sites are also included. Note that editing position III is even shared between some chloroplast *ndhA* mRNAs and the plant mitochondrial *ndh1* transcripts

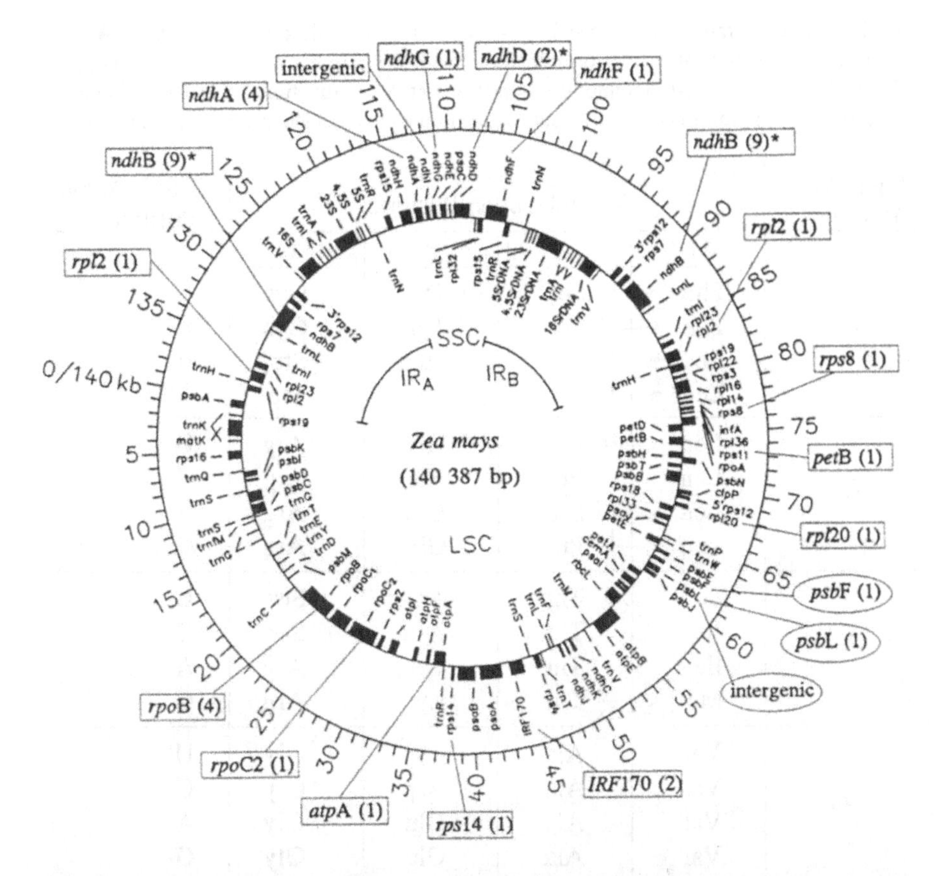

Fig. 2. Genomic location of editing sites identified in transcripts of higher plant plastids. The gene organization of the maize chloroplast genome is used as a reference system. The inverted repeat regions IR_A and IR_B, respectively, separate the circular plastid genome into a large (LSC) and a small (SSC) single copy region. Genes outside the circle are transcribed clockwise. Genes and intergenic regions in which editing sites have been detected in maize and other higher plant species are denoted by enlarged and framed gene symbols. Sites detected only in dicot species are marked by circled gene symbols. The numbers in parantheses following the gene symbols give the number of editing sites encoded in the respective gene. Numbers denoted by asterisks refer to the maximum number of editing sites in those cases where interspecific variation was detected.

confirms that editing of the chloroplast *psbF* transcript is an essential RNA processing step and thus provides the first direct proof for the biological significance of plant organellar RNA editing (Bock et al. 1994).

The recently completed nucleotide sequence of the maize plastome (Maier et al. 1995) provided an excellent system for a computer-aided screen for putative editing sites. More than 50% of the maize plastome-encoded mRNA sequences, in-

Table 2. Codon transitions caused by C-to-U editing of chloroplast transcripts. Arrows indicate the direction of the codon transitions with the number of observed events given above. Three threonine-to-methionine transitions create a functional translation initiation codon (Start), the fourth (Internal) results in the formation of an internal methionine codon.

First Codon Position	Second Codon Position				Third Codon Position
	U	C	A	G	
U	Phe ←1x—	Ser	→Tyr	Cys	U
	Phe ←3x—	Ser	Tyr	Cys	C
	Leu ←13x—	Ser	Stop	Stop	A
	Leu ←4x—	Ser	Stop	Trp	G
C	Leu ←1x—	Pro	1x His (→Tyr)	Arg	U
	Leu	Pro	His	Arg	C
	Leu ←5x—	Pro	Gln	Arg	A
	Leu ←1x—	Pro	Gln	Arg	G
A	Ile	Thr	Asn	Ser	U
	Ile	Thr	Asn	Ser	C
	Ile	Thr	Lys	Arg	A
	Met ← Start 3x / Internal 1x	Thr	Lys	Arg	G
G	Val	Ala	Asp	Gly	U
	Val	Ala	Asp	Gly	C
	Val	Ala	Glu	Gly	A
	Val	Ala	Glu	Gly	G

cluding all those that contain putative editing sites, were determined by cDNA analysis. Out of the more than 200 candidate sites, altogether 27 functional C-to-U editing sites could be verified (see Fig. 2). A rough estimate based on the alignments of all plastome-encoded polypeptides (similar to the alignment shown in Fig. 1 for the *ndh*A-encoded protein), taking into account also the bias for certain codon transitions (see below), supports the conclusion that the number of 27 sites identified is very close to the total number of sites encoded in the maize plastome. Given the overall number of codons in the maize plastome, the percentage of codons changed by editing is rather low (0.13%).

A strong bias for the second codon position becomes evident upon statistical evaluation of the codon transitions caused by plastid RNA editing (Table 2). In addition, editing shows a significant preference for certain codon transitions: UCA (Ser) to UUA (Leu) is by far the most frequent transition whereas others, such as

AC(U,C,A) to AU(U,C,A) and GCN to GUN have not been found thus far (Table 2). At present it is unclear whether or not these preferences are indicative of mechanistic constraints, e. g., inhibitory effects of purines at 5' neighboring positions. However, as outlined below, these biases were shown not to reflect a role of the chloroplast translation machinery in RNA editing.

The above-mentioned computer screen limits the detection of putative editing sites to coding regions, since the two criteria used were amino acid identity and codon transition frequencies. To test whether or not editing is restricted to protein coding sequences, some 5' and 3' untranslated regions as well as complete intergenic spacer regions of polycistronic transcripts were analyzed. This led to the detection of an editing site located 10 nucleotides upstream of the translation initiation codon of the *ndh*G reading frame from maize and rice (K. Neckermann, R. M. Maier and H. Kössel, unpublished data). It appears that the C-to-U transition at this site destabilizes an RNA secondary structure, which in turn may facilitate ribosome binding for initiation of translation. Thus, it is tempting to speculate that RNA editing in this case plays a regulatory role by unmasking a ribosome binding site and making it accessible for the chloroplast translation machinery. To date no editing events have been detected in intron sequences. Chloroplast structural RNAs seem also to be exempt from editing: neither the rRNAs nor the tRNAs analyzed so far contain any sequence deviations with respect to their DNA templates (unpublished results).

In spite of extensive cDNA sequencing, no alterations other than C-to-U transitions have been detected in any plastid RNA molecule, implying that editing in higher plant chloroplasts is restricted to conversions of cytosine to uracil. However, "reverse editing" (by U-to-C conversion) was shown recently to occur in plastids of the hornwort *Anthoceros formosae* (Yoshinaga et al. 1996).

Mechanistic Aspects of Plastid RNA Editing

The presence of a U residue instead of the genomically encoded C can be explained by alternative hypotheses: (1) cotranscriptional nontemplated U incorporation instead of C or (2) posttranscriptional modification of C to U. To distinguish between these two possibilities, processing intermediates of several chloroplast transcripts were analyzed. The maize *rpl2* gene contains an intron and is cotranscribed with the upstream *rpl23* gene. Editing in the *rpl2* initiator codon is virtually complete in spliced RNA molecules. The unspliced dicistronic precursor molecules, however, were shown to be only partially edited (Freyer et al. 1993). This finding provides circumstantial evidence for a posttranscriptional C-to-U conversion mechanism, although absolute proof can be only provided by demonstrating a precursor-product relation between the unedited/unspliced and the edited/spliced molecules.

In the case of the maize *petB* gene (which is cotranscribed with the downstream *petD*), complete editing was observed for all cDNAs, irrespective of whether they originated from unspliced, partially or fully spliced, and/or dicistronic or monocistronic RNA molecules (Freyer et al. 1993). Therefore, editing of the *pet*B message must be viewed as a very early step in mRNA processing which precedes both splicing and cleavage of polycistronic into monocistronic mRNAs. Furthermore, it can be concluded that *petB* editing is entirely independent of splicing and processing of polycistronic transcripts. This conclusion was further substantiated by a detailed analysis of transcripts encoded by the maize *ycf3* reading frame, which is split by two introns. In this case editing also precedes both intron removal and cleavage of polycistronic into monocistronic mRNAs (Ruf et al. 1994).

A high molecular weight transcriptionally active chromosome (TAC) consisting of DNA, nascent RNA, and protein can be isolated from various plastids (Hallick et al. 1976, Krupinska and Falk 1994; Reiss and Link 1985). The TAC is capable of elongating *in vivo* initiated transcripts *in vitro* (for review see Igloi and Kössel 1992). TAC-associated mRNAs were analyzed with respect to editing and splicing. Most of the investigated mRNAs were found to be already fully spliced and edited (unpublished data), implying that editing and splicing activities are closely associated with or even are integral parts of the TAC. The early and efficient conversion of a pre-mRNA into its edited form may protect the chloroplast from the production of aberrant polypeptides by premature translation of unedited transcripts.

The biochemical reaction mechanism underlying plastid RNA editing is still elusive. Cytosine can be chemically converted to uracil by a simple deamination reaction. RNA editing in higher plant mitochondria was recently shown to involve cytosine deamination (Yu and Schuster 1995; Blanc et al. 1995), which seems to be an attractive possibility for the editing mechanism in chloroplasts as well. Alternatively, transamination, base excision, or nucleotide excision mechanisms have to be considered.

Since chloroplast editing shows the above-mentioned strong preference for the second codon position as well as a bias towards certain types of codon transitions, speculations were put forward about a possible involvement of the plastid translational apparatus in the editing process. Analysis of RNA editing in the barley mutant *albostrians*, however, has disproved all those speculations. This mutant contains in its white sectors ribosome-free plastids (Börner et al. 1976; Hagemann and Scholz 1962), implying the absence of any translational activity. RNA editing was show not to be impaired at any of the sites active in the barley wild type transcripts (Zeltz et al. 1993). This finding clearly demonstrates that chloroplast editing is neither linked to nor dependent on the chloroplast translational apparatus. Furthermore, all the protein components of the chloroplast editing machinery must be encoded in the nuclear genome.

A major focus of current research is the identification of factors involved in plastid RNA editing. Using chloroplast transformation, the essential *cis*-acting determinants for editing of the initiation codon of the tobacco *psb*L transcript were shown to reside within a 98-nt fragment spanning the editing site (Chaudhuri et al. 1995). Expression of the chimeric *psbL* gene construct led to a significant decrease in the editing efficiency of the endogenous *psb*L mRNA, the editing of which is virtually complete in wild-type tobacco. This decrease appeared to be specific for the *psbL* site, other endogenous editing sites remaining unaffected. Thus it seems that transgene mRNA and *psbL* mRNA are competing for one and the same editing factor(s), which is present in rate-limiting concentrations (Chaudhuri et al. 1995). The nature of such a site-specific *trans*-acting factor, however, remains enigmatic. Small trans-acting RNA molecules termed guide RNAs have been shown to mediate editing site selection in kinetoplastid mitochondria (Blum et al. 1990; for review see Benne 1993). The 5' region of a guide RNA exhibits complementarity to the unedited precursor RNA, thus allowing hybridization and formation of an RNA-RNA duplex. Mismatches downstream of the annealed region direct the enzymatic reaction (in this case U insertions or deletions).

Several attempts have been undertaken to identify guide-RNA-like molecules in chloroplasts. Fractions of small RNAs were analyzed by hybridization, computer screens were devised to identify possible plastome-encoded guide RNAs, and a candidate guide RNA gene was tested for function in a knockout experiment (Bock and Maliga 1995). However, no functional guide RNA has been identified to date and the search for specificity factors continues.

Evolution and Phylogeny of RNA Editing in Chloroplasts

The alignment in Fig. 1 indicates that in the liverwort *Marchantia polymorpha*, the four amino acid residues restored by editing of the *ndh*A transcript in the higher plant species are already specified by the "correct" codon at the DNA level. This is also the case for all the other editing events in chloroplast transcripts, implying that the chloroplasts of the lower plant *Marchantia polymorpha* entirely lack an RNA editing apparatus. Editing seems to be also absent from the plastids of the algae investigated to date, as well as from cyanobacteria, the evolutionary progenitors of chloroplasts.

Whether editing is an ancient or a derived feature of plastid gene expression is still a matter of debate. The apparent absence of editing from lower plant plastids is not necessarily indicative of an "editing late" scenario. For example, editing can be also viewed as an evolutionary relic from the RNA world which was eventually lost in certain lineages of evolution but still persists in others. Like the old and still unsolved question of "introns early" or "introns late", the question of "editing early" or "editing late" cannot be answered without the accumulation of a much more complete set of phylogenetic data.

The *ndhB*-encoded transcript with its six editing sites is the most frequently edited mRNA in maize chloroplasts (Maier et al. 1992b). Thus *ndhB* seems to be a suitable candidate for studying the structural and functional conservation of chloroplast editing sites within the plant kingdom. The six maize editing sites were found to be conserved in the *ndhB* transcripts of barley and rice as well as in the dicotyledonous species tobacco (Freyer et al. 1995). Interestingly, several additional sites are observed for each of these three species. The additional sites show species-specific divergence, which, surprisingly, is even more extensive among the closely related graminean species. For example, two of the sites are shared between barley and tobacco but not between the two monocots barley and maize. A similar situation has been reported for the single editing site of the *pet*B transcript which is present in maize (Freyer et al. 1993) and tobacco (Hirose et al. 1994) but not in rice. Consequently, the presence or absence of individual editing sites is not a reliable criterion for evaluating phylogenetic relationships.

The poor evolutionary conservation of chloroplast editing sites can be readily explained if one assumes that the evolutionary divergence is caused by loss of preexisting sites rather than by acquisition of new sites. Whereas an acquisition model requires convergent evolution by independent creation of identical new sites (and of the respective specificity factors), existing editing sites can be easily eliminated in a certain lineage of plant evolution by simple C-to-T mutations at the DNA level. The latter scenario offers an evolutionary much more conceivable explanation.

To test whether or not the loss of an editing site is accompanied by the loss of the capacity to edit this site, a heterologous editing site was incorporated into the tobacco chloroplast genome by biolistic transformation. The site was taken from the spinach *psbF* gene and replaced the homologous tobacco sequence (which specifies the "correct" codon at the DNA level). The spinach *psbF* editing site remained unmodified in the transplastomic tobacco plants generated, indicating that the editing capacity of evolutionary eliminated editing sites is not maintained (Bock et al. 1994). In the absence of selective pressure the site-specific editing factor is eventually lost. Such a loss of a specificity factor may be also the case of one of the *rpo*B-encoded editing sites in barley. In spite of structural conservation, the C residue that undergoes editing to U in maize remains unchanged in barley (Zeltz et al. 1993). This apparent silencing of an editing site may indicate either that the codon transition from proline to leucine is silent with respect to protein function or that proline can be tolerated in the barley protein owing to a compensatory mutation somewhere else in the protein.

Comparison of Plastid and Plant Mitochondrial RNA Editing

Among the numerous RNA editing systems known to date (Table 1), editing in the plant mitochondrial compartment is certainly the most closely related to the plastid editing system. Striking parallels become evident upon comparison of the editing systems in the two plant organelles. Firstly, the only modifications found in higher plant chloroplast transcripts are C-to-U transitions, which are also observed as the vast majority of mitochondrial editing events, the exceptions being only a very few U-to-C conversions (Schuster et al. 1990). Such "reverse editing" events have been described recently also for the chloroplasts of a lower plant, the hornwort *Anthoceros formosae* (Yoshinaga et al. 1996). A second characteristic shared between plastid and mitochondrial editing is the strong preference for the second codon position as well as the bias towards certain codon transitions (Table 2). Furthermore, the apparent absence of editing from both plastids and mitochondria of the liverwort *Marchantia polymorpha* is also suggestive of an evolutionary link between the two organellar editing systems.

The comparatively low number of both plastid editing sites and edited transcripts, however, contrasts with the much higher editing frequencies of plant mitochondrial transcripts (approximately 2% of the nucleotides in mitochondrial protein coding regions are altered by RNA editing). In rare cases, structural RNA's in plant mitochondria are also subject to editing (Binder et al. 1994; Maréchal-Drouard et al. 1993), which seems not to be the case in plastids.

Sets of loose consensus sequences shared between editing sites encoded in mitochondrial and chloroplast transcripts were identified (Gualberto et al. 1990; Maier et al. 1992b; Zeltz et al. 1996) and led to the speculation that certain plastid and mitochondrial editing sites may be recognized by common *trans*-acting specificity factor(s). This is particularly conceivable in those cases where editing sites are present at homologous positions of homologous plastid and mitochondrial genes. For example, editing at site III of the chloroplast *ndhA* transcript results in a transition from a serine to a leucine codon. A mitochondrial editing event leads to the same codon transition in the homologous position of the *nad1*-encoded mRNAs from at least three plant species (Maier et al. 1992a).

In view of the above-mentioned similarities and of the existence of common sequence motifs shared between homologous and even nonhomologous transcripts of the two plant organelles, it appeared reasonable to assume that mitochondrial RNA sequences could be substrates for the editing machinery of chloroplasts and *vice versa*. To test mitochondrial editing sites for function in chloroplasts, transgenic tobacco plants were produced which carried *Petunia* mitochondrial *coxII* sequences. However, none of the seven editing sites active in mitochondria was active in chloroplasts (Sutton et al. 1995). This finding demonstrates that the editing machineries of the two plant organelles are not completely identical. At least some of the factors engaged in editing must be organelle-specific. This conclusion was further substantiated by a reciprocal study: a promiscuous plastid DNA fragment carrying the *rpoB* gene was shown to be evolutionary transferred to the mitochondrial genome of rice. All three editing sites active in chloroplasts were shown to be nonfunctional in mitochondria (Zeltz et al. 1996).

In spite of these differences, the many similarities described above are suggestive of common components and/or mechanistic steps in the two plant organellar editing systems. Thus it appears likely that the two systems arose from common evolutionary roots. How far these roots date back is one of the many challenging questions that remain to be answered.

Acknowledgements: Work in the authors' laboratory was supported by grants from the Deutsche Forschungsgemeinschaft (SFB 206 and 388; KO 253/13-1; MA 1592/2-2) and the Human Frontier Science Program Organization (grant RG-

437/94). We thank Marita Hermann and Maria Escobar-Friedrich for excellent technical assistance and Dr Gabor L. Igloi for critical reading of the manuscript. R.B. is the recipient of a fellowship from the Boehringer Ingelheim Fonds, Stuttgart, Germany. R.M.M. is recipient of a fellowship from the Deutsche Forschungsgemeinschaft. S.R. is supported by a fellowship from the Landesgraduiertenförderung Baden-Württemberg. F.A. is supported by fellowships from the Deutsche Akademische Austauschdienst and the University of Costa Rica.

References

Benne R (1993) RNA editing. The alteration of protein coding sequences of RNA. Ellis Horwood, Chichester

Benne R (1994) RNA editing in trypanosomes. Eur J Biochem 221: 9-23

Benne R, Van den Burg J, Brakenhoff J, Sloof P, Van Boom JH, Tromp MC (1986) Major transcript of the frameshifted *coxII* from trypanosome mitochondria contains four nucleotides that are not encoded in the DNA. Cell 46: 819-826

Binder S, Marchfelder A, Brennicke A (1994) RNA editing of tRNA(Phe) and tRNA(Cys) in mitochondria of *Oenothera berteriana* is initiated in precursor molecules. Mol Gen Genet 244: 67-74

Blanc K, Litvak S, Araya A (1995) RNA editing in wheat mitochondria proceeds by a deamination mechanism. FEBS Lett 373: 56-60

Blum B, Balakara N, Simpson L (1990) A model for RNA editing in kinetoplastid mitochondria : small "guide RNA" molecules transcribed from maxicircle DNA provide the edited sequence information. Cell 60: 189-198

Bock R, Maliga P (1995) *In vivo* testing of a tobacco plastid DNA segment for guide RNA function in *psbL* editing. Mol Gen Genet 247: 439-443

Bock R, Hagemann R, Kössel H, Kudla J (1993) Tissue-specific and stage-specific modulation of RNA editing of the *psbF* and *psbL* transcript from spinach plastids - a new regulatory mechanism? Mol Gen Genet 240: 238-244

Bock R, Kössel H, Maliga P (1994) Introduction of a heterologous editing site into the tobacco plastid genome: the lack of RNA editing leads to a mutant phenotype. EMBO J 13: 4623-4628

Bonnard G, Gualberto JM, Lamattina L, Grienenberger JM (1992) RNA editing in plant mitochondria. CRC Crit Rev Plant Sci 10: 503-524

Börner T, Schumann B, Hageman R (1976) Biochemical studies on a plastid ribosome-deficient mutant of *Hordeum vulgare*. In: Bücher T, Neupert W, Sebald S, Werner S (eds) Genetics and biogenesis of chloroplasts and mitochondria. Elsevier, Amsterdam, pp 41-48

Chaudhuri S, Carrer H, Maliga P (1995) Site-specific factor involved in the editing of the *psbL* mRNA in tobacco plastids. EMBO J 14: 2951-2957

Chen SH, Habib G, Yang CY, GU ZW, Lee BR, Weng SA, Silberman SR, Cai SJ, Deslypere JP, Rosseneu M, Gotto AM, Li WH, Chan L (1987) Apolipoprotein B-48 is the product of a messenger RNA with an organ-specific in-frame stop codon. Science 238: 363-366

Chen X., Kindle, KL, Stern DB (1995) The initiation codon determines the efficiency but not the site of translation initiation in *Chlamydomonas* chloroplasts. Plant Cell 7: 1295-1305

Covello PS, Gray MW (1989) RNA editing in plant mitochondria. Nature 341: 662-666

Freyer R, Hoch B, Neckermann K, Maier RM, Kössel H (1993) RNA editing in maize chloroplasts is a processing step independent of splicing and cleavage to monocistronic mRNAs. Plant J 4: 621-629

Freyer R, López C, Maier RM, Martin M, Sabater B, Kössel H (1995) Editing of the chloroplast *ndhB* encoded transcripts shows divergence between closely related members of the grass family (Poaceae). Plant Mol Biol 29: 679-684

Gott JM, Visomirski LM, Hunter JL (1993) Substitutional and insertional RNA editing of the cytochrome *c* oxidase subunit I mRNA of *Physarum polycephalum*.. J Biol Chem 268: 25483-25486

Gualberto JM, Lamattina L, Bonnard G, Weil JH, Grienenberger JM (1989) RNA editing in wheat mitochondria results in the conservation of protein sequences. Nature 341: 660-662

Gualberto JM, Weil JH, Grienenberger JM (1990) Editing of the wheat *coxIII* transcript : evidence for twelve C to U and one U to C conversions and for sequence similarities around editing sites. Nucleic Acids Res 18: 3771-3776

Guedeney G, Corneille S, Cuiné S, Peltier G (1996) Evidence for an association of *ndh*B and *ndh*J gene products and ferredoxin-NADP-reductase as components of a chloroplastic NAD(P)H dehydrogenase complex. FEBS Lett 378: 277-280

Hagemann R, Scholz F (1962) Ein Fall Gen-induzierter Mutationen des Plasmotyps bei Gerste. Züchter 32: 50-59

Hallick RB, Lipper C, Richards OC, Rutter WJ (1976) Isolation of a transcriptionally active chromosome from chloroplasts of *Euglena gracilis*. Biochemistry 15: 3039-3045

Hiesel R, Wissinger B, Schuster W, Brennicke A (1989) RNA editing in plant mitochondria. Science 246: 1632-1634

Hiratsuka J, Shimada H, Whittier R, Ishibashi T, Sakamoto M, Mori M, Kondo C, Honji Y, Sun CR, Meng BY, Li YQ, Kanno A, Nishikawa Y, Hirai A, Shinozaki K, Sugiura M (1989) The complete sequence of the rice (*Oryza sativa*) chloroplast genome: Intermolecular recombination between distinct tRNA genes accounts for a major plastid DNA inversion during the evolution of cereals. Mol Gen Genet 217: 185-194

Hirose T, Wakasugi T, Sugiura M, Kössel H (1994) RNA editing of tobacco *petB* mRNAs occurs both in chloroplasts and non-photosynthetic proplastids. Plant Mol Biol 26: 509-513

Hoch B, Maier RM, Appel K, Igloi GL, Kössel H (1991) Editing of a chloroplast mRNA by creation of an initiation codon. Nature 353: 178-180

Igloi GL, Kössel H (1992) The transcriptional apparatus of chloroplasts. CRC Crit Rev Plant Sci 10: 525-558

Janke A, Pääbo S (1993) Editing of a tRNA anticodon in marsupial mitochondria changes its codon recognition. Nucleic Acids Res 21:1523-1524

Kavousi M, Giese K, Larrinua IM, McLaughlin WE, Subramanian AR (1990) Nucleotide sequence and map positions of the duplicated gene for maize (*Zea mays*) chloroplast ribosomal protein L2. Nucleic Acids Res 18: 4244-4244

Kössel H, Hoch B, Maier RM, Igloi GL, Kudla J, Zeltz P, Freyer R, Neckermann K, Ruf S (1993) RNA editing in chloroplasts of higher plants. In A Brennicke, U Kück (eds) Plant mitochondria. VCH, Weinheim, pp 93-102

Krupinska K, Falk J (1994) Changes in RNA-polymerase activity during development and senescence of barley chloroplasts. Comparative analysis of transcripts synthesized either in run-on assays or by transcriptionally active chromosomes. J Plant Physiol 143: 298-305

Kudla J, Igloi G, Metzlaff M, Hagemann R, Kössel H (1992) RNA editing in tobacco chloroplasts leads to the formation of a translatable *psbL* messenger RNA by a C to U substitution within the initiation codon. EMBO J 11: 1099-1103

Kuntz M, Camara B, Weil JH, Schantz R (1992) The *psbL* gene from bell pepper (*Capsicum annuum*): plastid RNA editing also occurs in non-photosynthetic chromoplasts. Plant Mol Biol 20: 1185-1188

Mahendran R, Spottswood MR, Miller DL (1991) RNA editing by cytidine insertion in mitochondria of *Physarum polycephalum*. Nature 349: 434-438

Maier RM, Hoch B, Zeltz P, Kössel H (1992a) Internal editing of the maize chloroplast *ndhA* transcript restores codons for conserved amino acids. Plant Cell 4: 609-616

Maier RM, Neckermann K, Hoch B, Akhmedov NB, Kössel H (1992b) Identification of editing positions in the *ndhB* transcript from maize chloroplasts reveals sequence similarities between editing sites of chloroplasts and plant mitochondria. Nucleic Acids Res 20: 6189-6194

Maier RM, Neckermann K, Igloi GL, Kössel H (1995) Complete sequence of the maize chloroplast genome: Gene content, hotspots of divergence and fine tuning of genetic information by transcript editing. J Mol Biol 251: 614-628

Maréchal-Drouard L, Ramamonjisoa D, Cosset A, Weil JH, Dietrich A (1993) Editing corrects mispairing in the acceptor stem of bean and potato mitochondrial phenylalanine transfer RNAs. Nucleic Acids Res 21: 4909-4914

Neckermann K, Zeltz P, Igloi GL, Kössel H, Maier RM (1994) The role of RNA editing in conservation of start codons in chloroplast genomes. Gene 146: 177-182

Ohyama K, Fukuzawa H, Kohchi T, Shirai H, Sano T, Sano S, Umesono K, Shiki Y, Takeuchi M, Chang Z, Aota S, Inokuchi H, Ozeki H (1986) Chloroplast gene organization deduced from complete sequence of liverwort *Marchantia polymorpha* chloroplast DNA. Nature 322: 572-574

Petschek JP, Mermer MJ, Scheckelhoff MR, Simone AA, Vaughn JC (1996) RNA editing in Drosophila *4f-rnp* gene nuclear transcripts by multiple A-to-G conversions. J Mol Biol 259: 885-890

Powell LM, Wallis SC, Pease RJ, Edwards YH, Knott TJ, Scott J (1987) A novel form of tissue-specific RNA processing produces apolipoprotein-B48 in intestine. Cell 50: 831-840

Reiss T, Link G (1985) Characterization of transcriptionally active DNA-protein complexes from chloroplasts and etioplasts of mustard (*Sinapis alba* L.). Eur J Biochem 148: 207-212

Ruf S, Zeltz P, Kössel H (1994) Complete RNA editing of unspliced and dicistronic transcripts of the intron-containing reading frame *IRF170* from maize chloroplasts. Proc Natl Acad Sci USA 91: 2295-2299

Schuster W, Brennicke A (1994) The plant mitochondrial genome: physical structure, information content, RNA editing and gene migration. Annu Rev Physiol Plant Mol Biol 45: 61-78

Schuster W, Hiesel R, Wissinger B, Brennicke A (1990) RNA editing in the cytochrome b locus of the higher plant *Oenothera berteriana* includes a U to C transition. Mol Cell Biol 10: 2428-2431

Shimada H, Sugiura M (1991) Fine structural features of the chloroplast genome: Comparison of the sequenced chloroplast genomes. Nucleic Acids Res 19: 983-995

Shinozaki K, Ohme M, Tanaka M, Wakasugi T, Hayashida N, Matsubayashi T, Zaita N, Chunwongse J, Obokata J, Yamaguchi-Shinozaki K, Ohto C, Torazawa K, Meng BY, Sugita M, Deno H, Kamogashira T, Yamada K, Kusuda J, Takaiwa F, Kato A, Tohdoh N, Shimida H, Sugiura M (1986) The complete nucleotide sequence of the tobacco chloroplast genome: its gene organization and expression. EMBO J 5: 2043-2049

Sommer B, Kohler M, Sprengel R, Seeburg pH (1991) RNA editing in brain controls a determinant of ion flow in glutamate-gated channels. Cell 67: 11-19

Sutton CA, Zoubenko OV, Hanson MR, Maliga P (1995) A plant mitochondrial sequence transcribed in transgenic tobacco chloroplasts is not edited. Mol Cell Biol 15: 1377-1381

Svab Z, Maliga P (1993) High-frequency plastid transformation in tobacco by selection for a chimeric *aadA* gene. Proc Natl Acad Sci USA 90: 913-917

Svab Z, Hajdukiewicz P, Malgia P (1990) Stable transformation of plastids in higher plants. Proc Natl Acad Sci USA 87: 8526-8530

Thomas SM, Lamb RA, Paterson RG (1988) Two mRNAs that differ by two nontemplated nucleotides encode the amino coterminal proteins P and V of the paramyxovirus SV5. Cell 54: 891-902

Yokobori S, Pääbo S (1995a) Transfer RNA editing in land snail mitochondria. Proc. Natl. Acad Sci USA 92: 10432-10435

Yokobori S, Pääbo S (1995b) tRNA editing in metazoans. Nature 377: 490

Yoshinaga K, Iinuma H, Masuzawa T, Ueda K (1996) Extensive RNA editing of U to C in addition to C to U substitution in the *rbcL* transcript of hornwort chloroplasts and the origin of RNA editing in green plants. Nucleic Acids Res 24: 1008-1014

Yu W, Schuster W (1995) Evidence for a site-specific cytidine deamination reaction involved in C to U RNA editing of plant mitochondria. J Biol Chem 270: 18227-18233

Zeltz P, Hess WR, Neckermann K, Börner T, Kössel H (1993) Editing of the chloroplast *rpoB* transcript is independent of chloroplast translation and shows different patterns in barley and maize. EMBO J 12: 4291-4296

Zeltz P, Kadowaki K, Kubo N, Maier RM, Hirai A, Kössel H (1996) A promiscuous chloroplast DNA fragment is transcribed in plant mitochondria but the encoded RNA is not edited. Plant Mol Biol 31: 647-656

The Mobile Introns in Fission Yeast Mitochondria: A Short Review and New Data

B. Schäfer and K. Wolf
Institut für Biologie IV (Mikrobiologie) der Rheinisch-Westfälischen Technischen Hochschule Aachen, Worringer Weg, 52056 Aachen, Germany

Key words: *Aspergillus nidulans*, *Schizosaccharomyces pombe*, yeast, mitochondria, introns, endonuclease.

Introduction

The mitochondrial genome of fission yeast is remarkable because of its small size of only 19.43 kbp in strain 50 of the Leupold collection (Fig. 1). Nevertheless, this genome contains two group I introns and one group II intron (Paschke et al. 1993). The construction of an intronless mitochondrial genome (Schäfer et al. 1991) yielded a small genome of only 14.62 kbp. The construction of strains with mitochondrial genomes harboring just one intron provides the opportunity to study the transmission of this special intron using the intronless mitochondrial genome as recipient. The two introns in the gene encoding the largest subunit of cytochrome c oxidase *(cox1)* belong to group I, the intron in the gene for apocytochrome b is a group II intron. In this contribution we shall briefly summarize the characteristics of these introns, focusing mainly on their mobility. We further report on a first attempt to characterize an intron-encoded endonuclease. Finally, we discuss the possibility of horizontal transfer of a group I intron between microorganisms and from the mitochondria to the nucleus.

The Mobile Group I Intron 1 in the *cox1* Gene

The rationale of a genetic experiment to demonstrate intron mobility is as follows. A strain with a mitochondrial genome harboring just one intron is used as the donor in this cross, and a strain with an intronless mitochondrial genome as recipient. The two strains are marked with different nuclear auxotrophic mutations and different mitochondrial drug resistance markers during meiosis. In a zygote, mitochondria fuse, and mitochondrial genomes recombine and segregate, yielding ascospores which may be mitochondrially homo- or heteroplasmic. By replica-plating colonies on drug-containing media, transmission of the mitochondrial alleles conferring resistance or sensitivity to a drug can be measured. The average of a large number of crosses reveals transmission values of a given mitochondrial drug resistance marker around 50%. Transmission of the intron, however, deviates from this value. In the case of *cox1* intron 1 between 85 and 90% of the progeny contains the intron, which can be measured by restriction enzyme analysis of mitochondrial DNA of the progeny and hybridization with intron probes. This type of experiment shows that the intron present in the *cox1* gene very frequently moves into the intronless allele of the *cox1* gene. This process has been termed "intron homing" (Wolf 1994).

The *cox1I1* Homing Endonuclease

Mobile group I introns are widespread, being found in organelle, bacteriophage, and nuclear genomes (Lambowitz and Belfort 1993). The basic mechanism of group I intron homing seems to be the same in all cases examined. The mobile

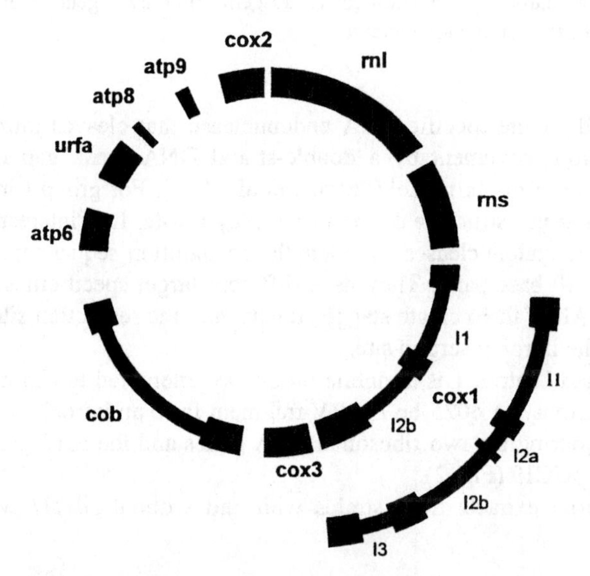

Fig. 1. The mitochondrial genome of *Schizosaccharomyces pombe.* The inner circle represents the gene map of strain *50* from the Leupold collection, the outer circle the *cox1* gene from strain *EF1*. Thick bars represent exons, thin bars introns. *rnl* and *rns* encode large and small ribosomal RNAs, *cox1*, *2*, and *3* are the genes for subunits of cytochrome c oxidase, while *atp6, 8*, and *9* code for subunits of ATPase, *cob* is the gene for apocytochrome b, and *urf a* represents an unassigned reading frame which is very likely involved in mitochondrial protein synthesis and genome integrity. tRNA genes are not indicated. I1, I2a, I2b, and I3 are group I introns in the *cox1* gene. Strain 50 has a group II intron in the *cob* gene, whereas strain *EF1* harbors a continuous cob gene

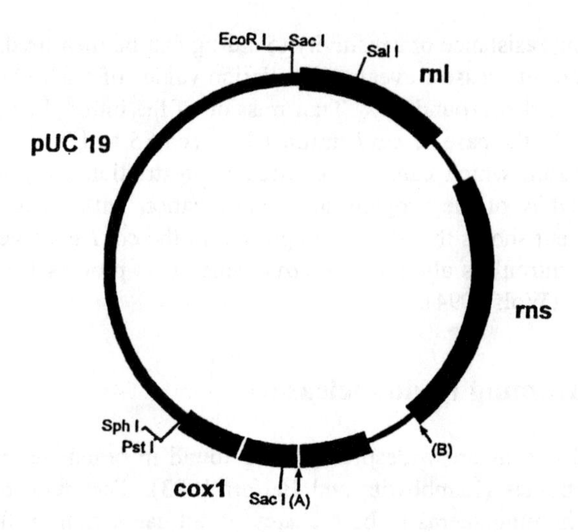

Fig. 2. Restriction map of the hybrid vector pP3E5-2. This vector contains a 4025-bp *Eco*RV fragment harboring the continuous *cox1* gene and the *rns* gene. A and B are cutting sites for the mitochondrial endonuclease

introns encode a site-specific DNA endonuclease that cleaves intronless alleles, initiating intron movement by a double-strand DNA break gap repair process according to the Szostak model (Szostak et al. 1983). For group I intron mobility neither splicing nor structure of the intron play a role. It is interesting that these intron-encoded endonucleases have lengthy recognition sequences, ranging from 14 to almost 40 base pairs. They have different target specificities and cleavage frequencies. All of them create staggered cuts, and the restriction site may or may not overlap the intron insertion site.

Since the *cox1* intron 1 is a mobile intron, we attempted to characterize its endonuclease activity. A 4025-bp *Eco*RV fragment from an intronless mitochondrial genome harbouring the two ribosomal RNA genes and the *cox1* gene was cloned in the vector pUC19 (Fig. 2).

Mitochondrial extracts from strains with and without *cox1I1* were prepared.

Table 1. Optimized conditions for endonuclease cut

Mitochondrial extract	20 µl
Tris-HCl, pH 7.5	25 µM
$MgCl_2$	25 µM
NH_4Cl_2	200 µM
Dithioerythrol	2.5 µM
Template	3 µg
Incubation time	15 min

Table 1 summarizes the optimized experimental conditions.

Under these conditions two endonuclease cuts were observed (Fig. 2), located as expected between exons 1 and 2 of the *cox1* gene (Fig. 2, A), the other downstream of the gene encoding the large ribosomal RNA (Fig. 2, B). It is astonishing that the cut at (B) is obviously more efficient than the expected cut at (A) (data not shown). Further optimization of the conditions and purification of the enzyme will be necessary. As controls we prepared mitochondrial extracts from a strain without *coxI1* and from a strain without mitochondrial DNA (*rho*^{O}) (Massardo et al. 1994), and we used carefully washed intact mitochondria from a strain with *coxI1*. In none of these cases was cutting activity observed.

Homing and Transposition of the Group II

Using the same approach as for intron *coxI1*, we have found that the *cob* intron is also mobile. In various crosses between a *cob* intron-containing donor and an intronless recipient, between 85 and 90% of the meiotic progeny carried the *cob* intron (Paschke et al. 1993).

In a recent paper (Schmidt et al. 1994) it has been shown that the *cob* intron is able to transpose to various sites in the intron (producing a twintron) and in the downstream non-coding sequence. Intron transposition is not very frequent, but integration events can be monitored using polymerase chain reaction (PCR). The underlying mechanism is a reverse splicing process followed by reverse transcription of the RNA. The results are summarized in Fig. 3.

It is now well documented that the group II introns found in yeasts and filamentous fungi bear striking similarities with retrovirus genomes (Michel and Lang

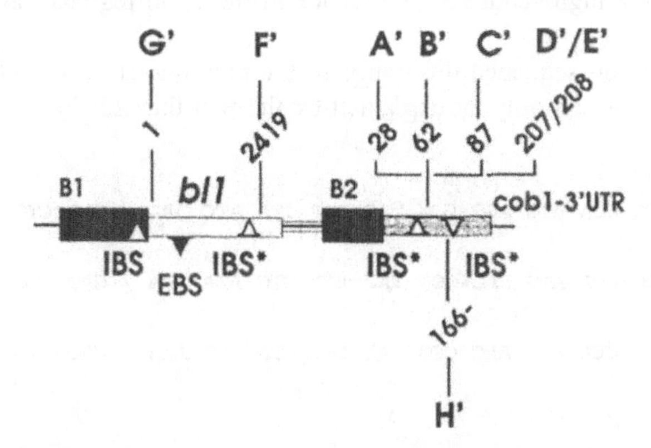

Fig. 3. Physical map of the *cob* gene with 3′ untranslated region. A′ - G′ indicate insertion sites in the sense strand, H′ marks an insertion of the *cob* intron in the antisense strand. B1 and B2 are the exons; IBS intron binding site, IBS* intron binding site-like sequence, EBS exon binding site. (Modified from Schmidt et al. 1994)

1985). Analysis of group II introns has revealed that each intron reading frame has the following conserved motifs in the order from 5′ to 3′ of the reading frame: (1) a protease-like motif, (2) a region found in reverse transcriptases of non-long-terminal-repeat-retrotransposons, (3) a reverse transcriptase-like region, (4) a region associated with maturase function, and (5) a zinc finger (see Moran et al. 1995 and references therein). Moran et al. (1995) demonstrated that efficient mobility of the group II introns *aI1* and *aI2* from *Saccharomyces cerevisiae* is dependent on the reverse transcriptase function and a putative DNA endonuclease.

The current model of group II intron mobility involves reverse transcription of an unspliced pre-mRNA, starting in an exon and ending in an exon. The cDNA integrates into the recipient mitochondrial DNA either directly or after synthesis of the second strand. This results in the insertion into the recipient allele of the intron plus flanking exon sequences. It is interesting to note that homing is accompanied by efficient but asymmetric co-conversion of the flanking exon sequences (Moran et al. 1995), which was also observed by Lazowska et al. (1994). The features of site-specific group II intron mobility are also applicable to the transposition to ectopic sites, which is much less frequent. This process, however, does not lead to conversion events (Schmidt et al. 1994).

A Footprint of Group II Intron Insertion in the *cob* Gene of *Schizosaccharomyces pombe*

The observation of co-conversion of flanking exons in the *cox1* exons of *Saccharomyces cerevisiae* has an interesting parallel in *Schizosaccharomyces pombe*. In our collection of wild-type isolates, the majority of strains have a continuous *cob* gene. In these strains, there is no sequence difference in the region where the intron is inserted in mosaic genes. Strains with an intron-containing *cob* gene, however, show a high sequence divergence in the exon regions flanking the intron, as shown in Fig. 4.

Since there is no sequence difference in the continuous gene, but only in the mosaic genes, this can only be explained by the fact that this intron is a later ac-

1) GAC AGA ATC CCA ATG **AGC** CCG **TAC** TAT **TTA** ATC AAA / GAT CTG ATC ACA ATA TTT

2) GAC AGA ATC CCA ATG **AAC** CCG **TAT** TAT **CTG** ATA AAA / GAT **TTG** ATA ACA ATA TTT

3) **GAT** AGA ATC CCA ATG **AAC** CCG **TAT** TAT **CTA** ATT AAA / **GAC** CTG ATT **ACG** ATA **TTC**

Fig. 4. Sequence of the *cob* exon region flanking the insertion point of the *cob* intron. The insertion of the intron is marked "/". 1 Strains *EF1, EF4, NCYC 132, NCYC 365* (without intron); 2 Strain *50* (Leupold-collection, with intron); 3 Strain *UCD-Fst 65-116* (with intron). The nucleotide differences are marked in bold print

quisition of the gene during evolution. The striking clustering of sequence differences around the intron insertion point could be the result of the co-conversion mechanism.

Intron-Encoded Proteins are Multifunctional

Both intron proteins, the *cox1I1* and the *cob* intron protein have more than one function. In previous papers we have shown that mutations in the open reading frame of *cox1I1* lead to a splicing deficiency (Schäfer et al. 1994). Furthermore, this intron-encoded protein is able to stimulate homologous recombination in *Escherichia coli* (Manna et al. 1991). Thirdly, it has the function of an endonuclease. In consequence, this protein is able to interact with DNA and RNA. All other known mobile introns encode either a maturase or an endonuclease. This is the first example of an intron encoding both functions.

Introns on the Move to the Nucleus

The picture of passive intron elements must be discarded in favour of a concept that mobile introns not only go home but also have the ability to transpose to non-allelic locations. There is a considerable body of evidence from other organisms than yeast that mitochondrial sequences can enter the nucleus and integrate in the chromosomes. There is evidence for a copy of (parts of) *cox1* intron 2 in the nuclear genome of fission yeast upstream of the *pho2* gene (S. Pietrikovski, personal communication). We have investigated this in a *rho*^{o} mutant of *Schizosaccharomyces pombe* to avoid possible contamination with mitochondrial DNA. A positive hybridization signal was observed, indicating that this intron is present. Sequence analysis will tell whether the entire intron and possibly also parts of the exons are integrated. This intron is of special interest, since it has been shown by Lang (1984) that there is evidence for horizontal transfer of this intron between *Aspergillus nidulans* and *Schizosaccharomyces pombe.*

Acknowledgements. We thank Mrs. A. Schreer for skilful technical assistence. Work was supported by the Deutsche Forschungsgemeinschaft

References

Lambowitz AM, Belfort M (1993) Introns as mobile genetic elements. Ann Rev Biochem 62:587-622

Lang BF (1984) The mitochondrial genome of the fission yeast *Schizosaccharomyces* pombe: highly homologous introns are inserted at the same position of the otherwise less conserved *cox1* genes in *Schizosaccharomyces pombe* and *Aspergillus nidulans.* EMBO J 3:2129-2136

Lazowska J, Meunier B, Macadre C (1994) Homing of a group II intron in yeast mitochondrial DNA by unidirectional co-conversion of upstream-located markers. EMBO J 13:4963-4972

Manna F, Massardo DR, Del Giudice L, Buonocore A, Nappo AG, Alifano P, Schäfer B, Wolf K (1991) The mitochondrial genome of *Schizosaccharomyces pombe*. Stimulation of intra-chromosomal recombination in *Escherichia coli* by the gene product of the first *cox1* intron. Curr Genet 19:295-299

Massardo DR, Manna F, Schäfer B, Wolf K, Del Giudice L (1994) Complete absence of mitochondrial DNA in the petite-negative yeast *Schizosaccharomyces pombe* leads to resistance towards the alkaloid lycorine. Curr Genet 25:80-83

Michel F, Lang B (1985) Mitochondrial class II introns encode proteins related to the reverse transcriptases of retroviruses. Nature 316:641-643

Moran JV, Zimmerly S, Eskes R, Kennell JC, Lambowitz AM, Butow RA, Perlman PS (1995) Mobile group II introns of yeast mitochondrial DNA are novel site-specific retroelements. Mol Cell Biol 15:2828-2838

Paschke T, Meurer P, Schäfer B, Wolf K (1993) Homing of mitochondrial introns in the fission yeast *Schizosaccharomyces pombe*. Endocytobiosis Cell Res 10:205-213

Schäfer B, Merlos-Lange A, Anderl C, Welser F, Zimmer M, Wolf K (1991) The mitochondrial genome of fission yeast. Inability of all introns to splice autocatalytically, construction and characterization of an intronless genome. Mol Gen Genet 225:158-167

Schäfer B, Wilde B, Massardo DR, Manna F, Del Giudice L, Wolf K (1994) A mitochondrial group I intron in fission yeast encodes a maturase and is mobile in crosses. Curr Genet 25:336-341

Schmidt WM, Schweyen RJ, Wolf K, Mueller MW (1994) Transposable group II introns in fission and budding yeast. Site-specific genomic instabilites and formation of group II IVS plDNAs. J Mol Biol 243:157-166

Szostak J, Orr-Weaver TL, Rothstein RJ, Stahl FW (1983) The double strand break repair model for recombination . Cell 33:25-35

Wolf K (1994) Mitochondrial introns in yeast - mobile genetic elements. Endocytobiosis Cell Res 10:55-63

Gene Transfer from the Zygomycete *Parasitella parasitica* to its Hosts: An Evolutionary Link Between Sex and Parasitism?

J. Wöstemeyer, A. Wöstemeyer, A. Burmester, and K. Czempinski
Lehrstuhl für Allgemeine Mikrobiologie und Mikrobengenetik, Friedrich-Schiller-Universität Jena, Neugasse 24, 07743 Jena, Germany

Key words: Parasitism, horizontal gene transfer, fungi, Zygomycetes.

Summary: *Parasitella parasitica* (synonyms: *P. parasitica* and *Mucor parasiticus*) is a heterothallic facultative mycoparasite which belongs phylogenetically to the *Mucor* group within the zygomycetes. It is a facultative parasite of several but not all mucoraceous fungi. The interactions between host and parasite present several peculiarities. (1) Infection of the host organism *Absidia glauca* especially is strongly mating type-specific. (+) strains of *P. parasitica* will infect exclusively (-) strains of *A. glauca*, whereas (-) strains of the parasite are required for the infection of (+) type hosts. We propose the hypothesis that host/parasite recognition is mediated by the same gamone system as is required for communication between sexual partners within mucorales (trisporic acid). (2) *P. parasitica* is a fusion biotroph. Infection is accompanied by a limited cytoplasmic continuum between the two partners, which allows invasion of the host by nuclei of the parasite. *P. parasitica* is not very aggressive. The host, *A. glauca*, survives the infection well and forms normal sporangiospores even close to infection sites. We were able to demonstrate by genetic and molecular techniques that genetic markers of the parasite are transferred into the host's nuclei at high rates. *Absidia* and *Parasitella* form a naturally occurring parasexual system with high efficiency. (3) Sexually compatible pairs of *P. parasitica* are not able to form zygospores. In order to reproduce sexually at least one of the partners has to grow parasitically. Measurements of trisporic acid levels in cocultures of *P. parasitica* with *A. glauca* led us to assume that interaction with a host organism is required for the cooperative synthesis of trisporic acid at a level high enough to induce sexual differentiation.

Introduction

Our studies on the transfer of genetic material associated with parasitic interactions were prompted by a stimulating publication dating back to the early 1920s, when the German botanist Hans Burgeff (1924) analyzed the sexual and parasitic interactions in *Mucor*-like fungi at the microscopic level. Among many other

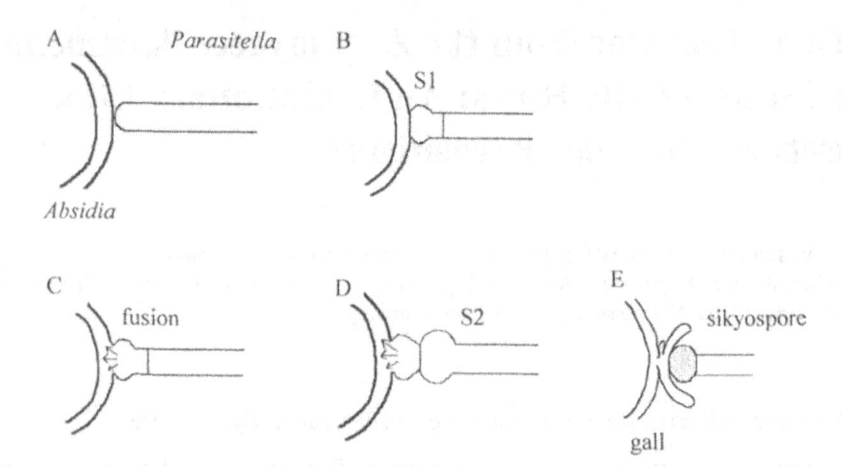

Fig. 1 A-E. Infection pathway of *P. simplex*. *P. simplex* grows towards the host **A** and forms the primary sikyotic cell (S1) and a septum **B**. The primary sikyotic cell fuses with the host **C**. A secondary sikyotic cell is formed **D**, which is surrounded by a gall from the host **E**. Finally *P. simplex* forms a sikyospore **E**

important observations he reported that the chemical principle inducing sexual development in these fungi seemed to be universal: a (+) mating type of a given species will induce early sexual reactions, i.e. zygophore formation, in many different species only if they belong to the (-) mating type. Burgeff also described the interaction between the facultative parasite *P. parasitica* and one of its possible host organisms, *A. glauca*. The most important observation was that the cell walls of both partners are dissolved in the contact zone. Cytoplasm and cell organelles from the parasites, including nuclei, invade the host's mycelium. Today we would classify mycoparasites of this kind as fusion biotrophs (Jeffries and Young 1994). In most instances the fusion zones between host and parasite are quite small, more in the range of plasmodesmata, and certainly not suitable for the transfer of organelles. In this respect the larger fusion areas produced by *P. parasitica*, the parasitic *Absidia* species, *A. parricida*, and another zygomycete, *Chaetocladium brefeldi*, are exceptions.

Burgeff's report stimulated us to look for permanent transfer of genetic material. Looking at the morphology of the infection (Fig. 1), a putative gene transfer is expected to be unidirectional. To begin with, both partners have the typical unseptate mycelium of mucoraceous fungi. After contact formation, *P. parasitica* forms a septum close to the infection site. Thus, it is highly unlikely that nuclei from the host mycelium will be transported into the parasite; transport in the opposite direction is much more likely. Another prerequisite for the formation of genetic recombinants between host and parasite is also met in the *Parasitella/Absidia* system: *A. glauca* forms sporangia with a normal amount of uninucleate spores even after infection. Thus, it may be expected that interspecific recombinant nuclei will appear in the subsequent vegetative spore generation and therefore will be amenable to analysis.

Once the prerequisites for a gene transfer are fulfilled and recognized, the design of appropriate experiments is easy. Cocultures of parasitically compatible (+) and (-) strains of host and parasite must be grown under conditions which allow the efficient formation of infection structures and sporulation of the host. In addition, dominant genetic markers that can be selected for in recombinants should be available in the donor, *P. parasitica*. This latter condition is fulfilled by using stable auxotrophic mutants of the recipient, *A. glauca*. Presumably most of the corresponding genes in a prototrophic donor will work in the recipient too, and will therefore be easily selectable. In such experiments, the donor organism will also sporulate. Therefore, it is necessary to distinguish between the two organisms as early during development as possible. *Absidia* and *Parasitella* can be distinguished on Petri dishes by morphological criteria. It is also possible to select for *Absidia* offspring by adding low levels of neomycin. *P. parasitica* is much more sensitive to this drug than *A. glauca* and will die preferentially.

In this paper we will summarize the key experiments that prove the genetic transfer from *P. parasitica* to its host *A. glauca*. We will also report on the progress that has been made towards checking the hypothesis that sexua lity and parasitism are linked at the level of partner recognition via the trisporic acid system.

Material and Methods

Most of our experiments were performed with the *Parasitella parasitica* strain CBS 412.66 (+) as a donor strain and several auxotrophic derivatives of *Absidia glauca* 101.48 (-) (Wöstemeyer and Brockhausen-Rohdemann 1987; Kellner et al. 1993; Wöstemeyer et al. 1990). For inoculum production all zygomycetes were grown on solid medium containing 15 g agar and 30 g malt extract per liter. The highest rates of infection structure formation were obtained on media containing maltose as carbon source (Burmester and Wöstemeyer 1994); glucose tends to suppress parasitic growth.

DNA was isolated essentially as described (Burmester and Wöstemeyer 1986) by equilibrium centrifugation of PEG precipitated DNA in ethidium bromide containing CsCl gradients. Isolation of mRNA (Vetter et al. 1994), cloning work (Czempinski et al. 1996), and transformation of *P. parasitica* (Burmester 1992) were performed as described previously.

Results

Gene Transfer from *P. parasitica* to *A. glauca*

We infected auxotrophic mutants of *A. glauca* with compatible wild-type *P. parasitica* and looked in the subsequent vegetative spore progeny for prototrophic *A. glauca* offspring. The percentage of prototrophic spores was in the range between 0.1% and 1.5% depending on the experiment. A mean value from nine independent experiments with different auxotrophy markers was 0.42%. The reversion rates of the recipient's mutants were always $< 10^{-8}$; for most mutants we saw never any revertants. This means in the worst case, that as little as 0.1% prototrophs is still at least 10^4 times more than could be expected on the basis of a reversion frequency of 10^8. The appearance of prototrophic *A. glauca* depends on coculturing compatible strains of host and parasite, which allow the formation of the typical infection structures (Fig. 1). Combinations of *A. glauca* and *P. parasitica* belonging to the same mating type will not develop infection structures, and indeed we did not observe prototrophic spores in such cases. Therefore, we conclude that the appearance of prototrophic derivatives is the consequence of a gene transfer from the parasite to its host, coupled with successful infection in the aerial mycelia of these fungi. We have christened these interspecies chimeras "pararecombinants," in an allusion to both parasexuality and parasitism.

In addition to the genetic line of evidence for this novel parasexual system, we can also prove the transfer of genes physically at the level of molecular biology by a simple model experiment. *P. parasitica* can be transformed with a plasmid containing the bacterial, Tn5 derived neomycin resistance gene under the control of a promoter from *A. glauca* originally belonging to the gene for the translational elongation factor EF1alpha (TEF). This plasmid propagates as an autonomously replicating plasmid and confers resistance against aminoglycoside antibiotics in both organisms (Burmester 1995). If a transformed *P. parasitica* strain is used for the infection, neomycin-resistant *A. glauca* derivatives can be found with reasonable frequency among the vegetative spore progeny. Southern blot analysis allows the neomycin resistance gene to be found in these derivatives (Kellner et al. 1993).

Stability of Genes Acquired via Infection

The efficiency of the *Parasitella/Absidia* parasexual system is quite high and reaches the percentage range among the progeny spores in some experiments. Another important question, especially with regard to the evolutionary impact of the system, is the mitotic stability of pararecombinants. In coenocytical mycelial fungi, stability rates cannot be measured as easily as in unicellular organisms like bakers' yeast. A reasonable method is to start a Petri dish culture with a single uninucleate spore and to wait until this culture has sporulated again. The progeny will normally be approximately 10^9 spores, which are all descendants of the primary nucleus. Measurements of stability in many different experiments through four or five subsequent sporulation cycles show that the amount of prototrophs normally ranges between 10 and 80%. Completely stable derivatives, 100% prototrophs, do occur, but are more the exception than the rule. Of course, these latter events cannot be distinguished from simple reversions. If we assume a division mode following an exponential model, a multiplication factor of 10^9 corresponds to 30 subsequent division cycles. A frequency of 10% prototrophs after such a growth cycle means that a fraction of 0.9 (90%) has lost the acquired genetic trait; this corresponds to 3% (0.9/30) per mitotic division (Kellner et al. 1993; Wöstemeyer et al. 1995). This is certainly rather unstable compared with the stability of normal chromosomal genetic markers, but it is reasonably stable if we compared with, for example, loss rates observed for derivatives of 2 µm yeast plasmids.

Communication Between Host and Parasite

The infection of *A. glauca* by *P. parasitica* depends on the mating type of the partners. Complementary combinations of (+) and (-) mating types are required for successful infection. This observation, which dates back to Burgeff (1924), and which has been supported by experiments with all *Parasitella* isolates available, has prompted us to speculate that the sexual gamone of *Mucor*-like fungi, trisporic acid, is also responsible for the successful interaction between host and parasite. According to a widely accepted model (van den Ende 1978; Werkman 1976), trisporic acid is synthesized from ß-carotene by the cooperative action of both mating types. A (+)-specific intermediate, 4-dihydromethyltrisporic acid, is passed on to the (-) mating type, which will convert it to trisporic acid. The (-) mating type produces a different intermediate, trisporin, which can only be converted to trisporic acid in the (+) mating type. Trisporic acid will then induce the first visible steps in sexual development in both mating types.

There are analogies between the sexual pathway and the parasitic interactions between *A. glauca* and *P. parasitica*. Both differentiation programs require the presence of complementary mating types, both lead to the production of a spore structure: zygospores in the case of sexual development, sikyospores as the final stage of the infection pathway. There is of course a fundamental difference: Whereas the zygospore contains genomes from both partners, the sikyospore is formed exclusively from the parasite's material. What happens if we look for the synthesis of trisporic acid in interspecies combinations of host and parasite? Whereas cocultures of the *P. parasitica* mating types produce only trace amounts of trisporic acid, complementary combinations of *P. parasitica* and *A. glauca* give rise to trisporic acid levels which are in the range of a mating type pair of *A. glauca*. This phenomenon, the deficiency of *P. parasitica* in producing trisporic acid and the ability of *A. glauca* to complement it, is presumably the biochemical reason for another interesting observation: *P. parasitica* will only be able to form zygospores if at least one of the partners grows parasitically. *P. parasitica* depends completely on a host organism for its sexual developmental pathway.

In order to prove the role of the trisporic acid system for both sexuality and parasitism, we have cloned and sequenced the gene for a key enzyme for the biosynthesis of trisporic acid (Czempinski et al. 1996). Experiments to construct gene disruption mutants at this locus have been started. The analysis of such mutants with respect to zygospore formation and the development of infection structures will answer the question.

Discussion

One important question with respect to the evolutionary impact of the parasitic gene transfer system *Parasitella parasitica/Absidia glauca* is: How important is this parasexual system for evolutionary processes? With our present knowledge this question cannot be answered: it is only possible to list the pros and contras. The gene transfer rates are very high. In addition, the host range of *P. parasitica* is quite broad: without aiming at completeness, we were able to demonstrate the transfer of a plasmid conferring resistance against neomycin to very different species within the infection spectrum, e.g., the biotechnically important heterothallic zygomycete *Rhizopus nigricans*. Even *Blakeslea trispora* (Choanephoraceae) is infected efficiently. However, our genetic data show that at least the biosynthesis genes that we have selected for in mutant recipients tend to be mitotically unstable. We do not know the reasons for this unexpected behavior. Perhaps *A. glauca* has developed mechanisms that select against the activity of duplicated genes such as RIP in *Neurospora crassa* or MIP in *Asobolus immersus* (Selker 1990). Perhaps most of the genetic material that enters the recipient's nuclei is propagated primarily in an extrachromosomal, autonomously replicating stage. This would explain the apparent instability that we have observed for prototrophy markers. We know that *A. glauca* wild types harbor an enormous amount of different circular plasmid DNAs, at least one of which codes for a protein product (Hänfler et al. 1992). In contrast to other filamentous fungi, zygomycetes have fewer difficulties in propagating extrachromosomal DNA. Should this be true for the DNA acquired via the parasitic pathway, permanent acquisition of a foreign piece of DNA would probably require integration into the chromosomal DNA or into one of the residing stable plasmids. Presently we have no idea how frequent such events will be. However, evolution does not need frequent events. It is sufficient for a newly acquired DNA fragment to be established permanently in the recipient. The genomes of zygomycetes have an extraordinarily high amount of repetitive DNA elements (at least 25% for *A. glauca*) (Wöstemeyer and Burmester 1986), many of which are interspersed throughout the genome. By intergeneric protoplast fusion between *P. parasitica* and *A. glauca* we showed that DNA elements, which have a low copy number in *P. parasitica*, may multiply enormously in the recombinant nuclei of fusion hybrids (A. Wöstemeyer, unpublished results). Such elements seem to be highly recombinogenic and have a tendency to spread within genomes. We can expect that the occasional introduction of such elements into a given genome, e.g., via the parasitic gene transfer system, leads to an increase in rearrangements. This may be interpreted in favor of the importance of pararecombinant formation at an evolutionary scale as well. Even if no immediate dramatic effects are observed, the mere acquisition of additional genetic material opens the possibility for evolutionary experiments.

A second important aspect of this novel interspecific gene transfer system is: Are the allusive relationships between sexuality and parasitism in *P. parasitica* and its hosts a rare exception in nature, or is there an ancient general concept that links sexuality to parasitism? With our present knowledge it is not possible to answer these questions conclusively. Some of the difficulties arise from uncertainties in the definition of sexuality. If we reserve 'sexuality' for the advanced method of reproduction of highly evolved eukaryotes, coupled to meiotic processes, then we would have to go back to very early processes in evolution until we find relationships. Speculations of this type are not within the scope of this paper. If, however, we accept interpreting "sexuality" essentially as gene transfer and the formation of genetic recombinants, then we can simply ask whether there are sexual or para-sexual systems other than *P. parasitica/A. glauca* which are associated with the parasitic lifestyle (Wöstemeyer et al. 1996). Among fungi, we know many different fusion biotrophs (Jeffries and Young 1994). It might be worth looking at those interactions for possible organelle - ideally nucleus - transfer. To our knowledge, the transfer of genes has been proven only in our laboratory for *P. parasitica* and *Absidia parricida* (A. Wöstemeyer, unpublished observations). *Chaetocladium brefeldi* is another good candidate, but experiments in this direction have not been performed.

Are there comparable systems in other eukaryotes? We are aware of only one interaction between a parasite and its host that includes the transfer of nuclei. Lynda Goff has described the complex interactions between the obligatory parasitic red alga *Choreolax* and its host *Polysiphonia*, another red alga, very carefully by light and electron microscopic techniques (Goff and Coleman 1984). During the infection process fusion bridges are formed at the interphase between the two organisms, allowing the transport of nuclei. According to this analysis the foreign nuclei are active in the foreign genetic background as well and might be involved in metabolic processes. However, it could not be analyzed whether genes from the parasite are established permanently in the host organism. This would require axenic cultures and the availability of appropriate mutants in the recipient or of dominantly selectable markers. What has been described for this interesting algal system, the physiological activity of imported nuclei, is lacking for *P. parasitica/A. glauca*. We know that occasionally genetic material of the parasite is added to the chromosomal complement of the host, but we have no idea if *Parasitella*'s nuclei fulfill a role in formation of infection structures after their transfer. It is of course tempting to speculate that such nuclei are involved in determining the gall-like structures formed at the host's site shortly after contact formation (Fig. 1), but rigorous experimental testing of such an assumption is presently beyond the technical possibilities of this system.

Acknowledgements. The experiments for the natural gene transfer project were supported by the Bundesministerium für Forschung und Technologie (genetic aspects), Deutsche Forschungsgemeinschaft (cloning of the gene for dihydromethyl-trisporate-dehydrogenase), and Fonds der Chemischen Industrie (TA effects). We especially thank Prof. Herman van den Ende (Amsterdam), who taught us to work with the TA system.

References

Burgeff H (1924) Untersuchungen über Sexualität und Parasitismus bei Mucorineen. I. Bot Abh 4: 1-13

Burmester A (1992) Transformation of the mycoparasite *Parasitella simplex* to neomycin resistance. Curr Genet 21: 121-124

Burmester A (1995) Analysis of the gene for the elongation factor 1 alpha from the zygomycete *Absidia glauca*. Use of the promoter region for constructions of transformation vectors. Microbiol Res 150: 63-70

Burmester A, Wöstemeyer J (1986) Cloned mitochondrial DNA from the zygomycete *Absidia glauca* promotes autonomous replication in *Saccharomyces cerevisiae*. Curr Genet 10: 435-441

Burmester A, Wöstemeyer J (1994) Variability in genome organization of the zygomycete *Parasitella parasitica*. Curr Genet 26: 456-460

Czempinski K, Kruft V, Wöstemeyer J, Burmester A (1996) Purification of 4-dihydromethyl-trisporate dehydrogenase from *Mucor mucedo*, an enzyme of the sexual hormone pathway, and cloning of the corresponding gene. Microbiology 142: 2647-2654

Goff LJ, Coleman AW (1984) Transfer of nuclei from a parasite to its host. Proc Natl Acad Sci USA 81: 5420-5424

Hänfler J, Teepe H, Weigel C, Kruft V, Lurz R, Wöstemeyer J (1992) Circular extrachromosomal DNA codes for a surface protein in the (+) mating type of the zygomycete *Absidia glauca*. Curr Genet 22: 319-325

Jeffries P, Young TWK (1994) Interfungal parasitic relationships. CAB Int., Wallingford.

Kellner M, Burmester A, Wöstemeyer A, Wöstemeyer J (1993) Transfer of genetic information from the mycoparasite *Parasitella parasitica* to its host *Absidia glauca*. Curr Genet 23: 334-337

Selker EU (1990) Premeiotic instability of repeated sequences in *Neurospora crassa*. Ann Rev Genet 24: 579-613

Vetter M, Wöstemeyer J, Burmester A (1994) Characterization of specific cDNA clones of the zygomycete *Parasitella parasitica*, derived from mRNAs which are regulated by the pheromone trisporic acid. Microbiol Res 149: 17-22

Wöstemeyer A, Teepe H, Wöstemeyer J (1990) Genetic interactions in somatic intermating type hybrids of the zygomycete *Absidia glauca*. Curr Genet 17: 163-168

Wöstemeyer J, Brockhausen-Rohdemann E (1987) Inter-mating type protoplast fusion in the zygomycete *Absidia glauca*. Curr Genet 12: 435-441

Wöstemeyer J, Burmester A (1986 Structural organization of the genome of the zygomycete *Absidia glauca*: evidence for high repetitive DNA content. Curr Genet 10: 903-907.

Wöstemeyer J, Wöstemeyer A, Burmester A, Czempinski K (1995) Relationships between sexual processes and parasitic interactions in the host-pathogen system *Absidia glauca-Parasitella parasitica*. Can J Bot 73: Suppl 1, S243-S250

Wöstemeyer J, Wöstemeyer A, Voigt K (1996) Horizontal gene transfer in the rhizosphere: a curiosity or a driving force in evolution? Adv Bot Res 24: 399-429

Trans-Kingdom Conjugation as a Model for Gene Transfer From Endosymbionts to Nucleus During the Origin of Exogenosomal Organelles

K. Yoshida, K. Kamiji, A. Mahmood, T. Sekito and H. Ishitomi
Department of Biological Science, Faculty of Science, Hiroshima University, Higashi-Hiroshima 739, Japan

Key words: Bacterial sex, conjugation hypothesis, endosymbiosis, evolution, exogenosome, kinetic model, organelle origin, pressure of vectorial gene transfer, retroconjugation, sexduction, subunit compartmentation.

Summary: The described new conjugation hypotheses easily explain the pressure and mechanism of vectorial gene transfer from organelles' ancestors to the eukaryotic nucleus during the origin and evolution of mitochondria and chloroplasts. A kinetic model of exogenosomal organelle origin and evolution is also presented.

Introduction

Eukaryotic organelles of mitochondria and chloroplasts have a closer kinship with free-living bacteria than other organelles since they have their own DNA and bacteria-like structure. These characteristics have led to the hypothesis of an endosymbiont origin for mitochondria and chloroplasts (Gray 1992, Margulis and Sagan 1986, Yoshida 1995). The hypothesis postulates that both aerobic bacteria and cyanobacteria invaded ancestral eukaryotic cells and the resultant endosymbionts evolved into mitochondria and chloroplasts, respectively. This is supported by the ubiquity of endosymbionts in nature (Margulis 1993), the transplantable characteristics of isolated mitochondria (Gunge and Sakaguchi 1979, Yoshida 1979, Yoshida and Takeuchi 1980), and the usage of mitochondrial codon (Osawa et al. 1992) and molecular phylogenetic data based on ribosomal small subunit RNA (De Soete 1983, Sogin 1992). However, the genome sizes of present mitochondria and chloroplasts are as much as 2 to 10 % of typical bacterial genomes. This indicates that the prokaryotic bacterial endocytobionts transferred almost all their genes into the host nuclear genome in a unidirectional fashion during their evolution of organelles in the eukaryotic cell. Unfortunately, there is no clear explanation of the one-way (vectorial) transfer mechanism.

On the other hand, the sexual processes of prokaryotic bacteria, particularly conjugation, are believed to have evolved more than 3 billion years ago (Halvorson and Monroy 1985, King and Stansfield 1990, Margulis and Sagan 1986). Unfortunately, the conjugation has never been seriously considered with respect to the origin and evolution of exogenosomal organelles or eukaryotic cells (Dyer and Obar 1994, Gray 1992, Halvorson and Monroy 1985, Margulis and Sagan 1986, Michod and Levin 1988, Yoshida 1995).

In this paper, we describe novel conjugation hypotheses which explain how and why prokaryotic symbionts transferred almost all their genes into the host nuclear genome during the evolutionary process.

Hypothesis and Discussion

In bacteria, there are three gene transfer systems; conjugation, transformation, and transduction. In this paper, we primarily focus on conjugation since the latter two probably played a less important role in the origin and evolution of exogenosomes (see below). Like sexual reproduction in higher organisms, bacterial conjugation is believed to be confined within a single species. Recently, there have been several reports on promiscuous conjugation via a wide host range of plasmids beyond the species barrier in bacteria (Hirsh 1990). More recently, trans-kingdom conjugation between the bacterium *Escherichia coli* and the eukaryotic yeast *Saccharomyces cerevisiae* has been reported (Heineman and Sprague 1989, Nishikawa et al. 1990). *Agrobacterium*-mediated conjugation-like transfer of plasmid DNA to plants has also been known to occur in nature (Zambryski et al. 1989) These findings encouraged us to formulate new conjugation hypotheses for the origin of exogenosomal organelles which are primarily based on the phenomena of bacterial conjugation and trans-kingdom conjugation. Before introducing the hypotheses, we quickly review salient features of bacterial conjugation and trans-kingdom conjugation as an aid to understanding the hypotheses.

Bacterial Conjugation

Conventional bacterial sex is generally determined by sexual or conjugative plasmids. Usually, *oriT* (origin of conjugal transfer), *mob* (mobilization), and *tra* (transfer) genes on the plasmid are indispensable for conjugation. The most extensively studied conjugative plasmid is the F plasmid, which is endowed with the typical conjugative genes of *oriT*, *mob*, and *tra* (Frost et al. 1994, Willetts and Skurray 1987, Willkins and Lanka 1993). The latter two genes are *trans*-acting genes, so those genes can be functional even when on a separate helper plasmid. The bacterial cell which harbors the F plasmid becomes a male bacterial cell, while the cell lacking the F plasmid is designated a female. The mixture of male and female bacterial cells causes the conjugation: male cells fix female cells by sexual pili coded by *tra* genes and subsequently the conjugation tube between male and female cells is formed by the action of *tra* genes. The specific nuclease coded by the *mob* gene nicks one strand of the F plasmids at the *oriT* site. From this site, the transfer of single stranded plasmid DNA into the recipient occurs in the 5' to 3' direction. The synthesis of complementary DNA of the transferred strand and the retained strand DNA starts simultaneously. Finally, the complete plasmid is formed both in male and female cells. Thus, the transconjugant becomes a male if it receives a full copy of the F plasmid. In *E. coli* Hfr strain in which the F plasmid is reversibly integrated in the bacterial chromosome and excised via insertion sequences, the integrated F can mobilize the whole bacterial chromosome, although the mobilization is easily interrupted by shearing force (see Margulis and Sagan 1986, Willets and Skurray 1987, Willkins and Lanka 1993 for details). As the result of inaccurate excision, i.e., illegitimate recombination, Hfr sometimes yields F' which carries an additional small amount of chromosomal DNA. F', can transfer the DNA to another female *E. coli* with very high frequency. This is called sexduction.

Trans-Kingdom Conjugation: Evidence and Characteristics

In the field of plant pathology, *Agrobacterium*-induced tumorigenesis has been well known since the 1950s. *Agrobacterium* can transfer T-DNA flanked with right (RB) and left border sequence (LB) on Ti plasmid to dicot plants. *Vir* (virulence) genes on Ti plasmid have some roles corresponding to those of *mob* and *tra*. After the transmission, the oncogenes on T-DNA are integrated into plant genomes and subsequently induce plant tumor. This conjugation-like process has been established as one of the gene introduction techniques in plants. Recently, the promiscuous conjugation between prokaryote and eukaryote, which is called trans-kingdom conjugation, has been reported in *E. coli* and yeasts including *S. cerevisiae* (Heineman and Sprague 1989, Nishikawa et al. 1990,1992, 1993, Nishikawa and Yoshida 1995), *S. kluyveri* (Inomata et al. 1994, Nishikawa et al. 1993), *Kluyveromyces lactis*, *Pichia angusta*, *Pachysolen tannophilus* (Hayman and Bolen 1993), and *Schizosaccharomyces pombe* (Sikorski et al. 1990), and in *Agrobacterium tumefaciens* and *S. cerevisiae* (Sawazaki et al. 1995).

Figure 1 shows schematically a plausible molecular mechanism of trans-kingdom conjugation between bacteria and yeasts. The conjugation mechanism is believed to be essentially the same as in bacterial conjugation. The trans-kingdom conjugative plasmids require at least *oriT*, *mob* and *tra* genes as bacterial conjugative plasmids. Since *mob* and *tra* are *trans*-acting genes, these two genes can be harbored by a separate helper plasmid as in bacterial conjugation. Our trans-kingdom conjugative pAY plasmids have *oriT*, *oriV* (origin of vegetative replication), *mob*, *ARS* (autonomous replication sequence of *S. cerevisiae*), and antibiotic resistance or nutrient requirement marker genes. The conjugative results by these plasmids upon their transfer are summarized in Table 1. This clearly indicates that no transconjugant appeared without conjugative plasmids and/or helper plasmids (cross nos. 2, 4, 5, 15, 16, 18, 19). The transmitted plasmids are usually present as independent plasmids in transconjugants, since the transconjugant's plasmids are easily curable under the nonselective conditions (cross nos. 3, 7, 13, 14, 17). However, at the low rate of about 10^{-7} , the plasmids are sometimes integrated into the host chromosomes (cross no. 1). *ARS*-less plasmids, which are incapable of replicating in *S. cerevisiae* yeast, are transiently expressed and form an abortive microcolony of transconjugants (see cross no.1) (Nishikawa et al. 1992). This re-

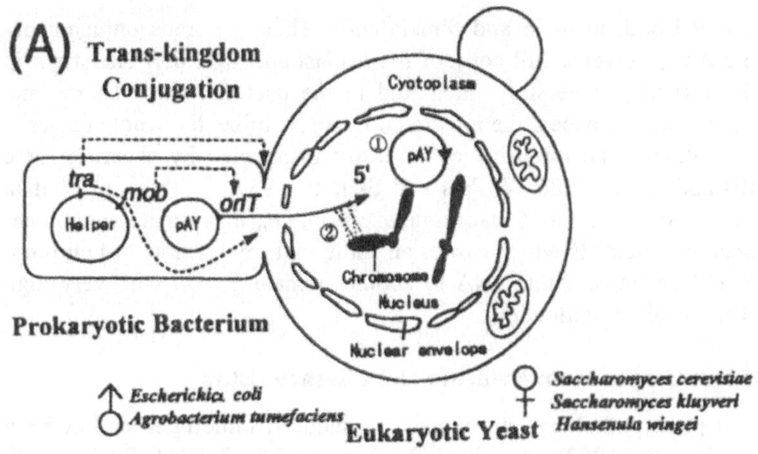
(A) Trans-kingdom
Conjugation
Cytoplasm
tra
mob
oriT
Helper
pAY
pAY
5'
Chromosome
Nucleus
Nuclear envelope
Prokaryotic Bacterium
Escherichia coli
Agrobacterium tumefaciens
Eukaryotic Yeast
Saccharomyces cerevisiae
Saccharomyces kluyveri
Hansenula wingei

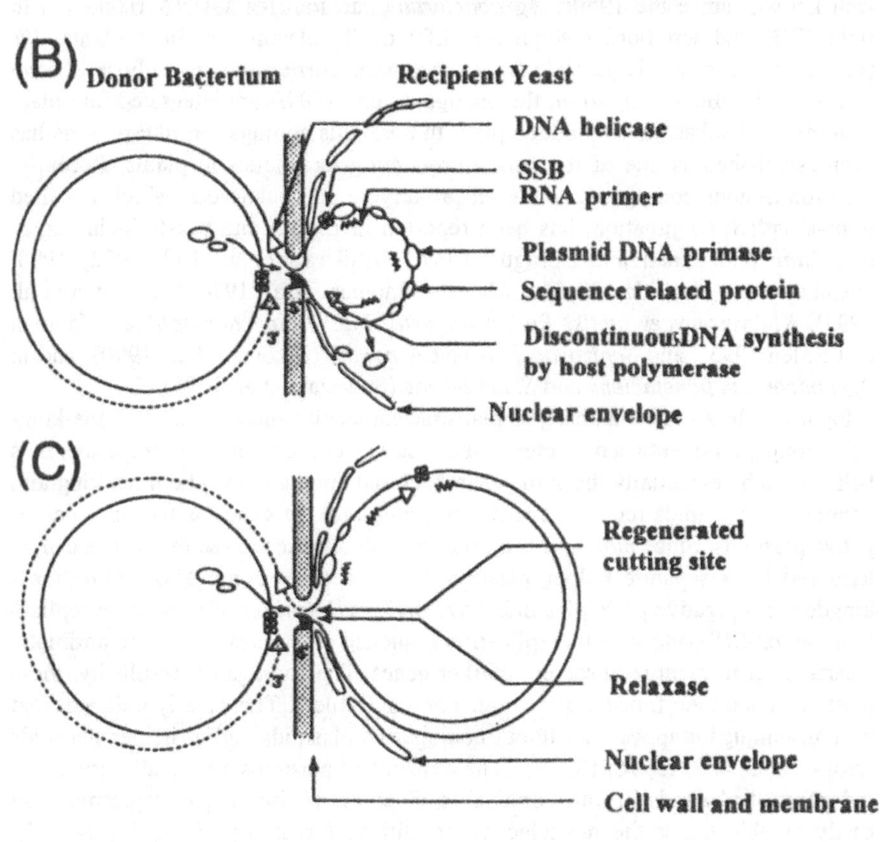
(B) Donor Bacterium
Recipient Yeast
DNA helicase
SSB
RNA primer
Plasmid DNA primase
Sequence related protein
Discontinuous DNA synthesis
by host polymerase
Nuclear envelope
(C)
Regenerated
cutting site
Relaxase
Nuclear envelope
Cell wall and membrane

Table 1. Trans-kingdom conjugation between bacteria and yeasts.[a]

No.	Crosses	Selection	Transconjugant numbers		
	Donor × Helper × Recipient		Curable	Uncurable	Abortive
***Escherichia coli* × *Saccharomyces cerevisiae* yeast**					
1	HB101(pAY201) × HB101(pRK2013) × YNN281	Ura	ND	0.7×10^{-7}	1.9×10^{-4}
2	HB101(pAY201) × YNN281	Ura	ND	ND	ND
3	HB101(pAY205) × HB101(pRK2013) ×YNN281	Ura	$2.0\times10^{-}$	NC	ND
4	HB101(pAY205) × YNN281	Ura	ND	ND	ND
5	HB101(---------) × HB101(pRK2013) × YNN281	Ura	ND	ND	ND
6	HB101(YRp7) × HB101(pRK2013) × YNN281	Ura	ND	ND	ND
7	HB101(pAY101, pAY215, pRH220) × YNN281	Trp	0.2×10^{-2}	NC	ND
		Leu	0.9×10^{-2}	NC	ND
		Trp Leu	0.2×10^{-2}	NC	ND
8	HB101(pAY-YAC-B) × HB101(pRH220) × YNN281	Ura	4.4×10^{-6}	NC	ND
9	HB101(pAY-YAC-E) × HB101(pRH220) × YNN281	Ura	2.3×10^{-6}	NC	ND
10	HB101(pAY-C22r, pRH220) × YNN281	Trp	1.4×10^{-3}	NC	NC
11	HB101(pAY-C22r, pRH220) × YNN281(pAY205)	Chy Km Tc	ND	ND	ND
12	YNN281	Ura	ND	ND	ND
***Escherichia coli* × *Saccharomyces kluyveri* yeast**					
13	HB101(pAY201) × HB101(pRK2013) × D1-15	Ura	1.0×10^{-7}	ND	ND
14	HB101(pAY205) × HB101(pRK2013) × D1-15	Ura	0.8×10^{-6}	ND	ND
15	HB101(pAY201) × D1-15	Ura	ND	ND	ND
16	HB101(pAY205) × D1-15	Ura	ND	ND	ND
***Agrobacterium tumefaciens* × *Saccharomyces cerevisiae* yeast**					
17	LBA1055(pAYSp-4) × HB101(pRK2013) × YNN281	Trp	0.7×10^{-6}	ND	ND
18	LBA1055(pAYSp-4) × HB101(------------) × YNN281	Trp	ND	ND	ND
19	LBA1055(------------) × HB101(pRK2013) × YNN281	Trp	ND	ND	ND
***Escherichia coli* × *Escherichia coli* (bacterial conjugation, control)**					
20	LE392(pAY101) × HB101(pRH210)	Ap Km	0.9	NC	ND

[a] Data from Inomata et al.(1994), Nishikawa et al.(1990,1992,1993) and Sawazaki et al. (1995). Transconjugant numbers are expressed by transconjugants/recipient. HB101 and LE392 are *E. coli* strains while LBA1055 is a *A. tumefaciens* strain. YNN281 and D1-15 are *S. cerevisiae* and *S. kluyveri* strains, respectively. Parentheses indicate the harboring plasmid; no parentheses and (---) both indicate absence of plasmid harboring. pAY series plasmids are trans-kingdom conjugative plasmids while pRH220 and pRK2013 are helper plasmids. „Selection“ indicates the first selection for transconjugants. Trp, Ura, Leu, Ap and Km indicate tryptophan, uracil, and leucine requirement, and ampicillin and kanamycin resistance, respectively. ND: Not detected, less than 10^{-9}; NC: not counted.

Fig. 1 A-C. Plausible molecular mechanism and summing up of trans-kingdom conjugation between bacteria and yeasts. **A** Trans-kingdom conjugation of prokaryote including *E. coli* and *A. tumefaciens* with eukaryotic yeasts including *S. cerevisiae*, *S. kluyveri* and *H. wingei*; **B** Plausible molecular mechanism of DNA replicative transfer near the start of conjugation; **C** that near the end of conjugation. **B** and **C** adapted from Wilkins and Lanka 1993

flects how transmission does not always depend on the plasmid's replicability in transconjugants. In other words, the *oriT*, *mob*, and *tra* system mobilizes any DNA following to *oriT* sequence. These conjugal events have been clearly verified by Southern hybridization and back-transformation (Nishikawa et al. 1990,1992,1993). The conjugation events between *E. coli* and *S. cerevisiae* in Table 1 (cross nos. 1-11) are schematically summarized in Fig. 1A (see also Nishikawa et al. 1992). Crosses 8 and 9 show the mobilization of YAC (yeast artificial chromosome) plasmids. Both pAY-YAC-B and pAY-YAC-E are present as a circular double-stranded plasmid in transconjugants, although pAY-YAC-B has *oriT* between telomeres (Mahmood et al. 1995). This fact indicates that the 5' terminus of transfer strand is not a free end but is bound with a nickase-like protein, as shown in Fig. 1B, C. An experimental design to gain further insight into whether a conjugative plasmid can move back from yeast to bacteria in the reverse direction during trans-kingdom conjugation. This possibility was also investigated; Table 1 shows the experimental result of trans-kingdom retroconjugation between *E. coli* and *S. cerevisiae*. If tetracycline- and kanamycin-resistant *E. coli* appears in cross no. 11, then the pAY205 plasmid of the yeast cell would be expected to be transferred into *E. coli* during trans-kingdom conjugation. However, the results clearly indicate that during trans-kingdom conjugation the transfer is always unidirectional, that is to say, the conjugative plasmid is transferred only from *E. coli* or *A. tumefaciens* to yeast cells, never in the reverse direction (see Yoshida 1995). pAY201 and pAY205, originally constructed for *S. cerevisia*, are successfully transferred into different species of yeast *S. kluyveri* by trans-kingdom conjugation and are quite functional in it (cross nos. 13, 14) (Inomata et al. 1994). Cross 17 shows successful conjugation between *S. cerevisiae* and *A. tumefaciens*, which is a relatively remote species from *E. coli* (Sawazaki et al. 1995).

Hypotheses: Gene Transfer Mechanism in the Origin of Exogenosomes

Our conjugation hypotheses on the gene transfer mechanism in the origin of organelles are based on the following assumptions: (1) the prokaryotic conjugation system was evolved well before the appearance of the eukaryotic cell; (2) trans-kingdom conjugation or sexduction may occur inside and outside a host cell. The former assumption has been generally accepted (Halvorson and Monroy 1985, King and Stansfield 1990, Margulis and Sagan 1986), while the latter assumption is introduced as a new hypothesis in this paper. However, circumstantial data including trans-kingdom conjugation described above support its possibility. Our two hypotheses (I and II) are schematically shown in Figs. 2 and 3, respectively. Hypothesis I postulates as follows. The bacterial candidates (ancestors) for exogenosomes transfer their genes into the nucleus of a host eukaryotic cell candidate (ancestor) (Fig. 2A) and the exogenosomal stage is subsequently established (Fig. 2B). The extra DNA copy is removed by self-recombination in the symbiotic bacteria as shown in Fig. 2C. After selection and stabilization, the symbionts finally

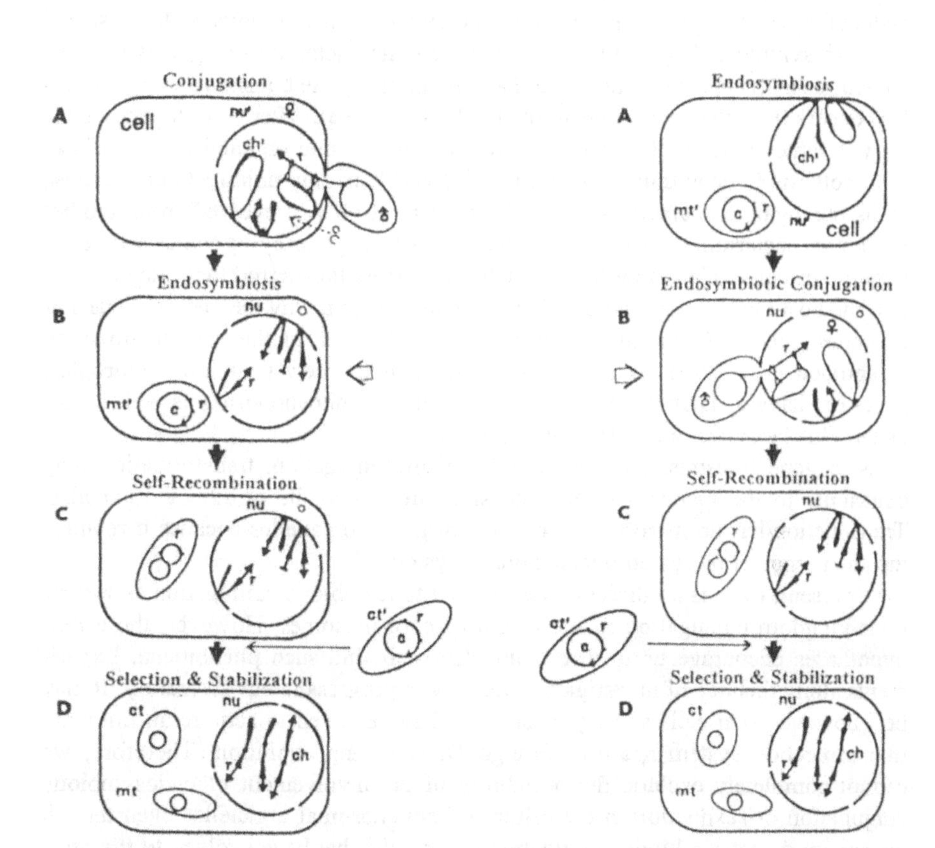

Fig. 2 A-D. Hypothesis I: Trans-kingdom conjugation involved in organelle origin. **A** DNA transfer into eukaryote ancestor by trans-kingdom conjugation together with transformation, resultant increase of genome size and development of nucleus and chromosome structures. **B** Establishment of endosymbiosis between the eukaryote ancestor and organelle ancestor. **C** Disintegration of duplicate genes in the symbiotic organelle ancestor by recombination. **D** Stabilization by selection and environmental stress. Ct, ct', mt, mt', nu, nu', ch, ch', T are chloroplast, chloroplast ancestor, mitochondrion, mitochondrion ancestor, nucleus, nucleus ancestor, chromosome, chromosome ancestor, and transfer region, respectively. The open arrow indicates an alternative position at which trans-kingdom conjugation may be involved

Fig. 3 A-D. Hypothesis II: Endosymbiotic conjugation involved in organelle origin. **A** establishment of endosymbiosis among organelle ancestral bacteria and eukaryote ancestor which increased the genome size and developed the nucleus and chromosomes. **B** Vectorial gene transmission by endocytobiotic conjugation or sexduction between symbiotic organelle ancestors and nucleus. **C** Gene integration into nuclear genome and disintegration by recombination. **D** Stabilization by selection and environmental stress. Labeling as in Fig. 2

evolved into organelles (Fig. 2D). On the other hand, hypothesis II postulates endocytbiotic conjugation as follows. The ancestor bacteria invade the host and establish symbiosis (Fig. 3A). The symbiotic male bacterium recognizes the host nucleus as a "female bacterium", so that conjugative gene transmission or sexduction occurs from the symbionts to the nucleus (Fig. 3B). After that, the extra gene copy is removed (Fig. 3C) and the symbionts evolve into organelles (Fig. 3D) as in hypothesis I. Gene transmission in both I and II is only unidirectional because of its conjugative characteristics. If the nucleus itself had evolved from another symbiotic bacterium, this would be more easy to explain by hypothesis II. Both hypotheses can explain how the organelles ancestors transferred their major genes into the host nuclear genome. Further, they also explain why the vectorial transfer of genes goes from organelles to nucleus and never in the reverse direction (Mahmood et al. 1995). This also indicates that the new symbiotic chloroplast ancestors have a capability to transfer the genes to mitochondrion as well as nucleus. This fact is favorable for our hypotheses.

As to gene transmission systems other than conjugation, transformation may contribute to the increase of genome size but less to the origin of organelles. Transduction has no important role in the origin of organelles because it requires the emergence of more complicated phage system.

At present, there is no direct evidence for endocytobiotic conjugation as well as trans-kingdom conjugation in the origin of exogenosomes. However, these new hypotheses encourage us to direct our efforts to find such phenomena. Experiments along this line of investigation are now in progress in our laboratory. It may be, however, that cells as they exist today have evolved to acquire an immune-like protection system against foreign DNA during evolution. Therefore, we cannot completely exclude the possibility of the involvement of endosymbiotic conjugation or sexduction in the origin of exogenosomal organelles even though we cannot detect symbiotic conjugation in the cells that have evolved to the present. Another approach to consolidate the hypothesis is to detect DNA sequence fossils, e.g., *oriT*, *mob*, and *tra* remaining in mitochondria and chloroplast DNAs, by Southern and computer analyses.

Hypothesis II easily explains the gene transfer mechanism as well as the pressure of vectorial gene transfer as described below, while hypothesis I does not require the assumption of endocytobiotic conjugation. However, it is highly probable that the processes of both hypotheses were involved in the origin and evolution of exogenosomal organelles.

Hypothesis on the Pressure of Gene Transfer During the Evolution of Organelles

Another important question is what causes exogenosomal organelles to transfer their genes to the nucleus. We speculate that by gene transfer these organelles have a better chance of selecting the best out of two or three gene products and evolve more quickly without causing mutation on their own haploid DNAs, as illustrated in Fig. 4. This mechanism by gene transfer and competitive mutational selection is very favorable especially for the evolution of functionally important

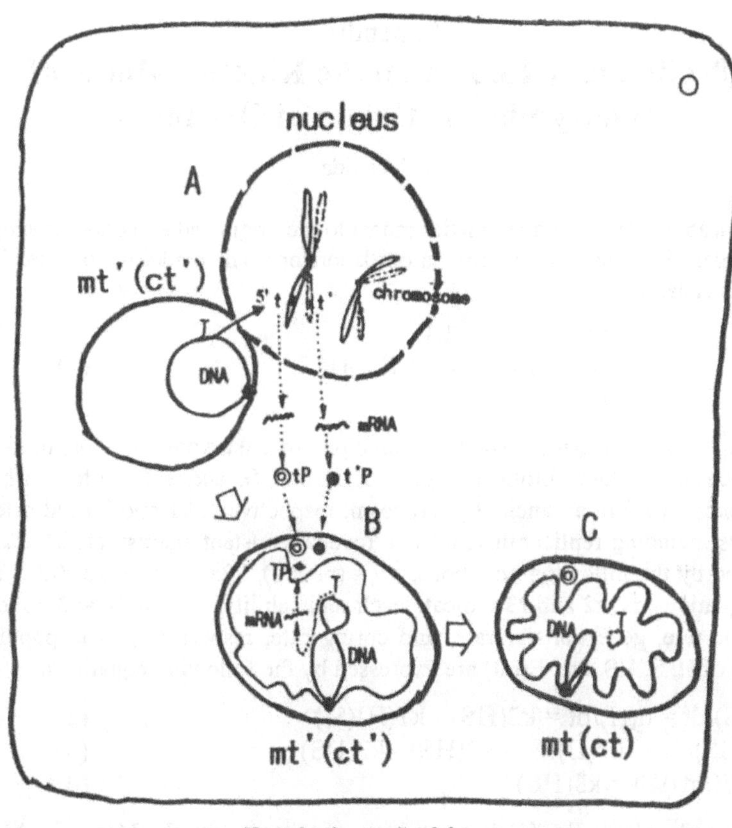

Fig. 4 A-C. Hypothesis III: Pressure of gene transfer during organelle evolution. **A** Conjugative transfer of organelle ancestral DNA into nuclear chromosomes on which mutation causes T to be t or t'. **B** Competition among T, t and t'. **C** Survival of t gene and loss of T gene. mt'(ct') and mt(ct) are mitochondrial (chloroplast) ancestors and mitochondrion (chloroplast) , respectively. TP, tP, and t'P are the gene products of genes T, t, and t', respectively. Dotted chromosomes are duplicative ones

proteins, because the organelles are believed to have lost their DNA repair system during evolution. The present compartmentation of ATPase subunits in organelles with their gene distribution may reflect this well (see, e.g., Lewin 1994, Yoshida 1995). For this advantage organelles may still retain the pressure of horizontal gene transfer even after loss of their conjugative characteristics and the emergence of eukaryotic sex (Yoshida 1995).

Acknowledgements: This study is supported in part by grants from the Electric Technology Research Foundation of Chougoku and the Ministry of Science, Culture and Education, Japan to K. Y.

Appendix: Preliminary Report on the Kinetics Model of Endosymbiotic Origin of Organelles

K. Yoshida

The application of the enzyme kinetics model to the origin and evolution of exogenosomal organelles is now in progress in our laboratory. The model is expressed by the following equation:

$$H + S \underset{k2}{\overset{k1}{\longleftrightarrow}} HS \overset{k3}{\longrightarrow} E(mt) \qquad (1)$$

Here, H, S, HS and E(mt) express populations of eukaryote ancestor, mitochondrial ancestor, endosymbiotic complex of H and S, and eukaryotes harboring mitochondria (mt) in an ancient microcosm, respectively. k1 and k3 indicate rate constants including replicability. k2 is a reverse constant against k1. k1 - k3 are expressed by the following equations. $k1 = (r1 \times a)$, $k2=(r2 \times b)$, $k3=(r3 \times c)$. In these equations, r1, r2 and r3 indicate each replicability while a, b, and c indicate symbiotic rate, gene transfer rate, and curing rate, respectively. The population changes of H, S, HS, and E(mt) are expressed by the following equations:

$$d(S)/dt = d(H)/dt = k2(HS) - k1(H)(S) \qquad (2)$$
$$d(HS)/dt = k1(H)(S) - k2(HS) - k3(HS) \qquad (3)$$
$$d(E(mt))/dt = k3(HS) \qquad (4)$$

When $r1= r2= r3= 1$, Eq. (1) is completely consistent with the Michaelis-Menten model. In this case, H acts as a catalyst to evolve mitochondria. The time course plot of Eqs. (2) and (3) is shown in Fig. A1. The abscissa indicates the population

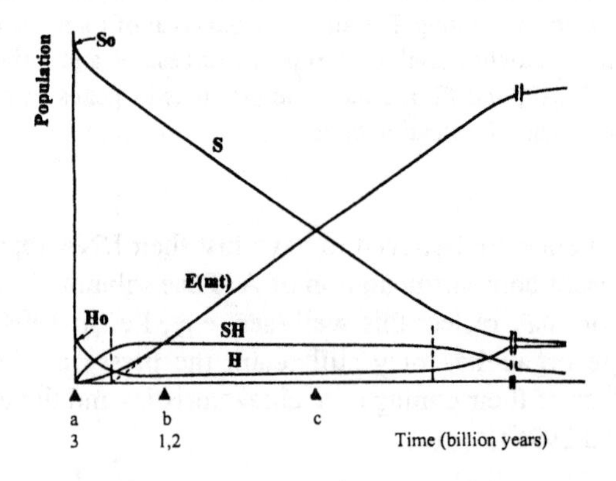

Fig. A1. Time course plot of Eq. (1). The times of the emergence of prokaryotic sex, mitochondrion, and adaptive radiation of eukaryotes are indicated by a, b, and c, respectively. Numbers are expressed in billion years ago

of H, S, HS and E(mt) during evolution time T of the ordinate. The area between the dashed lines indicates the stationary phase. After the stationary phase, the emergence of eukaryote and mitochondrion is established at the point of E(mt) over HS. Computer simulation and qualification of parameters a, b, c, r1, r2, and r3 are necessary for the improvement of Eqs. (1) to (4).

References

De Soete GA (1983) A least squares algorithm for fitting additive trees to proximity data. Psychometrika 48: 621-626

Dyer BD, Obar RA (1994) Tracing the history of eukaryotic cells. The Enigmatic Smile. Columbia University Press, New York

Frost LS, Ippen-Ihler K, Skurray RA (1994) Analysis of the sequence and gene products of the transfer region of the F sex factor. Microbiol Rev 58: 162-210

Gray MW (1992) The endosymbiont hypothesis revisited. Int Rev Cytol: 141: 233-357

Gunge N, Sakaguchi K (1979) Fusion of mitochondria with protoplasts in *Saccharomyces cerevisiae*. Mol Gen Genet: 170: 243-249

Halvorson HO, Monroy A (eds) (1985) The origins and evolution of sex. Alan R Liss, New York

Hayman GT, Bolen PL (1993) Movement of shuttle plasmids from *Escherichia coli* into yeasts other than *Saccharomyces cerevisiae* using trans-kingdom conjugation. Plasmid 30: 251-257

Heineman JA, Sprague GF Jr (1989) Bacterial conjugative plasmids mobilize DNA transfer between bacteria and yeast. Nature 340: 205-209

Hirsh PR (1990) Factors limiting gene transfer in bacteria. In: Fry JC, Martin JD (ed) Bacterial genetics in natural environments. Chapman and Hall, London, pp 31-40

Inomata K, Nishikawa M, Yoshida K (1994) The yeast *Saccharomyces kluyveri* as a recipient eukaryote in transkingdom conjugation: behavior of transmitted plasmids in transconjugants. J Bacteriol 176: 4770-4773

King RC, Stansfield WD (1990) "Sex". In: A dictionary of genetics, 4th edition, p 289, Oxford University Press, New York

Lewin B (1994) Genes V, p 738, Oxford University Press, Oxford

Lewin R (1984) No genome barriers to promiscuous DNA. Science 224: 970-971

Mahmood A, Kimura A, Takenaka M, Yoshida K (1995) The construction of mobilizable YAC plasmids and their behavior during trans-kingdom conjugation between bacteria and yeasts. Genet Anal 13: 25-32

Margulis L (1993) Symbiosis in cell evolution. WH Freeman, Boston

Margulis L and Sagan D (1986) Origin of Sex. Yale University Press, New Haven

Michod RE and Levin RE (eds) (1988) The evolution of sex. Sinauer Ass Inc, Sunderland

Nishikawa M, Inomata K, Sekito T, Suzuki K, Yoshida K(1993) Trans-kingdom conjugation as a symbiotic plasmid transfer between bacteria and yeasts. In: Sato S, Ishida M, Ishikawa H (eds) Endocytobiology V, pp 537-544, Tübingen University Press, Tübingen

Nishikawa M, Suzuki K, Yoshida K (1990) Structural and functional stability of IncP plasmids during stepwise transmission by trans-kingdom mating: promiscuous conjugation of *Escherichia coli* and *Saccharomyces cerevisiae*. Jpn J Genet 65: 323-334

Nishikawa M, Suzuki K, Yoshida K (1992) DNA integration into recipient yeast chromosomes by trans-kingdom conjugation between *Escherichia coli* and *Saccharomyces cerevisiae*. Curr Genet 21: 101-108

Nishikawa M, Yoshida K(1995) Trans-kingdom conjugation more frequently induces gene replacement than transformation does in *Saccharomyces cerevisiae* yeast.

Osawa S, Jukes TH, Watanabe K, Muto A (1992) Recent evidence for evolution of the genetic code. Microbiol Rev 56: 229-264

Sawazaki Y, Inomata K, Yoshida K (1995) The trans-kingdom conjugation between *Agrobacterium tumefaciens* and *Saccharomyces cerevisiae*, a bacterium and a yeast. Plant Cell Physiol 37: 103-106

Sikorski RS, Michaud W, Levin HL, Boeke JD, Hieter P (1990) Trans-kingdom promiscuity. Nature 345: 581-582

Sogin ML (1992) Comments on genome sequencing. In: Hartman H, Matsuno K (eds) The origin and evolution of the cell, pp 387-391

Willetts N, Skurray R (1987) Structure and function of the F factor and mechanism of conjugation. In: Neidhardt FC et al (eds) *Escherichia coli* and *Salmonella typhimurium*, vol 2, pp 1110-1113, American Society for Microbiology, Washington

Willkins B, Lanka E (1993) DNA processing and replication during plasmid transfer between gram-negative bacteria. In: Clewell DB (ed) Bacterial conjugation. Plenum Press, New York, pp 105-136

Yoshida K (1979) Interspecific and intraspecific mitochondria-induced cytoplasmic transformation in yeasts. Plant Cell Physiol 24: 851-856

Yoshida K (1995) Endosymbiotic conjugation: a new hypothetical mechanism for gene transfer to the host nucleus during the organelles origin. *Viva Origino* 23: 121-140

Yoshida K, Takeuchi I (1980) Cytological studies on mitochondria-induced cytoplasmic transformation in yeasts. Plant Cell Physiol 21: 497-509

Zambryski P, Tempe J, Schell J (1989) Transfer and function of T-DNA genes from *Agrobacterium* Ti and Ri plasmids in plants. Cell 56: 193-201

Chronobiology and Endocytobiology: Where do They Meet?

On the Evolution and Mechanisms of Eukaryotic Timekeeping

F. Kippert
Institute of Botany, University of Tübingen, 72076 Tübingen, Germany
Present address: Institute of Cell, Animal and Population Biology, University of Edinburgh, West Mains Road, Edinburgh EH9 3JT, Scotland

Key words: *Schizosaccharomyces pombe*, circadian clocks, ultradian clocks, timekeeping, evolution, cell division cycle, cellular signalling, stress response.

Introduction

In the Proceedings of Endocytobiology III, I had the opportunity to outline some ideas about how the evolution of cellular clocks may have occurred concomitantly with the evolution of eukaryotic cells (Kippert 1987). I speculated whether *internal timekeeping* might have been the initial selective advantage that cellular clocks provided to an endocytobiotic consortium developing into the early eukaryotic cell - a hypothesis which constituted a link between the fields of chronobiology and endocytobiology. Since then, there have been exciting developments in both fields. Much of the progress made in chronobiological research is of significance in evolutionary terms and is more ore less related to the field of endocytobiology. Thus, for Endocytobiology VI it may be timely to review this progress. The aim of the present review is to present briefly those recent findings that may be of relevance to our understanding of the evolution of cellular clocks, and discuss what this may tell us about today's oscillator mechanisms.

One major conclusion will be that eukaryotic timekeeping is in many respects different from that of prokaryotes. Studying a simple, unicellular eukaryote may be best suited to establishing its basic mechanisms. The fission yeast *Schizosaccharomyces pombe* is introduced as a model system for evolutionary chronobiology. Mutations in a number of genes which code for exclusively eukaryotic-type proteins were found to affect clock functions, indicating the advantage of this organism in the quest for conserved oscillator mechanisms in eukaryotes.

Rejecting the 'Eukaryotes Only' Dogma: Circadian Clocks in Prokaryotes

Most important to the issue of the evolution of cellular timekeeping mechanisms was the demonstration of circadian rhythms in prokaryotes. When I suggested an endocytobiotic origin of circadian rhythms, assuming that cellular clocks provided means of temporal coordination to an endocytobiotic consortium developing into the eukaryotic cell (Kippert, 1987), this was based on the fact that no circadian rhythms had been observed in any prokaryotic species. However, since then, the 'eukaryotes only' dogma has had to be rejected. Within a year, three reports described circadian rhythms in a particular group of cyanobacteria. These are characterised by their capability for photosynthesis and nitrogen fixation within single cells (for review see Bergman et al., 1997). Since nitrogenase is very sensitive to oxygen, it is important to have separation, spatial or temporal, of the two incompatible reactions. These species have taken the latter option, N_2 fixation occurs at night and is thus separated from photosynthesis. With these early cases, however, doubt remained (summarised by Kippert, 1991) as to whether the observed rhythms were the output of true circadian clocks.

Regarding the marine unicellular *Synechococcus* Miami BG043511 (Mitsui et al., 1986), it was questioned whether the reported rhythms were the output of a clock, or were related to a synchronised cell cycle. In photosynthetic organisms, it is obvious that the cell cycle can be synchronised to a light dark cycle (LD). If conditions are chosen such that the doubling time approaches 24 h, the subsequent 'free-run' will produce a rhythm with a 'circadian' period. Earlier studies on this organism were performed under such conditions and only recently has it been investigated at different temperatures or oxygenation levels (e.g. Mitsui et al., 1993). When growth rate was altered in these studies, the free-running period was found to change in parallel with values as short as 6 h and long as 30 h. This strongly suggests that the rhythm is cell cycle-related, and that the cell cycle is not under the control of a circadian clock in this organism. In general, in most cases where circadian rhythms of cell division have been sought in cyanobacteria, none have been found (see below, p. 173 ff).

A different problem is encountered with *Synechococcus* RF-1. In this intensively studied species (for review see Huang and Grobbelaar, 1995) with a much lower growth rate, it was the proximity of the 'free-running' period to 24 h which gave grounds for initial scepticism (Kippert, 1991). Since period was found to be close to 24 h even under conditions where a deviation, at least transiently, would have been expected, *i.e.* after temperature steps (Huang et al., 1990), these doubts remained. In order to detect a small but significant deviation of period it is necessary to monitor rhythms over an extended time span. Measuring alkalisation of the medium, rhythms could be followed for up to 2 weeks (Kippert, 1996c). Under a variety of combinations of temperature and light intensity, it was not possible to achieve a period significantly different from 24 h. A review of the literature reveals several more cases of periods indistinguishable from 24 h. This suggests that these species may have the capability to register some subtle changes in their environment not completely controlled in the experi-ments. Whether they employ this to synchronise their circadian clock to the geophysical 24 h cycle or to initiate some non-clock related physiological program remains an open question.

However, any ambiguity about the existence of true circadian rhythms in prokaryotes has vanished completely in recent years with the combined efforts of the groups of Golden, Johnson and Kondo studying *Synechococcus* sp. PCC7942 (for review see Johnson et al., 1996). Notably, this species, for which molecular genetic techniques are well established, is not capable of nitrogen fixation. In their systematic studies on circadian gene expression, these workers demonstrated that the rhythm of this species fulfils the canonical criteria for circadian rhythms (see Edmunds, 1988), i.e. entrainability, free-run under constant conditions, and temperature-compensation of the period. By using bacterial luciferase as a reporter for the rhythmic expression of the *psbA* gene coding for a protein of photosystem II, Kondo et al. (1993) were able to monitor a great number of colonies simultaneously. This enabled large-scale screening and thereby the isolation of a considerable number of clock mutants. Most important was the isolation of mutants in period length, ranging from 16 to 60 h (Kondo et al., 1994), which proved unequivocally the endogenous nature of the rhythmicity in this species. No information is available yet for possible oscillator mechanisms, but this should soon change with the characterisation of genes cloned by complementation of the clock mutants. Remarkably, about 75% of the mutants were rescued by an 8 kb fragment of wild-type *Synechococcus* DNA (Johnson et al., 1996). Further analysis of the information encoded on this fragment should give important insights into the circadian system of this organism.

In summary, it can be concluded, as Johnson et al. (1996) have put it, that 'at the present time, there is no doubt that at least some cyanobacteria have a *bona fide* circadian clockwork mechanism'. It has, however, to be kept in mind that there are also several studies where no circadian rhythms could be detected under appropriate experimental conditions (e.g. Gallon et al., 1991; for review see Bergman et al., 1997). This situation contrasts with their ubiquitous occurrence in eukaryotes and may force caution about equating the rhythms found in prokaryotes and eukaryotes. Hopefully, further searches will provide some indication about the distribution of circadian rhythms in prokaryotes and it will be important to see whether they can be detected in other orders of eubacteria and in archaebacteria. Alternatively, since the genomes of a gram-negative bacterium (Fleischman et al., 1995) and an archaebacterium (Bult et al., 1996) have now been completely sequenced, it will be most interesting to see whether genes affecting clock function in *Synechococcus* have homologues in these genomes.

What does the existence of prokaryotic circadian rhythms mean to the evolution of their counterparts in eukaryotes? Johnson et al. (1996) speculate whether the clock mechanism of a cyanobacterium-like prokaryote could have been passed by endosymbiosis to its host. It cannot be assumed that a cyanobacterial clock was the source of the timekeeping mechanisms of the eukaryotic nucleocytoplasm, because this would not explain circadian clocks in any heterotrophic eukaryote. On the other hand, it could well have contributed to the timekeeping in photoautotrophs. Hwang et al. (1996) have recently shown that in the alga *Chlamydomonas reinhardtii* there is a circadian rhythm in the transcription of chloroplast-encoded genes. It should be interesting to compare this rhythm with the rhythms of gene expression in cyanobacteria. This might provide some indication whether the chloroplast circadian system is of prokaryotic origin or whether the cyanobacterial clock has been lost during plastid evolution. There is only one organism where the issue of a wider contribution of the chloroplast compartment to circadian timekeeping has been addressed. In the algal flagellate *Euglena gracilis* it is possible to have bleached mutants which are completely devoid of plastids. Carré et al. (1989) and Schmidt and Balzer (1995) have studied two such strains and found the rhythms in photoautotrophic and heterotrophic cells to be remarkably similar. That the rhythms of the bleached mutant displayed faster damping under constant conditions (Schmidt and Balzer, 1995) may be a difference worth studying in more detail.

Is There a Typical Eukaryotic Clock?: The Story of *per* and *frq*

Having realised that circadian rhythms do exist in at least some prokaryotic species, the question is what the relationship of these rhythms is to those found in eukaryotes. Before this question can be addressed, we need to know whether there is such a thing as a 'typical eukaryotic circadian clock'. For three decades, circadian research was hampered by the intrinsic problems of what may be called the 'pharmacological approach': the agonists and antagonists used to perturb the oscillator have had at least ambiguous specificity. The state of the art 10 years ago is well reflected in the book by Edmunds (1988) which lists more than 30 very different models for the circadian oscillator that had been put forward by that time.

A breakthrough came from applying the molecular biological approach to the clock mutants of the classical objects of circadian research, the fruit fly *Drosophila melanogaster* and the bread mould *Neurospora crassa* (for reviews see Hall, 1995; Loros, 1995; Dunlap, 1996; Rensing, 1996). Both the *period* (*per*) gene of *D. melanogaster* and the *frequency* (*frq*) gene of *N. crassa* had been promising candidates since the isolation of the first mutants:
(1) the respective screens produced almost exclusively mutants affected at the *per* and *frq* loci;
(2) several alleles were found for each of them which showed either lengthening or shortening of the free-running period or apparent arhythmicity;
(3) in some of the alleles temperature-compensation, a distinguishing feature of the circadian oscillator, was affected. Both the predominance of *per* and *frq* alleles and the variety of different phenotypes made the gene products likely key players of the circadian oscillator in these two organisms. Rather disappointingly, when the genes were cloned, only a very limited sequence homology could be detected between them. However, the subsequent molecular characterisation revealed a picture that in its general features appears to be very similar for the two organisms. The favourite model at present is that of a negative feedback loop in which *per*/*frq* proteins control the transcription of their own mRNA. The substantial

evidence in support of this model has been extensively reviewed (Hall 1995; Loros 1995; Dunlap 1996; Rensing 1996), and *per* and *frq* have been assigned state variables of the circadian oscillator. From an evolutionary point of view, regarding the relationship to prokaryotic circadian clocks, it should be noted that there are two features of the proposed *per* feedback loop that would be typical of eukaryotes, these beeing the temporally regulated phosphorylation of the *per* protein and its transport into the nucleus.

There are, however, serious challenges to considering the *per* and *frq* feedback loops as universal and/or central oscillator mechanisms (for a discussion see also Hardeland 1994). Firstly, there is the observation that circadian rhythms can persist in enucleated specimens of the unicellular alga *Acetabularia mediterranea* (Sweeney and Hastings 1961), which clearly denies a central role to chromosomal DNA and nucleus in circadian timekeeping, at least in this species. Secondly, there are observations from *N. crassa* that clearly conflict with the suggested role of *frq* being a state variable of the circadian oscillator. In the *frq7* mutant, the oscillator is largely desensitised to the phase-shifting action of the cycloheximide, whereas protein synthesis itself is not (Dunlap and Feldman 1988). More importantly, the null mutants *frq9* and *frq10* still show circadian rhythms, although with much reduced stability and with very poor temperature compensation (Aaronson et al. 1994). Thirdly, one should in this context consider the existence of higher-frequency (ultradian) temperature-compensated rhythms that share characteristics and probably to some extent mechanisms with circadian rhythms (see p. 170 ff). Some of these ultradian clocks have periods that are definitely too short to envisage a feedback loop based on transcription. Among these we find, most notably, the ultradian rhythm of courtship song of *D. melanogaster,* with a period of around 1 min (for review see Kyriacou et al. 1993). The period of this rhythm is also determined by the *per* gene and is affected by mutations in a fashion that parallels their effects on the period of the circadian clock (see below). Also to the *per* gene relates a recent unexpected finding about the *per* homologue of the silkmoth *Anteraea pernyi.* Unlike its *Drosophila* counterpart, the *per* protein of *A. pernyi* never enters the nucleus of what are assumed to be the pacemaker neurons of this insect (Sauman and Reppert 1996). This means that a negative feedback loop as envisaged by the model for the fruit fly cannot operate in the silkmoth. Instead, Sauman and Reppert (1996) provide evidence that the temporal regulation is via antisense RNA. Thus, in these two homometabolous insects the roles of the respective *per* homologues in the circadian oscillator seem to be entirely different (for a discussion of the implications of this finding see also Hall 1996; Sassone-Corsi 1996). This is all the more surprising since the *per* homologue of *A. pernyi* is able to restore rhythmicity in *D. melanogaster* null mutants. Further compara-tive studies are now needed in order to find out what this puzzling finding means to both the evolution and the mechanism of circadian clocks.

Another potentially discomfiting aspect for the proponents of a rather simple feedback model, as realised by Hall (1995), is that mutations affecting the clock can have pleiotropic effects. For the *per* gene, this seems to be restricted to timekeeping processes. In the different alleles of *per* which result in (with respect to the circadian clock) in period lengthening, period shortening or apparent arrhythmicity, both courtship song periodicity (Kyriacou et al. 1993) and developmental timing (Kyriacou et al. 1990) are affected in a parallel fashion. In the ciliate *Paramecium bursaria*, mutations that alter circadian period length also affect the timing of maturation (Miwa and Yajima 1995) and several physiological processes such as the period of the contractive vacuole, swimming velocity, membrane potential and intracellular K^+ concentration (Tokushima et al. 1994). There are also two observations concerning ultradian clocks. Mutations in the *clk1* gene of the nematode *Caenorhabditis elegans*, besides altering the period of the ultradian rhythm in defecation, also affect processes diverse as embryonic and post-embryonic development, egg production rate, cell cycle period, life span, swimming and pumping cycles (Wong et al. 1995). The *GTS1* gene of the budding yeast *Saccharomyces cerevisiae*, overexpression of which lengthens the period of the ultradian clock of *Schizosaccharomyces pombe* (F. Kippert, unpublished), affects heat tolerance, sporulation, flocculation, fatty acid composition, cell cycle timing and life span (Yaguchi et al. 1996; Bossier et al. 1997).

So is it possible to fit the various observations into an overall picture of eukaryotic clock mechanisms? The answer is probably no, and a major reason for this is that the screens for clock mutants in eukaryotes have been far from saturating. Notably, almost all of the mutations isolated so far are semi-dominant (Hall 1995; Dunlap 1996). This does not, of course, mean that recessive mutations do not exist, but rather that they could not be detected in the way the screens were performed. Considering the fact that in the haploid yeast *S. pombe* 18 genes have already been identified, the mutation, deletion or overexpression of which cause an altered clock phenotype, it seems reasonable to assume that the number of genes not detected in those earlier screens may be considerable (see also below, p. 181 ff). Therefore, the hunt for genes involved in 'typical eukaryotic' clock mechanisms seems to be only beginning.

The Eukaryotic Cell as a Cellular Clockshop: Ultradian Clocks are Timekeeping Devices

When I suggested an endocytobiotic origin of eukaryotic clocks (Kippert 1987), I thought to provide explanation not only for the evolution of circadian clocks, but also for cellular clocks whose periods neither match the geophysical cycle nor have any other exogenous correlate. Because of this lack of any periodic external time cue to which synchronisation could mean adaptive advantage, it seemed reasonable to consider that non-circadian clocks could have provided some means of internal timekeeping used for the coordination of cellular functions. This might have been of importance in cells that had evolved to the compartmental complexity of the eucyte.

To distinguish cellular clocks *sensu stricto* from the plethora of other biological oscillations, I have suggested two characterising features (Kippert, 1992). First, since a clock is thought to serve as a timekeeping device, constancy of period is demanded under a variety of different conditions. The most obvious aspect is temperature compensation, ensuring that the clocks runs at the same speed at different (steady state) temperatures; but period should likewise not be affected by differences in nutrient supply, humidity, osmolarity, irradiation, etc. The second feature is that a clock is expected to control a variety of different cellular functions. While the demonstration of this can be taken as indication of a clock, its absence (when only one oscillating parameter has been observed) does not necessarily speak against it.

Early data on ultradian clocks were treated with scepticism, but studies in recent years have provided ample evidence for their existence (for review see Kippert 1997a). Table 1 lists those ultradian oscillations of unicellular eukaryotes for which temperature-compensation has been clearly demonstrated. In the cases of the two ciliates, period determination was exact enough and was carried out over a wide enough temperature range to enable determination of a Q_{10}. For both *Tetrahymena pyriformis* (Kippert 1996b) and *Paramecium tetraurelia* (Kippert 1996a) the value was 1.08, which is well within the range of those found for circadian clocks. Independence of period in different media has been shown for *P. tetraurelia* (Kippert 1996a) and *S. pombe* (Kippert 1997b; Kippert and Lloyd 1997a). Most remarkably, in the more than 100 experiments performed under very different conditions with a long period strain (wild-type Tübingen) of *S. pombe*, the free-running period was never outside the range of 40 - 44 min (Kippert 1997b). Taken together, these results suggest a general homeostasis of period length of these ultradian clocks which equals that of circadian clocks (Kippert 1997a).

In contrast to this striking stability of period under different environmental conditions, a considerable variability was found for different strains of *S. pombe*, ranging from 28 to 56 min (Kippert 1997c). While the periods of most strains cluster around 30 - 36 min, the strains with longer periods can probably be considered clock mutants as far as period is concerned. Here it has to be appreciated that all these strains are derived from the same original isolate. Mutations must therefore have been either spontaneous or introduced in the course of mutagenic procedures in the history of the strains. These remarkable differences in period length of different strains strongly suggest that period may be affected by a considerable number of genes, the effect of which can best be revealed by studying a haploid organism (see also below, p. 181 ff). In the strains where this has been looked for, the circadian period is affected in a parallel fashion, although to a lesser degree (Kippert 1997c), indicating that these microbial ultradian clocks share some aspects of their mechanisms with their circadian counterparts.

Although ultradian clocks have mostly been studied in eukaryotic microbes, there are two interesting cases of very short period (~ 1 min) clocks in invertebrates. The ultradian rhythm of the *D. melanogaster* courtship song has already been mentioned. It is well temperature-compensated, even in those *per* alleles where temperature-compensation of the circadian period is less efficient than in the wild-type (Kyriacou et al. 1993). Research on a *C. elegans* clock has commenced only recently; here, an ultradian rhythm in defecation with a 45 s periodicity is also well temperature-compensated. A number of mutants have been isolated with periods ranging from less than 20 s to up to 100 s, and temperature compensation is dramatically impaired in some of these mutants (Iwasaki et al. 1995).

In considering the examples of *E. gracilis*, *T. pyriformis* and *S. pombe*, we become aware that even unicellular eukaryotes have a remarkable complexity of temporal structures. In the ciliate and the yeast, a circadian clock and two ultradian clocks of different time ranges appear to take over alternative control over a variety of cellular functions, such as cell division and energy metabolism, and the growth rate determines which clock is in control (Kippert 1997a). On the other hand, in *E. gracilis*, the 8- or 60-min rhythms in motility, the 4.5 h rhythm in tyrosine aminotransferase activity and circadian rhythms in various other parameters occur simultaneously with each other (Balzer et al. 1989b; Lewandowski et al. 1995).

What comes to mind regarding this difference is that of three organisms *E. gracilis* is the only photoautotroph. To be noted in this context is the observation, actually the breakdown of another dogma in circadian research, that in the dinoflagellate *Gonyaulax polyedra* two distinct oscillators are operating within a single cell (for review see Roenneberg 1996). These two circadian clocks are normally in synchrony with each other, but can be desynchronised and then free-run with different periods. Thus we have two examples where in photoautotrophic organisms two or more cellular clocks are in operation simultaneously. This is certainly not enough to postulate a general pattern, but it seems rather tempting to

Table 1. Temperature-compensated ultradian rhythms in unicellular eukaryotes

Species	Rhythmic parameter	Period	Reference
Euglena gracilis	Motility	8 min	Lewandowski et al. (1995)
Tetrahymena pyriformis	Cell division, respiration	30 min	Kippert 1996b
Schizosaccharomyces pombe	Cell division, metabolism	40 min	Kippert and Lloyd (1995)
Euglena gracilis	Motility	60 min	Lewandowski et al. (1995)
Acanthamoeba castellanii	Respiration, protein synthesis	70 min	Lloyd et al. (1982)
Paramecium tetraurelia	Cell division, motility	70 min	Kippert (1996a)
Schizosaccharomyces pombe	Cell division, metabolism	4 h	Kippert (1997a, unpublished)
Euglena gracilis	TAT activity	4.5 h	Balzer et al. (1989a)
Tetrahymena pyriformis	TAT activity, cell division	5 h	Michel and Hardeland (1985)

Modified from Kippert, 1997a. If there are differences between strains, only one value is given.

speculate that the endocytobiotic cyanobacteria from which today's plastids evolved may have brought along an additional complexity of cellular timekeeping in photoautotrophic eukaryotes. It would now be most interesting to see whether when *E. gracilis* is deprived of its plastids it loses any of this complexity. In summary, the temporal order of the single eukaryotic cells is a very complex one, irrespective of whether different clocks represent alternative modes or operate simultaneously. To regard eukaryotic cells as a 'cellular clockshop' and study the relationships between the different oscillations may provide a valuable approach to addressing both the evolution and the mechanism of these clocks.

Cell Division Cycles and Cellular Clocks: The Eukaryotic Way of Multiplication

The cell division cycle is probably one of the most, if not *the* basic cellular function found to be under control of both circadian and ultradian clocks. The phenomenon of 'gating' of cell division has been known for a long time (for a review see Edmunds, 1988). However, only since the late 1980s has the interaction of the circadian clock with the cell division cycle of unicellular eukaryotes been studied in more detail. Circadian rhythms of cell division have been demonstrated in the photoautotrophs *Chlamydomonas reinhardtii* (Goto and Johnson 1995), *Euglena gracilis* (for review see Edmunds 1996) and *Gonyaulax polyedra* (Homma and Hastings 1989a), and the yeasts *Schizosaccharomyces pombe* (Kippert et al. 1991) and *Saccharomyces cerevisiae* (F. Kippert, unpublished). Gating of S phase was found for *E. gracilis* (Edmunds 1996), *G. polyedra* (Homma and Hastings 1989b), and *S. pombe* (F. Kippert, unpublished). The finding on *C. reinhardtii* is remarkable because there had been a long-standing debate about whether its cell cycle is under circadian clock control, but in their recent series of well designed experiments Goto and Johnson (1995) provided compelling evidence that this is indeed the case. This suggests that, wherever is observed under appropriate experimental conditions, the cell cycle of eukaryotes is under circadian control, one condition being that the interdivision time equals or exceeds the circadian period (circadian-infradian rule; see Edmunds 1988).

In remarkable contrast to this is the situation in cyanobacteria (as it is to be conceived at present). Early studies have been criticised because of inappropriate conditions (Kippert 1991), and indeed the cell cycle of *Synechococcus* BG043511 appears not to be under circadian clock control (see section, p. 168 ff). In *Synechococcus* RF-1 with its longer generation time, no circadian gating of cell division could be detected under conditions where various other rhythms were observed (Rojek et al. 1994; F. Kippert, unpublished). Ironically, the only convincing demonstration of gating of cell division (but not DNA synthesis) so far comes from *Synechococcus* PCC7942 with a growth rate higher than one division per day (Mori et al. 1996). Remarkably, these fast-growing cultures show pronounced circadian rhythms in gene expression (Kondo et al. 1997). Circadian rhythms of cell division or gene expression have never been observed in fast growing cultures of any unicellular eukaryote.

Regarding these differences in interplay between circadian clocks and the cell division cycle, it is worth considering the differences between prokaryotic and eukaryotic cell cycles in general. Considering the evolution of the eukaryotic cell cycle, Nasmyth (1975) points out that these differences are indeed fundamental, and goes on to suggest that the invention of the mechanisms of mitotic DNA segregation may have been a key step in eukaryotic evolution. At the very centre of the regulation of mitosis are cyclin-dependent kinase (CDKs), for which the *S. pombe cdc2* kinase is the prototype, and their associated cyclins (for review see Nigg, 1995; D'Urso and Nurse, 1995). Thus any clock control would be expected to interact here. Indeed, in *E. gracilis* both the tyrosine phosphorylation state of the *cdc2* homologue, and the protein level of a B-type cyclin oscillate (Edmunds 1996). In addition, a pronounced rhythm in the activity of its *cdc2* kinase has been observed in the dinoflagellate *Gambierdiscus toxicus* (van Dolah et al. 1995); the kinase was constitutively expressed, but activated only in a phase concurrent with the presence of mitotic cells.

In this context, it is interesting to note that CDK- and cyclin-like proteins exist in the non-dividing nervous tissue of both invertebrates (Krucher and Roberts 1994) and vertebrates (Lew and Wang 1995). Intriguingly, the concentration of the cyclin-like protein in the eye of the mollusc *Bulla gouldiana* was found to be affected by treatments that phase-shift the circadian rhythm (Krucher and Roberts 1994), leading the authors to suggest that the circadian system may be related to the biochemical mechanisms regulating the eukaryotic cell cycle. Here it is a most remarkable finding that the circadian changes in the phosphorylation state of the *E. gracilis cdc2* kinase persist in stationary phase (Edmunds 1996), indicating that this reversible tyrosine phosphorylation is coupled to the circadian oscillator, and perhaps suggesting that there is more to *cdc2* than regulation of the cell division cycle.

The phenomenon of gating of stages of the cell division cycle is not restricted to circadian rhythms but has also been observed with ultradian clocks. Gating of cell division had been indicated in studies with *Tetrahymena pyriformis* (Michel and Hardeland 1985; Kippert 1996b), and has recently been demonstrated for *S. pombe* (Kippert and Lloyd 1997b) and on the single cell level for *Paramecium tetraurelia* (Kippert 1996a). Gating of S phase by the ultradian clock has been shown for *S. pombe* (Kippert and Lloyd 1997b). Since *S. pombe* is one of the model organisms in cell cycle research (D'Urso and Nurse 1995), further analysis might be able to make use of the well characterised *cdc* mutants of this organism. By so doing, it might be demonstrated that the ultradian clock's control over S phase and mitosis are independent of each other, and thus gating takes place at least twice in the cell cycle of fission yeast (Kippert and Lloyd 1997b).

For the control over mitosis, it was already possible to identify a link on the output control pathway from the clock to the processes regulating entry into mitosis. The protein kinases and phosphatases counteracting in the regulating of the

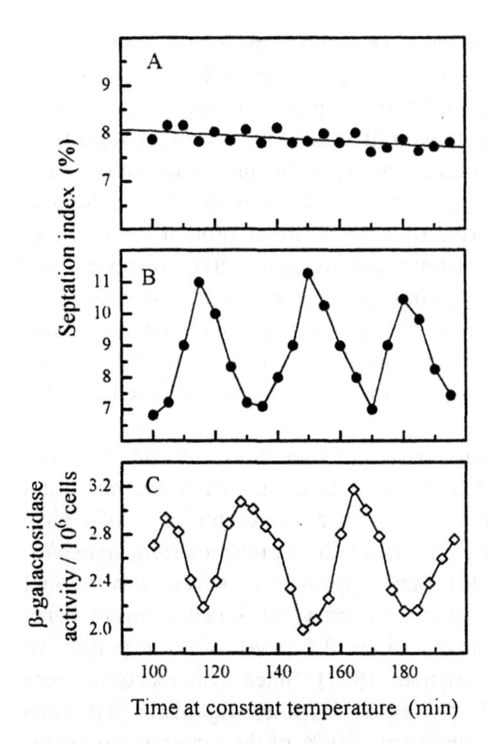

Fig. 1A-C. Gating of the entry into mitosis by the *wee1*⁺ protein kinase of *Schizosaccharomyces pombe*. **A** Absence of the ultradian rhythm in cell division (measured as septation index) in a *wee1* deletion mutant. **B** Restoration of the ultradian rhythm of cell division in this mutant by reintroduction of the *wee1*⁺ gene. **C** Gating of S phase by the ultradian clock in the deletion mutant (monitored as the expression of β-galactosidase under control of the S phase-specific *cdc18* promoter)

phosphorylation state of tyrosine 15 of the *S. pombe cdc2* kinase, which ultimately defines the entry into mitosis, are well known (D'Urso and Nurse 1995). Of these, the *wee1*⁺ protein kinase is absolutely essential for gating of mitosis to occur (Kippert, 1997d). As Fig. 1A shows, there is no rhythm in the septation index (the percentage of cells approaching division) in a *wee1* deletion mutant, but the rhythm can be restored by reintroducing the *wee1*⁺ gene (Fig. 1B). Gating of S phase is not impaired in the *wee1* deletion mutant (Fig. 1C), indicating that the clock remains unaffected in these cells. Since the general regulation of *cdc2* (and its homologue CDKs) is highly conserved (Nigg 1995), it will be most interesting to see whether this pathway of clock control is conserved as well; a study involving *wee1*⁺ homologues of other organisms is now under way. In evolutionary terms, it is important to keep in mind that we are dealing here with clock control over an exclusively eukaryotic regulatory network.

The Network Counts: Eukaryotic Signalling Cascades in Eukaryotic Clock Function

Since the title of this section promises a review and discussion of 'eukaryotic signalling cascades', it is necessary to clarify what eukaryotic signalling means.

Much of eukaryotic signalling involves second (and third) messengers and protein phosphorylation events. The question now is to what extent are they eukaryote-specific? Let us here consider only the evolution of protein phosphorylation as a regulatory device (but similar considerations will apply for most other aspects of cellular signalling). While serine/threonine and tyrosine phosphorylation have long been considered typical eukaryotic inventions in cellular regulation, the past few years have brought forward a series of reports about protein kinases and phosphatases in different classes of eubacteria and archaebacteria (for review see Zhang, 1996; Kennelly and Potts, 1996). Although their occurrence in some prokaryotes appears to be due to lateral gene transfer, and participation in cellular regulation could not be demonstrated in other cases, it is very likely that regulatory protein phosphorylation initially was not initially a eukaryotic invention (for discussion see Zhang 1996).

The question then remains do they play any major role in prokaryotic cell regulation, and how widespread are they? Some answer comes from the microbial sequencing projects where we now have at least the absolute minimum of a phylogenetically representative set. In the genome of the cyanobacterium *Synechocystis* sp. PCC6803 (Kaneko et al. 1996), there appear to be seven conventional (Hank-type) kinases, whereas in the genome of the yeast *Saccharomyces cerevisiae* (Goffeau et al. 1996), which contains about 2.5 times as many genes, we find 113 such kinases (Hunter and Plowman 1997). Since cyanobacteria seem most 'advanced' in this respect, and the search in other organisms has been largely negative (e.g. in *E. coli*, after sequencing 60 % of the genome, no candidate has yet be detected; Zhang 1996), it seems safe to conclude that (1) 'eukaryotic-type' kinases and phosphatases exist in prokaryotes; and that (2) in eukaryotes, they have reached a completely different level of ubiquity and complexity. Of particular significance is the observation that there are kinase subtypes which are not found at all in prokaryotes. These include, for example, the kinases of the MAP kinase pathways (for review see Waskiewicz and Cooper 1995), which in eukaryotes have attained such a level of complexity that Pelech (1996) fittingly coined the term 'cellular intranet'. It therefore seems justified to consider a eukaryotic-specific network of signalling, and discuss whether and how its constituents are involved in circadian and ultradian clock mechanisms. In doing this, we have to distinguish between the central oscillator mechanism and the input and output pathways. Regarding the universality of eukaryotic signal ling pathways, it comes as no surprise that they are involved in various ways in signal input (e.g. in entrainment to the LD), as well as output control pathways (bringing a plethora of cellular functions under control of the clock). These interactions have been reviewed by Anderson and Kay (1996) and Carré and Kay (1996).

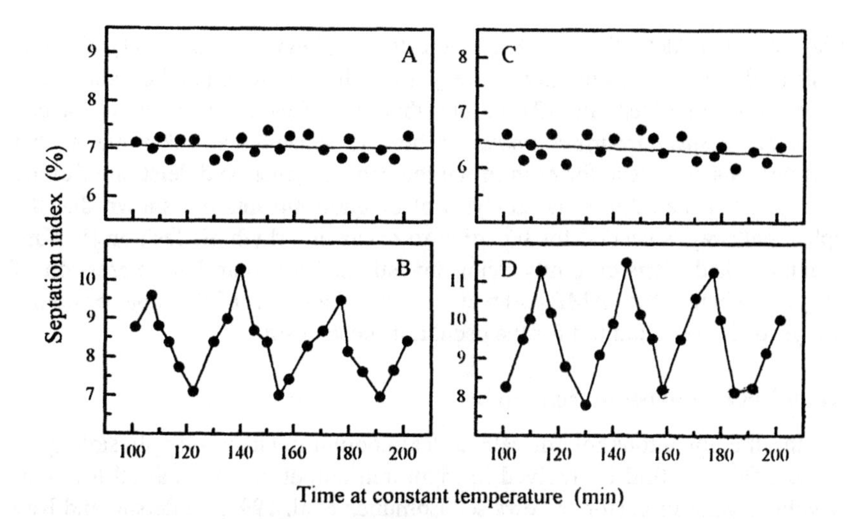

Fig. 2A-D. Absence of the ultradian rhythm of cell division (measured as septation index) in phospholipase C and phosphoinositide 3-kinase deletion mutants of *Schizosaccharomyces pombe*, and rescue of the rhythmic phenotype by transformation with the respective gene. **A, B**. phospholipase; **C, D**. phosphoinositide 3-kinase

cAMP/Protein Kinase A Signalling

Much earlier work employing the 'pharmacological approach' suggested the involvement of the cAMP pathway in the oscillator (for review see Edmunds 1988), but the conclusions drawn from these studies suffered from the insufficient specificity of the inhibitors used. A most important finding in recent years is that circadian oscillations in intracellular cAMP have been observed in an increasing number of organisms, from the microbes *Euglena gracilis* (Edmunds 1996) and *Paramecium bursaria* (Hasegawa et al. 1995) to the suprachiasmatic nucleus, the mammalian pacemaker (Prosser and Gillette 1991). In *E. gracilis*, this is brought about by circadian changes in the activities of both adenylate cyclase and phosphodiesterase (Edmunds 1996). There is good evidence that in *E. gracilis* cAMP and protein kinase A are involved in the output pathway which brings the cell division cycle under clock control (for review see Edmunds 1996). This is a particularly promising starting point for further analysis because it identifies a link between the central oscillator and a 'hand of the clock', i.e. the rhythmic activity of the *cdc2* kinase. It is suggestive that the observed fluctuations in cAMP level and kinase A activity participate in the temporal control over other cellular functions as well. On the other hand, there appears at the moment to be no compelling evidence for a role in the input pathway of any non-neuronal system.

That cAMP and protein kinase A may even be evolved in the central oscillator is indicated by an altered rhythmicity in the *dunce* mutant of *Drosophila melan-*

ogaster, which is defective in phosphodiesterase (Levine et al. 1994). Genetic analysis of the role of this pathway is very difficult because of (1) the redundancy of the enzymes involved, and (2) the fact that some function is essential for cell survival. *Schizosaccharomyces pombe* is here an exceptional eukaryote in that there is only a single gene for each of the relevant enzymes, and deletion of either of them is not lethal. An initial analysis of deletion mutants has shown that the complete pathway is needed for septation to occur in a rhythmic fashion (F. Kippert, unpublished). It is currently being investigated by controlled expression of these genes whether the cAMP pathway of *S. pombe* is part of the input or output pathways, or is even required for the oscillator mechanism.

Calcium/Phosphoinositide Signalling

Regarding the ubiquitous role of calcium/calmodulin in eukaryotic physiology, it is not surprising to find it involved in input and output pathways at all levels of eukaryotic organisation (for reviews see Edmunds et al. 1992; Anderson and Kay 1996; Carré and Kay 1996). In the plant model system *Arabidopsis thaliana* the molecular genetic analysis has now reached such a level that detailed insights into the role of calcium/calmodulin in synchronisation to the LD can be expected in the near future (Barnes et al. 1997). Recently, Johnson et al. (1995) made use of genetically engineered aequorin to measure calcium levels *in vivo* in both *A. thaliana* and *Nicotiana tabacum,* and found clear-cut circadian rhythms of cytoplasmic calcium. Interestingly, the time course corresponded reasonably well to the one previously postulated (Kippert 1987). Since the values were well within a physiologically relevant range, it can be assumed that calcium is at least involved in the circadian modulation of various cellular functions in these two plants. Whether it is also a component of the central oscillator remains to be determined.

An experimental hint at the involvement of calcium in the oscillator mechanism comes from a recent study on the role of the phosphoinositide cycle (intimately linked to calcium signalling; for review see Berridge 1993) in the ultradian clock of *S. pombe*. Lithium salts are probably the only agent for which a universal effect on circadian oscillators (by lengthening of the free-running period) has been observed so far (for review see Klemfuss 1992). This is also found for the circadian clock of *S. pombe* (F. Kippert, unpublished). With the ultradian clock the effect is particularly profound - the period is lengthened by about 50% (Kippert et al. 1997), which makes it a suitable model system for the analysis of the mode of action of lithium. Over the years a couple of different cellular targets of lithium have been suggested, but the most likely candidate at present is the enzyme inositol monophosphatase, eventually required to keep the phosphoinositide cycle running (for review see Atack et al. 1995). Controlled expression in *S. pombe* of the human gene, and of variants made by *in vitro* mutagenesis which are largely insensitive towards lithium inhibition, has provided unequivocal evidence that inositol monophosphatase is the target for the chronobiological effect of lithium (Kippert 1997e). This suggests that phosphoinositide signalling (and thereby probably calcium signalling) play a role at least in the ultradian oscillator mechanism. This supposition is substantially corroborated by the absence of rhythmicity in deletion mutants for two further enzymes involved in phosphoinositide signalling. Both the phospholipase C and the phosphoinositide 3-kinase deletion mutants show arrhythmic septation (Fig. 2A,C). Rhythmicity can be restored by reintroducing the respective genes (Fig. 2B,D). Further analysis will aim to elucidate whether inositol signalling defines a state variable of the oscillator or whether its activity is only required as a prerequisite to keep the oscillator working.

Protein Phosphorylation/Dephosphorylation

Regarding protein phosphorylation and dephosphorylation in general, we still have to rely largely on the pharmacological approach with its ambiguous interpretations, but in this case it may be less problematic because different inhibitors give a fairly coherent picture. In *Gonyaulax polyedra* protein phosphorylation/dephosphorylation appear to be involved in both the input pathway and the oscillator itself (Comolli et al. 1994, 1996). A general inhibitor of protein phosphorylation blocked light-induced phase shifting and, depending on the concentration, slowed down and eventually stopped the circadian oscillator (Comolli et al. 1994). Different inhibitors of serine/threonine phosphatases caused either period lengthening or phase delays (Comolli et al. 1996). Kinase inhibitors were also found to lengthen the free-running period of the circadian pacemaker in the *Bulla gouldiana* eye (Roberts et al. 1989). The period lengthening indicates that protein phosphorylation is part of the oscillator. Using *S. pombe* as the experimental system, it is now possible also to apply the molecular genetic approach to the study of the role of protein kinases and phosphatases in clock functioning. It has already be shown that a MAP kinase pathway is involved in the ultradian rhythm in septation, since any single deletion mutant down the pathway has lost this rhythmicity (F. Kippert, unpublished). It remains to be established whether the input pathway, output pathway or central oscillator is affected or whether the cascade may even be required at more than one point. MAP kinase pathways may be of particular interest in the study of circadian rhythms because many of the agents causing phase-shifts of circadian rhythms at comparable concentrations induce MAP kinase signalling (see next section, this page). In conclusion, there is now such ample evidence for the participation of eukaryotic signalling in eukaryotic clock functions that this should no longer be neglected in any modelling of feedback loops for the circadian oscillator.

The Importance of Being Alert: Clocks, Endocytobiosis and Stress

One further aspect to be discussed is a possible relationship between circadian rhythms and the heat-shock or, more generally, the stress response, and the role that heat-shock proteins (HSPs) may play in clock mechanisms. These relationships can be considered twofold. Firstly, circadian rhythms in heat tolerance have been observed in yeast and plants (Kippert 1989; Rikin 1992; and references therein). For a variety of organisms there are also reports of circadian rhythms in either the constitutive levels or inducibility of HSPs (for review see Rensing and Monnerjahn 1996). A circadian rhythm in the expression of a major HSP70 has been observed in the two yeasts *Saccharomyces cerevisiae* and *Schizosaccharomyces pombe* (F. Kippert, unpublished). Employing the *E. coli* β-galactosidase as a reporter gene, rhythms in *SSA1* expression (a *S. cerevisiae* HSP70 with both constitutive and heat-inducible expression) could be demonstrated. The finding is remarkable in two respects. Firstly, the phase relationship observed for the expression of the *SSA1* gene fits well to that of the rhythms in heat tolerance in *S. pombe* (Kippert, 1989) and *S. cerevisiae* (F. Kippert, unpublished). The second observation relates to the conservation of clock mechanisms. Since the construct shows comparable rhythms in *S. cerevisiae* and the heterologous host *S. pombe*, the mechanism of circadian clock control over expression of a major HSP70 has apparently been conserved between these two only distantly related yeasts.

It seems reasonable to assume an adaptive significance of such rhythms, that of preparing the organism to withstand stresses which it will encounter only during a certain part of the daily cycle. Primarily we have to envisage the differences in temperature (sometimes dramatic), depending on the habitat, and we have also to consider the potentially damaging UV portion of light. In addition, there are other fluctuations that accompany the cyclical changes in light and/or temperature which also cause stress to the organism (e.g. changes in osmolarity and the concentrations of free radicals and oxidising agents). The advantage of being prepared at the time when (or before) it is needed is not to be underestimated.

The other aspect of the relationship between the stress response and the circadian clock is related to phase shifting. I have pointed out earlier that many agents that result in phase shifts also elicit the stress response (Kippert 1987). This calls into question the specificity of the drugs used in many phase shift experiments. For instance, such diverse treatments as heat shock, UV light, sodium arsenite, oxidative stress and anisomycin, which all phase shift circadian rhythms (Edmunds 1988), also induce MAP kinase cascades (Woodgett et al. 1996). The case of anisomycin is of particular significance because this protein synthesis inhibitor has been used frequently in phase shifting experiments to demonstrate a role of protein synthesis in the circadian oscillator. Interestingly, in the eye of the mollusc *Aplysia californica* two proteins whose concentrations are affected by different phase shifting treatments (such as light, serotonin and cAMP) were identified as HSPs (Koumenis et al. 1995). There is at present no experimental evidence that either the induction of stress proteins or the induction of the MAP kinase pathway plays a role in phase-shifting, but it seems an attractive hypothesis worth following up.

Likewise, the role of HSPs in endocytobiosis deserves attention. As Jeon (1995) has pointed out, we have to assume that any association that has led to a stable cellular consortium may in the beginning have caused some stressful condition to the partners. Thus, when we look at recent endocytobioses, it is not a surprise to find elevated levels of HSPs in endocytobionts, for instance in the X-bacteria of *Amoeba proteus* (for review see Jeon 1995 and this volume). Then we should also consider that many HSPs are expressed at a constitutive level and serve as molecular chaperones for multiple purposes, e.g. in protein folding, storage and transport (for reviews see Ellis 1996). In particular their role in protein transport may have been of crucial significance for evolving endocytobiotic associations. When genetic information is transferred from one compartment to the other, there is an absolute requirement for mechanisms to ensure that the gene product can be transported back into the original compartment. Well fitting in this context is the observation that the number of chaperone genes has increased dramatically in eukaryotes as compared to both eubacteria and archaebacteria (Pennisi 1996). In addition, they have acquired multiple additional functions in eukaryotic cells (Ellis 1996); and some of these chaperones are also involved in signal transduction events (Pennisi 1996). We may speculate that in the early days this could have been employed for signalling between the partners in the endocytobiotic consortium developing into the eukaryotic cell.

The Advantage of Being Simple: *S. pombe* as a Model in Evolutionary Chronobiology

In the preceding sections, I have reviewed a series of recent significant advances in chronobiological research that may shed some light both on the evolution of cellular clocks and on their mechanism(s). However, it has also become apparent that these results have left many questions unanswered as well as creating new ones. In this last section I would like to advertise the fission yeast *Schizosaccharomyces pombe* as an experimental system in which many of these questions can be addressed. *S. pombe* is one of the eukaryotic model systems (for review see Hayles and Nurse 1992); it has well established classical and molecular genetics and has been used in pioneering studies in various areas of cell biology. Now, some of its characteristics may prove valuable for chronobiological analysis as well.

S. pombe is haploid, a fact which allows the isolation of recessive mutants in simple screens. That the clock mutants isolated in other eukaryotic organisms so far are mostly semi-dominant tells us something about the screen, but nothing about the nature of mutations affecting the clock. Some of the mutants discussed in previous sections, e.g. those in the cAMP/protein kinase A pathway and the MAP kinase pathway, would not have been detected (as loss of function mutants) in a screen performed with a diploid organism. The remarkable differences in period length of the ultradian clock (and to a lesser extent of the circadian clock) found for different strains of *S. pombe* (p. 171 f) strongly indicate that a considerable number of clock-affecting genes are yet to be discovered.

Furthermore, since *S. pombe* can been grown on defined media under a variety of different conditions, it will be possible to isolate clock-related mutations that become effective only under limiting conditions. Good examples of this are the auxotrophic mutations in *Neurospora crassa* affecting fatty acid and phospholipid composition which show their dramatic effect on the circadian period only as long as the organism is not grown on rich medium (for review see Coté et al. 1996).

S. pombe is a very simple eukaryotic organism. This means that it has evolved the full basic repertoire of eukaryotic-specific proteins but many of these are not yet essential for survival (at least under laboratory conditions). The example of the cAMP/protein kinase A pathway being required for the ultradian rhythm in cell division has been given above (in section, p. 176 ff); by studying *S. pombe* we now have the perhaps unique opportunity of elucidating the role this pathway has in clock functioning. On the other hand, the signalling machinery in particular has not yet reached the level of redundancy that we experience with higher eukaryotes where many regulatory proteins occur in multigene families whose members have overlapping functions (where one often experiences the situation that deletion of a single gene is without effect, whereas deletion of all of them is lethal). Taken together, being both complete and simple may be a major advantage of this organism. This should help to investigate the minimum requirement of cellular functions that is essential for a normal working of the clock - an aspect which is also most important in evolutionary terms.

Another question in evolutionary chronobiology is that of the conservation of clock mechanisms. *S. pombe* is an excellent host for the expression of heterologous proteins, which is a prerequisite for any complementation studies. Given the example of the unexpected differences in the regulation of the *per* homologues in two insect species (see p. 168 f), the significance of complementation studies is obvious. Attempts to find homologues from higher eukaryotes that can complement clock mutations in *S. pombe,* or affect the clock when overexpressed, should provide important clues about the degree of conservation of clock mechanisms.

The ultradian clock of *S. pombe* is now the best characterised oscillation of its kind. Since there is considerable evidence that circadian and ultradian clocks share at least some aspects of their mechanisms (see Kippert, 1997a), ultradian clocks may serve well as a model system for their circadian counterparts. The shorter time scale of ultradian experiments allows additional approaches for which circadian clocks may be less suited, e.g. experiments employing the induction and repression of clock-related genes. The relationship between the different clocks is an important issue in regard to both the evolution of these clocks and their mechanism(s). Of particular informative value here may be the analysis of what happens in the cell when it switches from one periodicity to another as growth rate changes.

In the present review I have stressed repeatedly that eukaryotic timekeeping mechanisms apparently involve eukaryote-specific proteins and have discussed their relationship to the cell division cycle, cellular signalling pathways and the cellular stress response. It therefore seems appropriate to end with some recent results on an *S. pombe* protein which appears to constitute a link between all these aspects. In the course of my systematic analysis of potential 'clock gene' candidates among already characterised *S. pombe* genes, the *wis2*$^{+}$ gene has been iden-

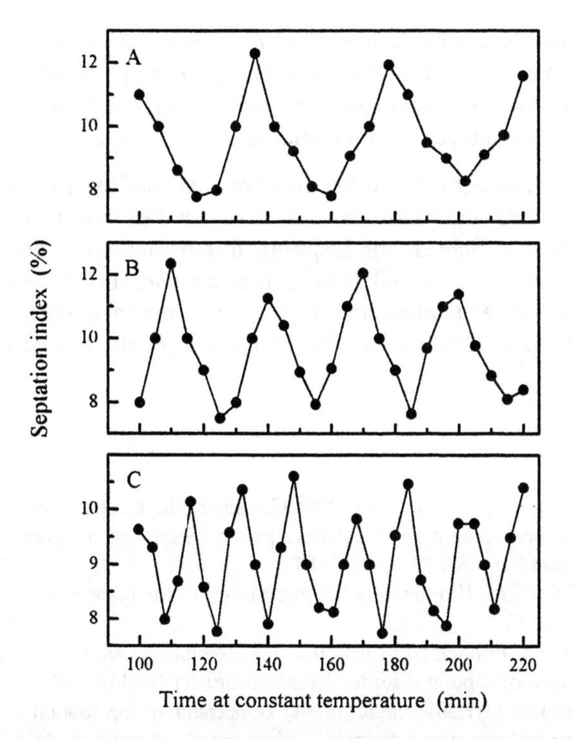

Fig. 3A-C. Effect of the level of transcription of the *wis2*$^{+}$ gene on the period of the ultradian rhythm of cell division in *Schizosaccharomyces pombe.* **A** Deletion mutant; **B** Wild type level of expression; **C** Strong overexpression of the *wis2*$^{+}$ gene

tified as affecting the period of the ultradian clock. A deletion mutant in the *wis2*$^{+}$ gene had a long period of 44 min which is rather outside the normal range (Fig. 3A). I therefore tested whether re-introduction of the gene affected period, and found that it indeed brought the period back to a rather wild type-like 30 min (Fig. 3B). Strong overexpression then resulted in a further dramatic shortening of the period to 17 min (Fig. 3C). Thus, the case of *wis2*$^{+}$ represents the first observation of a gene whose expression level directly affects the period of a cellular clock.

The character of the *wis2*$^{+}$ gene is such that that it exemplifies the possible interactions between clock function and various aspects of cellular physiology: (1) the gene was originally isolated in a cell cycle-related screen (Weisman et al. 1996); (2) the gene product is a cyclophilin, a class of molecules which act as molecular chaperones; it is expressed at constitutive level but superinduced by heat shock; (3) cyclophilins of this particular subtype are involved in cellular signalling (Pennisi, 1996). Considering the multiplicity of these interactions, the further analysis of how *wis2*$^{+}$ expression affects the ultradian clock promises to be an exciting piece of clock research. The proteins which interact with the *wis2*$^{+}$ gene product are now prime candidates for further clock-related functions. It seems quite likely that *S. pombe*, as one of the most recent systems in chronobiology, will soon become the one with the highest number of identified 'clock genes' with known cellular function - a perspective which strongly recommends its value in both evolutionary and molecular chronobiology.

Acknowledgements. I am grateful to the many members of the yeast community who provided the strains and plasmids used in my studies described here. I am grateful to Prof. H.E.A. Schenk for inspiring discussions on endocytobiology during my time in Tübingen, as well as his patient support during the preparation of this review. The critical reading and discussion of the manuscript by Harriet McWatters, Paul Hunt, and Prof. Murdoch Mitchison is greatly appreciated.

References

Aaronson BD, Johnson KA, Dunlap JN (1994) Circadian clock locus *frequency*-protein encoded by a single open reading frame defines period length and temperature compensation. Proc Natl Acad Sci USA 91: 7683-7687

Anderson SL, Kay SA (1996) Illuminating the mechanism of the circadian clock in plants. Trends Plant Sci 1: 51-57

Atack JR, Broughton HB, Pollack SJ (1995) Inositol monophosphatase - a putative target for Li^+ in the treatment of bipolar disorder. Trends Neurosci 18: 343-349

Balzer I, Neuhaus-Steinmetz U, Hardeland R (1989a) Temperature-compensation in an ultradian rhythm of tyrosine aminotransferase activity in *Euglena gracilis*. Experientia 45: 476-477

Balzer I, Neuhaus-Steinmetz U, Quentin E et al. (1989b) Concomitance of circadian and circa-4- hour ultradian rhythms in *Euglena gracilis*. J Interdiscipl Cycle Res 20: 15-24

Barnes SA, McGrath RB, Chua NH (1997) Light signal transduction in plants. Trends Cell Biol 7: 21-26

Bergman B, Gallon JR, Rai N, Stal LJ (1997) N_2 fixation by non-heterocystous cyanobacteria. FEMS Microbiol Reviews 19: 139-186

Berridge MJ (1993) Inositol trisphosphate and calcium signaling. Nature 361: 315-325

Bossier P, Goethals P, Rodrigues-Pousada C (1997) Constitutive flocculation in *Saccharomyces cerevisiae* through overexpression of the *GTS1* gene, coding for a 'GLO'-type Zn-finger containing protein. Yeast, *in press*

Bult CJ, White O, Olsen GJ et al. (1996) Complete genome sequence of the methanogenic archaeon, *Methanococcus jannaschii*. Science 273: 1058-1073

Carré IA, Kay SA (1996) Mechanisms of input and output in circadian transduction pathways. In: DP Verma, Signal Transduction in Plant Growth and Development. Springer, New York, pp 231-247

Carré IA, Oster AS, Laval-Martin DL, Edmunds LN Jr (1989) Entrainment and phase-shifting of the circadian rhythm of cell division by light in cultures of the achlorophyllous ZC mutant of *Euglena gracilis*. Curr Microbiol 19: 223-229

Comolli J, Taylor W, Hastings JW (1994) An inhibitor of protein phosphorylation stops the circa- dian oscillator and blocks light-induced phase shifting in *Gonyaulax*. J Biol Rhythms 9: 13-26

Comolli J, Taylor W, Rehman J, Hastings JW (1996) Inhibitors of serine/threonine phosphoprotein phosphatases alter circadian properties in *Gonyaulax* polyedra. Plant Physiol 111: 285-291

Coté G, Lakin-Thomas PL, Brody S (1996) Membrane lipids and circadian rhythms in *Neurospora crassa*. In: T Vanden Driessche (ed) Membranes and Circadian Rhythms. Springer, New York, pp 13-46

Dunlap JC (1996) Genetic and molecular analysis of circadian rhythms. Annu Rev Genet 30: 579-601

Dunlap JC, Feldman JF (1988) On the role of protein synthesis in the circadian clock of *Neurospora crassa*. Proc Natl Acad Sci USA 85: 1096-1100

D'Urso G, Nurse P (1995) Checkpoints in the cell cycle of fission yeast. Curr Opin Genet Dev 5: 12-16

Edmunds LN Jr (1988) Cellular and Molecular Bases of Biological Clocks. Models and Mechanisms for Circadian Timekeeping. Springer, New York

Edmunds LN Jr (1996) Cross-talk between clocks: regulation of cell division cycles by circadian oscillators. In: T Vanden Driessche, Membranes and Circadian Rhythms. Springer, New York, pp 95-124

Edmunds LN Jr, Carré IA, Tamponnet C, Tong J (1992) The role of ions and second messengers in circadian clock functions. Chronobiol Internat 9: 180-200

Ellis RJ (ed) (1996) The Chaperones. Academic Press

Fleischman RD, Adams, MD, White, O et al. (1995) Whole-genome random sequencing and assembly of *Haemophilus influezae* RD. Science 269: 496-512

Gallon JR, Hashem MA, Chaplin AE (1991) Nitrogen fixation by *Oscillatoria* spp. under autotrophic and photoheterotrophic conditions. J Gen Microbiol 137: 31-39

Goffeau A, Barr BG, Busse R et al. (1996) Life with 6000 genes. Science 274: 546-567

Goto K, Johnson CH (1995) Is the cell division cycle gated by a circadian clock? The case of *Chlamydomonas reinhardtii*. J Cell Biol 129: 1061-1069

Hall JC (1995) Tripping along the trail to the molecular mechanisms of biological clocks. Trends Neurosci 18: 230-240

Hall JC (1996) Are cycling gene products as internal zeitgebers no longer the zeitgeist of chronobiologists? Neuron 17: 199-802

Hardeland R (1994) Periodic gene expression as an element of cellular oscillators? In: R Hardeland (ed) Cell Biological Problems in Chronobiology. University of Göttingen, Göttingen, pp 6-12

Hasegawa K, Tsukahara Y, Shimamoto M et al. (1995) A mechanism regulating circadian changes in the resting membrane potential in *Paramecium*. Biol Rhythm Res 26: 398

Hayles J, Nurse P (1992) Genetics of the fission yeast *Schizosaccharomyces pombe*. Annu Rev Genet 26: 373-402

Homma K, Hastings JW (1989a) Cell growth kinetics, division asymmetry and volume control at division in the marine dinoflagellate *Gonyaulax polyedra:* a model of circadian clock control of the cell cycle. J Cell Sci 92: 303-318

Homma K, Hastings JW (1989b) The S phase is discrete and is controlled by the circadian clock in the marine dinoflagellate *Gonyaulax polyedra*. Exp Cell Res 182: 635-644

Huang T-C, Grobbelaar N (1995) The circadian clock of the prokaryote *Synechococcus* RF-1. Microbiology 141: 535-540

Huang T-C, Tu J, Chow T-J, Chen T-H (1990) Circadian rhythm of the prokaryote *Synechococcus* RF-1. Plant Physiol 92: 531-533

Hunter T, Plowman GD (1997) The protein kinases of budding yeast: six score and more. TIBS 22: 18-22

Hwang S, Kawazoe R, Herrin DL (1996) Transcription of *tufA* and other chloroplast-encoded genes is controlled by a circadian clock in *Chlamydomonas*. Proc Natl Acad Sci USA 93: 996-1000

Iwasaki K, Liu DWC, Thomas JT (1995) Genes that control a temperature-compensated ultradian clocks in *Caenorhabditis elegans*. Proc Natl Acad Sci USA 92: 10317-10321

Jeon KW (1995) The large, free-living amoeba: wonderful cells for biological studies. J Eukaryot Microbiol 42: 1-7

Johnson CH, Knight MR, Kondo T et al. (1995) Circadian oscillations of cytosolic and chloroplastic free calcium in plants. Science 269: 1863-1865

Johnson CH, Golden SS, Ishiura M, Kondo T (1996) Circadian clocks in prokaryotes. Mol Microbiol 21: 5-11

Kaneko T, Sato S , Kotani H et al. (1996) Sequence analysis of the genome of the unicellular cyanobacterium *Synechocystis* sp. strain PCC6803. II. Sequence of the entire genome and assignment of potential protein-coding regions. DNA Res 3: 109-136

Kennelly PJ, Potts M (1996) Fancy meeting you here! A fresh look at 'prokaryotic' protein phosphorylation. J Bacteriol 178: 4759-4764

Kippert F (1987) Endocytobiotic coordination, intracellular calcium signaling, and the origin of endogenous rhythms. In: Lee JJ, Fredrick JF (eds) Endocytobiology III, Ann NY Acad Sci 503: 476-495

Kippert F (1989) Circadian control of heat tolerance in stationary phase cultures of *Schizosaccharomyces pombe*. Arch Microbiol 151: 177-179

Kippert F (1991) Essential clock proteins/circadian rhythms in prokaryotes - what is the evidence? Bot Acta 103: 2-4

Kippert F (1992) Ultradian and circadian clocks - two sides of one coin? J Interdiscipl Cycle Res 23: 192-196

Kippert F (1996a) An ultradian clock controls locomotor behaviour and cell division in isolated cells of *Paramecium tetraurelia*. J Cell Sci 108: 867-873

Kippert F (1996b) The temperature-compensated clock of *Tetrahymena*: Oscillations in respiratory activity and cell division. Chronobiol Intern 13: 1-13

Kippert F (1996c) Long-term recordings of a novel circadian rhythm in *Synechococcus* RF-1 reveal an exact 24 hours period. Abstract # 253, 1st Eur Phycol Congr Cologne, Germany

Kippert F (1997a) Temperature-compensation of ultradian rhythms: a general homeostasis of period length identifies ultradian clocks as timekeeping devices. Submitted

Kippert F (1997b) The ultradian clock of *Schizosaccharomyces pombe* shows a general homeostasis of period length under diverse growth conditions. Submitted

Kippert F (1997c) Genetic variability of period length of the *Schizosaccharomyces pombe* ultradian clock: the case of a haploid organism. Submitted

Kippert F (1997d) Activity of the *wee1*$^{+}$ protein kinase is required for ultradian clock control over mitosis in *Schizosaccharomyces pombe*. In preparation

Kippert F (1997e) Inositol monophosphatase is the target for the period lengthening effect of lithium on the ultradian clock of *Schizosaccharomyces pombe*. Submitted

Kippert F, Lloyd D (1995) A temperature-compensated ultradian clock ticks in *Schizosaccharomyces pombe*. Microbiology 141: 883-890

Kippert F, Lloyd D (1997a) Rhythms in respiration, fermentation, and medium acidification: outputs of an ultradian clock in fast growing *Schizosaccharomyces pombe*. Submitted

Kippert F, Lloyd D (1997b) The ultradian clock of *Schizosaccharomyces pombe:* timing of cell cycle stages in fast growing cells. Submitted

Kippert F, Ninnemann H, Engelmann W (1991) Photosynchronization of the circadian clock of *Schizosaccharomyces pombe*: Mitochondrial cytochrome *b* is an essential component. Curr Genet 19: 103-107

Kippert F, Diebold S, Feil K (1997) Lithium dramatically slows down the ultradian clock of *Schizosaccharomyces pombe*. Submitted

Klemfuss H (1992) Rhythms and the pharmacology of lithium. Pharmacol Ther 56: 53-78

Kondo T, Strayer CA, Kulkarni RD et al. (1993) Circadian rhythms in prokaryotes: luciferase as a reporter gene of circadian gene expression in cyanobacteria. Proc Natl Acad Sci USA 90: 5672-5676

Kondo T, Tsinoremas NF, Golden SS et al. (1994) Circadian clock mutants of cyanobacteria. Science 266: 1233-1236

Kondo T, Mori T, Lebedeva NV et al. (1997) Circadian rhythms in rapidly dividing cyanobacteria. Science 275: 224-228

Koumenis C, Nunez-Regueiro M, Raju U et al. (1995) Identification of three proteins in the eye of *Aplysia*, whose synthesis is altered by serotonin. J Biol Chem 270: 14619-14627

Krucher NA, Roberts MH (1994) Identification of CDK- and cyclin-like proteins in the eye of *Bulla gouldiana*. J Neurobiol 25: 1200-1206

Kyriacou CP, Oldroyd M, Wood J, et al. (1990) Clock mutations alter developmental timing in *Drosophila*. Heredity 64: 395-401

Kyriacou CP, Greenacre ML, Thackeray JR, Hall JC (1993) Genetic and molecular analysis of song rhythms in *Drosophila*. In: MW Young (ed) Molecular Genetics of Circadian Rhythms. Marcel Dekker, New York, pp 171-193

Levine JD, Casey CI, Kalderon DD, Jackson, FR (1994) Altered circadian pacemaker functions and cyclic AMP rhythms in the *Drosophila* learning mutant *dunce*. Neuron 13: 967-974

Lew J, Wang JH (1995) Neuronal cdc2-like kinase. TIBS 20: 33-37

Lewandowski MH, Domoslawski J, Balzer I et al. (1995) Demonstration of temperature compensation for ultradian rhythms of dark motility in *Euglena gracilis*. In: R Hardeland (ed) Cellular Rhythms and Indoleamines. University of Göttingen, Göttingen, pp 59-70

Lloyd D, Edwards SW, Fry JC (1982) Temperature-compensated oscillations in respiration and cellular protein content in synchronous cultures of *Acanthamoeba castellanii*. Proc Natl Acad Sci USA 79: 3785-3788

Loros JJ (1995) The molecular basis of the *Neurospora* clock. Semin Neurosci 7: 3-13

Michel U, Hardeland R (1985) On the chronobiology of *Tetrahymena*. III. Temperature compensation and temperature dependence in the ultradian oscillation of tyrosine aminotransferase. J interdiscipl Cycle Res 16: 17-23

Mitsui A, Kumazawa S, Takahashi A et al. (1986) Strategy by which nitrogen-fixing cyanobacteria grow photoautotrophically. Nature 323: 720-722

Mitsui A, Suda S, Hanagata N (1993) Cell cycle events at different temperatures in aerobic nitrogen-fixing marine unicellular cyanobacterium *Synechococcus* sp. strain Miami BG 043511. J Mar Biotechnol 1: 89-91

Miwa I, Yajima H (1995) Correlation of circadian rhythms with the length of immaturity in *Paramecium bursaria*. Zool Sci 12: 53-59

Mori T, Binder B, Johnson CH (1996) Circadian gating of cell division in cyanobacteria growing with average doubling times of less than 24 h. Proc Natl Acad Sci USA 93: 10183-10188

Nasmyth K (1995) Evolution of the cell cycle. Phil Trans R Soc Lond B 349: 271-281

Nigg EA (1995) Cyclin-dependent protein kinase: key regulators of the eukaryotic cell cycle. BioEssays 17: 471-480

Pelech SL (1996) Signalling pathways: kinase connections on the cellular intranet. Curr Biol 6: 551-554

Pennisi E (1996) Expanding the eukaryote's cast of chaperones. Science 274: 1613-1614

Prosser RA, Gillette MU (1991) Cyclic changes in cAMP concentration and phosphodiesterase activity in a mammalian circadian clock studied in vitro. Brain Res 568: 185-192

Rensing L (1996) Genetics and molecular biology of circadian clocks. In: PH Redfern, B Lemmer (eds) Handbook of Experimental Pharmacology. Physiology and Pharmacology of Biological Rhythms. Springer Verlag, *in press*

Rikin A (1992) Circadian rhythm of heat resistance in cotton seedlings: synthesis of heat-shock proteins. Eur J Cell Biol 59: 160-165

Roberts MH, Bedian V, Chen Y (1989) Kinase inhibition lengthens the period of the circadian pacemaker in the eye of *Bulla gouldiana*. Brain Res 504: 211-215

Roenneberg T (1996) Complex circadian system of *Gonyaulax*. Physiol Plant 97: 733-737

Rojek R, Harms C, Hebeler M, Grimme LH (1994) Cyclic variations of photosynthetic activity under nitrogen-fixing conditions in *Synechococcus* RF-1. Arch Microbiol 162: 80-84

Sassone-Corsi P (1996) Circadian rhythms - same clock, different works. Nature 384: 613-614

Sauman I, Reppert SM (1996) Circadian clock neurons in the silkmoth *Anteraea perny* - novel mechanisms of period protein regulation. Neuron 17: 889-900

Schmidt I, Balzer I (1995) Light perception and circadian rhythms in *Euglena gracilis*. In: R Hardeland (ed) Cellular Rhythms and Indoleamines. University of Göttingen, pp 71-78

Sweeney BM, Haxo FT (1961) Persistence of a photosynthetic rhythm in enucleated *Acetabularia*. Science 134: 1361-1363

Tokushima H, Okamoto K-I, Miwa I, Nakaoka Y (1994) Correlation between circadian periods and cellular activities in *Paramecium bursaria*. J Comp Physiol A 175: 767-772

Van Dolah FM, Leighfield TA, Sandel HD, Hsi CK (1995) Cell division in the dinoflagellate *Gambierdiscus toxicus* is phased to the diurnal cycle and accompanied by activation of the cell cycle regulatory protein, CDC2 kinase. J Phycol 31: 395-400

Waskiewicz AJ, Cooper JA (1995) Mitogen and stress response pathways: MAP kinase cascades and phosphatase regulation in mammals and yeast. Curr Opin Cell Biol 7: 798-805

Weisman R, Creanor J, Fantes P (1996) A multicopy suppresor of a cell cycle defect encodes a heat shock-inducible 40 kDa cyclophilin-like protein. EMBO J 15: 447-456

Wong A, Boutis P, Hekimi, S (1995) Mutations of the *clk-1* gene of *Caenorhabditis elegans* affect developmental and behavioral timing. Genetics 139: 1247-1259

Woodgett JR, Kyriakis JM, Avruch J et al. (1996) Reconstitution of novel signalling cascades responding to cellular stresses. Phil Trans R Soc Lond B 151: 135-142

Yaguchi S-I, Mitsui K, Kabawata K-I et al. (1996) The pleiotropic effect of the *GTS1* gene product on heat tolerance, sporulation and the life span of *Saccharomyces cerevisiae*. Biochem Biophys Res Commun 218: 234-237

Zhang, C-C (1996) Bacterial signalling involving eukaryotic type protein kinases. Mol Microbiol 20: 9-15

1.3 Protein Import into Cell Organelles - Exogenosomes and Endogenosomes

Evolution of Protein Sorting Signals

G. von Heijne
Department of Biochemistry, Stockholm University, 106 91 Stockholm, Sweden

Summary: Sorting signals route proteins to the correct subcellular compartment and define classes of evolutionarily conserved protein motifs. Their surprisingly low degree of sequence conservation suggests that partially functional sorting signals may arise continuously during evolution, and thus that existing proteins may be continually tested in new compartments. Intertaxonic combination events such as those involved in the establishment of mitochondria and chloroplasts have necessitated the evolution of new classes of organellar import signals, and have resulted in interesting hybrids between preexisting and newly created sorting signals.

Introduction

In the mammalian cell, the most important primary protein sorting events lead to the secretory pathway, to mitochondria, to the nucleus, and to peroxisomes (von Heijne 1996). While the secretory pathway delivers proteins to all compartments between the ER and the plasma membrane as well as to the lysosome, some proteins can also be directly translocated across the plasma membrane (Muesch et al. 1990) or into lysosomes and vacuoles (Seguireal et al. 1995). In plant cells, the chloroplast is another important organelle that imports most of its proteins from the cytoplasm.

Endosymbiotic events are evident both in the similarity between different classes of sorting signals and in the evolutionary conservation of certain components of the various protein translocation machineries.

The Secretory Pathway

Strong homologies exist between components of the protein translocation systems found in the inner membrane of bacteria, the ER membrane of higher cells, and the thylakoid membrane of chloroplasts (Schatz and Dobberstein 1996), suggesting a common origin. In addition, the signal peptides that target proteins to these translocation machineries are related in overall design: a positively charged N-terminal region, a central hydrophobic stretch, and a more polar C-terminal segment (von Heijne et al. 1989; von Heijne 1990). In all three systems, the signal peptide is removed by signal peptidases with similar substrate specificities (Dalbey and von Heijne 1992). It has also been found that signal peptides may function in more than one of these systems (Talmadge et al. 1980; Wiedmann et al. 1984; Seidler and Michel 1990).

It is thus clear that the bacterial inner membrane translocation system can be viewed as an evolutionary precursor to the ER and thylakoid systems. The latter case is particularly interesting, since plastid-encoded proteins like cytochrome *f* have a "classical" signal peptide that targets the thylakoid (Rothstein et al. 1985; Anderson and Gray 1991), whereas nuclear-encoded thylakoid proteins have composite targeting signals where a stroma-targeting transit peptide precedes the thylakoid-targeting signal peptide (von Heijne et al. 1989).

Mitochondrial Import

So far, none of the components identified in the mitochondrial outer and inner membrane translocation machineries (Ryan and Jensen 1995) have known homologues in other organelles. The mitochondrial protein import machinery thus seems to be unique. Likewise, mitochondrial targeting peptides are distinct from the secretory signal peptides, both in terms of overall design and in terms of their cleavage sites. In general, they are rich in basic residues and tend to form amphipathic α-helices in detergent micelles or lipid bilayers (Roise 1993), and they are cleaved from the mature protein by matrix-localized proteases with substrate specificities that are very different from the signal peptidases (Ou et al. 1994; Branda and Isaya 1995).

The only known link between the protein-sorting machinery of mitochondria and those involved in secretion and thylakoid import is provided by two related proteases in the intermembrane space that cleave sorting signals on certain proteins located in this compartment (Nunnari et al. 1993). Not only is the substrate specificity of one of these two proteases similar to the signal peptidases, but the signals that are removed also have a central hydrophobic stretch (von Heijne et al. 1989). There is an ongoing debate over whether nuclear-encoded intermembrane space proteins are sorted directly to their final location (the so-called "stop-transfer" model) or whether they reach the intermembrane space via the matrix (the "conservative sorting" model) (Rospert et al. 1994; Gärtner et al. 1995; Ono et al. 1995; Rojo et al. 1995).

Chloroplast Import

Not many components of the import machinery in the outer and inner envelope membranes of chloroplasts are known, and those that have been identified have no detectable sequence homology to any of the other translocation systems discussed here (Alefsen et al. 1994; Soll 1995). As noted above, however, the thylakoid protein import machinery is clearly related to the bacterial inner membrane translocation system, although studies of the energetic requirements of thylakoid import suggest that there are additional variations on the pathway not found in bacteria (Robinson and Klösgen 1994).

Higher plant stroma-targeting transit peptides are distinct both from secretory signal peptides and mitochondrial targeting peptides: they are enriched in hydroxylated amino acids (and a number of transit peptides have been shown to be phosphorylated *in vitro* (Waegemann and Soll 1996)) and generally lack acidic residues (von Heijne et al. 1989). Whether a particular secondary structure is induced upon contact with the chloroplast surface is unclear. A high affinity for certain galactolipids found only in chloroplast membranes has been suggested to provide part of the targeting specificity (van't Hof et al. 1991).

Certain algae such as *Chlamydomonas reinhardtii* seem to have evolved transit peptides that look more like mitochondrial targeting peptides than higher plant transit peptides (Franzén et al. 1990). Whether this means that these organisms also have components in their chloroplast import machinery related to the mitochondrial import system is not known.

Conclusion

From what we know today, it seems that the secretory machinery of prokaryotes was the first protein targeting/translocation system that evolved, and that parts of this system have survived in endosymbiotic organelles such as chloroplasts an mitochondria. The import machineries that deliver nucelar-encoded proteins into these organelles do not seem to be related, however, suggesting independent evolutionary origins.

References

Alefsen H, Waegemann K, Soll J (1994) Analysis of the chloroplast protein import machinery. J Plant Physiol 144: 339-345

Anderson CM, Gray J (1991) Cleavage of the precursor of pea chloroplast cytochrome-f by leader peptidase from *Escherichia coli*. FEBS Lett 280: 383-386

Branda SS, Isaya G (1995) Prediction and identification of new natural substrates of the yeast mitochondrial intermediate peptidase. J Biol Chem 270: 27366-27373

Dalbey RE, von Heijne G (1992) Signal peptidases in prokaryotes and eukaryotes - a new protease family. Trends Biochem Sci 17: 474-478

Franzén LG, Rochaix JD, von Heijne G (1990) Chloroplast transit peptides from the green alga *Chlamydomonas reinhardtii* share features with both mitochondrial and higher plant chloroplast presequences. FEBS Lett 260: 165-168

Gärtner F, Bömer U, Guiard B, Pfanner N (1995) The sorting signal of cytochrome b(2) promotes early divergence from the general mitochondrial import pathway and restricts the unfoldase activity of matrix Hsp70. EMBO J 14: 6043-6057

Muesch A, Hartmann E, Rohde K, Rubartelli A, Sitia R, Rapoport TA (1990) A novel pathway for secretory proteins. Trends Biochem Sci 15: 86-88
Nunnari J, Fox TD, Walter P (1993) A mitochondrial protease with two catalytic subunits of nonoverlapping specificities. Science 262: 1997-2004
Ono H, Gruhler A, Stuart RA, Guiard B, Schwarz E, Neupert W (1995) Sorting of cytochrome b(2) to the intermembrane space of mitochondria - Kinetic analysis of intermediates demonstrates passage through the matrix. J Biol Chem 270: 16932-16938
Ou WJ, Kumamoto T, Mihara K, Kitada S, Niidome T, Ito A, Omura T (1994) Structural requirement for recognition of the precursor proteins by the mitochondrial processing peptidase. J Biol Chem 269: 24673-24678
Robinson C, Klösgen RB (1994) Targeting of proteins into and across the thylakoid membrane - A multitude of mechanisms. Plant Mol Biol 26: 15-24
Roise D (1993) The amphipathic helix in mitochondrial targeting sequences. In: Epand RM (ed) The amphipathic helix. CRC Press, Boca Raton, pp 257-283
Rojo EE, Stuart RA, Neupert W (1995) Conservative sorting of F0-ATPase subunit 9: Export from matrix requires Delta pH across inner membrane and matrix ATP. Embo J 14: 3445-3451
Rospert S, Müller S, Schatz G, Glick BS (1994) Fusion proteins containing the cytochrome b2 presequence are sorted to the mitochondrial intermembrane space independently of hsp60. J Biol Chem 269: 17279-17288
Rothstein S, Gatenby A, Willey D, Gray J (1985) Binding of pea cytochrome f to the inner membrane of *Escherichia coli* requires the bacterial secA gene product. Proc Natl Acad Sci USA 82: 7955-7959
Ryan KR, Jensen RE (1995) Protein translocation across mitochondrial membranes: What a long, strange trip it is. Cell 83: 517-519
Schatz G, Dobberstein B (1996) Common principles of protein translocation across membranes. Science 271: 1519-1526
Seguireal B, Martinez M, Sandoval IV (1995) Yeast aminopeptidase I is post-translationally sorted from the cytosol to the vacuole by a mechanism mediated by its bipartite N-terminal extension. EMBO J 14: 5476-5484
Seidler A, Michel H (1990) Expression in *Escherichia coli* of the psbO gene encoding the 33 kd protein of the oxygen-evolving complex from spinach. EMBO J 9: 1743-1748
Soll J (1995) New insights into the protein import machinery of the chloroplast's outer envelope. Bot Acta 108: 277-282
Talmadge K, Stahl S, Gilbert W (1980) Eukaryotic signal sequence transports insulin antigen in *Escherichia coli*. Proc Natl Acad Sci USA 77: 3369-3373
van't Hof R, Demel RA, Keegstra K, de Kruijff B (1991) Lipid peptide interactions between fragments of the transit peptide of Ribulose-1,5-bisphosphate carboxylase oxygenase and chloroplast membrane lipids. FEBS Lett 291: 350-354
von Heijne G (1990) The signal peptide. J Membr Biol 115: 195-201
von Heijne G (1996) Targeting signals for protein import into mitochondria and other subcellular organelles. In: Hartl F-U (ed) Advances in molecular and cell biology. Vol 43, JAI Press
von Heijne G, Steppuhn J, Herrmann RG (1989) Domain structure of mitochondrial and chloroplast targeting peptides. Eur J Biochem 180: 535-545
Waegemann K, Soll J (1996) Phosphorylation of the transit sequence of chloroplast precursor proteins. J Biol Chem 271: 6545-6554
Wiedmann M, Huth A, Rapoport TA (1984) Xenopus oocytes can secrete bacterial betalactamase. Nature 309: 637-639

Protein Import Into Peroxisomes

R. Erdmann and W.-H. Kunau
Ruhr-Universität Bochum, Institut für Physiologische Chemie, Abteilung Zellbiochemie, 44780 Bochum, Germany

Key words: Peroxins, peroxisomal targeting signals, peroxisome biogenesis.

Introduction

Peroxisomes are small, single membrane-bound organelles occurring ubiquitously in eukaryotic cells. They do not contain DNA and lack an independent protein synthesizing machinery. The number, size, protein composition, and biochemical functions of these organelles vary between cell types and in response to environmental conditions. Because of the multiplicity of peroxisomal biochemical functions, peroxisomes can best be described as "multipurpose" organelles. Common to most peroxisomes is the ß-oxidation of fatty acids and the consumption of H_2O_2 by catalase. More specialized functions include the synthesis of cholesterol, bile acid, and ether lipids in mammals, the glyoxylate cycle in plants and fungi, methanol oxidation in yeasts, and glycolysis in trypanosomatids (van den Bosch 1992). In humans, defects of structure and/or function of peroxisomes give rise to a group of genetically distinct, usually fatal inborn errors, the peroxisomal disorders (Lazarow and Moser 1989). Their discovery about a decade ago triggered the dramatic increase of interest of cell biologists in this organelle.

Originally, peroxisomes were thought to derive by budding from the endoplasmic reticulum (Novikoff and Shin 1964; De Duve and Baudhuin 1966). The widely accepted view now is that the organelles arise by budding or fission from preexisting peroxisomes (Lazarow and Fujiki 1985), although evidence for de novo synthesis of peroxisomes has been presented which would not be consistent with this model (Sulter et al. 1993; Waterham et al. 1993; Baerends et al. 1996). The ability to direct proteins accurately and efficiently to different compartments is a fundamental feature of eukaryotic cells and requires an elaborate delivery system. In this respect peroxisomes are similar to mitochondria, chloroplasts, and the nucleus in that their luminal proteins are synthesized on free cytosolic ribosomes and posttranslationally imported into the organelles (Lazarow and Fujiki 1985; Subramani 1992). In the last decade remarkable progress has been made in the elucidation of the mechanisms underlying peroxisomal protein import which to summarize is the purpose of this review.

Peroxisomal Targeting Signals

The intracellular sorting of proteins is mediated by defined signal sequences on the target protein and the recognition of such signals by specific signal recognition factors which direct the protein to its destination. A major contribution to our understanding of peroxisomal protein import was the discovery of two peroxisomal targeting signals (PTSs) which were shown to target proteins to the peroxisomal matrix (Subramani 1992).

PTS1, first discovered in firefly luciferase by Gould et al. (1987), comprises the C-terminal three amino acids of the majority of peroxisomal matrix proteins and consists of species-specific variations of the tripeptide consensus Ser-Lys-Leu. The majority of the known peroxisomal matrix proteins are targeted to the organelles by a PTS1, although different variants of the SKL motif are tolerated by the different species (Gould et al. 1990a, b). In mammalian cells, allowable substitutions are A or C for S, R or H for K, and M for L, but these variants sort with different efficiencies. In some yeasts C-terminal tripeptides NKL, ARF, or AKI can substitute for SKL (Aitchison et al 1991).

PTS2, first found by Swinkels et al. (1991), resides within the N-terminal first 20-30 amino acids of a number of peroxisomal matrix proteins. Alignment as well as site directed mutagenesis of PTS2 containing proteins led to the PTS2 consensus sequence Arg-Lys/Ile-5x-His-Leu (Swinkels et al. 1991; Osumi et al. 1991; Osumi et al. 1992; Glover et al. 1994; Erdmann 1994, Faber et al. 1944). Interestingly, in analogy to mitochondrial protein import, the signal sequence of some PTS2 containing proteins, including the mammalian peroxisomal thiolases, is cleaved upon import and a signal sequence processing protease has recently been localized to the peroxisomal matrix. However, in contrast to the mitochondrial import pathway, the ability to process the signal sequence seems not to be a prerequisite for peroxisomal protein import (Gietl et al. 1994; Authier et al. 1995).

Because several peroxisomal matrix proteins lack a consensus PTS1 or PTS2 and others have been reported to be sorted by internal regions which do not resemble PTS1 or PTS2 (Small et al. 1988, Kragler et al. 1993), the existence of additional peroxisomal targeting signals has to be taken into consideration. However, the recent finding that proteins may be imported into peroxisomes as dimers or homomultimers (see below) opens the possibility that proteins might also get imported as heteromultimers. Proteins lacking a PTS might then hitch a ride into peroxisomes by association with PTS1- or PTS2-containing proteins. A sequence feature of the interaction with a PTS-containing protein might then mistakenly be defined as a peroxisomal targeting signal.

There is considerable evidence that targeting of peroxisomal integral membrane proteins seems to be PTS1- and PTS2-independent. Thus, the existence of special peroxisomal membrane targeting signals (mPTS) may be anticipated, but so far no consensus sequence for a peroxisomal membrane targeting signal has been defined. However, PMP47 from *Candida boidinii* has been shown to be targeted to the peroxisomal membrane by a 20 amino acid hydrophilic loop separating two transmembrane segments (Dyer et al. 1996). In contrast, the N-terminal 45 amino acids of Sc-Pex3p which comprise the putative membrane span of this protein are sufficient to target reporter proteins to peroxisomes and anchor them into the membrane (Krause 1995).

Putative Components of the Peroxisomal Protein Import Machinery

A major breakthrough in our understanding of peroxisome biogenesis at the molecular level came from genetic approaches in yeast species such as *Saccharomyces cerevisiae*, *Hansenula polymorpha*, *Pichia pastoris*, and *Yarrowia lipolytica* as well as Chinese hamster ovary (CHO) cells and fibroblasts from patients suffering from peroxisomal disorders (Zoeller and Raetz 1986; Erdmann et al. 1989; Tsukamoto et al. 1991; Shimozawa et al. 1992; Gould et al. 1992; Waterham et al. 1992; Nuttley et al. 1993; Lazarow 1993). These genetic approaches have been the primary route to the identification of proteins essential for peroxisome biogenesis. The diversity of experimental systems led to a profusion of names for peroxisome assembly genes and proteins, including the acronyms PAS, PAF, PER, PAY, PEB, and PMP and an even greater array of numbering systems. Recently we unified the nomenclature for these factors for peroxisome biogenesis and from now on, all proteins involved in peroxisome assembly will be defined as peroxins with PEX being the gene acronym (Distel et al. 1996). Although a precise function has not yet been elucidated for most of these proteins, some of them are characterized by defined sequence motifs which might provide a first clue to their specific role in peroxisome biogenesis. The so far known peroxins are summarized in Table 1.

Pex1p and **Pex6p** (Erdmann et al. 1991; Heyman et al. 1994; van der Klei et al. 1996; Voorn-Brouwer et al. 1993; Spong and Subramani 1993; Nuttley et al. 1994; Tsukamoto et al. 1995), belong to the family of AAA proteins (ATPases associated with a variety of cellular activities (Kunau et al. 1993). Members of this protein family are characterized by the presence of one or two AAA domains, homologous 200 amino acid modules which contain the consensus sequence for ATP binding.

Pex2p, **Pex10p**, and **Pex12p** are characterized by the presence of a characteristic zinc finger motif, the C3HC4 "ring" finger (Tsukamoto et al. 1991; Shimozawa et al. 1992; Erdmann and Kunau 1992; Tan et al. 1995; Kalish et al. 1995 Liu et al. 1996). These motifs are likely to be involved in protein-protein interactions; however, the corresponding partner proteins have not yet been identified.

Pex3p is an integral peroxisomal membrane protein which contains short luminal N-termini followed by one transmembrane segment while the major part of the proteins face the cytosol (Höhfeld et al. 1991; Baerends et al. 1996).

Pex4p belongs to the family of ubiquitin conjugating enzymes (UBC) (Wiebel and Kunau 1992; Crane et al. 1994; van der Klei et al. 1996), suggesting that at least one ubiquitination reaction may be required for peroxisome biogenesis.

Table 1. Summary of peroxemblins

Gene	Peroxin Characteristics	Authors
PEX1	117-127 kDa; belongs to the family of AAA-ATPases; contains two AAA domains; intracellular localization not yet detected	Erdmann et al. (1991); Heyman et al. (1994)
PEX2	35-52 kDa; contains characteristic C_3HC_4-zinc finger motif, integral peroxisomal membrane protein	Tsukamoto et al. (1991); Shimozawa et al. (1992); Berteaux et al. (1995); Waterham et al. (1996)
PEX3	51-52 kDa; integral peroxisomal membrane protein	Höhfeld and Kunau (1991); Baerends et al. (1996); Wiemer et al. (1996)
PEX4	21 to 24 kDa, ubiquitin conjugating protein, associated with the peroxisomal membrane	Wiebel and Kunau (1992); Crane et al. (1994)
PEX5	64-69 kDa; contains at least six TPR motifs; PTS1 recognition factor; localized to the cytosol as well as to the peroxisomal membrane and matrix	McCollum et al. (1993); van der Leij et al. (1993); Dodt et al. (1995); Wiemer et al. (1995); Fransen et al. (1995); van der Klei et al. (1995); Nuttley et al. (1995); Szilard et al. (1995)
PEX6	112-127 kDa; belongs to the family of AAA-ATPases; contains two AAA-domains; has been localized to the cytosol	Spong and Subramani. (1993); Voorn-Brouwer et al.(1993); Nuttley et al. (1994); Tsukamoto et al. (1995); Yahraus et al. (1996)
PEX7	42 kDa; contains seven WD40 motifs; PTS2 recognition factor; localized to the cytosol as well as to the peroxisomal membrane and matrix	Marzioch et al. (1994); Zhang and Lazarow (1995)
PEX8	71-81 kDa; contains both a C-terminal PTS1 and an N-terminal PTS2; has been localized to the peroxisomal matrix and inner aspects of the peroxisomal membrane	Waterham et al. (1994); Liu et al. (1995); Rehling (1996)
PEX9	42 kDa; integral peroxisomal membrane protein	Eitzen et al. (1995)
PEX10	34-48 kDa; integral peroxisomal membrane protein; contains C3HC4 zinc finger motif, suggested to be involved in peroxisome proliferation or lumen formation	Tan et al. (1995); Kalish et al. (1995)
PEX11	27-32 kDa; peroxisomal membrane protein, involved in peroxisome proliferation; deficiency results in giant peroxisomes	Erdmann and Blobel (1995); Marshall et al. (1995); Moreno et al. (1995)
PEX12	48 kDa; contains a degenerated C3HC4 zinc finger motif.	Kalish et al. (1996)
PEX13	43 kDa; C-terminal SH3-domain, membrane receptor for the PTS1 recognition factor; putative docking protein for peroxisomal protein import	Elgersma et al. (1996); Erdmann and Blobel (1996); Gould and Crane (1996)
PEX14	38 kDa; peripheral peroxisomal membrane protein.	Komori (1997); Albertini et al. (1997)
PAS9	23 kDa; localized at the cytoplasmic surface of the peroxisome	Huhse (1995)
PAS12	40 kDa; localized in the cytosol as well as the cytosolic surface of peroxisomes. The protein contains a C-terminal consensus sequence for farnesylation	Götte (1995)
PAS21	43 kDa; peroxisomal integral membrane protein	Elgersma (1995)
PAS22	48 kDa; cytosolic DnaJ-homologue	Elgersma (1995)

Pex5p and **Pex7p** function as recognition factors for PTS1 and PTS2 respectively (Marzioch et al. 1994; Zhang and Lazarow 1995; McCollum et al. 1993; van der Leij et al. 1993; Brocard et al. 1994; Terlecky et al. 1995; van der Klei et al. 1995; Dodt et al. 1995; Wiemer et al. 1995; Fransen et al. 1995; Nuttley et al. 1995; Szilard et al. 1995). Pex5p, the PTS1 recognition factor is characterized by the presence of seven tetratricopeptide repeats (TPR) (Goebl and Yanagida 1991; Lamb et al. 1995), and Pex7p, the PTS2 recognition factor, contains at least six WD40 domains (Neer et al. 1994), both features well suited to protein-protein interactions.

Of all peroxins characterized so far, only the **Pex8p** (Waterham et al. 1994; Liu et al. 1995; Rehling (1996); Rehling et al., (1996a) contains an obvious PTS. Surprisingly, the protein contains both a C-terminal PTS1 and an N-terminal PTS2.

Pex9p is an integral peroxisomal membrane protein of 42 kDa with no similarity to other proteins (Eitzen et al. 1995).

Pex11p is a peroxisomal membrane protein of 27 kDa. Deficiency of the protein does not result in a protein import defect. However, cells deficient in Pex11p are characterized by the appearance of giant peroxisomes, and an involvement of this peroxin in peroxisome proliferation or parceling has been suggested (Erdmann and Blobel 1995; Marshall et al. 1995).

Pex13p, an integral peroxisomal membrane protein of 43 kDa, is characterized by the presence of an C-terminal src-homology 3 (SH3) domain and has been reported to function as a docking protein for peroxisomal matrix protein import (Elgersma et al. 1996; Erdmann and Blobel 1996: Gould et al. 1996).

Pex14p is a peripheral membrane protein of 38 kDa with no significant similarity to other proteins, which because of its multiple interactions with other peroxins such as Pex5p, Pex7p, and Pex13p may play a central role in peroxisomal protein import (Komori et al. 1997; Albertini et al. 1997).

The following peroxins have not yet been defined after the new nomenclature.

Pas9p does not contain any characteristic sequence feature. This 23-kDa protein is found at the cytoplasmic surface of the peroxisomal membrane (Huhse and Kunau 1995).

Pas12p is localized in the cytosol as well as at the cytoplasmic surface of peroxisomes (Götte 1995). This 40-kDa protein contains a C-terminal consensus sequence for farnesylation.

Pas21p is an integral membrane protein of 43 kDa with no significant sequence similarities to other proteins (Elgersma 1995).

Pas22p belongs to the DNA-J protein family (Elgersma 1995). In contrast to other peroxins, deficiency of this protein only results in a partial import defect for peroxisomal matrix proteins.

Although most of the peroxins mentioned above are promising candidates for components of the peroxisomal protein import machinery, for only the two PTS recognition factors Pex5p and Pex7p has this actually been proven. These proteins were shown to interact directly with PTS1 and PTS2 respectively (McCollum et al. 1993; Brocard et al. 1994; Terlecky et al. 1995; Dodt et al. 1995; Wiemer et al. 1995; Fransen et al. 1995; Rehling et al. 1996b; Zhang and Lazarow 1996) and their deficiency results in characteristic partial import phenotypes consistent with their function as specific targeting signal recognition factors. An attempt to identify further components by their interaction with these two proteins in the two-hybrid system was recently undertaken by Kunau and coworkers. Interestingly, both PTS recognition factors interacted with a number of peroxins and in addition with each other. According to the two-hybrid results, Pex8p, Pas9p, Pex13p, and Pex14p are likely to act in concert with the two signal recognition factors to target peroxisomal matrix protein to peroxisomes and translocate them across the membrane (Huhse 1995; Huhse and Kunau 1995; Rehling 1996, Rehling et al. 1996a,b; Erdmann and Blobel 1996; Albertini et al. 1997). However, as the two-hybrid system is for several reasons prone to false positive results (Bartel et al. 1993), these findings have to be interpreted with caution. Independently performed *in vitro* binding studies have only so far confirmed the physical interaction of the two signal recognition factors and the association of the PTS1 recognition factor with Pex13p (Rehling et al. 1996b; Elgersma et al. 1996; Erdmann and Blobel 1996; Gould et al. 1996). Surprisingly, cells deficient in Pex13p are defective not only in PTS1-dependent but also in PTS2-dependent protein import, although no interaction of Pex13p and the PTS2 recognition factor was detected in the two-hybrid system. However, this result might be explained by the fact that Pex13p also interacts with Pex14p, which binds the PTS2 recognition factor in the two-hybrid system (Albertini et al. 1997). Thus, Pex13p might not only be a membrane receptor for the PTS1 recognition factor, but might function as a general docking protein for peroxisomal matrix protein import. The observed multiple interactions of peroxins might indicate the involvement of heteromeric complexes in peroxisomal import, however, they also could reflect the existence of a protein import cascade involving transient interactions of peroxins.

Mechanism of Peroxisomal Protein Import

The widely accepted view on peroxisomal protein import is that in analogy to the protein import into mitochondria and chloroplasts, peroxisomal membrane and matrix proteins are synthesized of free ribosomes and posttranslationally imported into the organelles (Lazarow and Fujiki 1985). However, in contrast to mitochondria and chloroplasts, peroxisomes can translocate prefolded, even oligomeric proteins across the peroxisomal membrane. These include albumin cross-linked to PTS1 peptides (Wendland and Subramani 1993; Walton et al. 1992; Hill and Walton 1995), DHFR fusion proteins complexed with aminopterin (Häusler et al. 1996), dimeric thiolase (Glover et al. 1994), trimeric chloramphenicol acetyltransferase (McNew and Goodman 1994), and even disulfide-bonded IgG molecules and 9 nm gold decorated with PTS1-resembling peptides (Walton and Subramani 1995). How peroxisomes accommodate such large structures has not yet been solved. These results suggest the existence of large pores in the peroxisomal membrane, but evidence for such large structures is still missing. Alternatively, protein import by pinocytotic or endocytotic invagination of the peroxisomal membrane has been considered, but this idea is not supported by experimental evidence either. The latter idea also raises new questions regarding the fate of the invaginated membranes (Rachubinski and Subramani 1995; McNew and Goodman 1996).

A first step towards elucidating the mechanism of the topogenesis of peroxisomal proteins is the intracellular localization of components of the peroxisomal protein import machinery. Surprisingly, the recognition factor for PTS1 exhibits different species-specific intracellular locations, ranging from cytosolic, peroxisomal membrane-associated, to intraperoxisomal (McCollum et al. 1993; van der Lej et al. 1993; Brocard et al. 1994; Terlecky et al. 1995; van der Klei et al. 1995; Dodt et al. 1995; Wiemer et al. 1995; Fransen et al. 1995; Nuttley et al. 1995; Szilard et al. 1995). Equally puzzling results have emerged on the localization of the PTS2 recognition factor. Within the same species, this protein has been localized to the peroxisomal matrix, but another group found the protein to be primarily cytosolic and partially associated with the peroxisomal membrane (Marzioch et al. 1994; Zhang and Lazarow 1995, 1996; Rehling et al. 1996b). Pex13p, the putative docking protein for peroxisomal protein import, is an integral peroxisomal membrane protein (Elgersma et al. 1996; Erdmann and Blobel 1996; Gould et al. 1996). Pas9p and Pex14p, both of which interact with the PTS1 recognition factor, are peripheral peroxisomal membrane proteins (Albertini et al. 1997). In addition, Pas9p is also found in the cytosol. Pex8p, which interacts with the PTS1 recognition factor as well, is found in the peroxisomal matrix, but is also associated with the inner layer of the peroxisomal membrane (Waterham et al. 1994; Liu et al. 1995; Rehling 1996).

Although puzzling at first, the observed multiple subcellular localization of the signal recognition factors may give a clue to the peroxisomal protein import mechanism. The assumption that the PTS recognition factors might be multipresent is supported by the fact that the peroxins which were found to interact with the recognition factors are also found in either the cytosol, at the peroxisomal membrane, or in the peroxisomal matrix. The data collected so far are consistent with the following model: a putative heteromeric signal recognition complex consisting of the two signal recognition factors might bind proteins which are to become peroxisomal in the cytosol and direct them to peroxisomes by docking to a putative Pex13p, Pex14p, Pas9p complex at the peroxisomal membrane (Fig. 1). From there (a) the recognition complex might transfer the proteins to be imported to the translocation machinery and then shuttle back to the cytosol or (b) the complex might be partly disassembled but the signal recognition factors remain bound to the proteins during the import process, therefore entering the peroxisomal matrix and after release of the proteins shuttling back to the cytosol (Fig. 1). As an alternative, (c), identical or similar signal recognition factors might be present in both compartments; the cytosolic PTS recognition factors deliver the pro-

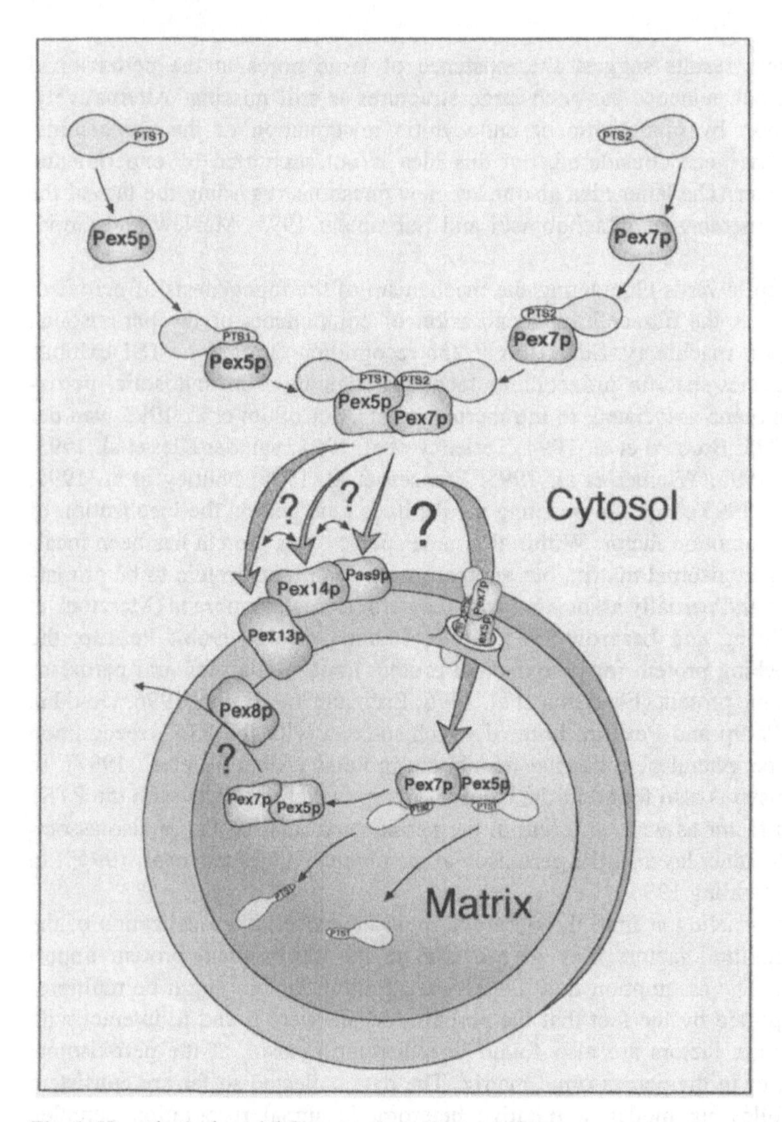

Fig. 1. Hypothetical model for peroxisomal protein import

teins to be imported to the translocation site, the intraperoxisomal ones take them from there. Suggestions b) and c) are not only supported by the fact that it is hard to see how a peroxisomal signal recognition factor can be found in the peroxisomal lumen without linkage to the biological function of the protein, but also by the matrix and membrane localization of the signal recognition factor binding peroxin Pex8p. An intriguing hypothesis might be that Pex8p is involved in the recycling of the PTS1 recognition factor to the cytosol. Which of these models reflects the *in vivo* situation has yet to be elucidated.

Although a body of data has been accumulated within the last few years, the principal mechanisms underlying peroxisomal protein import are still unclear (Rachubinski and Subramani 1995). The ultimate challenge is still elucidation of how folded proteins traverse the peroxisomal membrane. Nevertheless, one thing can already be concluded from what we know: the mechanism of peroxisomal protein import seems to differ from all we currently know about protein import into other organelles.

References

Aitchison JD, Murray WW, Rachubinski RA (1991) J Biol Chem 266: 23197-23203

Albertini M, Rehling P, Erdmann R, Girzalsky W, Kiel JAKW, Veenhuis M, Kunau WH (1997) Cell 89: 83-92

Authier F, Bergeron JJM, Ou WJ, Rachubinski RA, Posner BI, Walton PA (1995) Proc Natl Acad Sci USA 92: 3859-3863

Baerends RJS, Rasmussen S, Hilbrands RE, van der Hiede M, Faber KN, Ruevekamp PTW, Kiel JAKW, Cregg JM, van der Klei IJ, Veenhuis M (1996) J Biol Chem 271: 8887-8894

Bartel P, Chien CT, Sternglanz R, Fields S (1993) Biotechnology 14: 920-924

Berteaux-Lecellier V, Picard M, Thompson-Coffe C, Zickler D, Panvier-Adoutte A, Simonet JM (1995) Cell 81: 1043-1051

Brocard C, Kragler F, Simon MM, Schuster T, Hartig A (1994) Biochem Biophys Res Commun 204: 1016-1022

Crane, DI, Kalish JE, Gould SJ (1994) J Biol Chem 269: 21835-21844

De Duve, Baudhuin (1966) Physiol Rev 46: 323

Distel B, Erdmann R, Gould SJ, Crane DI, Cregg JM, Dodt G, Fujiki Y, Goodman JM, Just WW, Kiel JAKW, Kunau WH, Lazarow PB, Mannaertz GP, Moser HW, Osumi T, Rachubinski RA, Roscher A, Subramani S, Tabak HF, Tsukamoto T, Valle D, van der Klei I, van Veldhoven PP, Veenhuis M (1996) J Cell Biol 135: 1-3

Dodt G, Bravermann N, Wong C, Moser A, Moser HW, Watkins P, Valle D, Gould SJ (1995) Nature Gen 9: 115-125

Dyer JM, McNew JA, Goodman JM (1996) J Cell Biol 133: 269-280

Eitzen GA, Aitchison JD, Szilard RK, Veenhuis M, Nuttley WM, Rachubinski RA (1995) J Biol Chem 270: 1429-1436

Elgersma Y (1995) University of Amsterdam, thesis

Elgersma Y, Kwast L, Klein A, Voorn-Brouwer Van den Berg M, Metzig B, America T, Tabak HF, Distel B (1996) J Cell Biol 135: 97-109

Erdmann R (1994) Yeast 10: 935-944

Erdmann R, Blobel G (1995) J Cell Biol 128: 509-523

Erdmann R, Blobel G (1996) J Cell Biol 135: 111-121

Erdmann R, Kunau WH (1992) Cell Biochem Funct 10: 167-174

Erdmann R, Wiebel FF, Flessau A, Rytka J, Beyer A, Fröhlich KU, Kunau WH (1991) Cell 64: 499-510

Faber KN, Keizer-Gunnink I, Pluim D, Harder W, Ab G, Veenhuis M (1994) FEBS Lett 357: 115-120

Fransen M, Brees C, Baumgart E, Vanhooren JCT, Baes M, Mannaerts GP, Van Veldhoven PP (1995) J Biol Chem 270: 7731-7736

Gietl C, Faber KN, Van der Klei IJ, Veenhuis M (1994) Proc Natl Acad Sci USA 91: 3151-3155
Glover JR, Andrews DW, Rachubinski RA (1994) Proc Natl Acad Sci USA 91: 10541-10545
Glover JR, Andrews DW, Subramani S, Rachubinski RA (1994) J Biol Chem 269: 7558-7563
Goebl M, Yanagida M (1991) Trends Bioch Sci 16: 173-176
Gould SJ, Kalish JE, Morell JC, Bjorkman J, Urquhart AJ, Crane D (1996) J Cell Biol 135: 85-95
Gould SJ, Keller GA, Hosken N, Wilkinson J, Subramani S (1990) J Cell Biol 108: 1657-1664
Gould SJ, Keller GA, Subramani S (1987) J Cell Biol 105: 2923-2931
Gould SJ, McCollum D, Spong AP, Heyman JA, Subramani S (1990) Yeast 6: 613-638
Häusler T, Stierhof YD, Wirtz E, Clayton C (1996) J Cell Biol 132: 311-324
Heyman JA, Monosov E, Subramani S (1994) J Cell Biol 127: 1259-1273
Hill PE, Walton PA (1995) J Cell Sci 108: 1469-1476
Höhfeld J, Veenhuis M, Kunau WH (1991) J Cell Biol 114: 1167-1178
Huhse B (1995) PhD Thesis, Bochum, Germany
Huhse B, Kunau WH (1995) Cold Spring Harbor Symp Quant Biol 60: 657-662
Kalish JE, Theda C, Morrell JC, Berg JM, Gould SJ (1995) Mol Cell Biol 15: 6406-6419
Komori M, Rasmussen S, Kiel JAKW, Baerends RJS, Cregg JM, van der Klei IJ, Veenhuis M (1997) EMBO J 16: 44-53
Kragler F, Langeder A, Raupachova J, Binder M, Hartig A (1993) J Cell Biol 120: 665-673
Krause T, Ruhr-University Bochum, Germany, thesis
Kunau WH, Beyer A, Franken T, Götte K, Marzioch M, Saidowsky J, Skaletz-Rorowski A, Wiebel FF (1993) Biochimie 75: 209-224
Lamb JR, Tugendreich S, Hieter P (1995) Trends Bioch Sci 20: 257-259
Lazarow PB (1993) Trends Cell Biol 3: 89-93
Lazarow PB, Fuyiki Y (1985) Ann Rev Cell Biol 1: 489-530
Lazarow PB, Moser HW (1994) In: The Metabolic Basis of Inherited Disease. CR Shriver et al (ed) New York, McGraw-Hill. 7th ed 1479-1509
Liu H, Tan X, Russell KA, Veenhuis M, Cregg J (1995) J Biol Chem 270: 10940-10951
Liu Y, Gu KL, Dieckmann CL (1996) Yeast 12: 135-143
Marshall PA, Krimkevich YI, Lark RH, Dyer JM, Veenhuis M, Goodman J (1995) J Cell Biol 129: 345-355
Marzioch M, Erdmann R, Veenhuis M, Kunau WH (1994) EMBO J 13: 4908-4918
McCollum D, Monosov E, Subramani S (1993) J Cell Biol 121: 761-774
McNew JA, Goodman JM (1994) J Cell Biol 127: 1245-1257
Moreno M, Lark R, Campbell KL, Goodman JM (1995) Yeast 10:1447-1457
Neer EJ, Schmidt CJ, Nambudripad R, Smith TF (1994) Nature 371:297-300
Novikoff AB, Shin WY (1964) J Mikros Oxford 3:187-206
Nuttley WM, Brade AM, Eitzen GA, Veenhuis M, Aitchison JA, Szilard RK, Glover JR, Rachubinski RA (1994) J Biol Chem 269: 556-566
Nuttley WM, Brade AM, Gaillardin C, Eitzen GA, Glover JD, Aitchison JD, Rachubinski RA (1993) Yeast 9: 507-517
Nuttley WM, Szilard RK, Smith JJ, Veenhuis M, Rachubinski RA (1995) Gene 160: 33-39
Osumi T, Tsukamoto T, Hata S (1992) Biochem Biophys Res Commun 186: 811-818
Osumi T, Tsukamoto T, Hata S, Yokota S, Miura S, Fuyiki Y, Hijikata M, Miyazawa S, Hashimoto T (1991) Biochem Biophys Res Commun 181: 947-954
Rachubinski RA, Subramani S (1995) Cell 83: 525-528
Rehling (1996) PhD Thesis, Bochum, Germany

Rehling P, Albertini M, Kunau WH (1996a) Ann N Y Acad Sci USA, 804: 34-46
Rehling P, Marzioch M, Niessen F, Wittke E, Veenhuis M, Kunau WH (1996b) EMBO J 15: 2901-2913
Shimozawa N, Tsukamoto T, Suzuki Y, Orii T, Shirayoshi Y, Mori T (1992) Science 255: 1132-1134
Small GM, Szabo LJ, Lazarow PB (1988) EMBO J 7: 1167-1173
Spong AP, Subramani S (1993) J Cell Biol 123: 535-548
Subramani S (1992) J Membrane Biol 125: 99-106
Sulter GJ, Vrieling EG, Harder W, Veenhuis M (1993) EMBO J 12: 2205-2210
Swinkels BW, Gould SJ, Bodnar AG, Rachubinski RA, Subramani S (1991) EMBO J 10: 3255-3262
Szilard RK, Titorenko VI, Veenhuis M, Rachubinski RA (1995) J Cell Biol 131: 1453-1469
Tan X, Waterham HR, Veenhuis M, Cregg JM (1995) J Cell Biol 128: 307-319
Terlecky SR, Nuttley WM, McCollum D, Sock E, Subramani S (1995) EMBO J 14: 3627-3634
Tsukamoto T, Hata S, Yokota S, Miura S, Fujiki Y, Hijikata M, Hashimoto T, Osumi T (1994) J Biol Chem 269: 6001-6010
Tsukamoto T, Miura S, Fujiki Y (1991) Nature 350: 77-81
Tsukamoto T, Miura S, Nakai T, Yokota S, Shimozawa N, Suzuki Y, Orii T, Fujiki Y, Sakai F, Bogaki A, Yasumo H, Osumi T (1995) Nature Gen 11: 395-401
Van den Bosch H, Schutgens RBH, Wanders RJA, Tager JM (1992) Annu Rev Biochem 61: 157-197
Van der Klei I, Hilbrands RE, Swaving GJ, Waterham HR, Vrieling EG, Titorenko VI, Cregg JM, Harder W, Veenhuis M (1995) J Biol Chem 270: 17229-17236
Van der Klei IJ, Veenhuis M (1996) Ann NY Acad Sc 804: 773-774
Van der Leij I, Franse MM, Elgersma Y, Distel B, Tabak HF (1993) Proc Natl Acad Sci USA 90: 11782-11786
Voorn-Brouwer T, Van der Leij I, Hemrika W, Distel B, Tabak H (1993) Biochim Biophys Acta 1216: 325-328
Walton PA, Gould SJ, Feramisco JR, Subramani S (1992) Mol Cell Biol 12: 531-541
Walton PA, Hill PE, Subramani S (1995) Mol Biol Cell 6: 675-683
Waterham HR, De Vries Y, Russell K, Xie W, Veenhuis M, Cregg JM (1996) Mol Cell Biol 16: 2527-2536
Waterham HR, Titorenko VI, Haima P, Cregg JM, Harder W, Veenhuis M (1994) J Cell Biol 127: 737-749
Waterham HR, Titorenko VI, Swaving GJ, Harder W, Veenhuis M (1993) EMBO J 12: 4785-4794
Waterham HR, Titorenko VI, Van der Klei IJ, Harder W, Veenhuis M (1992) Yeast 8: 961-972
Wendland M, Subramani S (1993) J Cell Biol 120: 675-685
Wiebel FF, Kunau WH (1992) Nature 359: 73-76
Wiemer EA, Nuttley WM, Bertolaet BL, Li X, Francke U, Wheelock MJ, Anne UK, Johnson KR, Subramani S (1995) J Cell Biol 130: 51-65
Zhang JW, Lazarow PB (1995) J Cell Biol 129: 65-80
Zhang JW, Lazarow PB (1996) J Cell Biol 132: 325-334
Zoeller RA, Raetz CRH (1986) Proc Natl Acad Sci USA 83: 5170-5174

Membrane Transport of Proteins: A Multitude of Pathways at the Thylakoid Membrane

R. B. Klösgen, J. Berghöfer, and I. Karnauchov
Botanisches Institut der Ludwig-Maximilians-Universität, Menzinger Strasse 67, 80638 München, Germany

A characteristic feature of the eukaryotic cell is its compartmentalization and the presence of organelles. Each organelle provides particular functions for the cell and accordingly houses a unique set of proteins. Only a minority of these proteins are synthesized in the organelles themselves, because only mitochondria and plastids contain residual genomes and gene expression machineries, due to their endocytobiotic origin in free-living prokaryotes. However, even these genomes code for only a fraction of the proteins required, so the majority of mitochondrial and plastidic proteins are also encoded by nuclear genes and synthesized in the cytosol. Consequently, the eukaryotic cell had to develop mechanisms that ensure the intracellular transport of all these organelle proteins to and into their appropriate target organelles.

Intracellular protein sorting, however, is only a first step in the biogenesis of an organelle protein. Since some organelles are compartmentalized inside as well, mechanisms are often required not only for specific intracellular but also for intraorganellar sorting. In chloroplasts, even the envelope consists of two membranes, and in addition, there is the internal thylakoid membrane system that encloses a luminal space and separates it from the surrounding stroma. A nuclear-encoded protein from the thylakoid lumen thus needs to pass three membranes, the two envelope membranes plus the thylakoid membrane, on its way to its final destination.

Such transport processes have been extensively studied in recent years using proteins from the thylakoid membrane. Nuclear-encoded thylakoid proteins are synthesized in the cytosol as precursors with NH_2-terminal transit peptides which are capable of specifically recognizing receptor structures on the surface of the plastid and thus initiate the import of the protein into the organelle. According to their properties, two classes of transit peptides can be distinguished (Fig. 1). One class consists of chloroplast import signals which mediate the transport of the attached proteins only across the envelope membranes. These transit peptides are entirely removed by an endopeptidase of the chloroplast stroma and are not involved in the subsequent intraorganellar sorting, which is mediated in these instances by uncleaved signals present in the mature parts of the polypeptide chains. The second class of chloroplast transit peptides consists of bipartite targeting signals which carry two translocation signals in tandem. At the NH_2-terminus, there is again the chloroplast import signal which in most cases is removed by the stromal processing peptidase as well. This processing leads to the formation of proteins that still carry a fragment of the respective transit peptide, the so-called thylakoid-transfer domain, and are thus intermediate in size between the precursor and the mature polypeptides. These thylakoid-transfer domains provide the signals for transport across the thylakoid membrane and are proteolytically removed on the luminal side of these membranes.

Multiple Protein Transport Pathways Into and Across the Thylakoid Membrane

The thylakoid-transfer domains of all bipartite transit peptides analyzed so far have a common, conserved structure. They are positively charged at the NH_2-terminus, carry a hydrophobic core segment, and end in a polar COOH-terminal region with small side chain residues at positions -3 and -1 of the terminal processing site (von Heijne et al. 1989), i.e., they resemble strongly prokaryotic signal peptides which mediate the export of proteins from the bacterial cytosol into the periplasmic space. In spite of this structural conservation, however, proteins carrying bipartite transit peptides are translocated across the thylakoid membrane by at least three different and independent pathways (Fig. 1). Each pathway is specific only for a subset of thylakoid proteins (summarized in Robinson and Klösgen 1994; Klösgen 1997).

One of these pathways depends for its function on the presence of ATP and a stromal factor with homology to the bacterial SecA protein (see below). This Sec-dependent pathway is utilized by plastocyanin, the 33 kDa protein of the oxygen-evolving complex and subunit F of photosystem I (Hulford et al. 1994; Karnauchov et al. 1994; Robinson et al. 1994). In contrast, proteins like the 16-kDa and 23-kDa subunits of the oxygen-evolving complex, photosystem I subunit N, and photosystem II subunit T, require neither ATP nor a stromal factor for their thylakoid translocation, but depend instead strictly on the transthylakoidal proton gradient (Mould et al. 1991; Klösgen et al. 1992; Nielsen et al. 1994; Kapazoglou et al. 1995). This translocation route is therefore called the ΔpH-dependent pathway. The third translocation mechanism for proteins carrying bipartite transit peptides was originally discovered for subunit CFo-II of plastidar ATP synthase (Michl et al. 1994), but is apparently utilized also by the photosystem II subunit W (Lorkovic et al. 1995). CFo-II was shown to integrate spontaneously into the thylakoid membrane, i.e., it depends neither on the presence of nucleoside triphosphates or stromal factors, nor on the proton gradient across the thylakoid membrane (Michl et al. 1994). Even protease treatment of the thylakoid membrane does not prevent the subsequent integration of the protein (Robinson et al. 1996), which suggests that this process is also independent of any proteinaceous membrane receptor.

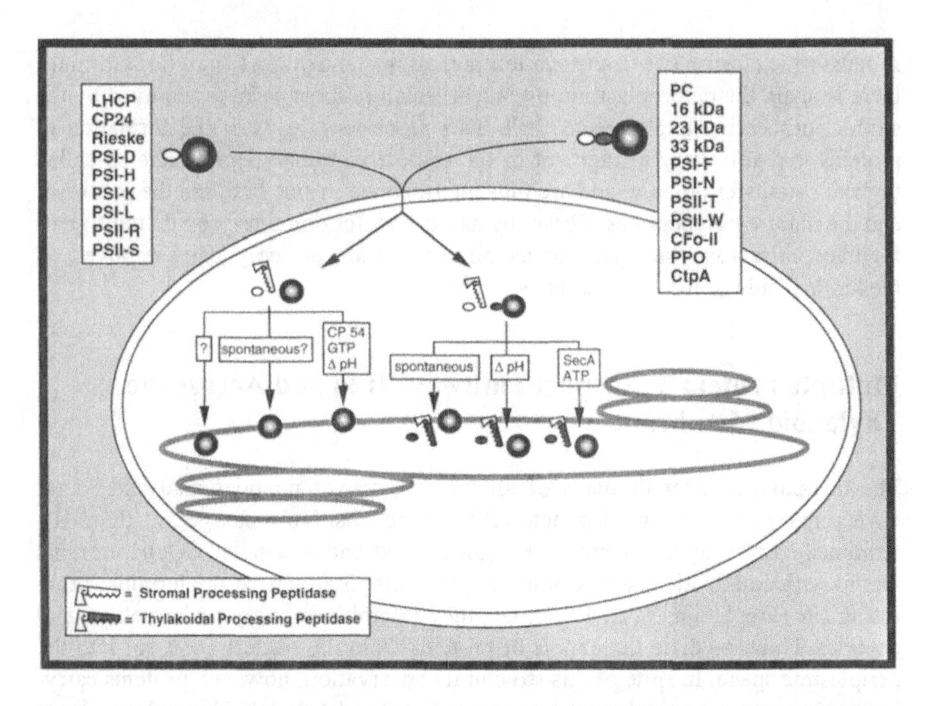

Fig. 1. Model of the protein translocation pathways at the thylakoid membrane. The chloroplast import signal in the transit peptides of nuclear encoded chloroplast proteins is indicated in white, the thylakoid-transfer domain present in bipartite transit peptides is shown in gray. Likewise, are the stromal and the thylakoidal processing peptidases indicated in white and gray, respectively

Not so much is known about the thylakoid transport pathways from those proteins that are targeted by internal, uncleaved signals in their mature parts. As yet, the only translocation mechanism that has been characterized in detail is that of LHCP, the apoprotein of the light-harvesting complex associated with photosystem II. Integration of LHCP into the thylakoid membrane is again different in mechanistic terms from the pathways described above, because it requires not only the transthylakoidal proton gradient but depends in addition on the presence of GTP and a stromal factor called CP54 (see below).

Assays for the Analysis of Protein Transport in Chloroplasts

Two experimental systems which are referred to as *in organello* and *in thylakoido* assays are widely used for the analysis of protein transport across the thylakoid membrane. In *in organello* assays, *in vitro* synthesized, radiolabeled precursor proteins are incubated with intact, isolated chloroplasts and their fate in the different compartments of the chloroplast is subsequently monitored. *In thylakoido* assays are similar, except that the precursor proteins are incubated with isolated thylakoids rather than intact chloroplasts so that the thylakoid translocation of a protein can be analyzed independently of the preceding transport across the chloroplast envelope. Moreover, in *in thylakoido* assays protein transport can be studied in the absence of stroma or nucleoside triphosphates, which is obviously not possible in intact chloroplasts.

In both assays, the effect of antimetabolites on the transport of proteins can be determined. For two of the pathways described above specifie inhibitors are available. Sodium azide inhibits the activity of the SecA protein (Oliver et al. 1990; Pugsley 1993) and thus affects the Sec-dependent protein transport, whereas nigericin dissipates the proton gradient across the thylakoid membrane and by this means inhibits the transport of proteins depending on the transthylakoidal ΔpH. In both instances, the proteins affected in thylakoid translocation are still efficiently imported into the chloroplast but accumulate in the stroma as intermediate processing products.

A complementary approach to distinguish the different translocation pathways is the competition experiment. In such experiments, the transport of a given protein is analyzed in the presence of a second protein that is added in excess. Such analyses have confirmed that the differences in the transport mechanisms described above are correlated with independent translocation routes across the thylakoid membrane (Cline et al. 1993; Karnauchov et al. 1994; Michl et al. 1994; Robinson et al. 1994). In contrast, at the chloroplast envelope, all stroma or thylakoid proteins analyzed so far compete with each other for import into the organelle (Cline et al. 1993; Rohling and Klösgen, unpublished results), which strongly suggests that a single, common translocation apparatus is involved in the import of most, if not all, chloroplast proteins into the organelle.

Influence of Transit Peptide and Mature Protein on the Choice of the Transport Pathway

Which part of a precursor protein determines its translocation route across the thylakoid membrane? This question has been studied in particular for those proteins that are translocated by the Sec- or the ΔpH-dependent pathway, since in both instances hydrophilic passenger proteins are transported across the thylakoid membrane into the luminal space. Most of these studies involved chimeric precursors which are composed of transit peptides and mature parts from different precursor proteins. Analysis of these chimeras in *in organello* and *in thylakoido* experiments showed that solely the transit peptide is capable of determining the translocation pathway (Robinson et al. 1994; Schmidt and Klösgen, unpublished results). All chimeras consisting of a transit peptide derived from a ΔpH-dependent precursor and a mature protein from the Sec-dependent group were translocated exclusively by the ΔpH-dependent pathway and, *vice versa*, all reciprocal chimeras were transported along the Sec-dependent translocation route. In neither case was an influence of the passenger protein on the choice of the pathway observed. Thus, the thylakoid-transfer domain of a bipartite transit peptide must contain all the signals that are required for the specific interaction of the protein with the corresponding thylakoidal protein transport machinery.

The observation that exclusively the transit peptide dictates the thylakoid translocation pathway for a hydrophilic protein is remarkable, because, as outlined above, the structure of the thylakoid-transfer domains of all bipartite transit peptides is to a large extent conserved. It is therefore of obvious interest to find out how these structurally similar targeting signals can distinguish between different transport routes. Mutations which have been introduced into the thylakoid-transfer domains of selected precursor proteins have shown that one important feature for specific pathway recognition is the type of positive charge that is found upstream of the hydrophobic core segment in such thylakoid-transfer domains (Chaddock et al. 1995). In all transit peptides of the ΔpH-type, this charge is provided by a twin pair of arginine residues, whereas the "twin-arginine motif" is lacking from all transport signals specific for the Sec-dependent pathway. Introduction of a twin pair of arginines into a Sec-type transit peptide or, *vice versa*, substitution of these arginine residues by lysines in a ΔpH-type transit peptide always resulted in a drastic decrease of the thylakoid transport efficiency (Chaddock et al. 1995). However, since in neither case was a mutated protein misdirected onto the alternative pathway, it must be concluded that these residues represent only part of the signal that is required for the determination of the translocation route. So far, the other components of this signal are still unknown.

It should be noted in this context that a passenger protein, though not influencing the translocation pathway, can in addition demand certain physiological conditions during its passage across the thylakoid membrane. In particular, proteins targeted by the Sec pathway often also require an intact transthylakoidal proton gradient for efficient translocation. This behavior is not restricted to chimeric proteins, but is for example also observed with the authentic precursor of the 33-kDa subunit from the oxygen-evolving complex. This protein, though targeted exclusively by the Sec pathway, is efficiently translocated across the membrane only in the presence of a transthylakoidal ΔpH, an effect that is particularly pronounced at low ATP concentrations in the stroma (Mant et al. 1995). The reason for this demand remains so far unknown but it was suggested that it might be related to the unfolding of the protein which is presumed to be required for its membrane transfer.

Components of the Thylakoidal Protein Translocation Machineries

An alternative approach to the characterization of the thylakoid translocation pathways is the isolation of components of the respective transport machineries. So far, we have isolated cDNA clones from spinach for two components of the

Sec-dependent transport apparatus, namely the translocation ATPase SecA (Berghöfer et al. 1995) and SecY, a constituent of the translocation pore in the thylakoid membrane (Berghöfer and Klösgen 1996). The genes for SecA and SecY are located in the nucleus of the plant cell and both proteins are synthesized in the cytosol as precursors with NH_2-terminal transit peptides for their transport into the chloroplast.

The precursor of spinach SecA of 117 kDa is efficiently imported into chloroplasts in *in organello* experiments. Within the organelle, the protein accumulates as a processing product of approximately 110 kDa, predominantly in the stroma but to some extent also associated with the thylakoid membrane (Berghöfer et al. 1995) which corresponds well with its assumed function as a stromal factor that is involved in protein targeting to the thylakoid membrane. On the amino acid level, the chloroplast protein shows almost 50% identity to SecA from *E. coli*, i.e., it is highly conserved between the prokaryote and the organelle. The amino acid identity of spinach SecY and its homolog from *E. coli* is lower (approximately 33%). However, its structure is apparently conserved, since for the spinach protein ten transmembrane spans are also predicted (Berghöfer and Klösgen 1996). The transit peptide of spinach SecY is approximately 110 residues in size which is unusually long but might be a requirement caused by the extraordinary hydrophobicity of this protein.

As yet, only a single component from any of the other translocation pathways has been characterized at the molecular level. This component is a protein from the chloroplast stroma required for the integration of LHCP into the thylakoid membrane (Li et al. 1995). It was named CP54 (Franklin and Hoffman 1993) due to its homology with the 54-kDa subunit of the cytosolic signal recognition particle involved in protein transport across the membrane of the endoplasmic reticulum. In contrast, the composition of the translocation machinery for the ΔpH-dependent pathway is still completely unknown.

Phylogenetic Aspects

What is the phylogenetic origin of the various translocation pathways at the thylakoid membrane? One of them, the Sec-dependent pathway, was obviously inherited from prokaryotes, because its mechanism, which depends on the presence of ATP and the SecA protein, is essentially identical to that of protein secretion into the periplasmic space of *E. coli* (Pugsley 1993). In coincidence with the endocytobiotic origin of chloroplasts from prokaryotic ancestors, the proteins translocated by this pathway, i.e., plastocyanin, the 33-kDa protein of the oxygen-evolving complex and subunit F of photosystem I, have also been found in cyanobacteria, the closest relatives of these descendants of original endocytobionts known today. In contrast, none of the proteins translocated by the ΔpH-dependent pathway has yet been found in cyanobacteria and, accordingly, a transport mechanism that depends exclusively on a transmembrane proton gradient has not been described in prokaryotes so far. It has therefore been suggested that this pathway might have been developed only in chloroplasts, possibly in parallel to the appearance of the new components of the photosynthetic machinery (Michl et al. 1994).

Why was it necessary to develop a new mechanism for protein translocation across the thylakoid membrane? Independent translocation pathways require at least in part additional transport machineries and thus can be assumed to be more demanding in terms of energy than a single, common transport apparatus. One possible reason would be saturation of the existing translocation route due to a large number of transported proteins, but since in bacteria a much higher number of proteins are translocated on the Sec-dependent route, this explanation appears unlikely. Alternatively, parallel transport routes might facilitate greater flexibility of the thylakoid system in response to environmental alterations, such as variations in light conditions or water supply, but so far there is no experimental support for this hypothesis.

There are, however, observations which suggest that the structure of a translocated protein might in some instances have played an important role in the development of a new translocation pathway. Passenger proteins like plastocyanin, which in the authentic situation are targeted by the Sec-pathway, are also efficiently transported by the ΔpH-dependent pathway if they are combined with transit peptides from the respective translocation group. In contrast, reciprocal constructs, i.e., chimeras consisting of a Sec-type transit peptide and a ΔpH-type passenger protein, are in most instances translocated only with low efficiency across the thylakoid membrane. At the very least they are retarded in the stroma; in some instances their subsequent thylakoid translocation is completely blocked (Clausmeyer et al. 1993). A typical example is the chimera PC/23, which consists of the transit peptide from plastocyanin and the mature 23-kDa protein from the oxygen-evolving complex. This protein is efficiently imported into isolated chloroplasts and cleaved in the stroma to an intermediate size. The intermediate processing product is capable of binding to the thylakoid membrane, but it cannot be further translocated across this membrane (Clausmeyer et al. 1993). Recent experiments have shown that this membrane-bound protein gets stuck within the thylakoid membrane in a pathway-specific manner (Schmidt and Klösgen, unpublished results), and we assume that the cause of this translocation arrest can be found in the mature 23-kDa protein, which is apparently a poor substrate for translocation by the Sec pathway. Of course, this working hypothesis is still preliminary and needs to be confirmed by other groups and with additional passenger proteins. However, we think that even on the basis of the available results, there is justification for seriously considering the structure of a translocated protein as a possible driving force for the development and maintenance of new translocation pathways at the thylakoid membrane.

Acknowledgements. This work was supported by a grant of the Deutsche Forschungsgemeinschaft (Sonderforschungsbereich 184, project B20).

References

Berghöfer J, Karnauchov I, Herrmann RG, Klösgen RB (1995) J Biol Chem 270: 18341-18346
Berghöfer J, Klösgen RB (1996) Plant Physiol 112: 863 (PGR 96-090)
Chaddock AM, Mant A, Karnauchov I, Brink S, Herrmann RG, Klösgen RB, Robinson C (1995) EMBO J 14: 2715-2722
Clausmeyer S, Klösgen RB, Herrmann RG (1993) J Biol Chem 268: 13869-13876
Cline K, Henry R, Li C, Yuan J (1993) EMBO J 12: 4105-4114
Franklin AE, Hoffman NE (1993) J Biol Chem 268: 22175-22180
Hulford A, Hazell L, Mould RM, Robinson C (1994) J Biol Chem 269: 3251-3256
Kapazoglou A, Sagliocco F, Dure L (1995) J Biol Chem 270: 12197-12202
Karnauchov I, Cai D, Schmidt I, Herrmann RG, Klösgen RB (1994) J Biol Chem 269: 32871-32878
Klösgen RB (1997) J Photochem Photobiol B: Biology 38: 1-9
Li X, Henry R, Yuan J, Cline K, Hoffman NE (1995) Proc Natl Acad Sci USA 92: 3789-3793
Lorkovic ZJ, Schröder WP, Pakrasi HB, Irrgang K-D, Herrmann RG, Oelmüller R (1995) Proc Natl Acad Sci USA 92: 8930-8934
Mant A, Schmidt I, Herrmann RG, Robinson C, Klösgen RB (1995) J Biol Chem 270: 23275-23281
Michl D, Robinson C, Shackleton JB, Herrmann RG, Klösgen RB (1994) EMBO J 13: 1310-1317
Nielsen VS, Mant A, Knoetzel J, Lindberg Müller B, Robinson C (1994) J Biol Chem 269: 3762-3766
Oliver DB, Cabelli RJ, Dolan KM, Jarosik GP (1990) Proc Natl Acad Sci USA 87: 8227-8231
Pugsley AP (1993) Microbiol Rev 57: 50-108
Robinson C, Cai D, Hulford A, Brock IA, Michl D, Hazell L, Schmidt I, Herrmann RG, Klösgen RB (1994) EMBO J 13: 279-285
Robinson C, Klösgen RB (1994) Plant Mol Biol 26: 15-24
Robinson D, Karnauchov I, Herrmann RG, Klösgen RB, Robinson C (1996) Plant J 10: 149-155
von Heijne G, Steppuhn J, Herrmann RG (1989) Eur J Biochem 180: 535-545

Analysis of Mitochondrial and Chloroplast Targeting Signals by Neural Network Systems

G. Schneider, J. Schuchhardt, A. Malik, J. Glienke, B. Jagla, D. Behrens, S. Müller, G. Müller, and P. Wrede
Freie Universität Berlin, Universitätsklinikum Benjamin Franklin, Institut für Medizinische/Technische Physik und Lasermedizin, AG Molekulare Bioinformatik, Krahmerstrasse 6-10, 12207 Berlin, Germany

Key words: Feature extraction, Kohonen network, Perceptron, sequence analysis.

Summary: Targeting sequences of nuclear-encoded mitochondrial and chloroplast proteins were investigated for characteristic sequence features. To evaluate the applicability of different neural network systems for feature extraction we employed a supervised and an unsupervised network training algorithm for targeting sequence analysis. Sets of sequences with less than 30% pairwise sequence identity were used in the experiments. Several physicochemical amino acid properties were used for sequence encoding. It turned out that the properties "refractivity", "volume", "polarity", and "hydrophobicity" are suited for separation of chloroplast and mitochondrial targeting sequences. Prediction accuracy of the neural networks for separation of chloroplast and mitochondrial sequences yielded correlation coefficients around 0.7 in both training and test set. The sequences were encoded only by their mean property values, which were obtained from averaging over the first 20 residues of the precursors. To locate possible targeting signals in the precursor sequences Kohonen networks were used. These systems were able to identify several characteristic patterns of chloroplast and mitochondrial targeting sequences. The predominant role of the distribution of arginines in mitochondrial sequences is substantiated by our findings. Chloroplast sequences seem to be characterized by stretches containing high contents of alanine, serine, and threonine. Putative locations of the targeting signals were found using the Kohonen networks for prediction of the features extracted. Analysis of the FNR precursor from *Cyanophora paradoxa* served as an example.

Introduction

Protein import into cell organelles is accomplished by means of targeting mechanisms (Pugsley 1989). In most cases the targeting signals of nuclear-encoded mitochondrial and chloroplast proteins are located in the N-terminal parts of the precursor sequences (Keegstra and von Heijne 1992; Schatz 1993). Having arrived at its destination the targeting sequence is cleaved off by special peptidases, resulting in the mature functional protein (Hawlitschek et al. 1988; Kalousek et al. 1988; Smeekens et al. 1990). It is a challenging task to define characteristic features of targeting signals and to infer possible implications for the evolution of targeting sequences. Secretory signals are the best understood targeting signals so far. Their main features are a positively charged N-terminal region, a hydrophobic core region, and a cleavage-site region with specific patterns of amino acids (von Heijne 1985). Similar signals are found in the precursors of nuclear-encoded mitochondrial proteins of the inter-membrane space (IMS) and in the precursor sequences of nuclear-encoded chloroplast proteins located in the thylakoid space (Halpin et al. 1989; von Heijne et al. 1989). This observation suggests that these proteins were secretory or membrane proteins of free-living endocytobionts (von Heijne et al. 1989; Schneider et al. 1992). As a result of intertaxonic combination (ITC) most of the genes coding for mitochondrial or chloroplast proteins (about 80%) were transferred to the nucleus (Sitte 1993). As a consequence, new targeting signals must have been developed in the course of evolution to guide these nuclear-encoded proteins to their destinations inside the organelles (Fig. 1). Apart from general findings relating to the overall composition of the targeting sequences, and some detailed knowledge of amino acid patterns around the protease cleavage-sites, not very much is known about the actual targeting signals of nuclear-encoded mitochondrial matrix and chloroplast stromal proteins (Arretz et al. 1991; Gavel and von Heijne 1990a, 1990b; Hendrick et al. 1989; Keegstra and von Heijne 1992; Roise and Schatz 1988; Schneider et al. 1995a). Except for charged residues and small stretches of hydrophobic amino acids their sequences are not conserved (von Heijne 1994). The aim of our work was threefold: (1) to derive sequence features that allow a distinction of nuclear-encoded mitochondrial and chloroplast proteins on the basis of the N-terminal parts of their precursor sequences, (2) to identify possible targeting signals, and (3) to locate characteristic features of the targeting sequences in the precursor. Three different types of artificial neural filter systems were used for this purpose: perceptrons (Minsky and Papert 1988; Rosenblatt 1962), multi-layer feedforward networks which were trained by an evolutionary algorithm (supervised training) (Hertz et al. 1991; Rechenberg 1973; Rumelhart et al. 1986), and Kohonen networks (unsupervised training) (Kohonen 1977, 1982, 1990).

Material and Methods

Data Preparation. From the PIR 1-International data base, release 41 (Barker et al. 1992) the precursor sequences of 55 nuclear-encoded mitochondrial and 43 chloroplast proteins from different organisms were selected. Only nuclear-encoded sequences with less than 30% pairwise identity were used in the experiments. The precursor sequences were restricted to the length of 50 residues

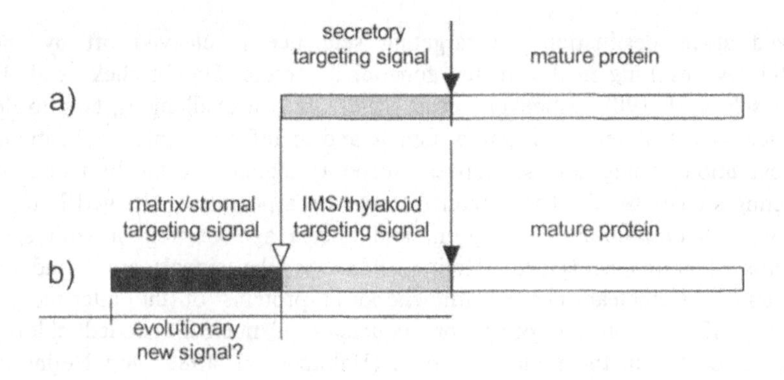

Fig. 1a, b. Targeting signals in precursor sequences of mitochondrial and chloroplast proteins. **a** Secretory protein or plasma membrane protein of the free-living endocytobiont prior to intertaxonic combination. The targeting sequence is shown in grey. **b** Intertaxonic combination led to the development of matrix/stromal targeting signals (black). The former secretory signal of the endocytobiont now serves as the mitochondrial inter-membrane space (IMS) or chloroplast thylakoid space targeting signal. Black arrow: Processing site of a membrane-bound signal peptidase; white arrow: matrix/stromal signal peptidase processing site

beginning with the N-terminus since the targeting signals are expected to be located within this region.

The main task to be performed by the neural networks is feature extraction. All amino acid sequences were described by a set of physicochemical parameters (Table 1). We hoped that this alternative representation of sequences reveals characteristic features that allow classification of chloroplast and mitochondrial targeting sequences. For the experiments the scales were normalized to [-1,1].

For supervised neural network training all 98 sequences were separated randomly 7:3 into a training and a test set. The training set consisted of 70 sequences (38 mitochondrial, 32 chloroplast), the test set contained 28 sequences (17 mitochondrial, 11 chloroplast). Feature extraction by Kohonen networks required the generation of sequence windows. Overlapping nine-residue segments were generated by the sliding-window technique.

Supervised Network Training. For the classification of sequence patterns by neural networks optimized by a supervised training algorithm the precursor sequences were restricted to the length of 20 residues beginning with the most N-terminal residue. The mean values of amino acid properties x_i were used as the input of the networks:

$$x_i = \frac{1}{20}\sum_{n=1}^{20}\xi_n \quad ,$$

Table 1. Amino acid properties used for sequence encoding. I hydrophobicity (Engelma et al. 1986), II hydrophobicity (Cornette et al. 1987), III volume (Zamyatnin 1972 IV volume (Harpaz et al. 1994), V bulkiness (Jones 1975), VI polarity (Jones 1975 VII hydrophilicity (Hopp and Woods 1981), VIII refractivity (Jones 1975)

Amino acid	I	II	III	IV	V	VI	VII	VIII
A	1.6	0.32	88.6	90.1	11.50	0.0	-0.5	4.34
C	2.0	5.48	108.5	113.2	13.46	1.48	-1.0	35.77
D	-9.2	-4.59	111.1	117.1	12.82	3.38	3.0	13.28
E	-8.2	-4.39	138.4	140.8	14.45	3.53	3.0	17.56
F	3.7	2.06	189.9	193.5	19.80	0.35	-2.5	29.40
G	1.0	0.00	60.1	63.8	3.40	0.0	0.0	0.00
H	-3.0	-0.61	153.2	159.3	13.69	51.60	-0.5	21.81
I	3.1	4.55	166.7	164.9	21.40	0.13	-1.8	19.06
K	-8.8	-4.75	168.6	170.0	15.71	49.5	3.0	21.29
L	2.8	2.34	166.7	164.6	21.40	0.13	-1.8	18.78
M	3.4	1.60	162.9	167.7	16.25	1.43	-1.3	21.64
N	-4.8	-2.89	117.7	127.5	11.68	49.7	0.2	12.00
P	-0.2	-3.16	122.7	123.1	17.43	1.58	0.0	10.93
Q	-4.1	-2.70	143.9	149.4	13.57	49.9	0.2	17.26
R	-12.3	-1.04	173.4	192.8	14.28	52.0	3.0	26.66
S	0.6	-1.96	89.0	94.2	9.47	1.67	0.3	6.35
T	1.2	-0.65	116.1	120.0	15.77	1.66	-0.4	11.01
V	2.6	3.78	140.0	139.1	21.57	0.13	-1.5	13.92
W	1.9	2.15	227.8	231.7	21.67	2.1	-3.4	42.53
Y	-0.7	0.98	193.6	197.1	18.03	1.61	-2.3	31.53

where ξ_n is the property value ξ at sequence position *n*. Three different types of feed-forward networks were trained by a (1,λ)-evolution strategy (Rechenberg 1973; Schneider and Wrede 1993; Schneider et al. 1994, 1995b): simple perceptrons, three-layered feedforward networks, and four-layered feedforward networks (Hertz et al. 1991; Rumelhart et al. 1986). The numbers of neurons in the hidden layers were 3, 5, or 10. A sigmoidal transfer function was employed by the hidden-layer neurons and the single output neuron. For classification, the output value of the networks were converted to binary values. A value of 1 was interpreted as a classification of the actual input pattern being a mitochondrial sequence, a value of 0 denotes putative chloroplast sequences

Network training by evolutionary algorithms involves the cyclic generation and test of variable values (connection weights and bias values). Per generation λ=1000 variable vectors were generated and tested. Training was stopped after 100 generations (cycles). Every experiment was repeated ten times for statistical reasons. An adaptive learning rate control was used (Rechenberg 1973). A simple pseudo-code formulation of the algorithm of the (1,λ) evolution strategy is:

```
1 Initialize (υ⃗ , σ)
2 FOR 1 TO cycles
3     FOR 1 TO λ
4         generate new σ
5         generate new υ⃗
6         calculate f(υ⃗ )
7     select best (υ⃗ , σ)
```

where $\vec{v}$ is the vector of free variables which have to be optimized, and σ is the learning rate. During training the mean-square-error, *mse*, served as the objective function f($\vec{v}$) which was minimized:

$$mse = \frac{1}{\mathrm{N}} \sum_{i=1}^{\mathrm{N}} \left(y_i^{target} - y_i^{actual} \right)^2 ,$$

where N is the total number of patterns in the training set. The target output values were 1 for a mitochondrial sequence and 0 for a chloroplast sequence. Classification accuracy of the networks was evaluated by the correlation coefficient according to Matthews (1975):

$$corr = \frac{(PN)-(UO)}{\sqrt{(N+U)(N+O)(P+U)(P+O)}} .$$

Where P is the number of true-positive predictions made (correctly classified mitochondrial patterns), N is the number of true-negative predictions (correctly classified chloroplast patterns), U is the number of false-negative predictions made (incorrectly classified mitochondrial patterns), and O is the number of false-positive predictions (incorrectly classified chloroplast patterns). The value of *corr* ranges between -1 and 1. A value of 1 indicates perfect classification.

Unsupervised Network Training. A Kohonen network was used to identify characteristic sequence patterns in the data (Kohonen 1977, 1982, 1990). The basic idea of the unsupervised training algorithm is that the input patterns attract the network neurons. As a result, during the training process many neurons move towards regions of the input space covered by many input patterns, whereas regions of the input space with a low pattern density attract only some network neurons. This process is shown for a special case in Fig. 2. Here, a two-dimensional Kohonen network was trained on the mapping of a two-dimensional input space (in most applications of Kohonen networks the input space is, however, high-dimensional). The training patterns (small spots) satisfy a Gaussian distribution in this special example (Fig. 2). The task for the network was to change the neuron weights in such a way that the topology of the network resembles the distribution of the patterns. In Fig. 2 the neurons are shown as black spots, where adjacent neurons are connected by a line. At the beginning of network training the network was randomly initialized (Fig. 2a). After a short

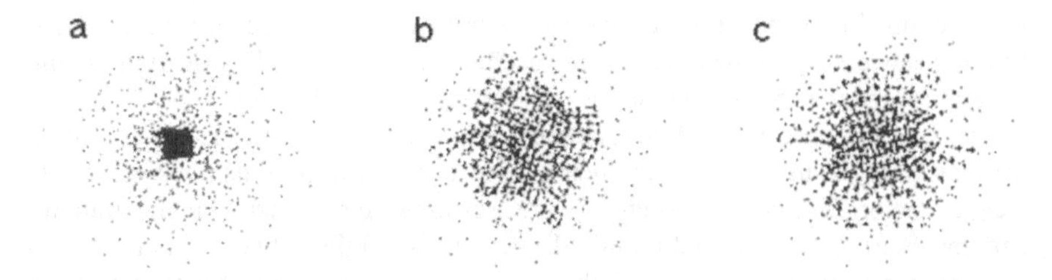

Fig. 2a-c. Training of a Kohonen network to approximate a gaussian distribution of patterns. **a** Randomly initialized state; **b** irregular intermediate state; **c** final network topology. *Small spots* represent input patterns, *large black spots* represent the network neurons which are connected by lines

period of time the network started to stretch out over the patterns (Fig. 2b). At the end of the training process the network found a possible geometry to approximate the distribution of the patterns (Fig. 2c). Most of the training patterns are located in the center of the input space, and the density of neurons is maximal in this region. The weight vectors can be analyzed to get an idea of characteristic features of the patterns located around a neuron. Additional details on self-organizing feature maps and the Kohonen algorithm can be found in the literature (Hertz et al. 1991; Ritter and Kohonen 1989; Zupan and Gasteiger 1993). An application to the identification of local motifs of three-dimensional protein structures has recently been described by us (Schuchhardt et al. 1995).

For targeting sequence analysis nine-residue windows generated by the "sliding-window" technique were encoded by four physicochemical property values per position, resulting in 36-dimensional input patterns. The properties "refractivity" (Jones 1975), "polarity" (Jones 1975), "hydrophobicity" (Engelman et al. 1986),

Table 2. Results of supervised perceptron training using a single property for sequence encoding. The *mse* values and their standard deviations averaged over ten runs are given. *mse:* mean square error

Codification	*mse* (Training data)	*mse* (Test data)
Hydrophobicity (Engelman et al. 1986)	$0.1828 \pm 5.3 \cdot 10^{-5}$	$0.1640 \pm 22.6 \cdot 10^{-5}$
Hydrophobicity (Cornette et al. 1987)	$0.2304 \pm 8.1 \cdot 10^{-5}$	$0.2413 \pm 13.3 \cdot 10^{-4}$
Volume (Zamyatnin 1972)	$0.1653 \pm 3.0 \cdot 10^{-5}$	$0.1425 \pm 5.3 \cdot 10^{-5}$
Volume (Harpaz et al. 1994)	$0.1572 \pm 4.9 \cdot 10^{-5}$	$0.1348 \pm 74.7 \cdot 10^{-5}$
Bulkiness (Jones 1975)	$0.2239 \pm 45.7 \cdot 10^{-5}$	$0.2071 \pm 52.1 \cdot 10^{-5}$
Polarity (Jones 1975)	$0.1886 \pm 7.3 \cdot 10^{-5}$	$0.1969 \pm 27.2 \cdot 10^{-5}$
Hydrophilicity (Hopp and Woods 1981)	$0.2346 \pm 13.2 \cdot 10^{-5}$	$0.2224 \pm 13.9 \cdot 10^{-4}$
Refractivity (Jones 1975)	$0.1650 \pm 13.8 \cdot 10^{-5}$	$0.1648 \pm 85.7 \cdot 10^{-5}$

and "volume" (Harpaz et al. 1994) were employed. These property scales were selected since they turned out to be useful for sequence classification in the experiments with perceptron systems (see Results, section below).

The neural network used to generate the feature map consisted of 10x10 neurons arranged in a plane. Each neuron was characterized by its position and its weight vector. The weight vectors were 36-dimensional, just like the training patterns. Starting from a random initialization in [-1,1] the network was trained to perform a nonlinear mapping of the input space. During the training process patterns were randomly selected from the pattern set and presented to the network.

Results

Classification of Mitochondrial and Chloroplast Protein Precursor Sequences by Supervised Network Training

To classify nuclear-encoded mitochondrial and chloroplast sequences on the basis of the N-terminal parts of their precursor sequences two types of artificial neural networks were trained: perceptrons, and multi-layered feedforward networks utilizing an evolutionary training algorithm. Several property scales were tested for their suitability for revealing characteristic targeting sequence features (Table 1). The *mse* served as the supervising function during network training, and the correlation coefficient according to Matthews (1975) was used to evaluate the prediction quality obtained with the trained networks. Table 2 summarizes the perceptron results obtained with a single property scale for sequence encoding. In general, network training converged and all sequence codifications led to reproducible results as indicated by the low standard deviations. The four most suited scales were "volume" (Harpaz et al. 1994), "refractivity" (Jones 1975), "volume" (Zamyatnin 1972), and "hydrophobicity" (Engelman et al. 1986).

The larger networks trained with the training data encoded by a single property did not lead to better results than did the simple perceptron systems (data not shown). In a second set of experiments several encoding properties were used to describe the data. Figure 3 shows the decision boundary represented by a perceptron trained with the training set which was encoded by two properties, "hydrophobicity" (Engelman et al. 1986) and "volume" (Zamyatnin 1972). Some of the sequences cannot be correctly classified even if a very complex non-linear separatrix is employed. This observation is reflected by the fact that the multi-layer networks were not able to significantly improve the Perceptron classification. A perceptron system trained with data encoded by the properties "volume" (Harpaz et al. 1994), "refractivity" (Jones 1975), "hydrophobicity" (Engelman et al. 1986), and "polarity" (Jones 1975) yielded a correlation coefficient $corr_{train} = 0.74$ and $corr_{test} = 0.77$. This was the best result obtained in our experiments. A typical training protocol is shown in Fig. 4.

From the experiments with supervised network training we conclude that:

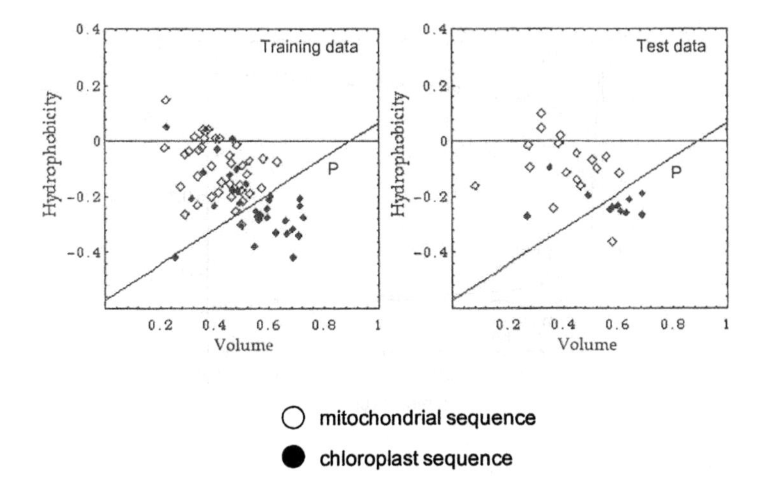

Fig. 3. The decision boundary P of a perceptron trained on the separation of mitochondrial and chloroplast precursor sequences. The 20 most N-terminal residues were encoded by their mean hydrophobicity (Engelman et al. 1986) and volume (Zamyatnin 1972)

- Mitochondrial and chloroplast targeting sequences differ in their overall physicochemical properties.
- Overall side-chain volume and hydrophobicity seem to be crucial for the distinction between the different types of targeting sequences.
- Characteristic differences are already found in the 20 most N-terminal residues.
- Perceptrons are useful for classification.
- Larger neural networks showed only slightly better prediction accuracy than perceptrons.
- Some sequences cannot be correctly classified either because some important sequence features have escaped our notice or because there is no significant difference between some of the chloroplast and some of the mitochondrial protein precursors.

Identification of Putative Targeting Signals by Kohonen Networks

To evaluate the findings made by supervised network training and to gain an idea of possible characteristic targeting signals in mitochondrial and chloroplast protein precursors a 10x10 Kohonen network was optimized using the entire data. Overlapping sequence windows each covering nine residues were used. Every position was encoded by four selected properties: "volume" (Harpaz et al. 1994), "refractivity" (Jones 1975), "hydrophobicity" (Engelman et al. 1986), and

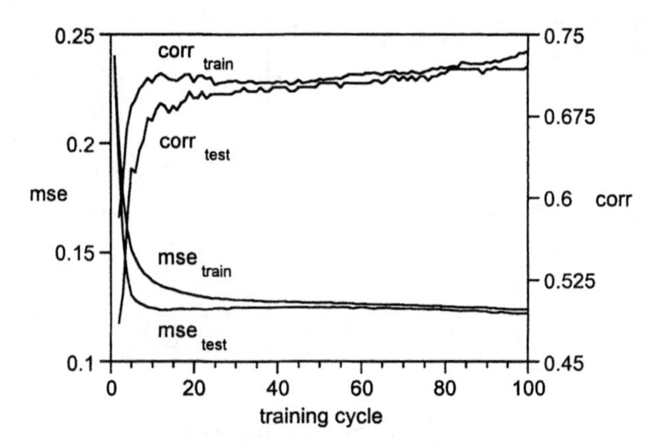

Fig. 4. Training protocol of a perceptron trained by a (1,200) evolution strategy. The input patterns were four-dimensional. *mse* Mean square error; *corr* correlation coefficient

"polarity" (Jones 1975). These properties were regarded as useful with respect to the results of perceptron training (see above). The feature map obtained is shown in Fig. 5. There are only a few neurons which cluster many sequences (white squares in Fig. 5). Most of the neurons contain only a few sequences, and several neurons do not represent any sequences (black squares in Fig. 5). The predominant neurons with many sequence windows are (8/7), (1/4), and (1/3).

Analysis of the sequences clustered by these neurons reveals that there are always chloroplast and mitochondrial sequences clustered together. Two characteristic neurons are described in more detail: neuron (1/9) contains 63% mitochondrial sequences and neuron (1/4) contains 65% chloroplast sequences. In these neurons 40 and 67 sequence fragments respectively are clustered.

Figure 6 displays 20 of the sequence windows clustered in these neurons. Figure 6a shows only mitochondrial sequences, Fig. 6b only chloroplast sequences. The mitochondrial sequences can be characterized by two conserved positively charged (mainly arginine) or polar residues (glutamic acid, aspartic acid) spaced three positions apart. The chloroplast sequences completely lack charged residues and are characterized by high contents of alanine, threonine and serine. To locate the fragments shown in Fig. 6 within the precursor sequences all mitochondrial fragments clustered in neuron (1/9) are highlighted by a grey box in Fig. 7, and all chloroplast fragments clustered in neuron (1/4) are highlighted by a grey box in Fig. 8.

The mitochondrial fragments seem to be randomly distributed among the first 50 positions of the precursors, whereas a slight tendency to being located near the N-terminus is observed for the chloroplast fragments. However, only 60% of the chloroplast precursors contain the motif of neuron (1/4), and only 51% of the

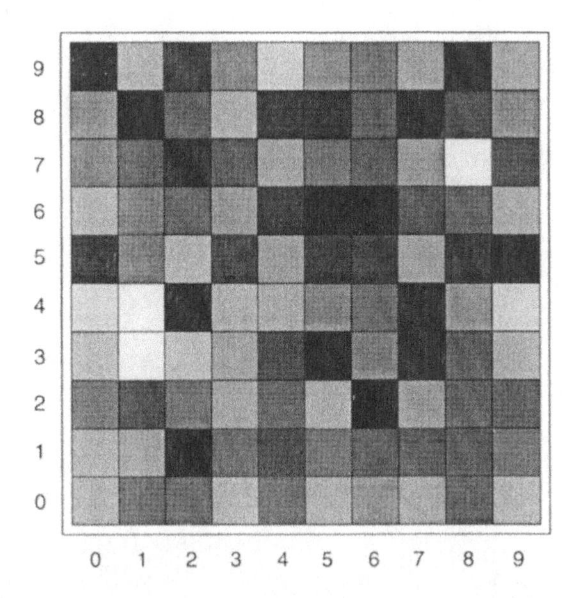

Fig. 5. Kohonen feature map giving a density plot of the sequence windows clustered by the 10x10 neurons. *Light squares* many sequences; *dark squares* few sequences

mitochondrial precursor sequences contain the motif of neuron (1/9). Other neurons clustering many sequence fragments of one kind, either mitochondrial or chloroplast, also show characteristic distributions of charged residues for the mitochondrial sequences and blocks of uncharged, hydroxylated residues for the chloroplast sequences.

From the experiments with the Kohonen networks we conclude:

- Kohonen networks are a well-suited method for the identification of local motifs in amino acid sequences sharing little sequence identity.
- Mitochondrial and chloroplast protein precursor sequences reveal common motifs.
- Both targeting signals contain characteristic distributions of mainly positively charged residues and blocks of small hydroxylated amino acids.
- Stretches containing many alanine, serine, and threonine residues seem to be predominant in chloroplast protein precursors, whereas positive charges are more frequently found in mitochondrial protein precursors than in chloroplast protein precursors.

Discussion

The task of sequence classification was surprisingly well performed by the perceptron systems, although only mean amino acid property values of the 20

a)	b)
PLRTAPQPP	STAAPPAAL
VARPAVRVA	TPVTSSSSL
PGQLPLRTP	ASLSATTTV
PARLVAQSS	ATVTSSAAV
VLRTPTRIG	SALAVTASI
ASKPTLNDV	ATASLSSTL
LPRCAPNVV	ASASVAAPL
LARTCLRST	SLPVSVTTL
ASRLASQMA	AATASLSST
LGNPSLQIP	AALSTSSTT
LGRLVPKLG	STLSSLSST
APQPPLRAF	ASIASSVAV
GSRTSMKDS	VTSSAAVAI
CARPLGRLV	VSVTTLPSF
TSNTPGHFS	STMALSSST
LVQGFLQPA	SLSTIASPI
VPHVEINGI	VSLSTIASP
TLKSCTRYL	APLTSALAV
TTRADPRGL	AASVMLSSV
FSRPGIRLL	PLTSALAVT

Fig. 6. Sequence fragments clustered by neurons of the Kohonen network. **a** Twenty mitochondrial sequences of neuron (1/9); **b** 20 chloroplast sequences of neuron (1/4). Conserved positive charges or polar residues are highlighted by a grey box in **a**

```
S00680  MAAAAASLRGVVLGPRGAGLPGARARGLLCSARPGQLPLRTPQAVALSSK  Homo sapiens
CCBY1H  MFSNLSKRWAQRTLSKSFYSTATGAASKSGKLTQKLVTAGVAAAGITAST  Saccharomyces cerevisiae
A27187  MLARTCLRSTRTFASAKNGAFKFAKRSASTQSSGAAAESPLRLNIAAAAA  Neurospora crassa
ZPBY    MLRNGVQRLYSNIARTDNFKLSSLANGLKVATSNTPGHFSALGLYIDAGS  Saccharomyces cerevisiae
A36442  MLNRFRPARLVAQSSRCLPLTRARAGPLPVNNARTLATRAAAVNTKEPTE  Neurospora crassa
O4RTV3  MAVLSRMRLRWALLDTRVMGHGLCPQGARAKAAIPAALRDHESTEGPGTG  Rattus norvegicus
CBBY2   MLKYKPLLKISKNCEAAILRASKTRLNTIRAYGSTVPKSKSFEQDSRKRT  Saccharomyces cerevisiae
DEBYPA  MLAASFKRQPSQLVRGLGAVLRTPTRIGHVRTMATLKTTDKKAPEDIEGS  Saccharomyces cerevisiae
S28950  CNSADWPAALVQGFLQPAVDDASQKRRVAHFTFQPDPESLQYGQTQKMNL  Rattus norvegicus
DEBYPX  MLSAISKVSTLKSCTRYLTKCNYHASAKLLAVKTFSMPAMSPTMEKGGIV  Saccharomyces cerevisiae
OTBY5A  MLRNTFTRAGGLSRITSVRFAQTHALSNAAVMDLQSRWENMPSTEQQDIV  Saccharomyces cerevisiae
OTHU5A  MLGAALRRCAVAATTRADPRGLLHSARTPGPAVAIQSVRCYSHGSQETDE  Homo sapiens
OWASN   MSHSRIISQLFVDALVLSTISFRPKSFKFFNIYILALPMASLRSVLKSQS  Emericella nidulans
OWASG   MPSPLRTAPQPPLRAFHNPPALRRLYSSTSHSAATPATSPFAPRHLLSIA  Aspergillus nidulans
XXRTAC  MAALAVLHGVVRRPLLRGLLQEVRCLGRSYASKPTLNDVVIVSATRTPIG  Rattus norvegicus
PWBOE   ARRAPSRDIMVAYWRQAGLSYIRYSQICAKAVRDALKTEFKANAMKTSGS  Bos taurus
LWNCA   MASTRVLASRLASQMAASAKVARPAVRVAQVSKRTIQTGSPLQTLKRTQM  Neurospora crassa
QYCHGM  MFWLRGGAQSCRGGETEDRMQRGMWGVGLARRRLSTSLSALPAAARDFVE  Gallus gallus
SYRTCA  MTRILTACKVVKTLKSGFGLANVTSKRQWDFSRPGIRLLSVKAQTAHIVL  Rattus norvegicus
UFRT    MNRAFCLLARSRRFPRVPSAGAVLSGEAATLPRCAPNVVRMASQNSFRIE  Rattus norvegicus
CSNCM   MFGPRHFSVLKTTGSLVSSTFSSSLKPTATFSCARAFSQTSSIMSKVFFD  Neurospora crassa
SYNCLM  MPLICARPLGRLVPKLGASLRPVLSSHAASPRRPVGVALEQHLGTESWKR  Neurospora crassa
YFBYAM  MEVTSMFLNRMMKTRTGLYRLYSTLKVPHVEINGIKYKTDPQTTNVTDSI  Saccharomyces cerevisiae
COBY10  MSYFPRTYAHLMRNVLAHNKGNIYLQIGTQLHDTQIKIRFNGVRYISRNH  Saccharomyces cerevisiae
R5BY31  MFGPFKLTSPVAGGLLWKIPWRMSTHQKTRQRERLRNVDQVIKQLTLGLH  Saccharomyces cerevisiae
R5BY32  MNSLIFGKQLAFHKIVPTTAIGWLVPLGNPSLQIPGQKQLGSIHRWLREK  Saccharomyces cerevisiae
R6BYM7  MWNPILLDTSSFSFQKHVSGVFLQVRNATKRAAGSRTSMKDSAGRRLGPK  Saccharomyces cerevisiae
TWBYM1  MIRTSSILKNCNYRYIHCIHRCLLNEANLKDRKTHNVERVSNEKTFEQAL  Saccharomyces cerevisiae
```

Fig. 7. Location of the mitochondrial sequence fragments clustered in neuron (1/9). The fragments are highlighted by grey boxes. Left column: PIR identifiers; Right column: origin of the proteins

```
CCKM6R   MLQLANRSVRAKAARASQSARSVSCAAAKRGADVAPLTSALAVTASILLT   Chlamydomonas reinhardtii
CFPM     MDRELSNLPNLIVEIFRIKDCTMQTRNAFSWIKKEITRSISVLLMIYIIT   Pisum sativum
FEPM1    MATTPALYGTAVSTSFLRTQPMPMSVTTTKAFSNGFLGLKTSLKRGDLAV   Pisum sativum
CUQH     MATVTSSAAVAIPSFAGLKASSTTRAATVKVAVATPRMSIKASLKDVGVV   Silene pratensis
DEMZMC   MGLSTVYSPAGPRLVPAPLGRCRSAQPRRPRRAPLATVRCSVDATKQAQD   Zea mays
DEZMMX   MLSTRTAAVAASASPASPWKLGGRSEGGASCDGCRTYRNTLRRRAAPAKV   Zea mays
DEZMG3   MASSMLSATTVPLQQGGGLSEFSGLRSSASLPMRRNATSDDFMSAVSFRT   Zea mays
DEPMNB   MATHAALASTRIPTNTRFPSKTSHSFPSQCASKRLEVGEFSGLKSTSCIS   Pisum sativum
TVWTGC   MASTAAPPAALVARRAASASVAAPLRGAGLAAGCQPARSLAFAAGADPRL   Triticum aestivum
PWSPD    MAALQNPVALQSRTTTAVAALSTSSTTSTPKPFSLSFSSSTATFNPLRLK   Spinacia oleracea
PWSPG    MACSLSFSSSVSTFHLPTTTQSTQAPPNNATTLPTTNPIQCANLRELRDR   Spinacia oleracea
PWNTG    MSCSNLTMLVSSKPSLSDSSALSFRSSVSPFQLPNHNTSGPSNPSRSSSV   Nicotiana tabacum
RKSZSJ   MATGAGAGAATVVSAFTGLKSTAQFPSSFKMSNAAAEWEQKTTSNGGRVR   Pinus thunbergiana
RKPMS    NTDITSNGERVKCMQVWPPIGKKKFETLSYLPPLTRDQLLKEVEYLLRKG   Pisum sativum
YCRP     MASFSFFGTIPSSPTKASVFSLPVSVTTLPSFPRRRATRVSVSANSKKDQ   Brassica napus
IBEG     MYCGRYETIGETRGNSLNVFIGAAAGFVAAVALINSGLATSFYSTPVRAV   Euglena gracilis
R5PM9    MASSTLSSLSSTPLQHSFADNLKTCSQFPNKSSGFMVFAQKKTKKTRKII   Pisum sativum
R5PM18   MALLCFNSFTTTPVTSSSSLFPHPTANPISRVRIGLPTNCLKGFRILTPI   Pisum sativum
R5PM25   MASVSSIFGCGVSMAPNSSLRNKAIRTERRSACGGLLIECSSRPQKKSTA   Pisum sativum
HHPM21   MAQSVSLSTIASPILSQKPGSSVKSTPPCMASFPLRRQLPRLGLRNVRAQ   Pisum sativum
AYBH     MAHCLAAVSSFSPSAVRRRLSSQVANVVSSRSSVSFHSRQMSFVSISSRP   Hordeum vulgare
AYSP     MASLSATTTVRVQPSSSSLHKLSQGNGRCSSIVCLDWGKSSFPTLRTSRR   Spinacia oleracea
F1SP4    MASIASSVAVRLGLTQVLPNKNFSSPRSTRLVVRAAEEAAAAPAAASPEG   Spinacia oleracea
F1SP5    MAAATASLSSTLLAPCSSKQPQPQQQHQHQQLKCKSFSGLRPLKLNISSN   Spinacia oleracea
F2MU10   MAASVMLSSVTLKPAGFTVEKTAARGLPSLTRARPSFKIVASGVKKIKTD   Arabidopsis thaliana
CDTO3C   MATSTMALSSSTFAGKAVKLSPSSSEITGNGRVTMRKTATKAKPASSGSP   Lycopersicon esculentum
```

Fig. 8. Location of the chloroplast sequence fragments clustered in neuron (1/4). The fragments are highlighted by grey boxes. Left column: PIR identifiers; Right column: origin of the proteins

most N-terminal residues of the precursor sequences were used for network training. Around 80% correct predictions were made (correlation coefficients around 0.7). This result is in accordance with previous observations by other groups who found an overall difference in hydrophobicity and charge of these targeting sequences (Claros 1995; Keegstra and von Heijne 1992; Popot and de Vitry 1990; Pugsley 1989; von Heijne 1994). Future experiments with supervised training might include position-specific information, e.g. by training networks with sequence windows encoded by position-specific amino acid properties.

The Kohonen network was able to extract locally conserved sequence motifs in a set of sequences with less than 30% pairwise sequence identity. A multiple alignment program applied to the same task was not able to focus on these small motifs (data not shown, see Data preparation section above). Therefore, we recommend the application of Kohonen networks to local alignment tasks. A detailed analysis of their suitability is currently being performed by us (in preparation). Another attracting feature of the Kohonen networks is the fact that sequence profiles can easily be derived from a cluster of sequence fragments represented by a neuron of the feature map. In Fig. 8 examples of such fragment clusters are shown. Sequence profiles derived from these collections of aligned fragments can be tested for their predictive value by a data base screen (Gribskov et al. 1987). It might well be that self-organizing feature maps provide techniques for the detection and analysis of distantly related proteins. Kohonen networks have already been successfully applied to the detection of protein families (Férran and Ferrara 1992).

One application of the trained Kohonen networks is the identification of characteristic motifs in amino acid sequences. Analysis of the only hitherto known nuclear-encoded cyanoplast sequence from *Cyanophora paradoxa* (Schenk 1991),

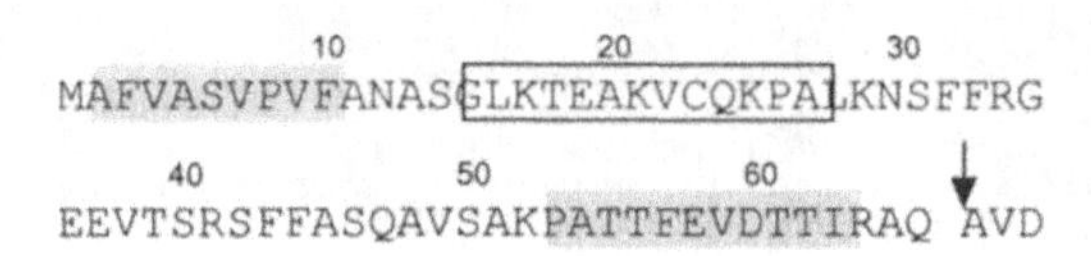

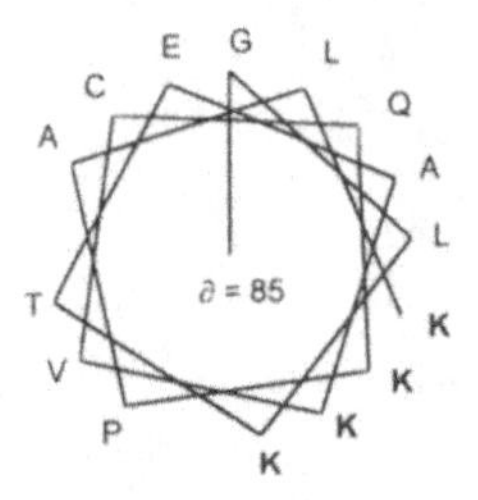

Fig. 9. Sequence analysis of the targeting sequence of the nuclear-encoded cyanoplast protein FNR from *Cyanophora paradoxa*. Top: the sequence fragments covered by neuron (1/4) are indicated by grey-shaded boxes; the fragments covered by neuron (1/9) are marked by the open box in the central part. The arrow denotes the putative peptidase cleavage-site. Bottom: A helical-wheel projection of the central part of the targeting sequence. A cluster of lysine residues is printed in boldface. The putative helix has a rotation angle of $\partial=85^{\circ}$

the FNR precursor, with the Kohonen network shows some remarkable results (Fig. 9). Three distinct blocks can be identified within the targeting sequence: (1) an N-terminal uncharged stretch of amino acids which is predicted as a characteristic chloroplast feature by the network; (2) a central, possibly amphiphilic portion, predicted as a mitochondrial feature; (3) a C-terminal portion which is very similar to the N-terminal part of the targeting sequence. This observation might lead to the conclusion that this targeting sequence is a modified mitochondrial targeting sequence. The modifications might be the N-terminal and the C-terminal part. Whether this is a generally valid statement concerning chloroplast targeting sequences cannot be decided from this single observation. However, it is in accordance with observations made by G. von Heijne and coworkers, who postulated a mosaic structure for some chloroplast targeting sequences (von Heijne et al. 1989). The Kohonen network classified the central sequence fragment as belonging to the motif of neuron (1/9) which represents a characteristic mitochondrial targeting sequence feature (Fig. 9). This sequence fragment forms an amphiphilic helix in the helical wheel plot (Schiffer and Edmundson 1967). Amphiphilic helices are postulated to be an important mitochondrial targeting signal besides an overall positive charge (Gavel et al. 1988; von Heijne 1986). We conclude that neuron (1/9) of the Kohonen network focused on the recognition of amphiphilic helical structures.

In future investigations we will try to extract specific features of targeting sequences for different types of chloroplasts (Bryant 1992). A further challenging task will be the analysis of mitochondrial and chloroplast targeting signals in higher plants using artificial neural networks. We think that these systems provide helpful tools for the detection of weak similarities between amino acid sequences.

Acknowledgements. Gunnar von Heijne, Rüdiger Cerff, Albin Nadine, and Uwe Hobohm are thanked for valuable discussions. Hainfried E. A. Schenk and W. Löffelhardt are thanked for leaving us the FNR precursor sequence to work with. This research was financially supported by the Bundesministerium für Bildung und Forschung (BMBF-DETHEMO project) and by a grant from the Fonds der Chemischen Industrie (to G.S.).

References

Arretz M, Schneider H, Wienhues U, Neupert W (1991) Processing of mitochondrial precursor proteins. Biomed Biochim Acta 50: 403-412

Barker WC, George DG, Mewes HW, Tsugita A (1992) The PIR-International protein sequence database. Nucleic Acids Res 20: 2023-2026

Bryant DA (1992) Puzzles of chloroplast ancestry. Curr Biol 2: 240-242

Claros MG (1995) MitoProt, a Macintosh application for studying mitochondrial proteins. Computer Applic Biosci 11: 441-447

Cornette JL, Cease KB, Margalit H, Spouge JL, Berzofsky JA, DeLisi C (1987) Hydrophobicity scales and computational techniques for detecting amphipathic structures in proteins. J Mol Biol 195: 659-685

Engelman DA, Steitz TA, Goldman A (1986) Identifying nonpolar transbilayer helices in amino acid sequences of membrane proteins. Annu Rev Biophys Biophys Chem 15: 321-353

Férran EA, Ferrara P (1992) Clustering proteins into families using artificial neural networks. Computer Applic Biosci 8: 39-44

Gavel Y, von Heijne G (1990a) Cleavage site motifs in mitochondrial targeting peptides. Protein Eng 4: 33-37

Gavel Y, von Heijne G (1990b) A conserved cleavage-site motif in chloroplast transit peptides. FEBS Lett 261: 455-458

Gavel Y, Nielsson L, von Heijne G (1988) Mitochondrial targeting sequences: why 'non-amphiphilic' peptides may still be amphiphilic. FEBS Lett 235: 73-177

Gribskov M, McLachlan AD, Eisenberg D (1987) Profile analysis: detection of distantly related proteins. Proc Natl Acad Sci USA 84: 4355-4358

Halpin C, Elderfield PD, James HE, Zimmermann R, Dunbar B, Robinson C (1989) The reaction specificities of the thylakoid processing peptidase and *Escherichia coli* leader peptidase are identical. EMBO J 8: 3917-3921

Harpaz Y, Gerstein M, Chothia C (1994) Volume changes on protein folding. Structure 2: 641-649

Hawlitschek G, Schneider H, Schmidt B, Tropschug M, Hartl FU, Neupert W (1988) Mitochondrial protein import:identification of a processing peptidase and of PEP, a processing enhancing protein. Cell 53: 795-806

Hendrick JP, Hodges PE, Rosenberg LE (1989) Survey of amino-terminal proteolytic cleavage sites in mitochondrial precursor proteins: leader peptides cleaved by two matrix proteases share a three-amino acid motif. Proc Natl Acad Sci USA 86: 4056-4060

Hertz J, Krogh A, Palmer RG (1991) Introduction to the Theory of Neural Computation. Addison-Wesley. Redwood City, CA

Higgins DG, Bleasby AJ, Fuchs R (1994) CLUSTAL V: improved software for multiple sequence alignment. Comput Appl Biosci 8: 189-191

Hurt EC, van Loon PGM (1986) How proteins find mitochondria and intramitochondrial compartments. Trends Biochem Sci 11: 204-207

Jones DD (1975) Amino acid properties and side chain orientation in proteins: a cross correlation approach. J Theor Biol 50: 167-183

Kalousek F, Hendrick JP, Rosenberg LE (1988) Two mitochondrial matrix proteases act sequentially in the processing of mammalian matrix enzymes. Proc Natl Acad Sci USA 85: 7536-7540

Keegstra K, von Heijne G (1992) Transport of proteins into chloroplasts. In: Herrmann RG (ed) Cell Organelles. Springer-Verlag, Wien/New York, pp 353-370

Kohonen T (1977) Self-Organization and Associative Memory. Springer-Verlag, Berlin

Kohonen T. (1982) Self-organized formation of topologically correct feature maps. Biol Cybern 43: 59-69

Kohonen T (1990) The self-organizing map. IEEE 78: 1464-1480

Matthews BW (1975) Comparison of the predicted and observed secondary structure of T4 phage lysozyme. Biochim Biophys Acta 405: 442-451

Minsky ML, Papert S (1988) Perceptrons. MIT Press. Cambridge, MA

Popot JL, de Vitry C (1990) On the microassembly of intergral membrane proteins. Annu Rev Biophys Biophys Chem 19: 369-403

Pugsley AP (1989) Protein Targeting. Academic Press, New York

Rechenberg I (1973) Evolutionsstrategie - Optimierung technischer Systeme nach Prinzipien der biologischen Evolution. Frommann-Holzboog, Stuttgart

Ritter H, Kohonen T (1989) Self-organizing semantic maps. Biol Cybern 61: 241-254

Roise D, Schatz T (1988) Mitochondrial presequences. J Biol Chem 263: 4509-4511

Rosenblatt F (1962) Principles of Neurodynamics. Spartan Books, New York

Rumelhart DE, McClelland JL, The PDB Research Group (eds) (1986) Parallel Distributed Processing. MIT-Press. Cambridge, MA

Schatz G (1993) The protein import machinery of mitochondria. Protein Science 2: 141-146

Schenk HEA (1993) *Cyanophora paradoxa*: anagenetic model or missing link of plastid evolution? Endocytobiosis Cell Res 10: 87-106

Schiffer M, Edmundson AB (1967) Use of helical wheels to represent the structures of proteins and to identify segments with helical potential. Biophys J 7: 121-135

Schneider G, Wrede P (1993) Development of artificial neural filters for pattern recognition in protein sequences. J Mol Evol 36: 586-595

Schneider G, Christmann H, Wrede P (1992) Protein targeting in the view of endocytobiology. Endocytobiosis Cell Res 9: 83-100

Schneider G, Schuchhardt J, Wrede P (1994) Artificial neural networks and simulated molecular evolution are potential tools for sequence-oriented protein design. Computer Applic Biosci 10: 635-645

Schneider G, Schuchhardt J, Wrede P (1995a) Peptide design in machina: development of artificial mitochondrial precursor cleavage sites by simulated molecular evolution. Biophys J 68: 434-447

Schneider G, Schuchhardt J, Wrede P (1995b) Development of simple fitness landscapes for peptides by artificial neural filter systems. Biol Cybern 73: 245-254

Schuchhardt J, Schneider G, Reichelt J, Schomburg D, Wrede P (1995) Classification of local protein structural motifs by Kohonen-networks. In: Schomburg D, Lessel U (eds) Bioinformatics: From Nucleic Acids and Proteins to Cell Metabolism. VCH, Weinheim, pp 85-92

Sitte P (1993) "Intertaxonic combination": introducting and defining a new term in symbiogenesis. In: Sato S, Ishida M, Ishikawa H (eds) Endocytobiology V. Tübingen University Press, Tübingen, pp 557-558

Smeekens S, Weisbeek P, Robinson C (1990) Protein transport into and within chloroplasts. Trends Biochem Sci 15: 73-76

von Heijne G (1985) Signal peptides: the limits of variation. J Mol Biol 184: 99-105

von Heijne G (1986) Mitochondrial targeting sequences may form amphiphilic helices. EMBO J 5: 1335-1342

von Heijne G (1994) Design of protein targeting signals and membrane protein engineering. In: Wrede P, Schneider G (eds) Concepts in Protein Engineering and Design. Walter de Gruyter Berlin, pp 263-279

von Heijne G, Steppuhn J, Herrmann RG (1989) Domain structure of mitochondrial and chloroplast targeting peptides. Eur J Biochem 180: 535-545

Zamyatnin AA (1972) Protein volume in solution. Prog Biophys Mol Biol 24: 107-123

Zupan J, Gasteiger J (1993) Neural Networks for Chemists. An Introduction. VCH, Weinheim

1.4 Metabolic Control and Ontogenetic Regulations Between Exogenosomes and Nucleus

Impact of Plastid Differentiation on Transcription of Nuclear and Mitochondrial Genes*

W. R. Hess, B. Linke, and T. Börner
Institut für Biologie, Humboldt- Universität, Chausseestr. 117, 10115 Berlin, Germany

Key words: *Hordeum vulgare* (albostrians mutant), chloroplast, plastid signal, gene regulation, mitochondria, transcription.

Summary: In plant cells the regulation of nuclear and chloroplast gene expression occurs in a coordinated manner. The transcription of certain nuclear genes has been shown to be affected by the state of chloroplast development. The signal molecules and transcription factors involved have not yet been identified. Mutant plants without a functional translational apparatus in their plastids, such as the barley *albostrians* mutant, provide a valuable experimental system for studying the effects of plastid differentiation on nuclear gene expression. Results of studies on this mutant are summarized. They indicate that the plastid has a much more complex influence on nuclear gene expression than was previously thought. The plastid was also observed to have an impact on the level of mitochondrial transcripts. The existence of several plastid-derived signal chains affecting the transcription of a large number of nuclear genes is proposed. These signal chains might not only respond to the developmental state of the plastids/chloroplasts, but also to changes in plastid metabolism and redox state as well as to oxidative stress. The data are discussed in the context of recent ideas about chloroplast-nuclear interactions.

Introduction

Eukaryotic cells have to coordinate the expression of genes in the different DNA-containing organelles. In higher plants, one regulatory component in the expression of nuclear genes is represented by the state of the plastid; "state" means in this context the developmental state, i.e., undifferentiated plastid vs chloroplast.

The observed effects of the plastid on gene activity in the nucleus are usually ascribed to a "plastid factor," "plastid signal," or "plastid signal chain" (Bradbeer

---

* Dedicated to Prof. Dr. Benno Parthier on occasion of his 65th birthday.

et al. 1979; Batschauer et al. 1986; Börner 1986; Oelmüller 1989; Taylor 1989; Mayfield 1990; Susek and Chory 1992; Gray et al. 1995). In this review we employ the term "plastid-derived signal chain," meaning a signal transduction chain that consists of several elements. These elements are transcription factors on the nuclear side, and on the plastid side they are the hitherto unidentified "plastid factor(s)." The plastid-derived signal chain may consist of just these two elements, plastid factor and transcription factor, but it is more likely that there are additional elements in between. The situation may be even more complex, since some of the elements of the plastid-derived signal chain may also participate in other signaling processes, and there may exist more than one plastid factor and more than one plastid-derived signal chain.

Compelling evidence for the existence of a signal transduction pathway originating in the plastids and determining the transcript level of certain nuclear genes was obtained chiefly from the study of three experimental systems, carotenoid-deficient mutants (Taylor 1989), plants treated with bleaching herbicides such as norflurazon, causing distinct photooxidative damage (Oelmüller 1989), and plastid ribosome-deficient mutants (Börner 1986, this work).

Among the few known higher plant mutants without plastid ribosomes, the barley mutant *albostrians* has been investigated especially well. Seedlings of *albostrians* barley can be totally white or green, but about 80% of them are white-green striped to a variable degree. This phenotype is caused by the nuclear recessive allele *albostrians* (*as*; Hagemann and Scholz 1962). The seedlings lack plastid ribosomes in cells of white leaf tissue, but contain normal amounts of plastid DNA (Hess et al. 1993, 1994a). In consequence of this defect, all plastid DNA-encoded proteins are missing and only traces of chlorophylls and carotenoids are detectable (Börner and Meister 1980; Hess et al. 1992). Although the ribosome-deficient plastids are transcriptionally active, introns are not spliced out from a distinctive set of chloroplast transcripts (Hess et al. 1994b; Hübschmann et al. 1996; Vogel et al. 1997a,b). This failure in splicing ensures that these plastids cannot assemble even trace amounts of functional ribosomes, because the unspliced transcripts code for a tRNA and for ribosomal proteins, which are also essential for translation.

Earlier studies demonstrated that nuclear-encoded proteins of the thylakoids (Börner et al. 1976) and Calvin cycle enzymes have drastically reduced levels in white leaves of *albostrians* barley (Bradbeer et al. 1979; Boldt et al. 1992). Since white and green leaves have the same nuclear genotype (*asas*) and differ only with respect to the presence of either normal chloroplasts or undifferentiated plastids, these data provided first genetic evidence for a signal deriving from the plastids and determining the different level of nuclear gene products in white and green leaves (Bradbeer and Börner 1978; Hagemann and Börner 1978; Bradbeer et al. 1979). The fact that white and green leaves of heterozygous plants (*Asas*) appear exactly like those of the *asas* plants provided further evidence that the plastids/chloroplasts are responsible for the observed alterations in nuclear gene expression (Hess et al. 1994c). We were able to show that the plastids affect the expression of nuclear genes in *albostrians* leaves at the level of transcription (Hess et al. 1994c).

Which Genes are Regulated by a Plastid-Derived Signal Chain?

The best studied nuclear genes affected by a plastid factor are those coding for Calvin cycle and thylakoid proteins, in particular *rbcS* and *lhcb1*. The transcription of these genes is repressed (or not activated) in the presence of undifferentiated plastids (Hess et al. 1994c). In white leaves of the *albostrians* mutant, the *lhcb1* transcript level is lowered to approximately 3% of its normal value (Börner and Hess 1993). Similar observations were made with several other experimental systems and genes (e.g., Oelmüller 1989; Taylor 1989; Gray et al. 1995). It appears advantageous for cells to be able to activate the expression of photosynthesis-related genes in the nucleus with the onset of chloroplast development and/or to repress them in cells that contain non-green plastids. Our data on nuclear gene transcription in *albostrians* barley, however, indicate that the presence of differentiated chloroplasts vs undifferentiated mutant plastids in cells also has an effect on genes not related to chloroplast functions at all.

One way to get an unbiased overview of nuclear gene expression is cDNA display (Liang and Pardee 1992). We subjected poly(A)+-RNA from white and green *albostrians* plants to reverse transcription and subsequent polymerase chain reaction using various primer sets (Fig. 1). Compared to the control, both reduced and enhanced transcript levels were observed in white *albostrians* plants. Several bands present in one sample were missing in the other. For more than 50% of all transcripts there were significant differences in abundance between green and white leaves. Hence

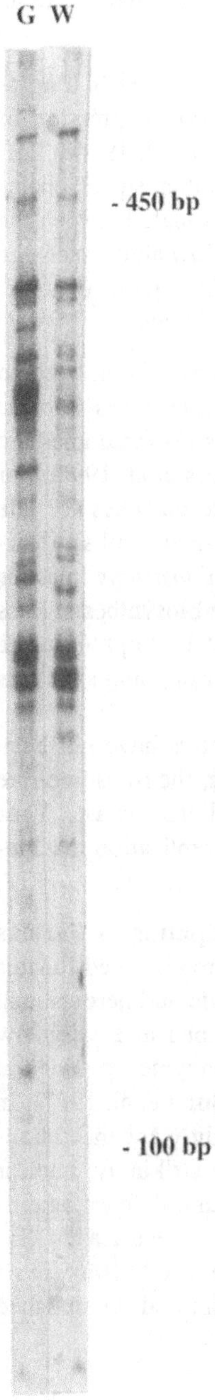

G W

- 450 bp

- 100 bp

Fig. 1. Different gene expression patterns in green (G) and white (W) leaves of the *albostrians* mutant of barley obtained by cDNA display according to Liang and Pardee (1992). Total RNA was reverse transcribed using the one-base anchored oligo-dT primer T_{11}-G followed by polymerase chain reaction using the T_{11}-G primer in combination with a random 10mer (5′-CTATCATGCG-3′). cDNAs were separated on a denaturing gel.

a broad spectrum exists of both up- and down-regulated nuclear genes in cells carrying the ribosome-deficient plastids. We obtained similar pictures with RNA from leaves bleached by the action of norflurazon (unpublished results).

The results of differential display were corroborated by Northern hybridization. The observations can be summarized as follows:

Nuclear Genes Coding for Chloroplast Proteins Directly Involved in Photosynthesis. Chloroplasts are essential for leaf-specific, high-level expression of genes such as *lhcb1* and the Calvin cycle enzymes (Hess et al. 1992, 1994c). Despite the probable interactions between plastid-derived and other signal chains within the plant cell (see below), it is often possible to discriminate between the effects of different signal transduction chains. Despite its very low abundance, the *lhcb1* mRNA is positively light-regulated as well as under the control of a circadian rhythm in white leaves of *albostrians* barley (Hess et al. 1994c).

Nuclear Genes for Chloroplast Proteins Involved in Processes Connected to or Directly Following Photosynthesis. The expression of these genes is often extremely low in the presence of undifferentiated mutant plastids (examples are enzymes of the glycolate pathway; Boldt et al. 1994, 1997; Hess et al. 1994c), or reduced to about 30-50% of their normal level (examples are enzymes for late steps of chlorophyll biosynthesis; Hess et al. 1992 and unpublished results). However, some are even induced in the presence of white *albostrians* plastids (examples are enzymes performing the first steps of porphyrin biosynthesis, Hess et al. 1992; Börner and Hess 1993; and unpublished results). Porphyrins are needed for both chlorophyll and heme biosynthesis, hence their ongoing synthesis in non-green tissue is not surprising.

Nuclear genes encoding house-keeping proteins of the plastids have not been tested yet for their transcription in *albostrians* barley. However, the ribosome-free plastids contain virtually the same amounts of DNA as normal chloroplasts. Thus at least the nuclear genes encoding the proteins of the plastid replication machinery appear to be normally expressed (Hess et al. 1993).

Nuclear Genes Encoding Non-Chloroplast Enzymes Participating in Various Metabolic Pathways. Diverse and in part opposite effects were observed for this group of genes. Genes coding for enzymes involved in cytosolic and peroxisomal processes that depend on photosynthetic activity are transcribed at a very low level in white leaves. Examples include nitrate reductase and enzymes involved in the glycolate pathway (Börner et al. 1986; Hess et al. 1994c; Boldt et al. 1997). In contrast, the presence of undifferentiated plastids causes a slightly enhanced transcript level for glycolytic enzymes (Hess et al. 1994c). Most strikingly, nuclear genes with a known or suggested function in stress response and defense against pathogens show much higher transcript levels in white than in green leaves. Examples are chalcone synthase, an Hsc70 protein, a homologue of the *Arabidopsis* Rps2-protein, ß-amylase (Dreier et al. 1995; Hess et al. 1994c; and unpublished results).

Mitochondrial Genes. The level of rRNAs and mRNAs transcribed from mitochondrial genes is about two- to fourfold higher in white than in green *albostrians* leaves. Examples are *rrn18, coxI, II, III, atp6,9, nad1,3, cob* (Börner and Hess 1993; and unpublished results).

Although this topic was not usually the primary focus of the investigations, the influence of a plastid-derived signal chain on non-chloroplast proteins has been described in other experimental systems, too. In norflurazon-treated plants, the disappearance of nitrate reductase mRNA under photodestructive conditions was detected (Oelmüller et al. 1988; Mohr et al. 1992). Similarly, an effect of the plastid state on gene expression was deduced for certain peroxisomal enzymes (Schwartz et al. 1992; see Oelmüller 1989 for older literature) and mitochondrial enzymes involved in glycine decarboxylation (Kim and Oliver 1990; Srinivasan and Oliver 1995). These peroxisomal and mitochondrial enzymes participate in photorespiration (glycolate pathway), i.e., in a process starting in the chloroplast with ribulose bisphosphate carboxylase. It appears reasonable that they would be regulated in the same way as the Calvin cycle (see Boldt et al. 1997). An unexpected observation was that even the morphology of palisade cells seems to be under the control of a plastid-derived signal chain (Börner and Förster 1981; Chatterjee et al. 1996; Keddie et al. 1996). The observation that so many nuclear and mitochondrial genes with entirely different functions are influenced in their activity by the plastid raises the question whether all these alterations are caused by a single plastid-derived signal chain.

Is There More Than one Plastid-Derived Signal?

Photodestruction of carotenoid-deficient chloroplasts and mutations like *albostrians* lead to a state of plastids which is different from etioplasts, amyloplasts, or other non-green plastids and may be described as "dedifferentiated" and "undifferentiated", respectively. The leaves bearing these proplastid-like organelles resemble "sink" rather than "source" organs. Lack of differentiated, active chloroplasts should have severe consequences for the metabolism of leaf cells. Such consequences could include the following: (1) altered levels of many metabolites including carbohydrates should occur. The concentration of carbohydrates is known to influence the transcription of nuclear genes (Sheen 1990; Koch 1996). (2) Certain phytohormones and other regulatory molecules may in dedifferentiated/undifferentiated plastids be produced only in very low amounts or not at all. For example, at least part of the biosynthesis of the stress-induced hormones abscisic acid (ABA) and methyl jasmonate takes place in chloroplasts (e.g., Bell and Mullet 1993; Bergey et al. 1996; Marin et al. 1996). White *albostrians* leaves are unable to synthesize stress-induced ABA (Quarrie and Lister 1984). (3) Light is known to cause oxidative stress in cells which are not protected by carotenoids. (4) The plastids of white leaves should differ from chloroplasts in their redox state. Thus, it is likely that certain metabolites, phytohormones, stress-inducing components, and constituents altered by the redox state represent elements of several signal chains which control the expression of nuclear genes and originate in the plastids, i.e., may be regarded as plastid factors and plastid-derived signal chains, respectively. However, levels of factors such as certain carbohydrates or phytohormones are perceived by the plant cell and consequently alter gene expression for reasons beyond signaling the state of plastid differentiation. Undoubted, some of the elements involved may be part of other signal chains. It remains to be determined whether and under what conditions these different signals and interactions are plastid-influenced in normal plants as well.

Indications for Links Between Signal Chains

A current observation is that signal transduction chains for light regulation and for the plastid developmental state are interlinked at some steps or may even share some elements in common. One parallel between recent data on the phytochrome signal transduction pathway and the expression of nuclear genes in green and white tissues of *albostrians* is interesting to note: genes that are stimulated via the phytochrome-Ca^{2+} pathway (Bowler et al. 1994) are repressed in the white tissue (photosynthesis genes), whereas the gene for chalcone synthase, which has been reported to be activated via the cGMP pathway (Bowler et al. 1994), is strongly induced in white *albostrians* plants. Thus these effectors may be shared by both signal transduction pathways.

One and the same promoter element may be used by different regulatory pathways including a plastid-derived one. The spinach *atpC* promoter possesses a single *cis* element mediating repression in darkness, in tissues with impaired plastids, and in roots (Bolle et al. 1996). On the other hand, this observation cannot be generalized, since in other promoters the functional dissection of light regulatory and developmentally regulating *cis* elements was possible, as in case of the pea *rbcS-3A* promoter (Kuhlemeier et al. 1989).

Most probably, further enlightenment about elements of plastid-derived signal chains is to be expected from the analysis of *Arabidopsis* mutants with light-uncoupled morphogenesis and gene regulation (*det*, *cop*, *doc*, *gun*, and *fus*; Susek et al. 1993; Li et al. 1994; Wei and Deng 1996). All these mutants affect the expression of positively light-regulated genes such as *rbcS* and *lhcb* . For example, the mutants *det2* (Chory et al. 1991) and *det3* (Cabrera et al. 1993) show constitutive expression of *rbcS* and *lhcb* in the presence of non-differentiated plastids. Interestingly, *det2* was recently identified as an enzyme involved in the biosynthesis of brassinosteroids (Li et al. 1996), adding a new candidate to the growing list of substances potentially involved in a plastid-derived signal chain. Very promising are the *gun* mutants, which are the result of a direct search for mutations uncoupling the expression of nuclear genes from the chloroplast state (Susek et al. 1993).

Conclusions

Effects of the postulated plastid-derived signal chain(s) have been elucidated by comparative studies on gene expression in leaves containing normally developed chloroplasts and leaves with severely impaired plastid differentiation. These defects are due to mutations blocking chloroplast development and gene expression (*albostrians* barley), to inhibition of plastid translation (Gray et al. 1995) and transcription (Rapp and Mullett 1991), or to photodestruction (Burgess and Taylor 1987; Oelmüller 1989). Since only the plastidic compartment should be affected in all these systems, it has been deduced that the postulated signal chain(s) originate(s) from the plastids/chloroplasts. If the low level of *lhcb1* transcription in white leaves is regarded as a consequence of a plastid-derived signal chain, the same must be done for all other differences between white and green leaves, including altered transcription of the chalcone synthase gene and of mitochondrial genes. We presume, therefore, that by its developmental, metabolic, and redox state the plastid influences the activity of many nuclear and mitochondrial genes. The impact of the plastid on mitochondrial genes may be an indirect one exerted via alteration of the activity of certain nuclear genes.

In a broad sense, phytohormones like ABA or jasmonate, metabolites like certain carbohydrates or heme, and even radicals produced within the plastids by photooxidation may consequently be regarded as "plastid factors" which initiate plastid-derived signal chains. The question is, therefore, to which cases the terms "plastid factor/signal" or "plastid-derived signal chain" should be applied. An earlier idea still discussed, was that there might exist a plastid gene encoding the "plastid factor" represented by a plastid RNA or protein that is specifically involved in the regulation of certain nuclear genes encoding proteins involved in photosynthesis and related processes (e.g., Bradbeer et al. 1979; Gray et al. 1995). Yet, the existence of such a regulatory plastid gene product has never been proven. We propose, therefore, to use the term "plastid factor" in a somewhat broader, though still traditional sense for any plastid component (not necessarily encoded by a plastid gene) initiating a "plastid-derived signal chain", yet restricted to those cases where the plastids are shown (or supposed) to signal their state to nuclear genes. The term "plastid factor" may become obsolete if the true nature of such a factor has been elucidated. There might exist more than one of such plastid-derived signal chains and plastid factors.

Currently, in contrast to the situation in plants, more details are known about intergenomic signaling from mitochondria to nuclear genes in yeast and mammals. A picture is emerging that clearly shows the existence of several different mitochondrial signals and signal chains, including heme, oxygen radicals, and metabolic signals, finally acting by a variety of transcription factors (Poyton and McEwen 1996). It would be surprising to find less diversity in the plant plastid-nuclear cross-talk.

Acknowledgements. We are grateful to Barb Sears for critical reading of the manuscript. This work was supported by a grant from the Deutsche Forschungsgemeinschaft, Bonn.

References

Batschauer A, Mösinger E, Kreuz E, Dörr I, Apel K (1986) The implication of a plastid-derived factor in the transcriptional control of nuclear genes encoding the light-harvesting chlorophyll a/b protein. Eur J Biochem 154: 625-634

Bell E, Mullet JE (1993) Characterization of an *Arabidopsis* lipoxygenase gene responsive to methyl jasmonate and wounding. Plant Physiol 103: 1133-1137

Bergey DR, Howe GA, Ryan CA (1996) Polypeptide signaling for plant defensive genes exhibits analogies to defense signaling in animals. Proc Natl Acad Sci USA 93: 12053-12058

Boldt R, Börner T, Schnarrenberger C (1992) Repression of the plastidic isoenzymes of aldolase, 3-phosphoglycerate kinase, and triosephosphate isomerase in the barley mutant "albostrians". Plant Physiol 99: 895-900

Boldt R, Koshuchowa S, Gross W, Börner T, Schnarrenberger C (1997) Decrease in glycolate pathway enzyme activities in plastids and peroxisomes of the *albostrians* mutant of barley (*Hordeum vulgare* L.). Plant Sci 124: 33-40

Boldt R, Pelzer-Reith B, Börner T, Schnarrenberger C (1994) Aldolases in barley (Hordeum vulgare L.): Properties and repression of the plastid enzyme in the plastome mutant "Albostrians". J Plant Physiol 144: 282-286

Bolle C, Kusnetsov VV, Herrmann RG, Oelmüller R (1996) The spinach *atpC* and *atpD* genes contain elements for light-regulated, plastid-dependent and organ-specific expression in the vicinity of the transcription start sites. Plant J 9: 21-30

Börner T (1986) Chloroplast control of nuclear gene function. Endocytobios Cell Res 3: 265-274

Börner T, Förster H (1981) Zum Einfluß des Plastoms auf Plastidenzahl und Zellform bei der Sorte "Mrs. Parker" von *Pelargonium zonale* hort. Wiss Z Univ Halle 30: 79-83

Börner T, Meister A (1980) Chlorophyll and carotenoid content of ribosome-deficient plastids. Photosynthetica 14: 589-593

Börner T, Schumann B, Hagemann R (1976) Biochemical studies on a plastid ribosome-deficient mutant of *Hordeum vulgare.* In: Bücher T, Neupert W, Sebald W, Werner S (eds) Genetics and Biogenesis of Chloroplasts and Mitochondria. Elsevier/North Holland, Amsterdam 1976, pp 41-48

Börner T, Mendel RR, Schiemann J (1986) Nitrate reductase is not accumulated in chloroplast-ribosome deficient mutants of higher plants. Planta 169: 202-207

Börner T, Hess WR (1993) Altered nuclear, mitochondrial and plastid gene expression in white barley cells containing ribosome-deficient plastids. In: Plant Mitochondria, Kück U, Brennicke A (eds) Verlag Chemie, Weinheim, Germany, pp 207-220

Bowler C, Neuhaus G, Yamagata H, Chua NH (1994) Cyclic GMP and calcium mediate phytochrome phototransduction. Cell 77: 73-81

Bradbeer JW, Börner T (1978) Activities of glyceraldehyde-phosphate dehydrogenase (NADP^{+}) and phosphoribulokinase in two barley mutants deficient in chloroplast ribosomes. In: Akoyunoglou G, Argyroudi-Akoyunoglou J (eds) Chloroplast Development. Elsevier/North-Holland, Amsterdam 1978, pp 727-732

Bradbeer JW, Atkinson YE, Börner T, Hagemann R (1979) Cytoplasmic synthesis of plastid polypeptides may be controlled by plastid-synthesized RNA. Nature 279: 816-817

Burgess DG, Taylor WC (1987) Chloroplast photooxidation affects the accumulation of cytosolic messenger RNA encoding chloroplast proteins in maize. Planta 170: 520-527

Cabrera y Poch HL, Peto CA, Chory J (1993) A mutation in the *Arabidopsis DET3* gene uncouples photoregulated leaf development from gene expression and chloroplast biogenesis. Plant J 4: 671-682

Chatterjee M, Sparvoli S, Edmunds C, Garosi P, Findlay K, Martin C (1996) DAG, a gene required for chloroplast differentiation and palisade development in *Antirrhinum majus.* EMBO J 15: 4194-4207

Chory J, Nagpal P, Peto CA (1991) Phenotypic and genetic analysis of det2, a new mutant that affects light-regulated seedling development in *Arabidopsis.* Plant Cell 3: 445-459

Dreier W, Schnarrenberger C, Börner T (1995) Light- and stress-dependent enhancement of amylolytic activities in white and green barley leaves: ß-amylases are stress-induced proteins. J Plant Physiol 145: 342-348

Gray JC, Sornarajah R, Zabron AA, Duckett CM, Khan MS (1995) Chloroplast control of nuclear gene expression. In: P Mathis (ed) Photosynthesis: from Light to Biosphere, vol III., Kluwer Acad Publ, pp 543-550

Hagemann R, Börner T (1978) Plastid ribosome deficient mutants of higher plants as a tool in studying chloroplast biogenesis. In: Akoyunoglou G, Argyroudi-Akoyunoglou JG (eds) Chloroplast Development. Elsevier/North Holland, Amsterdam pp 709-720

Hagemann R, Scholz F (1962) Ein Fall Gen-induzierter Mutationen des Plasmotyps bei Gerste. Der Züchter 32: 50-59

Hess WR, Schendel R, Rüdiger W, Fieder B, Börner T (1992) Components of chlorophyll biosynthesis in a barley albina mutant unable to synthesize d-aminolevulinic acid by utilizing the transfer RNA for glutamic acid. Planta 188: 19-27

Hess WR, Prombona A, Fieder B, Subramanian AR, Börner T (1993) Chloroplast *rps15* and the *rpoB/C1/C2* gene cluster are strongly transcribed in ribosome-deficient plastids: Evidence for a functioning non-chloroplast encoded RNA polymerase. EMBO J 12: 563-571

Hess WR, Hübschmann T, Börner T (1994a) Ribosome-deficient plastids of albostrians barley: extreme representatives of non-photosynthetic plastids. Endocytobios Cell Res 10: 65-80

Hess WR, Hoch B, Zeltz P, Hübschmann T, Kössel H, Börner T (1994b) Inefficient *rpl2* Splicing in Barley Mutants with Ribosome-Deficient Plastids. Plant Cell 6: 1455-1465

Hess WR, Müller A, Nagy F, Börner T (1994c) Ribosome-deficient plastids affect transcription of light-induced nuclear genes: genetic evidence for a plastid-derived signal. Mol Gen Genet 242: 305-312

Hübschmann T., Hess WR, Börner T (1996) Impaired splicing of the *rps12* transcript in ribosome-deficient plastids. Plant Mol Biol 30: 109-123

Keddie JS, Carroll B, Jones JDG, Gruissem W (1996) The DLC gene of tomato is required for chloroplast development and palisade cell morphogenesis in leaves. EMBO J 15: 4208-4217

Kim Y, Oliver DY (1990) Molecular cloning, transcriptional characterization, and sequencing of cDNA encoding the H-protein of the mitochondrial glycine decarboxylase complex in peas. J Biol Chem 265: 848-853

Koch KE (1996) Carbohydrate-modulated gene expression in plants. Annu Rev Plant Physiol Plant Mol Biol 47: 509-540

Kuhlemeier C, Strittmatter G, Ward K, Chua NH (1989) The pea *rbcS-3A* promoter mediates light responsiveness but not organ specificity. The Plant Cell 1: 471-478
Li HM, Altschmied L, Chory J (1994) *Arabidopsis* mutants define downstream branches in the phototransduction pathway. Genes Develop 8: 339-349
Li J, Nagpal P, Vitart V, McMorris TC, Chory J (1996) A role for brassinosteroids in light-dependent development of *Arabidopsis*. Science 272: 398-401
Liang P, Pardee AB (1992) Differential display of eucaryotic messenger RNA by means of the polymerase chain reaction. Science 257: 967-970
Marin E, Nussaume L, Queseda A, Gonneau M, Sotta B, Huguney P, Frey A, Marion-Poll A (1996) Molecular identification of zeaxanthin epoxigenase of *Nicotiana plumbaginifolia*, a gene involved in abscisic acid biosynthesis and corresponding to the ABA locus of *Arabidopsis thaliana*. EMBO J 15: 2331-2342
Mayfield SP (1990) Chloroplast gene regulation: interaction of the nuclear and chloroplast genomes in the expression of photosynthetic proteins. Curr Opinion Cell Biol 2: 509-513
Mohr H, Neininger A, Seith B (1992) Control of nitrate reductase and nitrite reductase gene expression by light, nitrate and a plastidic factor. Bot Acta 105: 81-89
Oelmüller R (1989) Photooxidative destruction of chloroplasts and its effect on nuclear gene expression and extraplastidic enzyme levels. Photochem Photobiol 49: 229-239
Oelmüller R, Schuster C, Mohr H (1988) Physiological characterization of a plastidic signal required for nitrate-induced appearance of nitrate and nitrite reductases. Planta 174: 75-83
Poyton RO, McEwen JE (1996) Crosstalk between nuclear and mitochondrial genomes. Annu Rev Biochem 65: 563-607
Quarrie SA, Lister PG (1984) Evidence of plastid control of abscisic acid accumulation in barley (*Hordeum vulgare* L.). Z Pflanzenphys 114: 295-308
Rapp JC, Mullet JE (1991) Chloroplast transcription is required to express the nuclear genes *rbcS* and *cab*. Plastid DNA copy number is regulated independently. Plant Mol Biol 17: 813-823
Schwartz BW, Daniel SG, Becker WM (1992) Photooxidative destruction of chloroplasts leads to reduced expression of peroxisomal NADH-dependent hydroxypyruvate reductase in developing cucumber cotyledons. Plant Physiol 99: 681-685
Sheen J (1990) Metabolic repression of transcription in higher plants. Plant Cell 2: 1027-1038
Srinivasan R, Oiver DJ (1995) Light-dependent and tissue-specific expression of the H-protein of the glycine decarboxylase complex. Plant Physiol 109: 161-168
Susek RE, Chory J (1992) A tale of two genomes: role of a chloroplast signal coordinating nuclear and plastid genome expression. Austr J Plant Physiol 19: 387-399
Susek RE, Ausubel FM, Chory J (1993) Signal transduction mutants of Arabidopsis uncouple nuclear CAB and RBCS gene expression from chloroplast development. Cell 74: 787-799
Taylor WC (1989) Regulatory interactions between nuclear and plastid genomes. Ann Rev Plant Physiol 40: 211-233
Vogel J, Hübschmann T, Börner T, Hess WR (1997a) Intron-internal RNA editing and splicing of *trnK-matK* precursor transcripts in barley plastids: support for MatK as an essential splice factor. J Mol Biol 270: 179-187
Vogel J, Hess WR, Börner T (1997b) Precise branch point mapping and quantification of splicing intermediates. Nucl Acids Res 25: 2030-2031
Wei N, Deng XW (1996) The role of the COP/DET/FUS genes in light control of Arabidopsis seedling development. Plant Physiol 112: 871-878

Glucose-6-Phosphate Dehydrogenase Isoenzymes from *Cyanophora paradoxa*: Examination of Their Metabolic Integration Within the Meta-Endocytobiotic System

T. Fester and H.E.A. Schenk
University of Tübingen, Botanisches Institut, Auf der Morgenstelle 1, 72076 Tübingen, Germany

Key words: *Cyanophora paradoxa*, Glaucocystophyta, glucose-6-phosphate dehydrogenase, isoenzymes, metabolic regulation, oxidative pentose phosphate pathway.

Summary: A partial characterization of the observed isoenzymes of the glucose-6-phosphate dehydrogenase (G6PDH) in *Cyanophora paradoxa* is given. The enzyme activities of G6PDH and 6-phosphogluconate-dehydrogenase detected in both cytosolic and cyanoplast cell compartments in comparison with biochemical properties of the isoenzymes of the G6PDH and nutritional characteristics of the flagellate allow speculation about the role of the cytosolic pentose phosphate cycle in *C. paradoxa*.

Introduction

The transfer of metabolic energy from photosynthetic endocytobionts to host cells is an important feature, perhaps the „raison d´être“, of such endocytobiotic systems. The photosynthetic endocytobiont generates large amounts of ATP (chemical energy) and NADPH (reducing power), compounds which the host cell urgently needs. Glucose-6-phosphate dehydrogenase is a keyenzyme at the beginning of the oxidative pentose-phosphate pathway (OPPP). Although the host OPPP is widely believed to be important for the production of NADPH by oxidation of organic compounds provided by the endocytobiont (most notably glycerinaldehyde-3-phosphate), Schnarrenberger et al. (1995) found that the cytosol of spinach leaves lacks the enzymes for the nonoxidative part of the OPPP. It remains an open question whether, despite these findings, the pathway is still capable of producing most of the NADPH needed by the host cell, or whether reducing power is transferred from the endocytobiont to the host cell by another mechanism. Because this transfer is essential for the establishment of the endocytobiotic relationship between photoautotrophic and heterotrophic organisms, a comparison of the OPPP of different such endocytobiotic systems seems to be important for our understanding of the development of metabolic integration between the two partners.

During intertaxonic combination (Sitte 1993), by gene transfer, recombination and development of the necessary protein import machinery, the endocytobiotic system changes via the stage of a meta-endocytobiotic system, a new taxonic entity (or, rather, a cascade of new taxonic entities), towards a complex cell, „the" recent eukaryotic cell. (A meta-endocytobiotic system is composed of the meta-host, the genomic hypersystem (GHRS) (Schenk 1993) and the meta-guest (Schenk 1994a), the genomic hyposystem (GHOS), now an exogenosome (Schenk 1992), a cell organelle with an evolutionary exogenous origin and thus the descendant of the original endocytobiont).

In the highly evolved (and on genomic level largely fused) meta-endocytobiotic system of higher plants G6PDH from the GHRS compartment is not subject to a sophisticated regulation mechanism, as is the isoenzyme from the GHOS, the chloroplast. The activity of this exogenosomal isoenzyme is tightly coupled to the redoxstate of the exogenosome by the thioredoxin-ferredoxin system. Most GHRS isoenzymes, by contrast, are not redox-regulated. (Anderson and Duggan 1976; Hilary 1972; Kaiser 1979; Mu et al. 1992).

Cyanophora paradoxa is an interesting meta-endocytobiotic system in which the original endocytobiont (endocytocyanobacterium = endocyanelle) has changed to an exogenosome (Schenk 1992a), in this case named cyanoplast or muroplast (Schenk 1994b). With respect to these plastids the new algal entity differs from modern unicellular algae in showing many morphologically „primitive" features. In 1977 however, the genome size of these cyano- (or muro-) plasts was found to be comparable to genomes from plant chloroplasts (Herdmann and Stanier 1977), and in 1986 Bayer and Schenk showed that over 80% of the exogenosomal proteins are coded on the nuclear genome of the genomic hypersystem (metahost) and, after expression, imported from the cytosolic compartment into the exogenosomal. Furthermore, we assume, as stated above, that apart from the amino acid exchange between the two compartments (Kloos et al. 1993; Schenk et al. 1987; Schenk 1990, 1992b) the import of triose phosphates from out the cyanoplasts (Schlichting and Bothe 1993) may be the most important metabolite translocation through the cyanoplast envelope. By looking at the isoforms of the important pentose phosphate cycle enzyme G6PDH in *Cyanophora paradoxa* we wanted to examine its role within the context indicated above.

Materials and Methods

Chemicals. Basic chemicals were obtained from E. Merck (Darmstadt, Germany), resins for chromatography from Pharmacia (Freiburg, Germany), with the exception of cibachron blue 3GA-agarose, which came from Sigma (Munich, Germany). Agents for electrophoresis came from Serva (Heidelberg, Germany) with the exception of APS, TEMED, and acrylamide/bisacrylamide, which were from

Bio-Rad. Enzyme substrates were all from Boehringer (Mannheim, Germany), and protease inhibitors were from Sigma, except aprotinin (Boehringer).

Organism, Growth and Harvest. Cyanophora paradoxa B 29.80, Pringsheim strain (Sammlung von Algenkulturen der Universität Göttingen, Pflanzenphysiologisches Institut, Göttingen, Germany) was grown and harvested as described by Zook and Schenk (1986).

Preparation of Extracts. Extracts from the complete organism were prepared as described by Fester et al. (1996), compartment-specific extracts as described by Deimel (1985), Zook and Schenk (1986), and Fester et al. (1996).

Purification of Cyanoplast G6PDH. Cyanoplast G6PDH was purified in a simple three-step procedure consisting of a batch process with DEAE-Sephadex A25 and chromatography on cibachron blue3GA-agarose and MonoQ (Fester et al. 1996). SDS-PAGE was performed under denaturing conditions (Laemmli 1970). Molecular mass markers were the proteins of the electrophoresis calibration kit for molecular weight determination of low-molecular-weight proteins from Pharmacia. Gels were stained either with Coomassie blue (Neuhoff et al. 1985) or with silver (Ansorge 1985).

Assessment of Different Biochemical Properties of the Compartment-Specific Isoenzymes. Ammonium sulfate precipitation and test for G6PDH activity are described by Fester et al. (1996), and analytical binding to the different chromatographic materials by the same authors (1997 in prep.).

Assessment of Kinetic Characteristics of the Isoenzymes. Examination of the different kinetic parameters is detailed by Fester (1993) and Völkle (1989), and inhibition by DTT by Fester (1993).

Results

Different Biochemical Properties of the two Isoenzymes

As can be seen from Table 1, there is a much greater amount of enzyme activity inside the cyanoplasts than in the cytosol. The precipitation with ammonium sulfate and the binding behaviour to different column materials (e.g., DEAE-sephadex A25 and A50) permit the inference that two G6PDH isoenzymes exist and that the cytosolic isoenzyme probably has a somewhat higher molecular mass and is more hydrophobic and/or less strongly negatively charged than the cyanoplast isoenzyme.

The greater amount in which the cyanoplast enzyme is present and its more hydrophilic properties, are the reasons why it is much easier to purify (Fester et al. 1996) than is the cytosolic enzyme in this organism.

Purification of the Glucose-6-Phosphate Dehydrogenase From the Cyanoplast Compartment

G6PDH from cyanoplasts of Cyanophora paradoxa (Fester et al. 1996) was isolated with a yield of 17% in a short, three-step-procedure (for purification see Table 2). The purification procedure started with cyanoplasts isolated from just freshly harvested algae (fresh weight 9 g). The purified cyanoplasts were lysed by treatment with lysozyme. The extract was subjected to a batch process with DEAE-Sephadex 25 and, after dialysis, to two consecutive chromatographies, first on Cibachron Blue 3GA-agarose, then on MonoQ. The specific activity of the purified enzyme was 120 U/mg; compared to crude extracts from cyanoplasts it was enriched by a factor of 1800. Using SDS-PAGE we determined a molecular mass of 59 kDa.

The best enrichment factor of the procedure is achieved by chromatography on the Cibachron Blue-material, because elution with 1 mM NADP is very selective. This step also ensures specifity for the cyanoplast compartment, because G6PDH from the cytosol does not elute with 1 mM NADP from Cibachron Blue material. The right enrichment factor can only be achieved if most of the phycobilisomes have been removed by the DEAE-A25 step before. Interestingly, on DEAE-A50, enzyme activity cannot be separated from that of phycobilisomes. Therefore the separation on DEAE-A25 would appear to be due to the larger size of the phycobilisomes in comparison to G6PDH. The phycobilisomes bind much less strongly to DEAE-A25 than to DEAE-A50.

Despite several attempts, we were unable to obtain an N-terminal sequence from the purified G6PDH. This is probably due to a blocked N-terminus, as has been reported for G6PDH from *Pichia jadinii* (Bergmann et al. 1991) and from potato (Graeve et al. 1994).

Table 1. *C. paradoxa,* G6PDH: different biochemical properties of the two isoenzymes

	Cyanoplast enzyme	Cytosolic enzyme
Activity in extracts	3.44 U/g fresh weight	0.5 U/g fresh weight
Precipitation with ammonium sulfate	40%-55% saturation	20% saturation
Binding to:		
DEAE-sephadex A25	+	-
DEAE-sephadex A50	+	+
Phenylsepharose	+	+
	Partial elution with low salt	Elution only with detergent
CBblue 3GA agarose	+	+
	Elution with NADP	No elution possible with NADP
2′,5′ADP-agarose	+	+
	Elution with NADP	Poor elution with NADP
MonoQ	Elution at 300-400 mM NaCl	Elution at 100-250 mM NaCl

Cbblue, Cibachron blue; + present; - absent

Table 2. *Cyanophora paradoxa*, purification of G6PDH from the cyanoplast compartment.

	raw extract	DEAE seph. A25	dialysis	CBblue 3GA	MonoQ
Activity (U)	25	18.9	18.3	5.8	4.2
Yield per step (%)		76	97	32	72
Overall yield (%)		76	74	24	17
Protein (mg)	370	12	12	0.1	0.03
Specific activity (U/mg)	0.068	1.74	1.53	60	120
enrichment per step		25.6	0.9	39.2	2
overall enrichment		25.6	23	900	1800

Different Kinetic Behavior of the two Isoenzymes

There are remarkable significant differences between the two isoenzymes (Table 3). The cytosolic isoenzyme shows only simple Michaelis-Menten kinetics and one broad pH optimum (maximum at pH 8.8). The kinetics of the cyanoplast isoenzyme however, cannot be linearized following Michaelis and Menten; rather, they show some kind of negative cooperativity and a smaller and lower pH optimum (pH 7.8) (see Fester et al. 1996) than the cytosolic enzyme.

The cyanoplast isoenzyme inhibited by DTT, can be reactivated by H_2O_2, whereas the activity of the cytosolic isoenzyme is affected by neither DTT nor H_2O_2. The G6PDH isoenzymes from *C. paradoxa* share these characteristics with G6PDH isoenzymes from green plants. In mixtures of cyanoplast and cytosolic extracts, the amount of G6PDH activity inhibited after treatment with DTT is directly proportional to the percentage of cyanoplast extract (Fester 1993).

Discussion

The different behaviors of the two isoenzymes in relation to DTT from the basis

Table 3. *Cyanophora paradoxa*: different kinetic behaviour of the two G6PDH isoenzymes

	Cyanoplast enzyme	Cytosolic enzyme
Inhibition by DTT	+ (Reversible)	-
pH Optimum	pH 7.8	pH 8.4 - 9.2 (8.8)
Kinetic in relation to NADP	Non-linear	$K_M = 0.015$ mM
Kinetic in relation to G6P	Non-linear	$K_M = 0.95$ mM
Mg^{++} dependence	Slight activation by 10 mM Mg^{++}	No influence on enzyme activity

for a method of determining the proportion of cyanoplast to cytosolic G6PDH in extracts from complete organisms. In case of *C. paradoxa* this can be used both for physiological experiments (quick measurement of the compartment-specific activities under various conditions) and as a quality control for cyanoplast preparations. So far there has been no method of ascertaining the absence of cytosolic contamination of cyanoplast extracts. Now the absence of G6PDH which cannot be inhibited by DTT might function as a quality control.

As stated above, G6PDH from cyanoplasts of *C. paradoxa* can be reversibly inhibited by DTT like G6PDH from other chloroplast and cyanobacterial sources. This is a strong indication that the enzyme is regulated by the ferredoxin-thioredoxin mechanism. G6PDH from cyanoplasts is the first such enzyme to be isolated, because so far only one other G6PDH has been isolated in plants, and this is from potato cytosol (Graeve et al. 1994).

The redox regulation mechanism ensures that oxidation of metabolites by the OPPP can provide NADPH as soon as photosynthesis is no longer capable of satisfying the needs of the organism. This is not only the case during darkness, but may also occur during nitrate assimilation/reduction, as was recently shown for *Chlamydomonas reinhardtii* (Huppe et al. 1992, 1994).

Kinetics from cyanoplast G6PDH for NADP and glucose-6-phosphate could not be linearized, but rather showed some kind of negative cooperativity (Fester et al. 1996). This might be an indication of an even more complex regulation mechanism for the cyanoplast enzyme.

The kinetic characteristics of the cytosolic enzyme (Fester et al. 1997 in prep.) accord well with data from modern plants too (Graeve et al. 1994). There is no influence of DTT or H_2O_2 on enzyme activity, the kinetics can be linearized according to Michaelis and Menten. The K_M values we measured (Table 3) can be compared with those from other cytosolic G6PDH. Thus, it seems there is no very sophisticated regulation mechanism for this enzyme, apart perhaps from the usual feedback inhibition by NADPH.

These observations suggest a minor role for cytosolic G6PDH and in fact at 10 U/g protein the cytosolic isoenzyme constitutes only about 13% (Fester et al. 1997 in prep.) of the total G6PDH activity. It was observed earlier (Provasoli and Pintner 1952, Pringsheim 1958) that *Cyanophora paradoxa* seems to be an obligate photoautotroph without any possibility of heterotrophic growth. For this reason it has not been possible so far to cultivate, for example, the cyanoplast free flagellate over a long period of time (Kies 1988). This indicates a strong metabolic integration of this complex organism.

Strong metabolic integration seems to be closely linked to the control of the redox state of the host compartment by the endocytobiont. A nice example of this connection has been described by Arillo et al. (1993). They compared two endocytobiotic systems of sponges and cyanobacteria, *Chondrilla nucula* and *Petrosia ficiformis*. *Chondrilla nucula*, which is unable to grow in the dark, looses control over its redox state soon after transfer to a dark place; *Petrosia ficiformis*, however, can provide enough reducing power by itself and can therefore colonize dark places as well. By leaving control over its own redoxstate to the endocytobiont, *Chondrilla nucula* has excluded itself from some habitats, in exchange, it has perhaps gained a more efficient system of energy transfer from the endocytobiont to the host cell.

Cyanophora paradoxa is an organism with apparently strong metabolic integration of the originally symbiotic partners. Mechanisms regulating the redox state of the GHRS compartment are crucial for the understanding of this integration. Because of the low activity of cytosolic G6PDH, the importance of the cytosolic OPPP from *Cyanophora paradoxa* seems to be doubtful, in analogy to the results from Schnarrenberger et al. (1995) for spinach leaves, where the cytosolic pathway is incomplete.

Schlichting and Bothe (1993), however, assume that ATP and NADPH are imported from the cyanoplasts in form of triosephosphates. They showed that cyanoplasts possess a phosphate translocator, transporting P_i, 3-PGA, and DHAP, and additionally a glucose carrier similar to plastids from higher plants. Glucose penetrates the inner cyanoplast membrane by facilitated diffusion at lower concentrations (< 2.5 mM) and by first order kinetics at higher concentrations (> 2.5 mM). Polyalcohols (such as glycerol or sorbitol) penetrate freely to an extent similar to ^3H_2O. In addition to these findings, Müller et al. (this volume) showed that under some physiological conditions *Cyanophora paradoxa* may accumulate large amounts of starch inside the host compartment. According to the observations of Schlichting and Bothe and of Müller et al. there has to be a cytosolic pathway for the generation of NADPH from the oxidation of carbohydrates. In addition to these results we found evidence for a high activity of cytosolic 6-phospho-gluconate dehydrogenase (6PGDH) - about 100 times more than the cytosolic G6PDH.

These findings seem to argue against a reduced importance of the cytosolic OPPP in *Cyanophora paradoxa*. Further experiments need to assess the activity of the other cytosolic OPPP enzymes as well as possible alternatives for the generation of NADPH within the host (GHRS) compartment of photosynthetic (meta-)endocytobiotic systems.

Acknowledgements. We thank Prof. Dr. H.-A. Bisswanger for providing an FPLC system (Pharmacia), Dipl. Biol. U. Gebhart for technical assistance regarding this system, and the Deutsche Forschungsgemeinschaft for financial support.

References

Anderson LE, Duggan JX (1976) Light modulation of glucose-6-phosphate-dehydrogenase. Partial characterization of the light inactivation system and its effects on the properties of the chloroplastic and cytoplasmic forms of the enzyme. Plant Physiol 58: 135-139

Ansorge W (1985) J Biochem Biophys Meth 9: 13-20

Arillo A, Bavestrello G, Burlando B, Sara M (1993) Metabolic integration between symbiotic cyanobacteria and sponges: A possible mechanism. Marine Biol (Berlin) 117: 159-162

Bayer M, Schenk HEA (1986) Biosynthesis of proteins in *Cyanophora paradoxa*: protein import into the endocyanelle analyzed by micro two-dimensional gel electrophoresis. Endocytobiosis and Cell Res 3: 197-202

Bergmann T, Jörnvall H, Wood I, Jeffery J (1991) Eukaryotic glucose-6-phosphate dehydrogenases: structural screening of related proteins. J Prot Chem 10 (1): 25-29

Deimel R (1985) Diplomarbeit, University of Tübingen

Fester T (1993) Diplomarbeit, University of Tübingen

Fester T, Völkle E, Schenk HEA (1996) Purification and partial characterization of the cyanoplast glucose-6-phosphate dehydrogenase in *Cyanophora paradoxa*. Endocytobiosis and Cell Res 11: 159-176

Graeve K, von Schaewen A, Scheibe R (1994) Purification, characterization, and cDNA sequence of glucose-6-phosphate dehydrogenase from potato (Solanum tuberosum L.). Plant J 5: 353-361

Herdmann M, Stanier RY (1977) The cyanelle: chloroplast or endosymbiotic prokaryote? FEMS Lett 1: 7-12

Hilary SJ (1972) Dithiothreitol: an inhibitor of glucose-6-phosphate dehydrogenase activity in leaf extracts and isolated chloroplasts. Planta (Berlin) 106: 273-277

Huppe HC, Vanlerberghe GC, Turpin DH (1992) Evidence for activation of the oxidative pentose phosphate pathway during photosynthetic assimilation of nitrate but not ammonium by a green alga. Plant Physiol (Bethesda) 100: 2096-2099

Huppe HC, Farr TJ, Turpin DH (1994) Coordination of chloroplastic metabolism in N-limited *Chlamydomonas reinhardtii* by redox modulation: II. Redox modulation activates the oxidative pentose phosphate pathway during photosynthetic nitrate assimilation. Plant Physiol (Rockville) 105: 1043-1048

Kaiser WM, Bassham JA (1979) Carbon metabolism of chloroplasts in the dark: oxidative pentose phosphate cycle versus glycolytic pathway. Planta 144: 193-200

Kies L (1988) The effect of penicillin on the morphology and ultrastructure of Cyanophora, Gloeochaete and Glaucocystis (Glaucocystophyceae) and their cyanelles. Endocytobiosis and Cell Res 5: 361-372

Kloos K, Schlichting R, Bothe H (1993) Glutamine and glutamate transport in *Cyanophora paradoxa*. Bot Acta 106: 435-440

Laemmli UK (1970) Cleavage of structural proteins during the assembly of the head of bacteriophage T4. Nature 227: 680-685

Mu, Hong, Ming-qu Li, Guang-Yao Wu, Xiang-Yu Wu (1992) Disactivation of glucose-6-phosphate-dehydrogenase by light-reduced thioredoxin. Acta Bot Sin 34: 37-42

Pringsheim EG (1958) Organismen mit blaugrünen Assimilatoren. Stud Plant Physiol (Praha) 165-184

Provasoli L, Pintner J (1952) Some interesting algal flagellates recently obtained in pure culture. News Bull Phycol Soc Amer 5: 7

Schenk HEA (1990) *Cyanophora paradoxa*: a short survey. In: P Nardon et al (eds) Endocytobiology IV. INRA, Paris, pp 199-209

Schenk HEA (1992a) *Cyanophora paradoxa*, identification and sequencing of nucleus-encoded cyanellar proteins. A proof for gene transfer. Endocytobiosis and Cell Res 8: 197-222

Schenk HEA (1992b) Cyanobacterial symbioses. In: A Ballows, HG Trüper, M Dwor-kin, W Harder, K-H Schleifer (eds) The Prokaryotes. A handbook of ecophysiology, isolation, identification, application. Springer, New York, vol 4, pp 3819-3854

Schenk HEA (1993) Some thoughts towards a discussion of terms and definitions in endocytobiology. In: S Sato, M Ishida, H Ishikawa (eds) Endocytobiology V, Tübingen University Press, Tübingen, pp 547-556

Schenk HEA (1994a) Glaucocystophyta model for symbiogenous evolution of new eukaryotic species. In: J Seckbach (ed) Evolutionary pathways and enigmatic algae, Kluwer, Dordrecht, pp 19-52

Schenk HEA (1994b) *Cyanophora paradoxa*: Anagenetic model or missing link of plastid evolution? Endocytobiosis and Cell Res 10: 87-106

Schenk HEA, Bayer MG, Maier T (1987) Nitrate assimilation and regulation of biosynthesis and disintegration of phycobiliproteids by *Cyanophora paradoxa*. Indications for a nitrogen store function of the phycobiliproteids. Endocytobiosis and Cell Res 4:167-176

Schlichting R, Bothe H (1993) The cyanelles (organelles of a low evolutionary scale) possess a phosphate-translocator and a glucose-carrier in *Cyanophora paradoxa*. Bot Acta 106: 428-434

Schnarrenberger C, Flechner A, Martin W (1995) Enzymatic evidence for a complete oxidative pentose phosphate pathway in chloroplasts and an incomplete pathway in the cytosol of spinach leaves. Plant Physiol (Rockville) 108: 609-614

Sitte P (1993) Intertaxonic combination: introducing and defining a new term in symbiogenesis. In: S Sato, M Ishida, H Ishikawa (eds) Endocytobiology V, Tübingen University Press, Tübingen, pp 557-558

Völkle E (1989) Diplomarbeit, University of Tübingen

Zook D, Schenk HEA (1986) Lipids in *Cyanophora paradoxa*. III. Lipids in cell compartments. Endocytobiosis and Cell Res 3: 203-211

The Phycobiliproteins Within the Cyanoplasts of *Cyanophora paradoxa* Store Carbon, Nitrogen, and Sulfur for the Whole Cell

N.E. Müller, O. Hauler and H.E.A. Schenk
University of Tübingen, Botanical Institute, Auf der Morgenstelle 1, 72076 Tübingen, Germany

Key words: *Cyanophora paradoxa,* Glaucocystophyta, cyanoplasts, cyanelles, phycobiliproteins, nitrogen storage, nitrogen deficiency, starch synthesis, sulfate deficiency, model calculation.

Summary: *Cyanophora paradoxa*, a protist and descendant of a cyanome, harbors photosynthetic active murein-enveloped cyanoplasts. Under normal culture conditions it preferentially uses cyanoplast located accessory phycobiliproteins for storage of assimilated carbon, instead of a starch pool located in its cytosol. However, during incipient nitrogen starvation, the direction of carbon flux in the cell changes: the assimilated photosynthetic carbon now becomes increasingly incorporated into the cytosolic starch, whereas phycobiliproteins are degraded. Sulfur deficiency shows a similar influence on carbon flux and biodegradation of phycobiliproteins. A model calculation prompts the assumption that assimilated nitrogen goes through an intracellular pool of metabolites with regulating properties (probably amino acids) before it is stored in the phycobiliproteins.

Introduction

The unicellular, hydrophotoautotrophic flagellate *Cyanophora paradoxa* (Korschikov 1924), phylum Glaucocystophyta (Kies and Kremer 1986), was for decades regarded as a symbiotic association, a so-called cyanome composed of a eukaryotic host with an endocytobiotic cyanobacterium (cyanelle, see Pascher 1929) as guest. Molecular biological investigations during recent years have led to the conclusion that the former symbiome was transformed by IITC (inequable intertaxonic combination) (Sitte 1990, 1993; Schenk 1993) via a cascade of gene transfers and new taxons to the recent complex eukaryotic cell. During this genomic transformation, the original cyanelle (of as yet unknown taxonomic position) has changed from an endocytobiont to an exogenosome, in this case a chloroplast-like cell organelle termed cyanoplast (Schenk 1990), which is characterized by the possession of a surrounding peptidoglycan wall (murein sacculus) (Schenk 1970; Heinz 1973; Aitken and Stanier 1979), by a variable pool size of bluegreen phycobiliproteins (PBP: allophycocyanin and C-phycocyanin) and other well conserved cyanobacterial properties on the one hand, and by a strongly reduced genome comparable to that of chloroplasts of higher plants on the other (Herdman and Stanier 1978; Löffelhardt and Bohnert 1994). Against this background, the question of to what extent a central nuclear control has been installed with regard to the catabolic (and anabolic) metabolism of the exogenosome deserves more attention. At first sight the organelle has anabolic functions similar to those of chloroplasts. Additionally to this property, in earlier experiments we observed degradation of phycobiliproteins (accessory chromoproteins of the cyanoplasts) caused by nitrogen starvation (Schenk et al. 1983, 1987). If the (nitrate-) nitrogen concentration in the medium is high enough, the phycobiliprotein content of the cell increases to an amount as high as or higher than the sum of all other cellular proteins within *C. paradoxa*. The genes responsible for the observed degradation of the phycobiliproteins are, surprisingly, not encoded on the cyanoplast DNA, but obviously on DNA of the eukaryotic nucleus (Schenk et al. 1987). Here we will describe first insights into some regulatory dependencies of nitrogen, sulfur, and carbon fluxes between cytosol and cyanoplasts.

Materials and Methods

Organism and Culture Conditions. *C. paradoxa* (Pringsheim strain 29.80, Culture Collection of Algae, University of Göttingen = UTEX 555) was cultivated as described earlier (Zook and Schenk 1986), the media for nitrogen and sulfate deficiency are described in Hauler and Schenk (1997, in preparation).

Quantitative Estimation of Pigments (Phycobiliproteins, Chlorophyll a, Carotenoids) and of relative cell number (RCN) was done by in vivo VIS spectroscopy (Schenk et al. 1983). Proteins were examined by the method according to Bradford (1976) and RNA (orcinol) and DNA (diphenylamine) by a modified Schmidt-Thannhäuser procedure (Munro and Fleck 1966). The cellular starch content was measured by the UV test of Boehringer Mannheim, the concentration of nitrate in the medium by HPLC (Hauler and Schenk 1997, in prep.), and the sulfate concentration following the method of Hwang and Dasgupta (1985) using the beryllon complex of barium. Probe preparation was described by Hauler (1993).

Model calculations. Typical data sets were carried on using the computational resources at the local university computer center and by PC-based routines using modified Runge-Kutta algorithm for fitting and splice-function approximation and interpolation to account for experimental restrictions on the total number of data. Experimental data were converted to nitrogen and sulfur contents per milliliter of culture based on the amino acid composition of phycobiliproteins, average nitrogen/sulfur content of nonpigment-associated proteins, base composition of nucleic acids, nitrogen contents of pigments, etc. Total nitrogen and sulfur contents and the amount of the metabolic nitrogen pool were compiled from these data.

Results and Discussion

Under advantageous growth conditions (about 4 - 5 mM NO_3^- in the culture medium, 3400 lx, 24 °C, and 4% CO_2 in air) *C. paradoxa* incorporates for storage most of the assimilated and surplus carbon into the phycobiliproteins. Some hours after the dilution of the cultures with fresh medium, the two (to four) cyanoplasts per cell have a distinct bluegreen colour, are large in diameter, and at this stage occupy more than a third of the cell volume. In this early proliferation stage only one to three small starch granules are visible within the cytosolic compartment of the eukaryote. Later, after 5 - 6 days, an increasing proportion of the phycobiliproteins are degraded (the content of PBP decreases) and the cyanoplasts begin to shrink, while the starch grains increase in number and size (Fig. 1).

The quantitative relations between PBP and starch concentrations are described by Hauler (1993). They documented how starch synthesis depends on the availability of nitrogen (nitrate) in the medium. If the cell has sufficient supplies of nitrogen, only a basic content of 2-5 pg starch per cell is present. When nitrogen starvation starts (4-5 days after the start of growth by dilution with new culture medium), increasing starch accumulation can be observed in the cytosol up to the 10th day (up to 35 pg/cell). At this time more than the half of the cell volume is filled up with starch grains (Fig. 1B). Clearly, these specifications depend on the original nitrate concentration in the medium: lower nitrate concentrations shorten

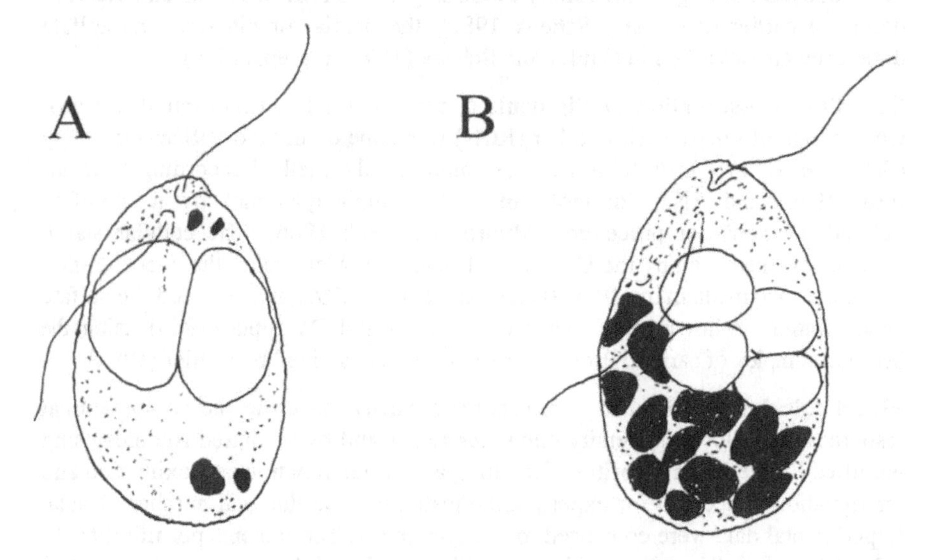

Fig. 1A, B. *C. paradoxa*: Two stages of cell anatomy dependent on availability of nitrogen in the medium. **A** Nitrogen saturation (about 4-5 mM nitrate; algal culture age: 1-2 days after dilution of the algal culture with new growth medium). **B** Nitrogen starvation (nitrate lower than 0.3 mM; algal culture age: about 6 days after dilution). Starch: black bodies in the cytosol after staining with Lugol solution ($KJ*J_2$ aq.)

the time for which the phycobiliproteins accumulate in the cyanoplasts (Schenk et al. 1987), while higher nitrate concentrations extend it.

The characteristics of the nitrate-PBP interdependency are as follows:

1. The nitrate uptake has zero-order kinetics, which forces the conclusion that the ammonium binding system is the rate limiting part for nitrate binding, and/or the photosynthetic part engaged in nitrate reduction is running at full speed. When, dependent on diminishing nitrate concentration in the medium, the nitrate import rate decreases, acceleration of the PBP synthesis rate decreases as well, and when the outside nitrate concentration reaches too low levels (< 0.3 mM), the PBP synthesis rate decreases. From this point on PBP synthesis must be driven by an intracellular pool of nitrogen-containing metabolites, as can be seen from Fig. 2. Again, we see a direct flow of nitrogen to the PBP.
2. When nitrate reduction is done, photosynthesis can utilize more energy equivalents in the Calvin-Benson cycle and hence builds starch molecules. Starch formation accelerates, although photosynthesis as measured by oxygen evolution decreases. This may also be seen phenomenologically: when the culture grows at maximum speed, degradation of PBP starts in order to free nitrogen. At this point the synthesis rate of starch starts to decrease as well, accounting for the need for carbon.
3. At the same time the production rate of PRO (all cell proteins except the PBP) decelerates. We think, that this is due to a flow of amino acids and metabolites

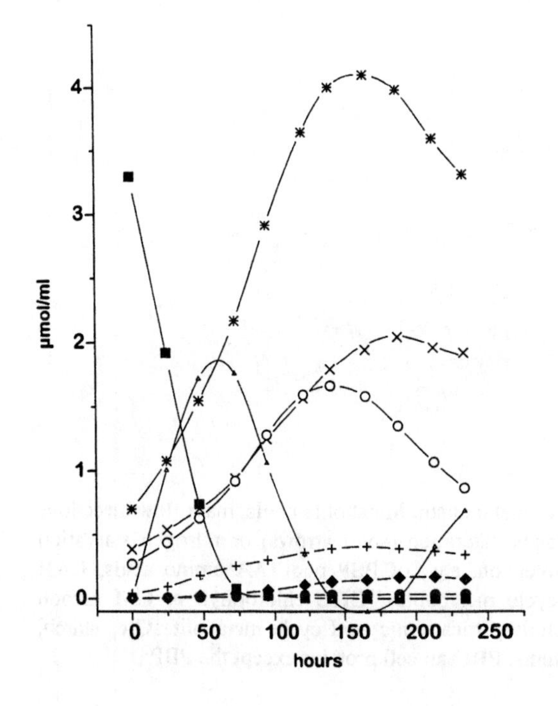

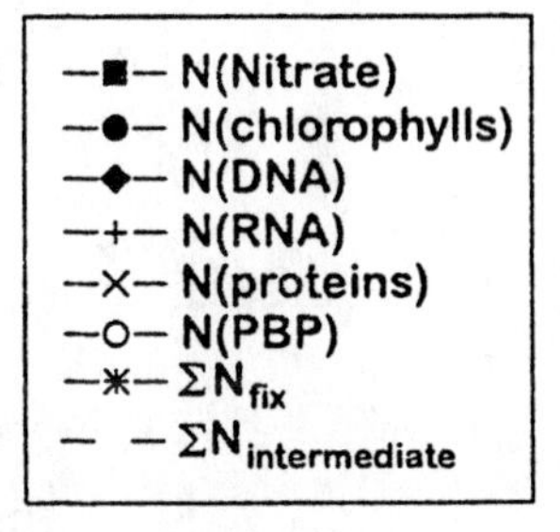

Fig. 2. *C. paradoxa:* Nitrogen Flux between different pools within the cell. Data calculated (in µmol/ml) from average of different experimental series; ■ nitrate concentration in the medium

from PBP to the PRO synthesis, which means flow of nitrogen, sulfur and carbon. During this period there is also more nitrogen liberated through PBP degradation than it seems to be subsequently reutilized in nitrogen-containing substances. Possibly this amount of PBP is degraded for a second reason, to liberate sulfur (see below).

4. When at last the PBP degradation rate decreases, starch will be utilized to substitute for carbon bodies. This helps to keep up the PBP content necessary to maintain membrane integrity for some time.

These few sentences show how cellular compounds could serve as dual-purpose components of the cell, mainly in the fields of control, metabolism, storage, and structure, e.g., like starch for storage or carotenes for structure. Evidence will be presented to show that the PBPs serve a dual purpose, for structure (of photosynthetic apparatus), and for storage of carbon, nitrogen, and sulfur. Experimental evidence was obtained by substitution of nitrate and sulfate to deficient cultures of *C. paradoxa* and by analysis of metabolic dynamic data. Comparing the dynamics of various major compound classes (chlorophyll, nucleic acids, proteins and starch) and some major metabolic pathways, very different characteristics can be observed. In the present case three levels are distinguishable: (1) the storage level of more or less slow activatable depots with peaking characteristics, (2) the input level with quick dynamic pools, and (3) the output level of controlled compounds (Fig. 2, 3).

Figure 4 compiles the most important rates of nitrate, PBP, PRO, chlorophyll, and starch versus time as fitted functions. First we focus on nitrogen caching. As under given culture conditions the cellular protein content PRO remains at ap-

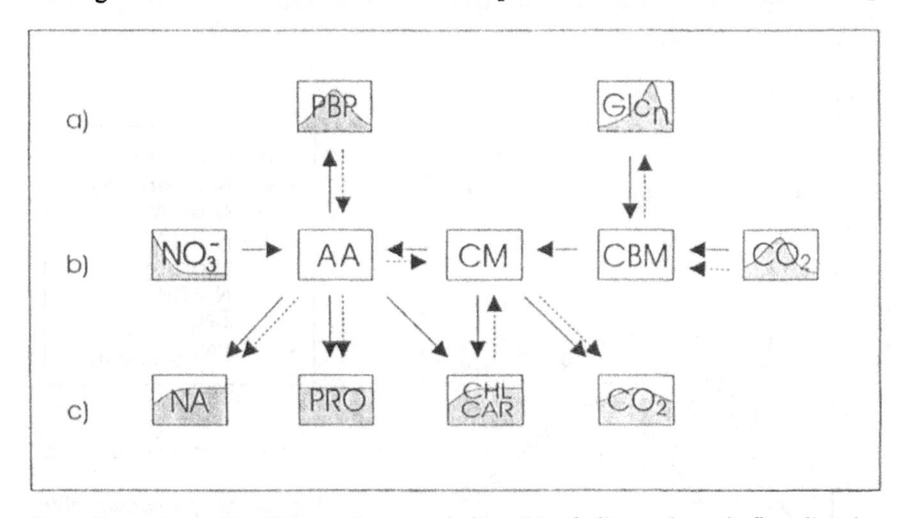

Fig. 3. Dynamic model of *C. paradoxa* metabolism. Metabolite pools, main flow directions of metabolites in conditions of nitrogen saturation (solid arrows) or nitrogen starvation (dotted arrows) and sink-source conversion, e.g., of PBP pool. **AA** amino acids, **CAR** carotenoids, **CBM** Calvin-Benson cycle metabolites, **CHL** chlorophyll a, **CM** carbon metabolites, e.g., carbonic acids including tricarbonic acid cycle metabolites**Glc_n** starch, **NA** nucleic acids, **PBP** phycobiliproteins, **PRO** all cell proteins except the PBP

proximately the same level throughout the first 7 days and therefore the PRO production rate corresponds strictly with the growth rate. The PBP synthesis rate and rate acceleration are higher than for PRO. This means that surplus nitrate-nitrogen is brought into the PBP with priority status. Candidates for this flow preference of the ammonium-capturing system (GS/GOGAT) are nitrogen-containing metabolites such as glutamine, glutamate, and alanine. (In very early, unpublished experiments with freed cyanoplasts we observed that within the group of soluble amino acids the glutamate has a photosynthetic ^{14}C incorporation rate five times higher than aspartate, alanine a rate four times higher, and both amino acids were lost into the medium at a rate of about 125 nmol/mg Chl×h maximum. Schlichting and Bothe (1993), finding that glutamate enters the cyanoplasts at an initial rate of about 60 and glutamine at about 360 nmol/mg Chl×h, suggested that glutamine is the nitrogen-compound which is supplied to the eukaryotic cytoplasm). One or several of these compounds are suspected of serving at the transcriptional and/or translational level in the regulation of biosynthesis and/or biodegradation of PBP. That is to say, surplus nitrate-nitrogen is cached within PBP.

To get a closer look at the role of sulfur in PBP metabolism we used depleted cultures and substituted nitrate and sulfate sequentially (Hauler 1993). There seem to be two kinds of effects; we will focus on metabolics first.

Sulfate deficient cultures which contain 0.13 mM sulfate (a quarter of the normally used concentration) show limited uptake and reduction of nitrate to a residual concentration of about a fourth of the initial concentration. Owing to a lack of sulfur the PBP are degraded to amino acids to allow growth of cell number. Subsequent addition of sulfate up to a concentration of 0.5 mM restores nitrate uptake and PBP production. However, cultures that are depleted of nitrate and sulfate do not respond to nitrate addition; in this case, although nitrate is present, PBP are

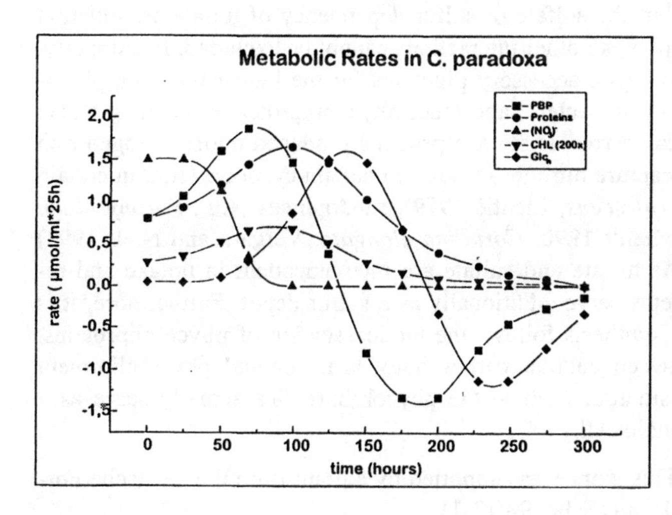

Fig. 4: *C. paradoxa:* Dynamic rates of different nitrogen and carbon fluxes (µmol/ml × 25 h): nitrogen in nitrate (NO_3^-), phycobiliproteins (PBP), cell proteins PRO (without PBP), and in chlorophyll, and carbon as glucose in starch. Symbols are for identification purposes only: ■ PBP, ● PRO, ▲ NO_3^-, ▼ chlorophyll (× 200), ◆ starch (based on glucose units) (× 0.002)

still degraded. The reason for this continuing degradation seems to be the flow of sulfur from PBP towards other metabolic and structural compounds, because sequential addition of sulfate restores nitrate uptake and PBP synthesis.

There may be another factor sharing responsibility for this effect, as can be seen from dynamic import rates. As previously mentioned, nitrate uptake is by zero-order kinetics, which shows the nitrate reduction and ammonium fixation to be rate-limiting. The coproduced photosynthetic ATP may accelerate the sulfate permease-driven import. When the nitrate import rate decreases, starch is accumulated and accounts for the consumption of ATP, which no longer accelerates sulfate import.

When no more nitrate is taken up, the nitrate-sulfate reduction interactions come to an end, in our opinion an indication that the PBP pool is the most important storage pool for nitrogen and sulfur in *C. paradoxa*.

Conclusions

Phycobiliproteins of *C. paradoxa* are coded on the DNA of the cyanoplasts and synthesized within this cellular compartment. The observation that the degradation genes of these proteins (surely, during an earlier symbiotic stage of the present complex cell, a property of the original endocytobiotic cyanobacterium, as shown for free-living modern cyanobacteria, e.g., *Synechococcus* sp.) (Collier and Grossman 1992) had come under the control of the eukaryotic nucleus (Schenk et al. 1987) seems to be of evolutionary interest. Phycobiliproteins serve as a cache for nitrogen available from the nitrate content of the given culture medium. Nitrate is only imported and reduced when a sufficient concentration of sulfate is present, and the reverse is also true. In consequence, nitrate assimilation stops if the sulfur concentration becomes minimal. Nitrate reduction may propagate an active sulfate import, but the sulfate or sulfur dependency of nitrate assimilation seems to be more complex, so other interactions cannot be excluded. In summary, phycobiliproteins are obligate accessory pigments for the light harvesting photosynthetic processes of cyanobacteria and other phycobiliprotein containing algae, but in addition they play a role as storage proteins produced under nitrogen-rich conditions in order to capture nitrogen, at least in certain cyanobacteria, in certain cryptophytes (e.g., *C. rufescens*, Lichtlé 1979), rhodophytes (e.g., *Porphyridium purpureum*, Levy and Gantt 1990; *Corallina elongata*, Vergara and Niell 1993) and in *C. paradoxa*. As nitrate and sulfate are interdependent in uptake and reduction, phycobiliproteins serve additionally as a sulfur depot. Furthermore, it is remarkable that starch synthesis follows the biodegradation of phycobiliproteins. Thus, the relation between cellular carbon-body turnover and phycobiliprotein concentrations reveal situations where the phycobiliproteins actually serve as a carbon depot for the whole cell.

Acknowledgements. This work was supported by a grant from the Deutsche Forschungsgemeinschaft, Bonn (Sche: 98/13-1).

References

Aitken A, Stanier RY (1979) Characterization of peptidoglycan from the cyanelles of *Cyanophora paradoxa*. J Gen Microbiol 112: 219-223

Bayer M, Schenk HEA (1986) Biosynthesis of proteins in *Cyanophora paradoxa*: Protein import into the endocyanelle analyzed by micro two-dimensional gel electrophoresis. Endocytobiosis and Cell Res 3: 197-202

Bradford MM (1976) Anal Biochem 72: 248-254

Collier JL, Grossman AR (1992) Chlorosis induced by nutrient deprivation in *Synechococcus* sp strain PCC 7942: not all bleaching is the same. J Bacteriol 174: 4718-4726

Deimel R (1985) Die Wirkung von Translationsinhibitoren auf die Proteinsynthese von *Cyanophora paradoxa*. Diplomarbeit, University of Tübingen

Fester T (1993) Glucose-6-phosphat Dehydrogenase bei *Cyanophora paradoxa*: Aufreinigung und Teilcharakterisierung des Isoenzyms der Cyanoplasten. Diplomarbeit, University of Tübingen

Fester T, Völkle E, Schenk HEA (1996) Purification and partial characterization of the cyanoplast glucose-6-phosphate dehydrogenase in *Cyanophora paradoxa*. Endocytobiosis and Cell Res 11: 159-176

Hauler O (1993) *Cyanophora paradoxa*. Regulation des Phycobiliproteidabbaus unter Stickstoffmangel und Sulfatmangelinduzierte Degradation der Phycobiliproteide. Diplomarbeit, University of Tübingen

Heinz G (1973) Versuche zur Isolierung der lysozymempfindlichen Stützmembran von *Cyanocyta korschikoffiana*, der Endocyanelle aus *Cyanophora paradoxa* Korsch. Diplomarbeit, University of Tübingen

Herdmann M, Stanier RY (1977) The cyanelle: Chloroplast or endosymbiotic prokaryote? FEMS Lett 1: 7-12

Huppe HC, Vanlerberghe GC, Turpin DH (1992) Evidence for activation of the oxidative pentose phosphate pathway during photosynthetic assimilation of nitrate but not ammonium by a green alga. Plant Physiol (Bethesda) 100: 2096-2099

Huppe HC, Farr TJ, Turpin DH (1994) Coordination of chloroplastic metabolism in N-limited *Chlamydomonas reinhardtii* by redox modulation: II. Redox modulation activates the oxidative pentose phosphate pathway during photosynthetic nitrate assimilation. Plant Physiol (Rockville) 105: 1043-1048

Hwang H, Dasgupta PK (1985) Spectrophotometric determination of trace aqueous sulfate using barium-beryllon II. Mikrochim Acta 1985 (I): 313-324

Kaiser WM, Bassham JA (1979) Carbon metabolism of chloroplasts in the dark: Oxidative pentose phosphate cycle versus glycolytic pathway. Planta 144: 193-200

Kies L (1988) The effect of penicillin on the morphology and ultrastructure of *Cyanophora, Gloeochaete* and *Glaucocystis* (Glaucocystophyceae) and their cyanelles. Endocytobiosis and Cell Res 5: 361-372

Kies L, Kremer BP (1986) Typification of the Glaucocystophyta. Taxon 35:128-133

Kloos K, Schlichting R, Bothe H (1993) Glutamine and glutamate transport in *Cyanophora paradoxa*. Bot Acta 106: 435-440

Korschikov AA (1924) Protistologische Beobachtungen. *Cyanophora paradoxa* n gen et sp Russ Arch Protistol 3: 57-74

Levy I, Gantt E (1990) Development of photosynthetic activity in *Porphyridium purpureum* (Rhodophyta) following nitrogen starvation. J Phycol 26: 62-68

Lichtlé C (1979) Effects of nitrogen deficiency and light of high intensity on *Cryptomonas rufescens* (*Cryptophyceae*). I. Cell and *photosynthetic* apparatus *transformations* and

encystment. Protoplasma 101: 283-299
Löffelhardt W, Bohnert HJ (1994) Molecular biology of cyanelles. In: DA Bryant (ed) The molecular biology of cyanobacteria. Kluwer, Dordrecht, pp 65-89
Munro HN, Fleck A (1966) The determination of nucleic acids. Methods Biochem Anal 14: 113-176
Pascher A (1929) Studien über Symbiosen. I. Über einige Endosymbiosen von Blaualgen in Einzellern. Jahrb Wiss Bot 71: 386-462
Pringsheim EG (1958) Organismen mit blaugrünen Assimilatoren. Stud Plant Physiol (Praha): 165-184
Provasoli L, Pintner J (1952) Some interesting algal flagellates recently obtained in pure culture. News Bull Phycol Soc Am 5: 7
Schenk HEA (1970) Nachweis einer lysozymempfindlichen Stützmembran der Endocyanellen von *Cyanophora paradoxa* Korsch Z Naturforsch 25b: 640-656
Schenk HEA (1990) *Cyanophora paradoxa*: a short survey, In: P Nardon et al (eds) Endocytobiology IV. INRA, Paris, pp 199-209
Schenk HEA (1992a) *Cyanophora paradoxa*, identification and sequencing of nucleus-encoded cyanellar proteins. A proof for gene transfer. Endocytobiosis and Cell Res 8: 197-222
Schenk HEA (1992b) Cyanobacterial symbioses. In: A Ballows, HG Trüper, M Dwor-kin, W Harder, K-H Schleifer (eds) The Prokaryotes. A handbook of ecophysiology, isolation, identification, application. Springer, New York, vol 4, pp 3819-3854
Schenk HEA (1993) Some thoughts towards a discussion of terms and definitions in endocytobiology. In: S Sato, M Ishida, H Ishikawa (eds) Endocytobiology V, Tübingen University Press, Tübingen, pp 547-556
Schenk HEA (1994a) Glaucocystophyta model for symbiogenous evolution of new eukaryotic species. In: J Seckbach (ed) Evolutionary pathways and enigmatic algae, Kluwer, Dordrecht, pp 19-52
Schenk HEA (1994b) *Cyanophora paradoxa*: Anagenetic model or missing link of plastid evolution? Endocytobiosis and Cell Res 10: 87-106
Schenk HEA, Hanf J, Neu-Müller M (1983) The phycobiliproteids in *Cyanophora paradoxa* as accessoric pigments and nitrogen storage proteins. Z Naturforsch 34c: 972-977
Schenk HEA, Bayer MG, Maier T (1987) Nitrate assimilation and regulation of biosynthesis and disintegration of phycobiliproteids by *Cyanophora paradoxa*. Indications for a nitrogen store function of the phycobiliproteids. Endocytobiosis and Cell Res 4: 167-176
Schlichting R, Bothe H (1993) The cyanelles (organelles of a low evolutionary scale) possess a phosphate-translocator and a glucose-carrier in *Cyanophora paradoxa*. Bot Acta 106: 428-434
Schnarrenberger C, Flechner A, Martin W (1995) Enzymatic evidence for a complete oxidative pentose phosphate pathway in chloroplasts and an incomplete pathway in the cytosol of spinach leaves. Plant Physiol (Rockville) 108: 609-614
Sitte P (1990) Phylogenetische Aspekte der Zellevolution. Biol Rundsch 28: 1-18
Sitte P (1993) Intertaxonic combination: introducing and defining a new term in symbiogenesis. In: S Sato, M Ishida, H Ishikawa (eds) Endocytobiology V, Tübingen University Press, Tübingen, pp 557-558.
Vergara JJ, Niell FX (1993) Effects of nitrate availability and irradiance on internal nitrogen constituents in *Corallina elongata* (Rhodophyta). J Phycology 29: 285-293
Völkle E (1989) Die Glucose-6-phosphat-dehydrogenase von *Cyanophora paradoxa* Teilcharakterisierung der kompartimentspezifischen Enzymformen. Diplomarbeit, University of Tübingen
Zook D, Schenk HEA (1986) Lipids in *Cyanophora paradoxa*, III Lipids in cell compartments. Endocytobiosis and Cell Res 3: 203-211

1.5 Molecular Evolution

Hypercycles in Biological Systems

M. Gebinoga
EVOTEC Biosystems GmbH, Grandweg 64, 22529 Hamburg, Germany

Summary: The idea of hypercycles as a theoretical tool to explain some aspects of early evolution was introduced by Manfred Eigen in 1971. Experimental data which show the relevance of hypercyclic coupled biosynthesis processes were gained nearly 20 years later. These experiments were done with phage-infected bacteria and show during the early stage of viral infection a hyperbolic increase of viral RNA synthesis rate which was a clear sign of hypercyclic coupled processes. The role of hypercyclic coupling in early molecular evolution is analogous to its role in viral evolution today: It provides evolutionary improvement of the phenotype by favoring the replication of its own genotype. Due to the growing implications of eukaryotic RNA viruses it is interesting and necessary to investigate these viruses considering the influence of hypercyclic organization structure. Better understanding may suggest possibilities for new antiviral strategies. Cell-free translation systems coupled with quasi viral amplification systems could help to establish pure *in vitro* systems with hypercyclic organization. Such systems on the one hand can give access to biotechnological synthesis of some proteins and on the other hand can act as a tool to investigate relevant processes of molecular evolution.

Introduction

How did life on earth arise? Why do we find the same basic equipment and the same genetic code in all organisms? Surely these are two questions which have been of importance to the whole field of biological sciences for a long time. At the time of Charles Darwin and his colleagues it was commonly thought that a living being, such as an animal or a plant, arise from the nonliving environment. These days we also consider the prebiotic steps and processes of molecular self-organization which are the connecting links between chemical and biological evolution. The more general question is, how did the first self-replicating structures (e.g., polynucleotides) develop and what conditions induced their ever-increasing complexity?

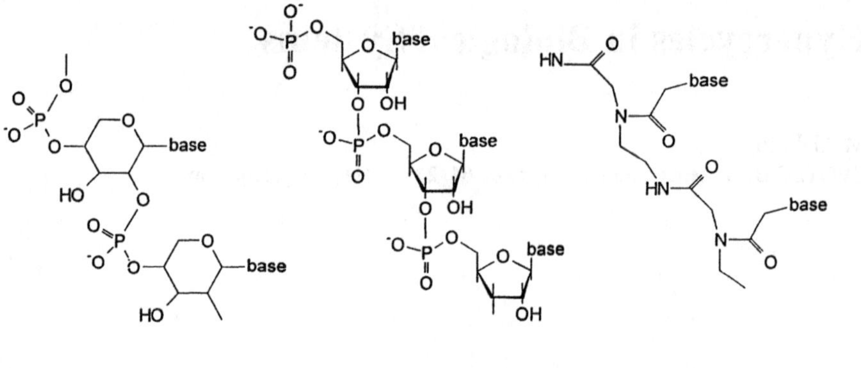

Pyranose-phosphate backbone of RNA Ribose-phosphate backbone of RNA Peptide-nucleic acid (PNA)

Fig. 1. Different possibilities of realizing a nucleic acid- like structure compared with normal RNA. An RNA with a pyranose-phosphate backbone was prepared by Eschenmoser et al. (Müller et al. 1990) and the PNA was synthesized by Nielsen et al. (1991)

Some early experiments in prebiotic evolution were done by Stanley L. Miller (1953), Sidney W. Fox (1965, 1969), and others. They have shown the possibility of synthesis of biomolecules by simple precursor molecules. What can happen when a mixture of energy-rich molecules, short polymers consisting of amino acids, nucleotides or other molecules comes together? A whole bundle of interactions must have existed between these different molecules. In the days of these first experiments the predominant opinion was that, firstly, proteins or protein-like molecules ("protenoids") existed and nucleic acids came later. The possibility that an RNA replication machinery preceded a protein synthesis apparatus was proposed by Carl Woese (Woese 1967), Lesley E. Orgel (Orgel 1968) and Francis H. Crick (Crick 1968) in the late 1960s, and with the discovery of ribozymes (Kruger et al. 1982) this theory became a commonly held view under the term "RNA world" of Walter Gilbert (1986).

Many experiments in prebiotic nucleic-acid chemistry were done in the laboratory of Orgel (Sulston et al. 1968, Lohrmann and Orgel 1973, Lohrmann 1975, Sleeper and Orgel 1979). However, there are some open questions in this field, especially about the generation of the stereospecific catalysis of the 5'-3'-bond between two nucleotides. Another interesting question is, why were nucleic acids with a sugar-phosphate backbone elected during the prebiotic evolution as carriers of genetic information? Eschenmoser (Müller et al. 1990) and Peter Nielsen (Nielsen et al. 1991) have shown that the common nucleobases can combine also with other sugars or without a sugar-phosphate backbone (Fig. 1) and that these polynucleotides also can replicate themselves.

Our interest is not focused on the questions above but deals with the processes of self-organization of the first polynucleotides. We are interested in the processes available for developing the very first machinery for self-replicating nucleic acids. Of course, we can never reproduce the same conditions prevailing nearly four-billion years ago, but we can ask what possible mechanism could build up such a molecular machinery for the generation, the storage, and the development of genetic information.

Theoretical Concepts of Molecular Evolution

Before we come to the concept of hypercycles, we would like to give a short introduction to two ideas which are closely related to the concept. Furthermore, they are powerful tools to handle some problems concerning molecular evolution. One of these tools is the concept of the quasi-species. It was Manfred Eigen who introduced these principles into the field of molecular evolution in 1971 (Eigen 1971) and has developed these theories with his coworkers and some colleagues in the following years. The sequence space, another important concept, was first introduced by Richard W. Hamming (Hamming 1980) to the theory of information. Ingo Rechenberg made the transfer from general coding theory to nucleic acids and proteins as carriers of information (Rechenberg 1973). An introduction of this concept to the theory of molecular evolution was also carried out by Eigen.

If we look at a polynucleotide chain, such as an RNA molecule, then it is obvious that it can carry information. Furthermore this molecule has the inherent ability to amplify itself with or without an enzyme. Without enzymatic aid this replication is not very accurate. Assuming that this specific self-replicating molecule can be replicated a few times until degradation destroys it, a whole family of mutated daughter molecules can also be replicated. If too many mutations occur, the population will consist of a random distribution of different sequences. If the information is lost, we can call such an event an "error catastrophy". If not too many mutations occur, then a population will arise with closely related molecules having a slightly different rate of replication. The molecule which replicates more rapidly is called the "master-sequence" by the competitors Eigen and Schuster and the whole population they call a "quasi-species" (Eigen and Schuster 1977, Eigen et al. 1988). The quasi-species is closely correlated with the so-called "wild-type" of a population of molecules or microorganisms. These quasi-species are not only a characteristic of RNA molecules but also of viruses and organisms. Exact determination of the wild-type distribution of phage Qβ demonstrated that only a small fraction of the sequence is identical to the wild-type sequence (Domingo et al. 1978). The overwhelming majority of all sequences represents the single and multiple error mutants of the wild-type sequence. Only the average over all sequences results once again in the wild-type sequence.

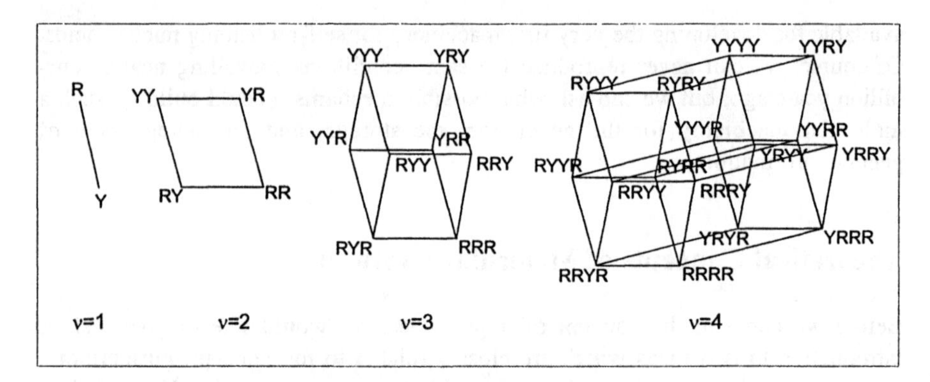

Fig. 2. The sequence space concept for polynucleotide chains consisting of pyrimidines (Y) and purines (R). The presented chain lengths of these binary sequences go from $\nu = 1$ to $\nu = 4$. The rapid increase in connectivity with increasing ν is clearly visible. The number of states increases with 2^{ν} in the binary case and with 4^{ν} for the AUCG code of a real RNA sequence

One problem occurred with this concept: What is the correct determination of distances between different mutants of a quasi-species distribution? To represent correctly the kinship distances between different sequences of the length ν we need a ν-dimensional point space, which we call the sequence space. Each sequence is presented by a point. In Fig. 2 it is easy to see why this point space can represent relationships between mutant sequences. The distances (Hamming distance) between two mutants of the quasi-species distribution are as short as possible. If we take any sequence then we get a one error mutant the distance of 1, a two error mutant the distance of 2, and so on. Overall, the sequence space can cover a hyper-astronomical number of sequences. By the way, it is clearly impossible to handle experimentally really large mutant distributions. It is only possible to handle very small parts of those. An RNA of 100 ribonucleotides represents $4^{100} \approx 10^{60}$ sequences. For the synthesis of all sequences you need the mass of 45-million stars ($9\text{x}10^{37}$ kg)! We can see that the sequence space of a real RNA distribution must be like a vast desert with only some very tiny oases, where the sequences of the quasi-species exist. Although the sequence space concept is a very powerful tool for general consideration of RNA sequences and mutant distributions, one problem exists. The sequence space concept can handle all mutants of a given sequence and a given length. On the other hand, it is an intrinsic characteristic of this general concept that it is not possible to deal with mutants which are generated by deletions or insertions. Nevertheless, the sequence space concept gives us an idea of what a quasi-species, e.g., of a virus population, is. Such a virus population is something like a very thin cloud or a net which occupies a few points in the sequence space.

An exact mathematical explication of the sequence space and the quasi-species concepts would be beyond the scope of this chapter. For a detailed description of these important concepts and the ideas based on those such as the shape space concept of Peter Schuster, we suggest research of the literature of Eigen, Schuster, and coworkers (Eigen 1971, 1985, 1986, Eigen and Schuster 1977, Eigen et al. 1988, Schuster et al. 1994).

General Characteristics of Hypercycles

A replication of the RNA quasi-species distribution with a few errors is equivalent to the possibility of amplifying the total information. For the replication rate, we can use the formula:

$$\frac{dx_i}{dt} = \left(E_i - E(t)\right) \cdot x_i$$

where x_i corresponds to the relative population number, E_i is the excess production of the self-replicating molecule i, and $E(t)$ corresponds to the average production of all self-replicating molecules (Eigen and Schuster 1977). Because of the erroneous replication, there are correct copies and some mutants with different rates of replication. Eigen has shown that the error rate of the information molecules is reciprocal to their chain length (Eigen 1971). In other words, the error-probability of the replication is also a complexity threshold for the information molecule. The maximal information content or the maximal chain-length of the sequence (ν_{max}) of such a quasi-species distribution depends on its error rate of replication. We can say:

$$\nu_{max} = \frac{\ln \sigma_m}{1 - q_m},$$

where ν_{max} is the number of symbols in the self-replicating molecule, σ_m is the selective superiority of the master sequence to the average of its mutant distribution, and $1\text{-}q_m$ corresponds to the average error rate per symbol. The arising question is, how can an enzyme-free replication of short polynucleotides, which are probably of the chain length of a tRNA, change to a catalyzed replication? For the coding of a small enzyme, larger polynucleotides are necessary. These RNAs would have a chain length approximately 3-5 times larger than the enzyme-free replicative molecules. But a larger polynucleotid needs a catalyzing enzyme for the correct replication of its information. Seemingly, it is a comparable paradox to the question which came first, the egg or the hen.

Eigen was the first to give an explanation to a possible mechanism of molecular evolution under the consideration of hypercyclic interactions. For the development of complex structures and at least protocells, three principles can be assumed (Eigen 1971, Eigen et al. 1980):

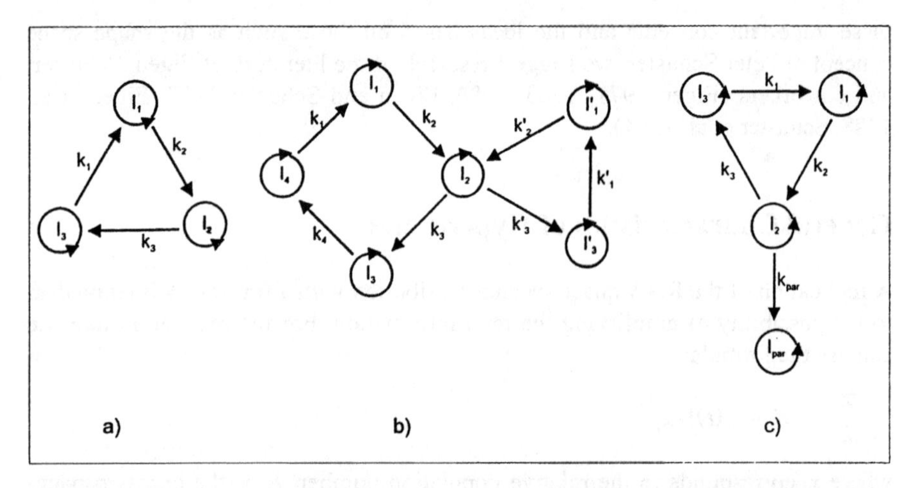

Fig. 3. Different hypercycles. **a** Hypercycle with three connected members. The autocatalytic unit I_1 gives catalytic support to the autocatalytic cycle I_2 and gets catalytic support itself from the encoded enzyme of the self-reproducible unit I_3. **b** Two joint hypercycles with one common member I_2 which is engaged in both hypercycles. **c** Three-membered hypercycle with a self-replicative, parasitic molecule species "par". The parasitic species gets support from species I_2, but does not give catalytic support to any species in the hypercycle. (After Eigen and Schuster 1977 and Hofbauer and Sigmund 1988)

1. There exists of a self-replicating molecule (e.g., RNA) which is capable of transmitting information with mutational errors from a parental strand to a daughter strand.
2. These molecules evolve in hypercyclic systems in which the whole information content is distributed over all members of the system (quasi-species). During this evolution, the phenotypic characteristics, such as reproductiveness, stability, and error proneness, improve.
3. Compartmentation on the one hand protects against parasitic infections and on the other hand allows a better coupling between the self-replicating molecules and their encoded translation products.

We agree with Eigen and Schuster (Eigen and Schuster 1977) calling these cycles of reactions hypercycles, which are hypercyclic in respect of their catalytic function. Such a catalytic hypercycle is a system of cyclic, connected self-replicative or autocatalytic units (Fig. 3).

John Maynard Smith has added some helpful criticism (Maynard Smith 1979) of problems which arise from the concept of hypercycles. One of the possible arising problems is the vulnerability of hypercycles to parasites. Parasites are here to be seen on a molecular level, i.e., molecules that are attached on a hypercycle and receive increased catalytic support but give no catalytic support to any member of the hypercycle (Fig. 3). Boerlijst and Hogeweg (Boerlijst and Hogeweg, 1991) have made computer simulations with a cellular automaton model. In this model they have used self-replicative molecules to simulate hypercyclic interactions. They have found that in incompletely mixed media, spiral structures emerged which made the hypercycles resistant to parasites. Furthermore, one can see these spiral structures as a very early form of a compartment. The following steps are a physical barrier (lipid membrane) against the environment. Compared with other catalytic cycles a hypercycle exhibits some special features which we would like to explain briefly. For a detailed description there are some publications which cover theory (Eigen 1971, Eigen and Schuster 1977; Eigen et al. 1980; Maynard Smith 1979; Boerlijst and Hogeweg 1991), mathematical methods (Eigen and Schuster 1977; Eigen et al. 1980b; Hofbauer and Sigmund 1988), and experiments (Eigen et al. 1991) of hypercyclic coupled systems.

A normal enzymatic reaction is an example of a simple catalytic cycle, where a constant concentration of the enzyme E catalyzes the transformation of a substrate S to the product P. The catalytic steps involve the intermediates of the enzyme with the substrate (ES) with the product (EP). After the transformation of the substrate to the product the enzyme can catalyze the reaction of the next substrate molecule. In general, the concentration of the substrate is largely excessive compared with the constant enzyme concentration, and in this case the reaction rate is of the order of zero (linear growth).

The well-known polymerase chain reaction (Saiki et al. 1985) is an example of a true self-reproductive process where the products catalyze their own replication and grow exponentially. Such an autocatalytic reaction can be written as:

$$dNTP \xrightarrow{DNA,Enzyme} DNA\,,$$

or, when the DNA concentration increases during the reaction:

$$dNTP + DNA \xrightarrow{Enzyme} 2\ DNA\,,$$

or, more formally (Eigen and Schuster 1977):

$$X \xrightarrow{I,E} I \quad \text{respective } X + I \xrightarrow{E} 2\,I\,,$$

where I is the information carrier and X corresponds to the precursor molecules (e.g., dNTPs). It is easy to see that in this case the reaction rate depends on the constant enzyme concentration and the growing DNA concentration. We have a first-order reaction with exponential growth of the DNA, because at first the enzyme concentration is much higher than the DNA concentration.

Now we can do the next step and present an autocatalytic system consisting of mutual growing nucleic acids and enzymes. The nucleic acid carries the information for the synthesis of the enzyme and the enzyme catalyzes the replication of

Table 1. The kinetic characteristics of catalytic systems

Normal enzymatic catalysis	PCR (autocatalytic reaction)	RNA replication with hypercyclic coupling to translation
$S \xrightarrow{E} P$	$DNA \xrightarrow{dNTP,E,primer} 2DNA$	$RNA \xrightarrow{NTP,E} 2RNA$
d[S]/dt ≈ 0, d[E]/dt = 0	d[DNA]/dt >0, d[E]/dt = 0, d[primer]/dt = 0, d[dNTP]/dt = 0	[RNA]/dt >0, d[E]/dt >0, d[NTP]/dt ≈0
[S] >> [E]	[DNA] < [E]	[RNA] ≈[E]
d[P]/dt = k [S][E] = k'	d[DNA]/dt = k [DNA][E] = k'' [DNA]	d[RNA]/dt = k [RNA][E] ≈ k [RNA]^2
$\int_{[P_0]}^{[P]} d[P] = \int_{t=0}^{t} kt$	$\int_{[DNA_0]}^{[DNA]} \frac{d[DNA]}{[DNA]} = \int_{t=0}^{t} kt$	$\int_{[RNA_0]}^{[RNA]} \frac{d[RNA]}{[RNA]^2} = \int_{t=0}^{t} kt$
$[P] = [P_0] \cdot kt$	$[DNA] = [DNA_0] \cdot e^{kt}$	$[RNA] = \frac{[RNA_0]}{1-kt}$
linear growth	exponential growth	hyperbolic growth of the RNA during a short time interval, as in viral infection cycles

S, substrate;
P, product;
E, enzyme;
k, k', k'', velocity constants;
t, time.

the information carrier. This is a true genotype-phenotype coupling. Eigen has named such systems "catalytic hypercycles", which connect autocatalytic units through a cyclic linkage. Table 1 shows the kinetic characteristics of the catalytic systems described above. The third case with the hyperbolic growth of the RNA is a simplified form but is also valid, if either the RNA concentration or the replicase concentration is small with respect to each other or the velocity constant.

This kinetic behavior of hypercyclic-coupled systems shows one very important characteristic of hypercycles. We are looking at a case where a hypercyclic-coupled system HC-a competes with another hypercycle HC-b. The hypercycle HC-b exhibits a "fitter" phenotype (e.g., shorter replication times), but is less concentrated than the first hypercycle HC-a. Under these circumstances, HC-a will win the competition regardless of the better fitness of its competitor HC-b. This is a once-only decision, where the hypercycle whose starting concentration was higher will win. This is also a possible explanation for the fact that after the formation of first living cells, one and only one genetic code and biochemistry existed. We know that slightly other forms of nucleic acids as information carriers are possible (e.g., PNAs), but they were never found in a natural system. This rigid decision pattern is well known for competing hypercycles. Once one hypercycle has overcome its competitors, no coexistence is possible.

Experimental Verification of Hypercycles

A simple example of a hypercyclic-organized process is the RNA-phage infection of a bacterial cell. The plus strand RNA of the infecting phage becomes translated by the ribosomal machinery of the host cell. One of the phage genome-instructed proteins associates with other host proteins to form a phage-specific RNA-replicase (e.g., Qβ-replicase). This enzyme exclusively recognizes some specific features of the viral plus and minus strand RNA and amplifies solely the viral RNA. The result is a burst of phage-RNA production. Because of the hypercyclic nature of the reaction, the increase of the RNA follows a hyperbolic growth law, until the metabolic supply of the host is exhausted. This was exactly what we found, analyzing the kinetics of viral RNA synthesis and viral protein synthesis in detail.

Hypercyclic Coupling During RNA Phage Development

In a bacteriophage-bacterial host system it was possible to observe a real hypercyclic coupling between the synthesis of viral RNA and the biosynthesis of the proteins encoded by the viral genome (Eigen et al. 1991) (Fig. 4). Although concentration profiles of the infection of bacteria by RNA phages were done by some groups, e.g., Viñuela et al. (Viñuela et al. 1967) and Cramer and Sinsheimer (Cramer and Sinsheimer 1971), a real in vivo measurement was not done, but was necessary for a clear distinction between different kinetic behavior of the viral synthesis processes.

For an exact estimation of the synthesis rate into the infected bacteria we tried to make the measurements using permeabilized bacteria (Moses and Richardson 1970). At first the bacteria became infected under exact defined conditions. After infection we took samples of the bacteria at fixed infection times for the estimation of the intracellular synthesis rates. For this approach we extracted the cytoplasmic membrane of the isolated bacteria after distinct incubation times with an appropriate organic solvent. After this treatment, the rest of the membrane allowed an exchange of nucleoside triphosphates and amino acids from the surrounding medium, while the high molecular weight of intracellular components

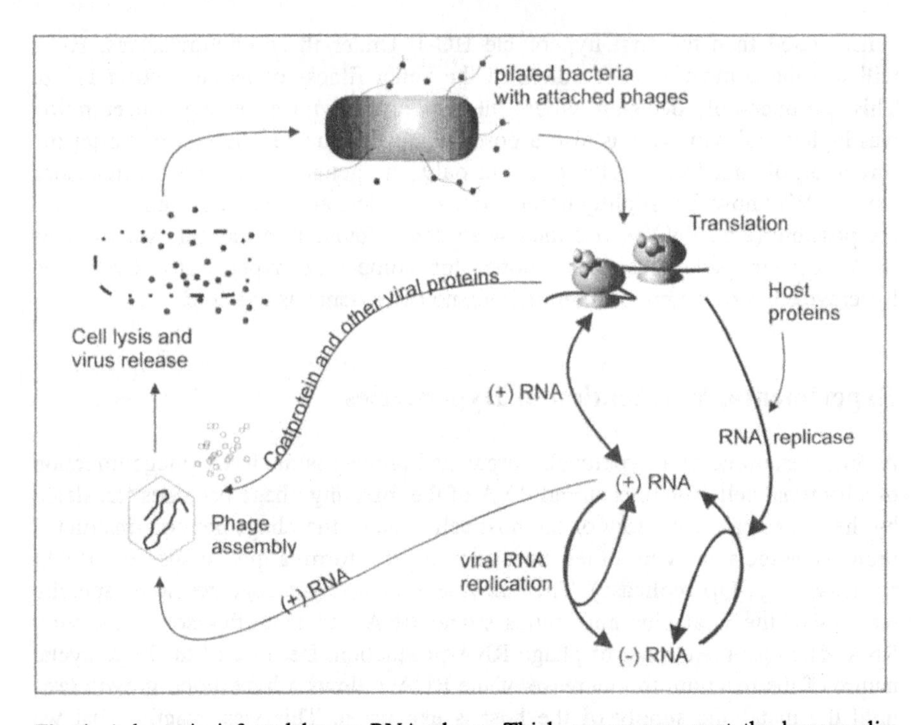

Fig. 4. A bacteria infected by an RNA phage. Thick arrows represent the hypercyclic-relevant interactions of viral components. The bacteria become infected by RNA phages which attach at the pili. The incorporated viral RNA becomes translated at the host ribosomes and first viral proteins will be synthesized. The viral RNA replicase consists of a viral subunit and host proteins and replicates solely the viral (+) and (-) RNA strand. (+) strand RNA and viral coat proteins come together and phage assembly starts. About 45 min after infection the bacterial cells lyse and release approximately 10^4 phage particles. The infective particles now can infect the next bacteria

was retained. The supplied nucleoside triphosphates and amino acids were radioactively labeled.

After preparation of the permeabilized, infected bacteria it was possible to estimate very precisely the synthesis rate of viral RNA and viral proteins during the whole infection cycle. Different phases of the nucleoside triphosphate incorporation curve can be distinguished. After a lag time of about 10 min a sudden increase in viral RNA synthesis was observed. The RNA synthesis rate reaches its maximum of more than 400 viral RNA molecules per minute and per cell at approximately 16-17 min after infection and then gradually decreases until the end of the infection with approximately 350 RNAs per minute and per cell. The estimation of the protein synthesis rate shows a strong similarity with the RNA synthesis rate. Lag time, the sharp increase in synthesis rate and the late phase cover the same time range. The rapid increase of RNA synthesis rate was not exponential because plotting doubling times against time after infection clearly shows a decreasing doubling time for the time of the increase. A characteristic of an exponentially growing population was the doubling of the population after constant time intervals. The hyperbolic growth phase of the viral RNA synthesis rate represents the first experimental evidence of hypercyclic coupling, as predicted by Eigen and Schuster (Eigen and Schuster 1977). The hyperbolic growth rapidly reaches its limits: the bacterial ribosomes become saturated with viral RNA, the RNA synthesis exhausts the bacterial cell, and the cell works now nearly only for the production of phages.

Hypercyclic Organization of Infection Cycles of RNA Viruses

We think that hypercycles, as well as their role in early evolution, also plays an important role in the infection cycle of RNA viruses and perhaps in other cellular processes. Besides the experimentally determined hypercyclic coupling in an RNA bacteriophage there is some evidence that also eukaryotic viruses use hypercycles during their reproductive phases into their hosts. Some RNA viruses have a very rapid life cycle and kill their host cells after a short time by cytopathic effects. The foot-and-mouth disease virus, Ebola virus, the influenza virus, and the vast majority of RNA viruses show such a behavior for lysing their host cells. On the other hand there are retroviruses (e.g., HIV-1) which exhibit a much slower time in infecting a cell. They also exhibit a nearly non-lytic life cycle. A possibility for these obvious differences is, on the one hand, a hypercyclic coupling without any obstacles for the majority of the RNA viruses and, on the other hand, a hypercyclic coupling which is under the control of repressors in the case of the retroviruses. Under these circumstances the retroviral infection machinery does not exhibit the over-exponential increase in one or more of its components.

A closer estimation of synthesis rates in eukaryotic viruses is of some interest in answering these questions, but there are experimental difficulties in preparing semipermeable eukaryotic cells comparable with that used in solvent-treated bacteria. Furthermore, animal viruses exhibit a different behavior in real infections compared with the defined infection of a cell culture. Last, but not least, it is easier to make such estimations with an RNA phage compared with a highly pathogenic virus (e.g., Ebola virus). Despite all of these difficulties there is a strong need to better understand infection cycles of RNA viruses. The known part of these emerging viruses (Morse and Schluederberg 1990) exhibits a clear and present danger for humans (e.g., Rift Valley fever virus, Lassa virus, HIV-1 and HIV-2, Filoviruses, etc.), and that the unknown RNA viruses are less harmful is very uncertain. Under these circumstances it would be very helpful to realize an *in vitro* system which is capable of simulating some general features of the regulation of RNA viruses. This *in vitro* system (quasi-viral system) can be

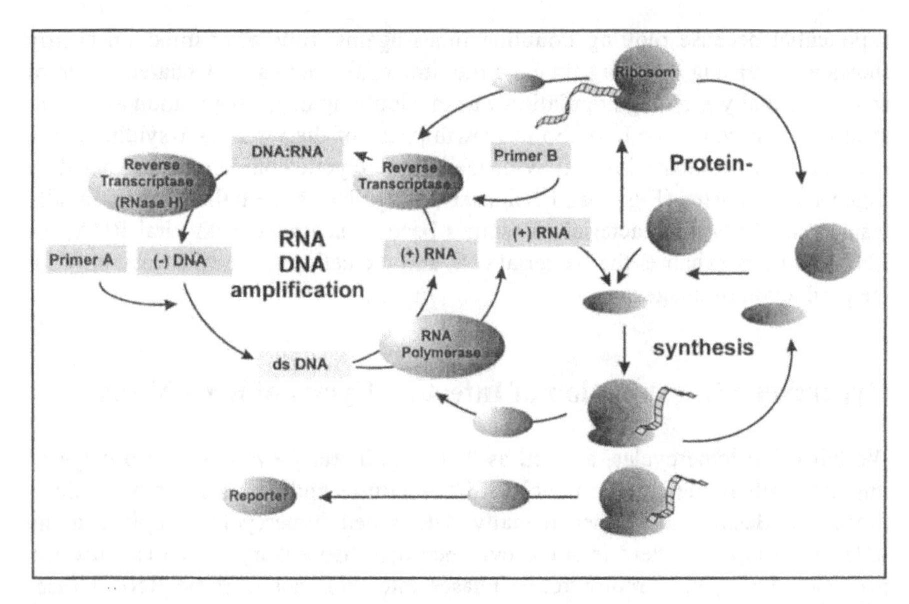

Fig. 5. Organization of a hypercyclic-coupled transcription-translation system. A 3SR-like reaction cycle produces RNA and DNA of the necessary proteins and the coding RNA will be translated by the protein-synthesizing part of the system. The *in vitro* synthesized RNA polymerase and reverse transcriptase amplify the RNA and the DNA. A supplemental RNA encodes a reporter protein or another protein and uses the amplification cycle as a "pillion rider"

designed in such a way that it does not exhibit such a great risk as the work with the whole virus.

Catalytic Hypercycles as Biotechnological Tools

In the development of modern biotechnology, it is important to build up new ways of gene shuffling to give the possibility of synthesizing proteins with new or better characteristics. Molecular evolution shows possibilities of how nature has overwhelmed obstacles during the history of living organisms and in the prebiotic phase of the very first self-reproducing entities. On the other hand, we have seen that nature has not used some paths,.which are of some interest, like the peptide nucleic acids (Nielson et al. 1991).

We have tried to establish an *in vitro* system of coupled transcription-translation which can work cell-free in such a sense that all of the components will be synthesized without a cellular environment. The results in the last years concerning nucleic acids amplification systems (Saiki et al. 1985, Joyce 1989, Guatelli et al. 1990) would make it possible to design the nucleic acid side of such a coupled system and then go forward to look for an efficient coupling strategy with a cell-free translation apparatus.

Our system should exhibit a form of hypercyclicy organization, like in a virus-infected cell or in a prebiotic environment (Fig. 5). In principle, such a transcription-translation system on the basis of a catalytic hypercycle can enable a synthesis of proteins without bacteria or other cells as carriers of the genetic information (Resto et al. 1992). Furthermore, there is a realization of such a coupled system as a quasi-viral system which will enable us to investigate interactions between viral components and parts of the host cell. Of special interest is the interaction between the necessary RNAs and the "pillion rider" RNA. We would like to see if the nucleic acid of the reporter protein show differences in the mutation rate compared with the enzyme coding nucleic acids. On the other hand, it is possible that the whole system does not use the "pillion rider" RNA, because it is easy to see that this arrangement is comparable with a hypercycle with a parasite. The "pillion rider" RNA receives catalytic support from the other gene products, but does not support any of them.

Conclusions

We have seen that hypercycles are not only a theoretical tool to answer questions about molecular evolution but also a real fact in virus development. Hypercyclic organization is a characteristic of RNA phages and, with a high probability, of RNA viruses in general. This last point enables us to use this hypercyclic coupling during virus development as a potentially powerful antiviral strategy.

In biotechnology there is the question of a possibility of expressing an information message of a nucleic acid in a protein message. The normal way of expressing an interesting gene uses its cloning into a vector construct which expresses the protein message in a bacterial or eukaryotic host cell. There are surely no fewer experimental problems with a totally cell-free system compared with a normal cloning system, but there are some clear advantages of such a technique. First, the system gives better access to regulation processes without disturbances by other cellular processes. A second point is the potential security problems around the gene technology. A cell-free system contains a less endangering potential because it does not need bacteria or cells for the steps where the cellular machinery synthesizes new proteins from possibly unknown nucleic acids. Of course, there is an uncertainty of potentially harmful new proteins or new nucleic acids. This known difficulty is under better control as an organism which can amplify itself and can grow on a large quantity of different culture media.

Acknowledgements: I thank Dr. Manfred Eigen and Dr. Peter Schuster for introduction into the field of molecular evolution, Dr. Frank Wirsching for valuable discussions and for comments on the manuscript, and Marlis Ude for her assistance in the preparation of the phage-infected bacteria.

References

Boerlijst MC, Hogeweg P (1991) Physica D 48: 17-28

Cramer JH, Sinsheimer RL (1971) J Mol Biol 62: 189-214

Crick FHC (1968) J Mol Biol 38: 367-379

Domingo E, Sabo D, Taniguchi T, Weissmann C (1978) Cell 13: 735-744

Eigen M (1971) Naturwissenschaften 58: 465-523

Eigen M (1985) Ber Bunsenges Phys Chem 89: 658-667

Eigen M (1986) Chemica Scripta 26B: 13-26

Eigen M, Biebricher CK, Gebinoga M, Gardiner Jr WC (1991) Biochemistry 30: 11005-11018

Eigen M, Gardiner Jr WC, Schuster PJ (1980) theor Biol 85: 407-411

Eigen M, McCaskill J, Schuster P (1988) J of Physical Chemistry 92: 6881-6891

Eigen M, Schuster P, Sigmund K, Wolff R (1980) BioSystems 13: 1-22

Eigen M, Schuster P. Naturwissenschaften (1977) 64: 541-565 (1978) 65: 7-41 (1978) 65: 341-369

Fox SW (1965) In: The Origin of Prebiological Systems and of their Molecular Matrices. New York, Academic Press

Fox SW (1969) Die Naturwissenschaften 56: 1-9

Gilbert W (1986) Nature 319: 618

Guatelli JC, Whitfield KM, Kwoh DY, Barringer KJ, Richman DD, Gingeras TR (1990) Proc Natl Acad Sci USA 87: 1874-1878

Hamming RW (1980) Coding and Information Theory, Prentice Hall

Hofbauer J, Sigmund K (1988) The Theory of Evolution and Dynamical Systems, Cambridge Univ. Press, Cambridge

Joyce GF (1989) Gene 82: 83-87

Kruger K, Grabowski PJ, Zaug AJ, Sands J, Gottschling DE, Cech TR (1982) Cell 31: 147-157

Lohrmann R, Orgel LE (1973) Nature 244: 418-420

Lohrmann RJ (1975) Mol Evol 6: 237-252

Maynard Smith J (1979) Nature 280: 445-446

Miller SL (1953) Science 117: 528-529

Morse SS, Schluederberg A (1990) J of Infectious Diseases 162: 1-7

Moses RE, Richardson CC (1970) Proc Natl Acad Sci 67: 674-681

Müller D, Pitsch S, Kittaka A, Wagner E, Wintner CE, Eschenmoser A (1990) Helv Chim Acta 73: 1410-1468

Nielsen PE, Egholm M, Berg RH, Buchardt O (1991) Science 254: 1497-1500

Orgel LE (1968) J Mol Biol 38: 381-393

Rechenberg I (1973) Evolutionsstrategie. Problemata Frommann-Holzboog, Stuttgart-Bad Cannstatt

Resto E, Iida A, Van Cleve MD, Hecht SM (1992) Nucl Acids Res 20(22): 5979-5983

Saiki RK, Scharf S, Faloona F, Mullis KB, Horn GT, Erlich HA, Arnheim N (1985) Science 230: 1350-1354

Schuster P, Fontana W, Stadler PF, Hofacker IL (1994) Proc Roy Soc Lond B Biol Sci 255: 279-284

Sleeper HL, Orgel LE (1979) J Mol Evol 12: 357-364

Sulston J, Lohrmann R, Orgel LE, Todd MH (1968) Proc Natl Acad Sci 59: 726-733

Viñuela E, Algranati ID, Ochoa S (1967) Eur J Biochem 1: 3-11

Woese C (1967) The evolution of the genetic code. In: The genetic code, pp 179-195, Harper & Row, New York

Evolutionary Optimization of Enzymes and Metabolic Systems

R. Heinrich
Humboldt-University Berlin, Institute of Biology, Theoretical Biophysics,
Invalidenstrasse 42, 10115 Berlin.

Introduction

Usually, the mathematical description of metabolic pathways concentrates on the simulation of steady states and time-dependent states of *variables* (concentrations of metabolic intermediates and fluxes) of these systems. Other quantities such as the kinetic constants of enzymes or the stoichiometric coefficients which define the topology of enzymic networks are considered as given *parameters* (i.e., they are inputs of the models). For these simulation models this distinction between variables and parameters is reasonable. Variations in the concentrations or fluxes may be experimentally observed in short time intervals, whereas the topology of the networks and the kinetic properties may change only very slowly or are even fixed during the lifespan of an organism.

In contrast to chemical systems of an inanimate nature, biochemical systems of living cells are the outcome of evolution. In the light of the Darwinian theory, one may state for biochemical systems that during evolution (1) new types of reactions were recruited by the cells leading to an increase in the complexity of biological organization, and (2) existing enzymic systems have adapted to environmental conditions. Both processes have been driven by mutation and natural selection. It seems, therefore, plausible to assume that contemporary metabolic systems have been developed by a stepwise improvement of their functioning.

Obviously, it would be a formidable task to follow in detail the origination and further development of metabolism during billions of years where living conditions have permanently changed and from which only few traces exist. On the other hand, it may be worth trying to explain the structural features of contemporary enzymic reaction systems on the basis of optimization principles. Certainly, evolution did not lead to a "global optimal state," but it is an experimental fact that mutations or other changes in the structure of present-day metabolism lead in most cases to a worse functioning of the cells.

In the literature, the following optimization principles are considered: (1) maximization of reaction rates and steady-state fluxes, (2) minimization of the concentrations of metabolic intermediates, and (3) minimization of transient times. Investigations concerning optimal stoichiometries and maximization of thermodynamic efficiencies have also been carried out (for a review, cf. Heinrich et al. 1991).

In the following quantitative treatments it is assumed that during evolution of cellular metabolism, some state function $\Phi(p,...,p_m)$ was maximized by variations of the system parameters p_j. Minimization problems may be transformed into such a maximization principle by considering $-\Phi = \max$. The parameters may enter the performance function Φ directly or via parameter-dependent concentrations S_i or fluxes J_i.

In optimization studies concerning metabolic systems, one has to take into account certain constraints which may be of different type. First, there are a number of physical constraints limiting the range of variations of kinetic parameters. Second, there are biological constraints which are often called *cost functions* (Reich 1983; Rosen 1986). The adequacy of optimization approaches depends essentially on the formulation of appropriate objective functions used to evaluate the fitness of a biological system. Generally, it seems to be difficult to derive the objective functions from more fundamental principles, such as the laws of physics. Accordingly, it is appropriate to derive them from heuristic arguments, and their validity should be judged by comparing theoretically predicted optimum properties with those of real systems.

Optimization of the Catalytic Properties of Single Enzymes

It has often been stressed that evolutionary pressure on the enzyme function was mainly directed toward maximization of catalytic activity, $v = max$, (Fersht 1974; Crowley 1975; Albery and Knowles 1976; Cornish-Bowden 1976; Brocklehurst 1977; Pettersson 1989; Heinrich and Hoffmann 1991). This hypothesis is strongly supported by the fact that the rates of enzymatically catalyzed reactions are typically 10^6-10^{12}-fold higher than those of the corresponding uncatalyzed reactions. Obviously, such high reaction rates may only be achieved if the kinetic properties of the enzymes fulfill certain requirements. It has been stated, for example, that enzymes with optimal catalytic activity are characterized by Michaelis constants close to the concentrations of their substrates *in vivo* (Hochachka and Somero 1973; Cornish-Bowden 1976). Other authors came to the conclusion that the K_m values tend to be large relative to the respective substrate concentrations (Crowley 1975). In contrast to these early studies which were mainly based on the most simple enzymatic two-step mechanism with the special assumption that the release of product from the enzyme-intermediate complex is irreversible, in the following a reaction mechanism is considered which involves two reversible binding processes of the substrate S and product P to the enzyme E and a reversible transformation of two enzyme-intermediate complexes (Scheme 1).

$$E + S \underset{k_{-1}}{\overset{k_1}{\rightleftharpoons}} ES \underset{k_{-2}}{\overset{k_2}{\rightleftharpoons}} EP \underset{k_{-3}}{\overset{k_3}{\rightleftharpoons}} E + P$$

Scheme 1

The steady-state reaction rate of this process may be expressed as

$$v = \frac{E_T}{D}(S \cdot q - P), \tag{1a}$$

with the thermodynamic equilibrium constant

$$q = \frac{k_1 k_2 k_3}{k_{-1} k_{-2} k_{-3}} \tag{1b}$$

and the denominator

$$D = \frac{1}{k_{-3}} + \frac{k_3}{k_{-2}k_{-3}} + \frac{k_2 k_3}{k_{-1}k_{-2}k_{-3}} + \left(\frac{k_1}{k_{-1}k_{-3}} + \frac{k_1 k_2}{k_{-1}k_{-2}k_{-3}} + \frac{k_1 k_3}{k_{-1}k_{-2}k_{-3}}\right)S + \left(\frac{k_2}{k_{-1}k_{-2}} + \frac{1}{k_{-1}} + \frac{1}{k_{-2}}\right)P. \tag{1c}$$

We are interested in those values of the elementary rate constants maximizing the absolute value $|v|$ of the reaction rate under the constraints of fixed values of the concentrations of the reactants and of the equilibrium constant q. According to Eq. (1), the reaction rate v is a homogeneous function of first degree of the elementary rate constants k_{+i}, that is,

$$v(\alpha k_i, \alpha k_{-i}) = \alpha v(k_i, k_{-i}), \tag{2}$$

with an arbitrary value of $\alpha>0$ For that reason, the rate v could be increased in an unlimited way when no constraints for the rate constants of the elementary reactions are imposed. According to quantum-mechanical and diffusional constraints, it is reasonable to take into account upper bounds on the individual rate constants upon optimizing the reaction rate, that is,

$$k_{\pm i} \le k_{\pm i,\max}\,. \tag{3}$$

Due to Eqs. (2) and (3), states of maximal activity have the property that one or more kinetic constants assume their maximal values. Because for q = constant, the numerator in Eq. (1a) is independent of the rate constants, and optimal states are characterized by those values of the rate constants minimizing the denominator D. Introducing normalized rate constants $(k_{\pm i}/k_{\pm i,\max} \rightarrow k_i)$, Eq. (3) may be rewritten as $k_i, k_{-i} \le 1$. The mathematical analysis shows that for the three-step

Table 1. Optimal solutions for the rate constants for the enzymic reaction depicted i Scheme 1 as functions of the concentration of the product for an equilibrium constant $q \geq 1$

Solution	k_1	k_{-1}	k_2	k_{-2}	k_3	k_{-3}
L_1	1	$\frac{1}{q}$	1	1	1	1
L_2	1	1	1	$\frac{1}{q}$	1	1
L_3	1	1	1	1	1	$\frac{1}{q}$
L_4	1	$\sqrt{\frac{P}{q}}$	1	1	1	$\sqrt{\frac{1}{Pq}}$
L_5	1	$\sqrt{\frac{S+P}{q(1+P)}}$	1	$\sqrt{\frac{1+P}{q(S+P)}}$	1	1
L_6	1	1	1	$\sqrt{\frac{2P}{q(1+S)}}$	1	$\sqrt{\frac{1+S}{2Pq}}$
L_7	$\sqrt{\frac{2q(1+P)}{S}}$	1	1	$\sqrt{\frac{2(1+P)}{Sq}}$	1	1
L_8	1	1	$\sqrt{\frac{2q(1+S)}{P}}$	1	1	$\sqrt{\frac{2(1+S)}{Pq}}$
L_9	1	$\sqrt{\frac{2(S+P)}{q}}$	1	1	$\sqrt{2q(S+P)}$	1
L_{10}	$k_1 = k_2 = k_3 = 1,\ k_{-1}^4 + k_{-1}^3 - \frac{Pk_{-1}}{q} - \frac{SP}{q} = 0,\ k_{-2} = \frac{P}{qk_{-1}^2},\ k_{-3} = \frac{1}{qk_{-1}k_{-2}}$					

mechanisms, ten different optimal solutions L_j are possible for a givenvalue of $q \geq 1$ (Heinrich and Hoffmann 1991). There are (1) three solutions with a submaximal value of one backward rate constant, (2) three solutions with submaximal values of two backward rate constants, (3) Three solutions with submaximal values of one backward rate constant and one forward rate constant, and (4) one solution with all backward rate constants being submaximal (cf. Table 1).

The optimal solutions L_j ($j \geq 4$) depend on the concentrations S and P. Therefore, Eq. (5) imposes various constraints on the allowed (S, P) values, depending on the type of the solution. For example, solution L_9 only exists if

$$S + P \leq \frac{1}{2q}, \tag{4}$$

and solution L_{10} determined by the fourth-order equation given in Table 1 only exists if

$$S \leq \frac{2q}{P} - 1, \; S \geq \frac{P}{q}, \; S \leq P\left(qP^2 + qP - 1\right). \tag{5a-c}$$

Equations (4) and (5) and analogous relations for the solutions $L_4,\ldots,L_8$ define, within the space of reactant concentrations, different subregions where the solutions L_j $(4 \leq j \leq 10)$ lead to rate constants fulfilling Eq. (3). The solutions L_1, L_2 and L_3 are independent of S and P and are, therefore, possible for all reactant concentrations. Some of these regions will overlap. Therefore, to make the solutions L_j unique functions of the reactant concentrations, one has to determine for given (S, P) values that solution which gives the highest enzymic activity.

In this way, one arrives at a unique subdivision of the (S, P)-plane into subregions R_j such that within region R_j solution L_j applies. In Fig. 1 these subregions are depicted for $q = 2$. From Table 1 and Figure 1, the following properties of the optimal solutions may be derived:

1. At very low substrate and product concentrations, optimal enzymic activity is achieved by improving the binding of S and P to the enzyme (solution L_9: *high (S, P)-affinity solution*).
2. When the substrate is present at a high concentration, it is weakly bound to the enzyme in the optimal state (solution L_7: *low S-affinity solution*). An analogous statement applies to the product (solution L_8: *low P-affinity solution*).
3. At variance with previous assumptions (e.g., Albery and Knowles 1976) an optimal enzymic activity is not compulsorily achieved by maximal values of the second-order rate constants.

The optimal values for the elementary rate constants are not only functions of S and P but also they depend on the equilibrium constant q. In particular, it follows from Eqs. (5a-c) that region R_{10} increases strongly in size with increasing values of the equilibrium constant q. One may conclude that for irreversible reactions $(q \to \infty)$, solution L_{10} becomes valid for all positive values of S and P. According to the central location of region R_{10} within the space of reactant concentrations, the corresponding solution L_{10} has been called the *central solution* (Wilhelm et al. 1994). For the reactant concentrations $S = P = 1$, which always belong to region R_{10}, the fourth-order equation given in Table 1 can be solved analytically. One obtains the two real solutions, $k_{-1} = \sqrt[3]{1/q}$ and $k_{-1} = -1$, and the two complex solutions $k_{-1} = \left(-1/2 \pm i\sqrt{3}/2\right)\sqrt[3]{1/q}$. As only the positive real solution is relevant, one may conclude that the optimal solution reads

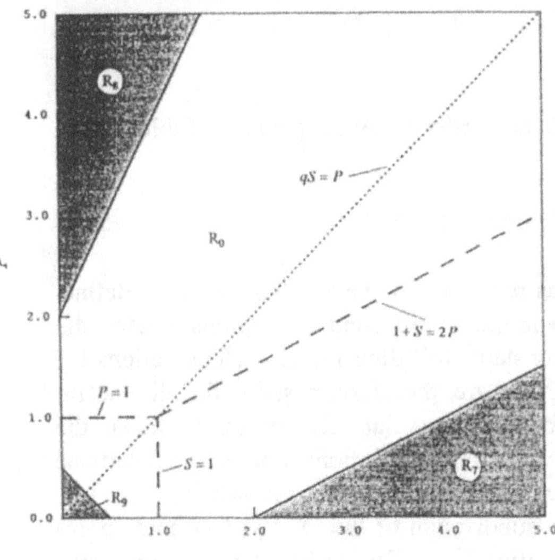

Fig. 1. Subdivision of the (S, P)-plane into subregions R_j corresponding to the ten solutions for optimal rate constants of the reversible three-step kinetic mechanism depicted in Scheme 1 for $q = 2$. The analytical expressions for the optimal rate as functions of the reactant concentrations S and P are given in Table 1. The vertices P_1, P_2, and P_3 of the central region R_{10} have the coordinates $(1/q^2, 1/q)$, $(2q - 1, 1)$, and $(1, q)$, respectively. Along the dotted line, $qS = P$ holds

$$k_{-1} = k_{-2} = k_{-3} = \sqrt[3]{\frac{1}{q}}\ ,\ k_1 = k_2 = k_3 = 1\ . \tag{6}$$

This solution shows some correspondence to that proposed by Stackhouse et al. (1985). In their *descending staircase* model, it was suggested that in the optimal state each of the three catalytic steps contributes equally to the equilibrium constant q. This optimization procedure for the catalytic activity of single enzymes has been applied to the reaction mechanisms of inorganic pyrophosphatase and of triosephosphate isomerase (Klipp and Heinrich 1994).

The kinetic Eq. (1) may be rewritten in the form of the reversible Michaelis-Menten equation

$$v = \frac{V_m^+ \frac{S}{K_{mS}} - V_m^- \frac{P}{K_{mP}}}{1 + \frac{S}{K_{mS}} + \frac{P}{K_{mP}}}\ , \tag{7a}$$

with the Michaelis constants

$$K_{mS} = \frac{k_2 k_3 + k_{-1} k_3 + k_{-1} k_{-2}}{k_1 \left(k_2 + k_3 + k_{-2}\right)}\ ,\ K_{mP} = \frac{k_2 k_3 + k_{-1} k_3 + k_{-1} k_{-2}}{k_{-3} \left(k_2 + k_{-1} + k_{-2}\right)}\ . \tag{7b,c}$$

Optimal values for these Michaelis constants are obtained by introducing $k_{\pm i}$ from Table 1 into Eqs. (7b,c). For the special case $S = P = 1$, one derives for $q = 1$:

$$\frac{S}{K_{mS}} = 1,\ \frac{P}{K_{mP}} = 1, \tag{8a,b}$$

and for $q > 1$:

$$\frac{S}{K_{\mathrm{mS}}} \cong 2\,,\ \frac{P}{K_{\mathrm{mP}}} \cong q^{-1/3} \tag{9a,b}$$

Equations (8a) and (9a) bear the interesting fact that the optimal Michaelis constant K_{mS} of the substrate is of the same order of magnitude as the substrate concentration S, irrespective of the equilibrium constant q. This gives a strong support to the hypothesis that there is mutual evolutionary adaptation of substrate concentrations and corresponding Michaelis constants (Lowry and Passonneau 1964; Cornish-Bowden 1976).

Optimization of Multienzyme Systems

Maximization of Steady-State Fluxes

The maximization of catalytic efficiencies as studied for single enzymes remains relevant also in the context of enzymic networks. Here, the difficulty arises that the concentrations of the intermediates are not fixed but depend on the kinetic parameters, which have changed during biological evolution. Moreover, due to the nonlinearity of most rate equations, the mathematical treatment is hampered by the fact that there are generally no explicit expressions for the parameter dependence of the performance functions.

Let us consider the case of a simple unbranched pathway of r nonsaturated enzymes where the steady-state flux J may be expressed analytically in the following way:

$$J = \frac{S\prod_{j=1}^{r} q_j - P}{\widetilde{E}\sum_{m=1}^{r}\frac{\widetilde{\tau}_m}{E_m}\left(1+q_m\right)\prod_{j=m+1}^{r} q_j}\,, \tag{10}$$

where S and P denote the concentrations of pathway substrate and of the end product, respectively, and $\widetilde{\tau}_j$ is the characteristic time of step j in a reference state with enzyme concentration $E_j = \widetilde{E}$ (Heinrich 1990). In the following, we are interested in those enzyme concentrations maximizing the steady-state flux J under the constraint

$$\sum_{j=1}^{r} E_j = E_{tot}\,, \tag{11}$$

which expresses the fact that the total enzyme concentration for a metabolic pathway is limited by the capacity of the living cell to synthesize proteins (Waley 1964). Using the method of Lagrange multipliers, the spectrum of optimal enzyme concentrations is determined by the condition

$$\frac{\partial}{\partial E_j}\left(J-\lambda\left(\sum_{m=1}^{r}E_m-E_{tot}\right)\right)=\frac{\partial J}{\partial E_j}-\lambda=0. \tag{12}$$

Introducing Eq. (10) into Eq. (12) yields

$$\frac{E_j}{E_k}=\sqrt{\frac{\tilde{\tau}_j(1+q_j)}{\tilde{\tau}_k(1+q_k)}\prod_{l=j+1}^{k}q_l}\,, \tag{13}$$

with $k>j$. This equation expresses the fact that in states of maximal steady-state activity poor catalysts should be present in high concentrations. However, the optimal distribution of enzyme concentrations also depends on the equilibrium constants. In the special case that all the enzymes have the same catalytic efficiency (i.e., $\tilde{\tau}_j=\tilde{\tau}$), Eq. (13) predicts for $q_j>1$ a monotonic decrease in the enzyme concentrations from the beginning toward the end of the chain.

Inserting Eq. (13) into Eq. (10), one arrives at an expression for the optimal flux which reads

$$J=\frac{E_{tot}\left(S\prod_{j=1}^{r}q_j-P\right)}{\tilde{E}\left\{\sum_{j=1}^{r}\sqrt{\tilde{\tau}_j(1+q_j)\prod_{m=j+1}^{r}q_m}\right\}^2}. \tag{14}$$

In the present case, where the reactions are described by linear rate equations, the optimal distribution of enzyme concentrations is independent of concentrations S and P of the initial substrate and the end product, respectively, of the pathway. This is no longer the case if saturation kinetics of the individual enzymes is taken into account (Heinrich and Hoffmann 1991). It is worth mentioning that optimization of the steady-state flux $J = max$ under the constraint of fixed total enzyme concentration (E_{tot} = constant) is mathematically equivalent to the problem of minimizing the total enzyme concentration at fixed steady-state flux (Heinrich et al. 1987; Brown 1991).

In the context of evolutionary optimization of metabolic pathways, considerations of the limited solvent capacity and the osmotic balance of living cells may play an important role (Atkinson 1969). The mathematical treatment may be extended to the case where an additional constraint concerning the upper limit for the total concentration of intermediates is introduced (Heinrich et al. 1987; Schuster and Heinrich 1991).

Optimal Stoichiometries

In the previous sections of this chapter the *optimization of kinetic parameters* has been considered. It may be argued that this kind of evolutionary optimization was nothing other than a fine-tuning which guaranteed the efficient interplay of enzymes within the pathways whose basic structure had evolved in a much earlier stage of evolution. The question arises of whether the special *topology of enzy-*

matic systems expressed by the molecular interactions may also be described as a result of an evolutionary optimization process. We are far from understanding in detail the origination of the different metabolic pathways observed in contemporary living cells. However, biochemists have rather clear ideas concerning the *temporal order* of the emergence of the main biochemical pathways (cf. Wald 1964; Hochachka and Somero 1973). On a more detailed level, the problem of the development of specific molecular interactions, as expressed by the stoichiometry of present-day metabolism, was probably closely related to that of the optimization of kinetic properties of enzymes. It was proposed that the evolution of metabolic pathways had involved the specialization of a smaller set of enzymes with less developed regulatory mechanisms and a much broader substrate specificity than the enzymes of present-day metabolism (Ycas 1974; Kacser and Beeby 1984). Such diversity may be regarded as necessary to make a metabolic system possible despite the limited gene content of primitive cells.

The relation between optimal kinetic and stoichiometric properties has been stressed in the pioneering work of Meléndez-Hevia (Meléndez-Hevia and Isidoro 1985; Meléndez-Hevia and Torres; 1988). Analyzing the stoichiometric structure of the nonoxidative phase of the pentose phosphate pathway, they came to the conclusion that the reduction in the number of reaction steps in the transformation of an initial substrate S into an end product P may be considered as a general principle of evolutionary optimization of metabolic pathways. In fact, as may be derived from the results presented in the previous section, the optimal flux through an unbranched chain of reactions will decrease with the increasing number of intermediate products if the total amount of available enzyme is limited.

Recently, Heinrich et al. (1997) has studied whether some structural properties of glycolysis may be understood on the basis of optimization principles. A remarkable feature of the stoichiometry of glycolysis is that it involves *ATP-consuming reactions*, despite the fact that its main biological function results in the *production of ATP*. It is, furthermore, striking that the two ATP-consuming reactions, hexokinase (HK) and phosphofructokinase (PFK), are located in the upper part, whereas the two ATP-producing reactions, phosphoglycerate kinase (PGK) and pyruvate kinase (PK), belong to the lower part of this pathway. Certainly, there are various constraints concerning the *chemical possibilities* of converting glucose into lactate, which are in favor of this special stoichiometric design. Beyond, it seems worthwhile to consider also the possible *kinetic advantages* of such a distribution of ATP-consuming and ATP-producing steps.

If, for simplicity, glycolysis is considered as an unbranched chain of reactions, the analysis may be based on Eq. (10) for the steady-state flux J (the glycolytic flux in this case). This formula can be applied also to chains with bimolecular reactions involving cofactors, if they are considered as external reactants.

In the above-mentioned paper, (Heinrich et al. 1997), ATP-producing sites are denoted as *P-sites* and the ATP-consuming sites as *C-sites*. Both types of reactions are called *coupling sites*. Reactions which are involved neither in ATP production

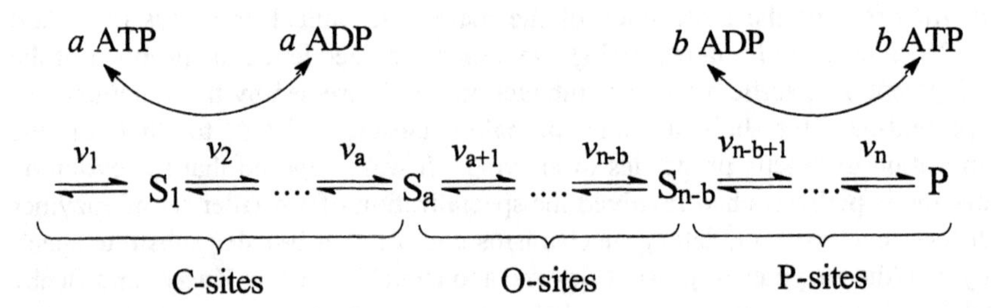

Scheme 2

nor in ATP consumption are denoted as *O-sites*. Denoting by a and b the number of C-sites and P-sites, respectively, the ATP-production rate is related to the glycolytic flux by $J_{ATP} = (b - a)J$. To identify the optimal structural design according to the principle $J_{ATP} = \max$, the kinetic properties of chains with different numbers and different locations of coupling sites are compared. Taking into account that a coupling of a reaction to ATP consumption or ATP production will change its thermodynamic properties, one may derive the following conclusions.

Theorem 1. (1) The replacement of an O-site by a C-site (i.e., $a \rightarrow a+1$) at any reaction increases the glycolytic rate J. (2) The replacement of an O-site by a P-site (i.e., $b \rightarrow b+1$) decreases J. This theorem points to the kinetic effects of a *change in the number* of coupling sites. The kinetic effects of a *variation in the location* of coupling sites at fixed numbers a and b are described by theorem 2.

Theorem 2. J as well as J_{ATP} are increased first by an exchange of a P-site at reaction i for an O-site at reaction m with $i < m$, and second by an exchange of a C-site at reaction j for an O-site at reaction m with $m < j$.

For the proof of these theorems, see Heinrich et al. (1997). From theorem 2 it follows that J_{ATP} becomes maximum when all P-sites are located at the lower end of the chain and all C-sites are located at the upper end of the chain (cf. scheme 2).

Taking theorem 1 into account, an optimum for the ATP-production rate J_{ATP} is obtained not only by proper localization of C- and P-sites at the two ends of the chain, but also by variation of their numbers a and b. Using realistic thermodynamic parameters, in particular $\Delta G^{0'}_{glyc} \cong -197\ \mathrm{kJ\ M^{-1}}$ (standard free energy change of the uncoupled interconversion of glucose into lactate) and $\Delta G_{ATP} \cong -50\ \mathrm{kJ\ M^{-1}}$ (free energy change of ATP-hydrolysis under physiological conditions), one obtains

$$d_{max} \cong \frac{\Delta G^{0}_{glyc}}{\Delta G^{0}_{ATP}} = 3.94\,, \qquad (15)$$

for the maximal number of ATP molecules produced for one molecule of ATP degraded (cf. Lehninger 1982). It may be shown that an optimum for the ATP production rate is achieved for $a = 1$ and $b = 3$ or $a = 2$ and $b = 4$. The latter optimum is in accord with the stoichiometric structure of glycolysis. There, the ATP-consuming reactions catalyzed by hexokinase and phosphofructokinase are located within the first part while the ATP-producing reactions catalyzed by the phosphoglycerate kinase and pyruvate kinase belong to the last part of glycolysis.

In the above analysis, the fact that real glycolysis is characterized by a splitting of C_6 compounds into two C_3 compounds at the aldolase reaction has been neglected. Introducing this feature does not change the main conclusions concerning the optimal location and the optimal number of ATP-consuming and ATP-producing steps derived for the linear model (cf. Heinrich et al. 1997).

The main result of the present investigation is that the optimization of kinetic properties favors pathways where the first steps are exergonic or coupled to exergonic processes (as ATP hydrolysis) and the subsequent steps are endergonic or coupled to endergonic processes (as ATP production). This result is in accordance not only with glycolysis but also with other metabolic systems. For example, the citric acid cycle starts with two exergonic reactions: (1) the citrate synthase reaction which involves hydrolysis of the energy-rich thioester bond of Acetyl-CoA, and (2) the isocitrate dehydrogenase. The subsequent reactions yield the energy-rich compound GTP and the redox equivalents NADH and $FADH_2$. The last reaction of the cycle, the malate dehydrogenase reaction, is very endergonic. Another example is gluconeogenesis, which starts by circumventing the pyruvate kinase step by two steps: the pyruvate carboxylase and the phosphoenolpyruvate carboxykinase, which both involve hydrolysis of either ATP or GTP. Similarly, the fatty acid oxidation is initiated by the fatty acid activation in an ATP-dependent acylation reaction to form fatty acyl-CoA. Further fatty acid oxidation yields NADH and $FADH_2$ which are reoxidized through oxidative phosphorylation to form ATP.

References

Albery WJ, Knowles JR (1976) Evolution of enzyme function and the development of catalytic efficiency. Biochemistry 15: 5631-5640

Atkinson DE (1969) Limitation of metabolite concentrations and the conservation of solvent capacity in the living cell. Curr Top Cell Regul 1: 29-43

Brocklehurst K (1977) Evolution of enzyme catalytic power. Biochem J 163: 111-116

Brown GC (1991) Total cell protein concentration as an evolutionary constraint on the metabolic control distribution in cells. J Theor Biol 153: 195-203

Cornish-Bowden A (1976) The effect of natural selection on enzymic catalysis. J Mol Biol 101: 1-9

Crowley PH (1975) Natural selection and the Michaelis constant. J Theor Biol 50: 461-475

Fersht AR (1974) Catalysis, binding and enzyme-substrate complementarity. Proc R Soc Lond B 187: 379-407

Heinrich R (1990) Metabolic control analysis: principles and application to the erythrocytes. In: A Cornish-Bowden, ML Cárdenas (eds) Control of metabolic processes. Plenum Press, New York

Heinrich R, Hoffmann E (1991) Kinetic parameters of enzymatic reactions in states of maximal activity. An evolutionary approach. J Theor Biol 151: 249-283

Heinrich R, Holzhütter H-G, Schuster S (1987) A theoretical approach to the evolution and structural design of enzymatic networks; linear enzymatic chains, branched pathways and glycolysis of erythrocytes. Bull Math Biol 49: 539-595

Heinrich R, Schuster S, Holzhütter H-G (1991) Mathematical analysis of enzymic reaction systems using optimization principles. Eur J Biochem 201: 1-21

Heinrich R, Montero F, Klipp E, Waddell TG, Meléndez-Hevia E (1997) Theoretical approaches to the evolutionary optimization of glycolysis; thermodynamic and kinetic constraint, Eur J Biochem 243: 191-201

Hochachka PW, Somero GN (1973) Strategies of Biochemical Adaptation, WB Saunders Company, Philadelphia

Kacser H, Beeby R (1984) Evolution of catalytic proteins. On the origin of enzyme species by means of natural selection. J Mol Evol 20: 38-51

Klipp E, Heinrich R (1994) Evolutionary optimization of enzyme kinetic parameters; the effect of constraints. J theor Biol 171: 309-323

Lehninger AL (1982). Principles of Biochemistry, Worth Publishers, New York

Lowry OH, Passonneau JV (1964) The relationships between substrates and enzymes of glycolysis in brain. J Biol Chem 239: 31-42

Meléndez-Hevia E, Isidoro A (1985) The game of the pentose phosphate cycle. J Theor Biol 117: 251-263

Meléndez-Hevia E, Torres NV (1988) Economy of design in metabolic pathways: further remarks on the game of the pentose phosphate cycle. J Theor Biol 132: 97-111

Pettersson G (1989) Effect of evolution on the kinetic properties of enzymes. Eur J Biochem 184: 561-566

Reich JG (1983) Zur Ökonomie im Proteinhaushalt der lebenden Zelle. Biomed Biochim Acta 42: 839-848

Rosen R (1967) Optimality Principles in Biology. Butterworths, London.

Schuster S, Heinrich R (1991) Minimization of intermediate concentrations as a suggested optimality principle for biochemical networks. I. Theoretical analysis. J Math Biol 29: 425-442

Stackhouse J, Nambiar KP, Burbaum JJ, Stauffer DM, Benner SA (1985) Dynamic transduction of energy and internal equilibria in enzymes: A re-examination of pyruvate kinase: J Am Chem Soc 107: 2757-2763

Wald G (1964) The origins of life. Proc Nat Acad Sci USA 52: 595-611

Waley SG (1964) A note on the kinetics of multi-enzyme systems. Biochem J 91: 514-517

Wilhelm T, Hoffmann-Klipp E, Heinrich R (1994) An evolutionary approach to enzyme kinetics: optimization of ordered mechanisms. Bull Math Biol 56: 65-106

Ycas M (1974) On earlier states of the biochemical system. J Theor Biol 44: 145-160

The Endocytobiological Concept of Evolution: A Unified Model

W. J. Schwemmler
Institut für Pflanzenphysiologie und Mikrobiologie, Freie Universität Berlin, c/o ZTL, Krahmerstrasse 6, 12207 Berlin, Germany

In memoriam to Prof. Dr. W. F. Gutmann († April 1997)

Summary: According to the synthetic evolution model, the most important evolutionary determinants are variation and selection. The biomechanical model offers an autonomous, self-regulating mechanism, but it can only preprogram the "species computer" endogenously, through genomic and/or structural mutations. The cybernetic model, in turn, postulates endogenous and exogenous flows of information, including even synthesis of novel genes. According to the now endocytobiological model, finally, main steps of evolution result from the interaction between the interspecific transfer of (master control) genes and the intraorganismic infrastructure. This will be demonstrated in detail in the "crocodilization" of the head of the cicada lantern carrier.

Cicades as Examples of Evolution Research

Along the banks of tropical rivers in South and Central America, we find the large cicada *Laternaria servillei* (lantern carrier). The head of this 10-cm-long insect extends forward to form a gigantic hollow, resonating chamber for amplification of its song (Fig. 1). This was formerly thought to contain a light organ, or symbiotic, bioluminescent bacteria. In fact, the organ has a strong resemblance to the head and snout of a crocodile with which the insect shares its habitat. The similarity extends to markings which resemble the lips and white teeth, nostrils and characteristic bulging eyes of the crocodile (Fig. 2). Two white dots within the eye spots even mimic the light reflex of a real eye. This mimicry is effective because the insect's predators are waterfowl which must always be on the alert for crocodiles. It has been suggested that the birds' vision is not sufficiently acute to determine readily whether their potential prey is a small animal close at hand or a large one at a distance. The crocodile markings on the cicada, therefore, are sufficient to trigger the birds' instinctive flight reaction. It can hardly be supposed that the convergent evolution of the lantern carrier mimicry was a random development, but how else can such strange examples of convergence be explained by current models of evolution? Even if the cicada's lantern was not selected because of its resemblance to a crocodile, but only because of its importance in sexual attraction (Ferrari 1993), the following discussion is still relevant.

Fig. 1. Head of the large cicada (the lantern carrier), showing a perfect imitation of a crocodile head. (Photo: W. Schwemmler 1995a)

Fig. 2. Head of a cayman species sharing the some living environment as the cicada. (Photo: Wendy 1993)

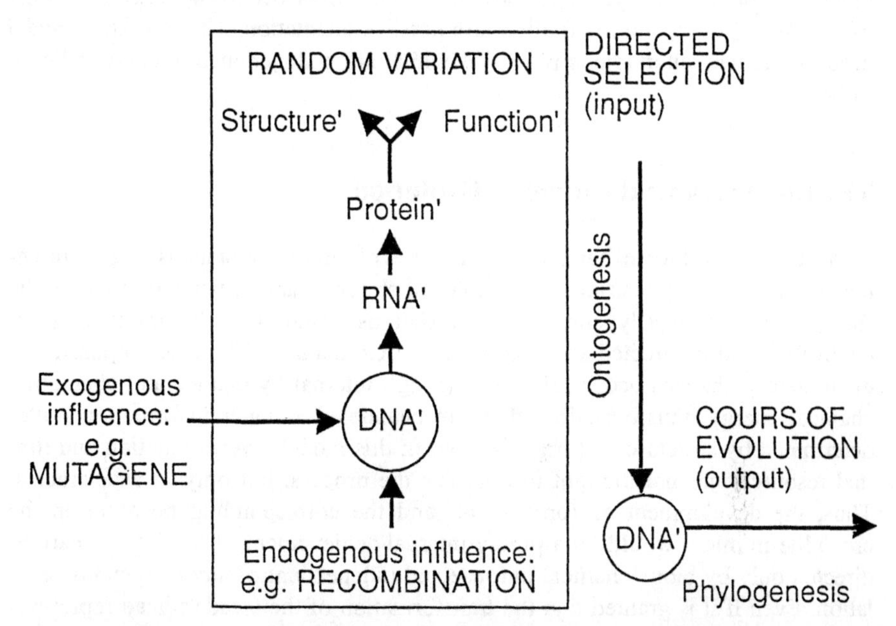

Fig. 3. According to the synthetic model of evolution (e.g., Mayr 1982), evolution occurs in principle through directed external and internal selection of chance mutants or recombinations. Following this hypothesis, the "species computer" does not possess any endogenous mechanism by which information from the environment is actively and directly translated into genetic adaptation

The Synthetic Model of Evolution

According to the synthetic model, the evolution of protective coloration is a classic example of the interaction of mutability, recombination, and selection: random micromutations or variations which heighten the mimicry effect, or mutations in higher-level regulatory genes are consistently selected until the wonder of a perfect crocodile simulation emerges (Fig. 3; Mayr 1982 a,b). As an obscure object of scientific curiosity, it now stands as an unresolved puzzle.

While in many simpler cases, it would seem quite reasonable to accept the contribution of mutation, recombination, and selection to the evolution of mimicry, here it seems more than questionable that these processes could be solely responsible. We have no means of explaining how the detailed and extremely specific information for the inborn structure, coloration, and instinctive behavior required for effective mimicry could have been transmitted from the environment into the genome through a long series of random, and thus non-directed, mutations and recombinations. After all, the species genome, as defined by neo-Darwinism, is relatively impervious to information acquired during the lifetime of the individual. The "Frankfurter Schule" (f.e. Gutmann 1995) has consciously established a model to counter the synthetic view of evolution as occurring largely through externally imposed, and internally more passive, adaptation. The Frankfurt-model proposes active, autonomous processes which are endogenously guided and controlled.

The Biomechanical Model of Evolution

According to the biomechanical model, the evolution of organisms is goal-oriented, and is driven by restrictive principles of structure and organization within the framework of primarily constructive variations, rather than by random, gene-controlled, and undirected variations (Fig. 4; Gutmann 1989). It is proposed that evolutionary changes occur primarily through internal hydraulics and chemomechanical energy transformation; they are channeled along transformation routes determined by structure and organization. In this model, overproduction and limited resources are not thought to optimize the process, but only to accelerate it. Thus, the development of form, color, and the corresponding behavior in the crocodile mimicry would be a purely internal achievement of the lantern carrier, directed only by biomechanical purpose and independent of any exogenous regulation. Even if it is granted that the transformation of the cicada's head represents an outgrowth of the „hydraulic head muscle tube“ resulting from an intraorganismic process of optimization initiated by an energy-transforming process of metabolism congruent with internal coherence and automobility (possibly even by an integral impetus from genetic variation), it would still be completely unclear by what concrete mechanism whole complexes of traits and functions of a foreign biohydraulic system could have been, and is still being, imitated in detail. This mystery increases when we remember that the insect has neither sensory nor intellectual perception of the situation which would allow it to protect itself as a constantly improving hydraulic energy machine from premature „biological recycling“ by the internally directed induction of protective form and coloration.

Admittedly, the „crocodilization“ of the cicada's lantern is a special and extreme example of evolution, but is it not precisely in such difficult cases that the explanatory value of a model has to prove itself? Might Professor Gutmann have sensed something of the dilemma faced by his overly abstract biohydraulic machines, which are exclusively programmed endogenously and completely immune to external stimuli, when he wrote: „It is not possible simply to translate traditional concepts of form into a constructional explanation, because our explanation is based on abstract principles, and not on a consideration of form.“ (Gutmann 1995). Could it be that the abstractions of his ontological and phylogenetic (re)constructions stand in a similar relationship to biology as the ideal gas laws do to chemistry: They can only be applied to real gases after the introduction of correction factors. If that were the case here as well, the biomechanical model of

THE BIOMECHANICAL EVOLUTION MODEL (Hydraulic theory)

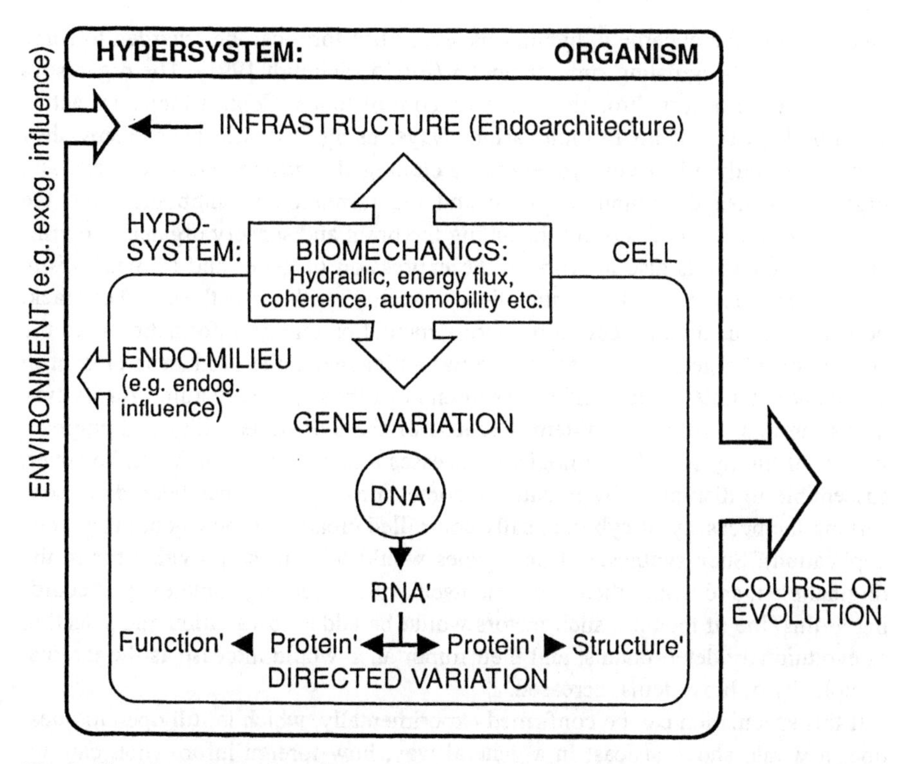

Fig. 4. The biomechanical model of evolution (e.g., Gutmann 1989) postulates an autonomous, self-regulating mechanism which is also able to vary and control the "species computer." However, this regulation is strictly endogenous, and occurs within the framework of mutations in the infrastructure and infrafunction which are directed by internal and external necessity of the construction. These may or may not be accompanied by genetic mutation; they occur as a result of an interaction between the total system (hypersystem) and its parts (hyposystems)

evolution might well make a decisive contribution to the foundation of a methodologically independent theoretical biology, since only the causally determined formation of types of basic body plans can be reconstructed from evolutionary history, but not the more or less random development of individual architectures.

It is apparent that neither the synthetic evolutionary model of a primarily exogenously directed biogenesis, nor the biomechanical model of an entirely endogenously determined and determining biohydraulics, is able to offer a satisfactory explanation for the phenomenon of mimicry. It is possible that the cybernetic model of evolution can offer a solution here, since it combines exogenous and endogenous aspects of both of the previous models, at least in a formal sense.

The Cybernetic Model of Evolution

According to the cybernetic hypothesis, every life form can be regarded to some extent as a self-regulating supercomputer (Fig. 5; Schmidt 1991). The organism's self-regulation occurs through cybernetic control loops. Control loops are active in all biological systems in multivarious ways: using feedback mechanisms, they control not only all of embryogenesis, including the differentiation of cells and organs, but also the immune system and the formation of antibodies, and the function of the nervous system, including the brain and sensory organs. The principle of feedback is also at work in the expression of genes and in metabolism, with its complex network of mutually interactive synthetic pathways. Feedback, however, means the impression of environmental or foreign information onto the species-specific genome. As soon as a particular metabolic product is present in sufficient amounts, it turns off the expression of those genes required for its synthesis through a feedback system. Because of the incredible utility and purposefulness of highly complex biological processes and systems, which are, however, susceptible to disruption by mutations, some thinking aloud has been done concerning the necessity of cybernetically controlled creation of new genes (f.e. gene duplication). Such synthesis of new genes would have to be provided for in the original plan, and would then complete itself with increasing complexity. According to this line of thought, such factors would be added to variation and selection as evolutionary determinants, and their importance would increase as the internal complexity of biosystems increased.

If this speculation can be confirmed experimentally, which is still open to question, it would show, at least in a general way, how foreign information can become incorporated into the own genome (species „biocomputer"). It would then be capable of being realized through goal-oriented gene synthesis, for example as an imitation of a crocodile. However, we still would not know who or what was the ultimate cause of this foreign programming. At first, it seems highly unlikely, even absurd, to suppose that the crocodile, or even the waterfowl, could have transmitted to the cicada subprograms of its physical traits or of its behavior-inducing stimuli, directly or indirectly. There are no interactions, or at least none which are observable, between the genome of the cicada, crocodile, and waterfowl, and for this reason, a priori there can be no synchronized control loops based on chains of interspecific and mutually induced structure formation, or even on instinct (AAM) or instinct-conditioning (EAAM) chains. There is also no possibility of an exogenously directed reactivation of a crocodile-convergent archegenetic program in the cicada. For one thing, the crocodile is phylogenetically much more recent than the cicada; for another, this convergence is only a superficial imitation of the crocodile's appearance, not of its actual genetic program or organ complexes. We have no other choice than to search for a more inclusive model, which can encompass the mutual integration of species-alien, genetic, and structurally inheritable programs.

THE CYBERNETIC EVOLUTION MODEL (Regulatory theory)

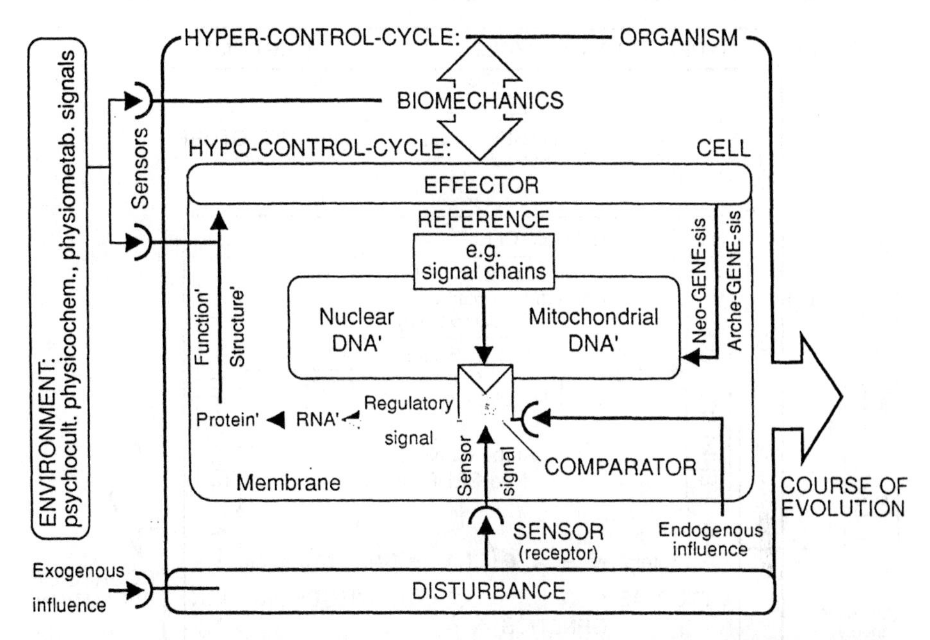

Fig. 5. The cybernetic model of evolution (e.g., Schmidt 1991) proposes that the direction of evolution is determined by self-regulation of each "species computer". This is achieved by cybernetic feedback loops beginning with those of gene expression and metabolism and extending to the resulting differentiation of structure and function; these include responses to environmental signals. In this case, both exogenous and endogenous flows of information interact, and may lead either to reactivation of archegenes (arche-gene-sis) or to synthesis of new genes (neo-gene-sis). However, the model gives no detailed indication whether or how genetic programs can be created, or exchanged between the "computers" of different species

The Endocytobiological Model of Evolution

The endocytobiological model (Fig. 6) is based upon the results of endocytobiological cell research conducted over the past ten years (reviewed by Schwemmler 1989, 1991, 1995b). This research has led to the now generally accepted thesis that eukaryotic cells arose through a series of phylogenetic endocytobioses (intracellular symbioses) between different species of prokaryotic cells. Taken to its final conclusion, this means nothing less than that the genetic programs of different species have been either incorporated into a hierarchy (symbiotic cell fusion), or partially exchanged (horizontal gene transfer), or even completely fused (genome/nuclear fusion). Even today, all three types of interspecific genetic transfer, partial or complete, can be observed. The compatibility of the different programs is ensured by the universality of the genetic code (genetic computer

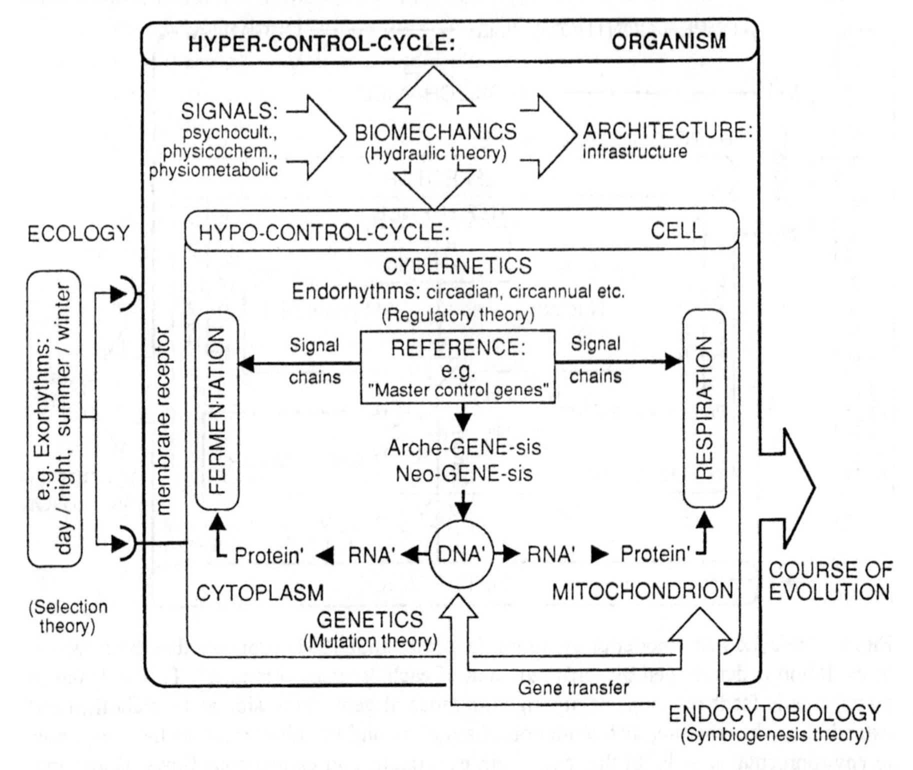

Fig. 6. The endocytobiological model of evolution (Schwemmler 1989, 1991, 1995b), finally, takes into account the fact that symbiogenesis, especially intracellular symbioses (endocytobioses), permit the exchange of genetic programs between different species, either completely or in part. This process occurs both phylogenetically and ontogenetically, and is based on the compatibility of the genetic programs of different species. If the host genome absorbs that of the endocytobiont (nuclear fusion), complete exchange occurs, but partial transfer may also occur (horizontal gene transfer). High-level regulatory genes control the interaction between cytoplasmic glycolysis and mitochondrial respiration (the latter is of symbiont origin); this interaction determines development in nearly all eukaryotes

language). Probably the most frequently occurring type, at any rate, more frequent than has previously been thought, is the exchange of simple genes („diskettes") between the computers of different species. This is the so-called "horizontal gene transfer," which is mediated by viruses and bacteria (episomes and plasmids). For example, plants synthesize the chromoprotein leghemoglobin, which is homologous to the hemoglobin in animals (Dayhoff 1972). This does not mean, of course, that plants arose from animals, but seems to suggest that interspecific gene transfer occurs.

The question can be raised of whether both phylogenetically and ontogenetically equivalent programs of high-level regulator genes were exchanged between species to then be developed in different, species-specific ways. Taking, for example, the simplest case of a viral disease, the same gene program induces both universal and unique symptoms in individuals with different organic infrastructures. Thus, the endocytogenetic model of evolution provides at least a general indication of how the structural, functional, and behavioral imitations of the lantern carrier might be the result of interaction between the interspecific execution of convergent regulator programs and the intraorganismic infrastructure, without, of course, being able to prove that this has occurred. The model might also indicate ways to explain the repeatedly postulated evolutionary jumps, such as the famous (and infamous) "hopeful monsters" of Goldschmidt (Gould 1977).

A Decisive Question

The average mammalian complement of DNA contains about $11x10^9$ nucleotide bases on each single strand, while that of bacteria comprises only about $2x10^6$ bases. In other words, in the evolution of human beings from bacteria, the maximum possible amount of information has been multiplied by a factor of about 5.500 (Fasold 1976). The decisive question in this context is how this enormous increase in information has been accomplished in the course of evolution. The synthetic model explains it as the result of mutations, such as alterations within and the duplication or loss of individual genes and chromosomes, or even the duplication of the entire genome (polyploidization). The cybernetic model explains it as primarily a matter of creation of new genes, while the endocytobiological model suggests it is derived, among other things, from horizontal interspecific gene transfer or cellular and nuclear fusion.

The widespread existence of atavism, appears to indicate that every species has collected a specific pool of archegenes, which are normally suppressed (cp. Ingber 1993). Conclusively, the further one follows this archetypical gene inventory back through evolution, the greater the similarity of the inventory will be in different species. If this is the case, each species would contain, in addition to its recent and active genetic inventory, a "closed box" of archegenes consisting of genes collected during important stages of its phylogenesis. These genes might be reactivated, either individually or in complexes, at any time when their repression was suspended by a suitable infrastructure, or by information from the environment (a process which might conceivably involve regulator genes). For example, mammals which have returned to an aquatic environment would not have reinvented streamlined forms, but would simply have reactivated their "fish" genes. However, the recent construction would no longer be achieved through the archetypical pathway, but through a modified path, since the infrastructure of the mammals has developed essentially out of the fish branch, presumably with additional structural genes coming into play.

These ideas receive experimental support in the work of Gehring (1985; cp. also Nüsslein-Volhard 1993) on the "homeobox". Indeed, current investigations of his research group at the Biological Center in Basel suggest that, for example, the structural principle of the eye, however differently nature might have embodied it in detail in different species, was developed only one time in the course of evolution. At the apex of the developmental cascade of eye morphogenesis it is argued that there is an overarching, homologous master control gene which might well be over 500 million years old. It controls a large number of different structural genes, which it guides directly or indirectly. As a result, it was possible for the master control gene to be exchanged functionally between insects and mammals (Halder et al., 1992).

An Attempt to Unify the Models of Evolution

As is well known, there are three forms of methodology in natural science: observation and description, experiment, and model building. They are closely related, since here, too, the more highly developed, historically more recent form subsumes the older. The increasing complexity becomes clear when we attempt to relate the different models to one another. Again, we shall borrow images and language from computer technology.

We begin with the observation that the lantern carrier's imitation of the crocodile is surely not accidental, but could (only?) have arisen from specific information about the environment. We find that the only evolutionary determinants offered by the synthetic model are variation and selection; there is no other mechanism for programming the "species computer" which could actively incorporate information about the environment into a directed adaptation to the environment. The biomechanical model does offer such an autonomous, self-regulating mechanism, but it can only preprogram, and, therefore, determine the species computer endogenously, through genomic mutations and/or structural mutations without genomic change. The cybernetic model, in turn, postulates endogenous and exogenous flows of information, up to and including synthesis of novel genes, but gives no indication of how these program exchanges between computers of different species might actually occur. Finally, according to the endocytobiological model, such ideas are developed through the new research field of endocytobiology: the interaction between interspecific transfer of a foreign gene/genome and the infrastructure of the organism is held responsible for the mimicry of the lantern carrier. However, who or what, in the end, coordinates such a mutual exchange of structural and genetic programs (in human beings, perhaps, psychological programs as well) between and within species to create such a meaningful whole, that one almost has the impression of orthogenesis? To this, the most decisive question of all, we must modestly admit we have, at least at present, no scientific answer (or perhaps there is one in chaos theory or in the hypothesis of self-organization?). Perhaps there are phenomena which are so complex that, given the finite nature of our „species computer", they simply will not yield to the vain attempt of our cognitive apparatus to understand them. In this case, it would be appropriate to reflect on these systematic limits to the ponderable.

Acknowledgments: Paul Wrede and Giesbert Schneider are thanked for inspiring discussions.

References

Dayhoff MO (1972) Atlas of protein sequence and structure. National Biochemical Research Foundation, Washington DC

Fasold H (1976) Bioregulation. Quelle & Meyer, Heidelberg

Ferrari M (1993) Farben im Tierreich. Tarnen, täuschen, überleben. Translation from the Italian. Krause B, Stütz Verlag, Würzburg

Gehring WJ (1985) The molecular basis of development. Scient Am 253: 137-149

Gould SJ (1977) The return of hopeful monsters. Natural History 86/6: 22-30

Gutmann WF (1989) Die Evolution hydraulischer Konstruktionen. Organismische Wandlung statt altdarwinistische Anpassung. Walter Kramer, Frankfurt

Gutmann WF (1995) Evolution von lebendigen Konstruktionen. EuS 6: 303-315

Halder G, Callaerts P, Gehring WJ (1995) Induction of ectopic eyes by targeted expression of the eyeless gene in Drosophila. Science 267: 1788-1792

Ingber DD (1993) The riddle of morphogenesis: A question of solution chemistry or molecular cell engineering? Cell 75: 1249-1252

Mayr E (1982a) Speciation and macroevolution. Evolution 36: 1119-1132

Mayr E (1982b) The growth of biological thought. The Belknap Press of Harvard University, Cambridge MA

Nüsslein-Volhard C (1993) Die Neubildung von Gestalten bei der Embryogenese von Drosophila. In: Wilke G (ed) Horizonte - Wie weit reicht unsere Erkenntnis heute? Wissenschaftliche Verlagsgesellschaft, Stuttgart

Schmidt F (1991) Fundamental principles of the theory of cybernetic evolution. Endocytobiosis and Cell Res 7: 109-129

Schwemmler W (1989) Symbiogenesis, a macromechanism of evolution.Progress towards a unified theory of evolution.Walter de Gruyter, Berlin, New York

Schwemmler W (1991) Symbiogenese als Motor der Evolution. Grundriß einer Theoretischen Biologie. Paul Parey, Berlin, Hamburg

Schwemmler W (1995a) Evolution: Obskures Objekt wissenschaftlicher Begierde. Wie „natürlich" ist das synthetische Evolutionsmodell? EuS 6: 343-346

Schwemmler W (1995b) Auf der Suche nach einer holistischen Evolutionsbetrachtung. In: Mey J, Schmidt R, Zibulla S (eds) Streitfall Evolution. S. Hirzel, Stuttgart

Wendy (1993) Tiere, die unseren Schutz brauchen. Ehapa Verlag, p 13

Giglio-Tos and Pierantoni: A General Theory of Symbiosis that Still Works

F. M. Scudo
Istituto di Genetica Biochimica ed Evoluzionistica (C.N.R.), Via Abbiategrasso 207, 27100 Pavia, Italy

Key words: Amphimixis, biomores (basibionts), biosphere, communities (biocoenoses), development, endocytobiosis, epigenetic inheritance, gender, genomic symbiosis, natural selection and fitness approximation to, norms of reaction, phylogeny, physiological symbiosis, probiosis, prokaryotes, stabilizing selection, sexuality, struggle for existence, taxonomy.

Summary: Giglio-Tos' (GT) theory explained basic features of life and its evolution on the premise that communities tend towards more mutualistic interactions, eventually reaching symbioses. Pierantoni's characterizations of bacterial and fungal symbioses corroborated a main prediction of this theory - the symbiotic origins of eukaryotic organelles - at a time when another - the switch from haplont to diplont gender systems - came into conflict with Mendelian genetics. The correct switch soon became evident - diplont gender systems derived by reuniting both determinants of haplont gender in the same haploid complement - easily correcting components of GT's theory affected by having assumed a different one. Soon GT's basic premises were amply justified by reformulations of older theories also in terms of mathematical models, such as showing that radically different equilibria can easily arise between the same host and symbiote or parasite (Kostitzin) or that complex communities, tightly bound by adversary interactions, tend to be more prone than others to collapse due to perturbations (Volterra). The smallish, at times very small, and ephemeral populations of more fragile communities can easily react to changes in conditions by different, largely haphazard selective modifications; each component species then changes mostly through selective replacements and hybridizations among its modified populations.

At variance from haplonts in which a single change in conditions of life is rather likely to directly fix previously rare alleles, in diplont, multicellular, interbreeding organisms a new selective regime will mainly modify common genotypic polymorphisms, eliminate part of them and select new ones. Novel plastic adjustements are thus easily modified also through polymorphisms, usually turning them into coexisting morphs which, later on, often "segregate" into distinct lineages. For various "Darwinisms", instead, even crossbreeding multicellular organisms would mostly change as a direct effect of rare, fitter variants being fixed by a single change in conditions, altough far too unlikely, slow and "costly" to account for fast change. According to these Darwinisms, the above positions would be wrong for reasons such as that their Malthusian struggles would be "typological", or their "group selection" would too easily result in symbioses often entailing "inheritance of acquired characters". The older theories now have a scarce following, being bitterly opposed by contemporary "Darwinisms", arriving at much the same conclusions as the older ones in more sophisticated ways, as well as by popular theories that claim being radically alternative to them. Much as it was a century ago, both Darwinisms and anti-Darwinisms now consider natural selection only in its most inefficient modes of operation, so that evolution would pose hard theoretical challenges according to the former while natural selection would be nearly powerless according to the latter. The older theories can easily account also for recent empirical knowledge, perhaps through more detailed specifications such as those on the origins of cells and coding by Cordón, de Duve and Ohnishi, namely cellular coding quite likely arose from collective modifications in communities of RNA organisms, commensual or parasitic and then symbiotic of protocells.

Introduction

Although having recently gained prominence in research, symbiosis failed to enter basic scientific education, still much influenced by the Synthetic Theory of Evolution (e.g. Zook 1994). This failure is only in part justified by the archaic and occasionally misleading terminology of symbiosis or by anglophone cultures having neglected it till not long ago (e.g. Sapp 1994), probably more so by deep, persisting disagreements on causal interpretations. Close symbioses might seem to lack a sound theoretical basis and this chapter aims to dispel this widespread impression by focalizing on little known theories. „Giglio-Tos and Pierantoni: A Problematic Theory and Its Demise“ outlines the naively axiomatized and mathematized theory on the evolutionary roles of symbioses which Giglio-Tos had introduced at the turn of the century mainly in a zoological context, and Pierantoni corroborated in some of its predictions. It also deals with various reasons why this theoretical-empirical body fell into disregard in the 1950s, a main one being its failure to integrate into more general developments of evolutionary ecology. These developments, taking place at much the same time in Italy, Russia, the USA and France, are sketched in „Theories on Evolution and Symbiosis“, which also elaborates upon their rejection by synthetic-type theories. Finally, „Origins of Prokaryotes: Two Main Alternatives“ will test these theoretical bodies through the very different scenarios they propose for the origins of the prokaryotes.

Many readers might find the theories here exposed hard to follow for two main reasons. One is that they deal with specific problems stated through technical terms, usually introduced in italics, with which they migh not be familiar; this is easily remedied by a glossary (p. 302). A far more serious obstacle is that,

Glossary of Unusual Terms (Origins are reported only when of rare usage.)

Amphimixis. Reproduction by union of gametes of different gender.

Apomixis. Production of a whole plant by a single vegetative cell, as for a cell of a bryophyte sporophyte resulting in a gametophyte.

Atavism. Particularly among vertebrates, a species rarely produces, under apparently normal conditions, individuals that look quite, or exactly like those of a closely related species or subspecies. The term is somewhat misleading since this "wrong" morph might in fact be the one further apart from the common ancestor.

Autosotery. Joint changes in a number of different genes, regardless of whether constitutive, epigenetic or just a matter of different induced expressions (Giglio-Tos).

Basibiont. Very small agglomeration, mostly of smallish peptides in solution, as originally due just to electromagnetic interactions (Cordón; cf. biomore).

Biocoenose. Several, distinctive animal forms more tightly associated than commonly in undisturbed nature, amomg themselves and with some equally distinctive plant forms.

Biomonade. Simple, cell-like association of biomores (Giglio-Tos).

Biomore. Very small agglomeration of smallish, unspecified biomolecules in solution, as originally due just to physical interactions (Giglio-Tos; cf. basibiont).

Biosphere. The parts of the earth, including its atmosphere, inhabited by living beings, if not interfered with by advanced human technologies (cf. noosphere).

Epigenetic inheritance. Non-constitutive modifications of genes or their expression, as for their expression being prevented through methylation of cytosines; as a rule these revert to the original through just one or two generations of amphimixis.

Genetic. Any entity in the sequence of developments re-establishing a genomic setup akin to the parent entity, in life cycles where reproduction results in entities that differ from one or both parents (Giglio-Tos).

Jordanion. (Lotsy) Denotes numerous variations, mostly in colours, within a species proper or Linneons, analogous to those Alexis Jordan discovered in Erophila (Draba) verna; as a rule these are interbreeding but not Mendelizing, and are usually interpreted by neo-Darwinists as swarms of species.

Mosaic (zoology). Developments which are for the most part completed before functional differentiation starts, so that each region develops almost independently of others; they mostly result in strict norms of reaction (as opposed to regulatory).

Noosphere. The parts of the earth and its living beings as modified by human technologies.

Norm of reaction. A precise functional morphology achieved in given conditions of life, which have no noticeable effect on its expression; when either of two, or more such reactions can occur, individual genotype specifies which one shall develop, depending upon conditions of life (as opposed to **reaction norm**).

Probiosis. Specific sets of entities or conditions that are prerequisites for the realization of other entities or conditions, such as floras are for herbivores (Giglio-Tos).

Reaction norm. An ample, unimodal phenotypic distribution as a response to variable conditions of life (as opposed to norms of reaction).

Regulatory (zoology). Developments through active interactions among different regions even before functional differentiation starts, so that the resulting traits are hardly affected by common genotypic variants but easily modified by conditions of life (as opposed to mosaic).

Stabilizing selection. According to traditional Darwinian theories, a selective process by which haphazardly variable phenes in multicellular organisms are standardized into one or more **norms of reaction**, largely regulated by genotypic polymorphisms. Darwinism-type theories assume, instead, that phenotypic invariance is achieved by fixing specific alleles to this end, while according to traditional theories these could only result in widely variable, "regulatory" phenes.

especially in the recent literature, references to older theories are often misleading. Thus the notion of fitness is usually regarded to be Darwin's main discovery, while in fact being a major component of Spencer's theory. For Darwin and his orthodox followers fitness is just a handy, although gross approximation to some effects of the struggle for existence at the level of individuals. A logical-sociological scheme (Fig. 1) attempts to alleviate such difficulties.

Giglio-Tos and Pierantoni: A Problematic Theory and Its Demise

Giglio-Tos' contributions were rooted in a radical epistemological shift collectively taking place at Turin's Zoology museum in the 1890s, at least in part motivated by the radical advances on animal and plant divergence. The raging debates on Malthusian struggles versus mutualism, or on Weismannism versus Lamarckism, were set aside to reformulate in terms of current knowledge the theories by the founding fathers, Lamarck, Geoffroy and Darwin. Apparently inspired by this circle, scholars such as the astronomer Schiaparelli in Milan and the mathematical physicist Volterra in Pisa (cf. Scudo 1992a) came out with positions wholly consistent with Giglio-Tos'. Later on and independently so did others, among whom Wright (e.g. 1982) and Schmalhausen (e.g. 1946) are of special relevance to this study. GT had thoroughly investigated the cytology, embryology and systematics of insects and vertebrates, with results at times conflicting with the established wisdom - e.g. exactly the same species as far as cytology could tell had rather different morphologies in different conditions of life, while species differing in so basic features as ploidy and mode of reproduction were morphologically indistinguishable. Such puzzles soon led him to regard both macromutations and the slow accumulation of minor hereditary changes as grossly insufficient to account for the divergence of taxa and the results of hybridization experiments. Major transformations would rather consist of all or none switches between different morphs through simultaneous changes in many genes, somehow depending upon conditions of life (autosotery); also later, however, (e.g. GT 1911), he did not attempt to specify this dependence. GT thus embarked on a theory (mainly 1900-1910) accounting for the major features of life from its very origins, drawing inspiration from the likes of Deleage, Roux, Altmann, Bertholet, Kekulé and Lothar Meyer. He qualified his mode of theorizing as analogous to studying the courses of rivers, which it would be futile to attempt justifying in all their details. These depend upon innumerable causal factors all of which, however, are easily qualified as special applications of the law of gravity (GT 1903: 6). GT's style of reasoning is, then, a peculiar blend of daring hypotheses and axiomatization of factual knowledge, to be sketched in its major lines without entering technicalities such as the mechanics of cell division.

Perhaps GT's most basic causal premise consists in ultimately reducing to chemical phenomena all processes of assimilation and reproduction characterizing autonomous life (1900). Even simple organic molecules can assimilate and

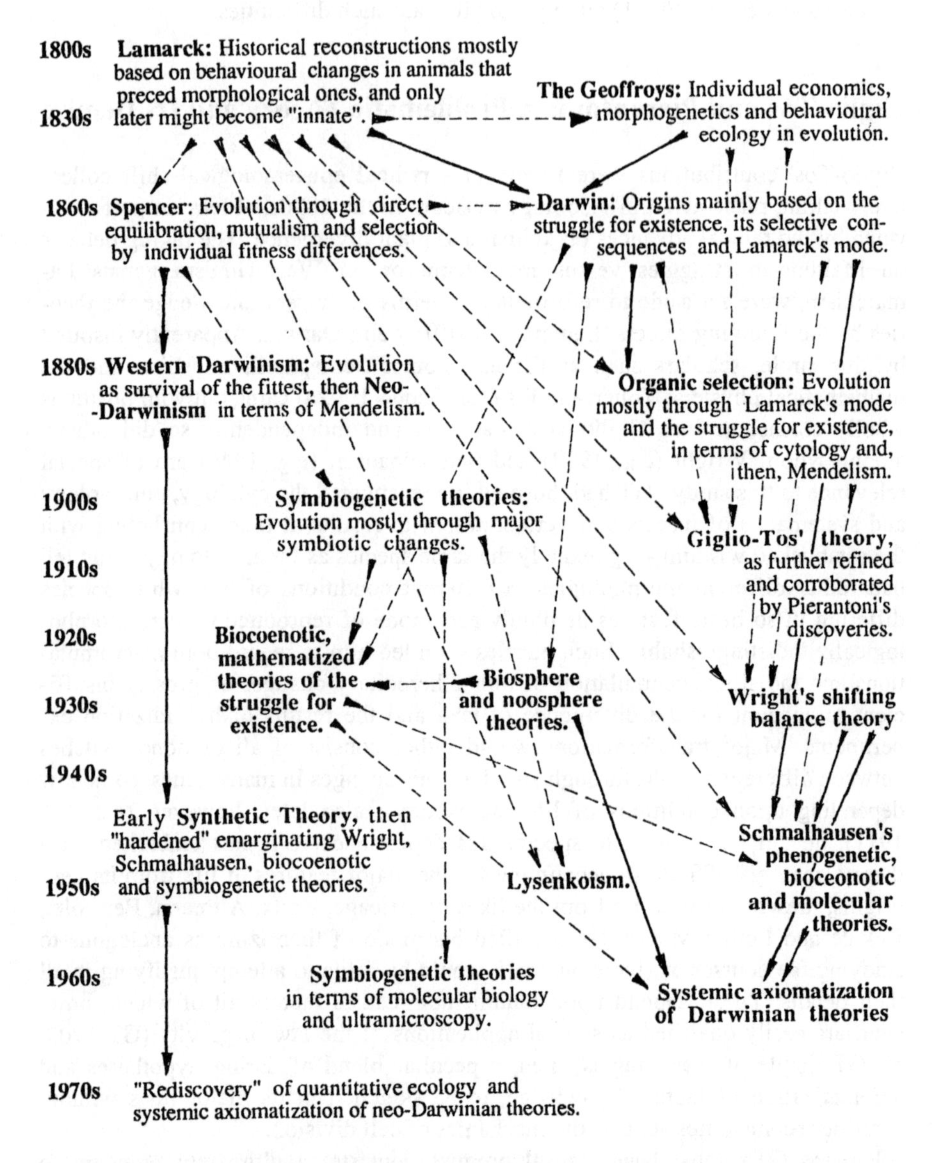

Fig. 1. Logical-sociological scheme. Effective, direct relationships among main theories on ecology and evolution (other than eliciting opposition) are displayed through arrows, solid ones when strong; these at times differ from claimed relationships.

reproduce, as suggested by a methylketone molecule turning into two acetic acid molecules by oxidation. Some simple organic molecules can no longer live autonomously since the proper conditions no longer exist, and some others never did since such conditions had never materialized. The ordinary compounds of organic chemistry would thus have easily originated, but this would not be enough to account for life. In its earliest form this would have consisted of special associations among different biomolecules with internal environments, the biomores, whose structures depended upon changes in their constituents. Close symbiotic associations of advanced biomores would be the earliest cells or biomonades, which paved the way to more complex symbiotic associations up to nucleated cell, namely special symbioses among non-nucleated ones. Most likely some vital characteristics of cellular plasmas are still due to complex, hierarchical symbioses among biomores, no longer capable of independent life and so deeply changed as to make it hard sorting them out as "individuals".

GT's analysis was and still is unique in ways such as jointly accounting for the development, a-sexual reproduction and evolution of so different entities as simple self-duplicating biomolecules and complex cells. All would be transformed through assimilation which would eventually result in doubling the parent entity, or in two "identical" entities differing from the parent , or in two different entities that also differ from the parent. In the last two cases a life cycle must also comprise a sequence of developments which regenerate the parent, through entities called genetic (a usage somewhat differing from the ordinary one of the term). The processes making up a life cycle will obviously differ in their chemical and mechanical details, such as the mode of the final splitting. Thus cell dieresis directly follows from duplicating all component elements so that, eventually, even a small displacement by one of them suffices to trigger it as a cascade process. Another basic premise by GT is that the origin of any new level of life is made possible or necessary by probiosis, namely the prior establishment of proper "habitat" conditions, such as land plants having made life possible for herbivores and these in turn for carnivores. Also complex multicellular organisms develop according to a few simple laws, still largely unspecified in their chemical and physical details (GT 1903). Thus chemical differentiation in the egg guarantees that already the first two blastomeres differ somewhat, and the locations of further chemical differences determine the modes of cellular evolution in subsequent, mostly asynchronous divisions. More and more cell types thus differentiate up to a limit phase, after which cellular heterogeneity usually decreases. Morphologies consist of foldings, evaginations and invaginations due to specific attractions or repulsions among proliferating cells, which critically depend upon doubling times of the most advanced cell types. Thus, if having a relatively long doubling time, the first cell of a given tissue will result in a local proliferation (polar development); even with a marginally shorter doubling time it could instead result in two symmetrical proliferations or in several, radially symmetrical ones if doubling time is still shorter. The development of individuals can thus be viewed as the evolution of special symbioses among cell types also at the genomic level (cf. below). On very different time scales these special symbioses among cell types would evolve in much the same modes as close symbioses among diverse organisms, through much the same mechanisms such as probioses.

Concerning heredity, GT (1905) starts by rejecting the common belief that the primary function of cross-breeding would be to favor the joint selection of hereditary units, by bringing them together in the same individual - i.e. if having this effect it would rather be just an accessory consequence (cf. "More Technical Comparisons Between Theories2).Since cells generally change also genomically in development, the primary function of cross-breeding by gender-differentiated gametes (amphimixis) is to reinstate in a germinal lineage much the same genomic constitution as in the parental cells. Nucleated cells started on this path by gender-differentiating a small fraction of their genes, enough to segregate into either of two polarities; more differentiations resulted in egg-sperm dimorphisms, also specified in the haplophase by segregation. The most primitive metazoans and different plants originated by switching to specify gametic dimorphisms in the diplophase, as in true hermaphrodites or "environmental" gender. Further cytological differentiation was imposing upon these organisms an ever more extensive gender-differentiation, to the point that it became problematic for both genomic genders to fully coexist in the same individual. The two complements thus partially separated so that females would lack the male counterparts for a fraction of their female genes, males likewise lacking the female counterparts for a fraction of their male genes. According to GT the partial gender differentiation of genomes in protists would correspond to the two complements veritably fusing after syngamy, and then re-differentiating. In high multicellular forms their whole genomes would be gender differentiated, namely sperm cells would only have male genes and eggs only female ones and the two complements would only juxtapose after amphimixis. Then, as a rule, only such "odd" sex-determining fractions of higher genomes could have a Mendelian inheritance with disjunction (GT, 1915; cf. also Grassi 1925). When proposed, this interpretation appeared to well relate known features in cytology with the radically different gender differentiations in the protists and in the higher multicellular forms, but it was not for long (cf. below). As Darwin had maintained, for GT preventing self-fertilization would only be an accessory consequence of diplont gender, regardless of how specified. On the other hand, parthenogenetic or vegetative modes of multiplication would not be a veritable re-production in the sense above, but just unchanging repetitions of the same genomic setup.

In 1909 Pierantoni discovered standard animal endocytobioses, independently rediscovered by Sulc a year later and soon confirmed by Buchner and others (e.g. Pierantoni 1940: 29-30): The enigmatic tissues of peculiar colorations (pseudovitellus) relatively common among invertebrates consist of exceptionally large, roundish cells (mycetomes), packed with bacterial or fungal symbiotes directly transmitted to the progeny. While algal and mycorrhizal symbioses were already reasonably well known, this breakthrough soon allowed to characterize innumerable cases of close, standardized microbial symbioses in metazoans and protists. A major step in understanding the symbiotic structure of living matter was thus achieved since the systematic distributions of these symbioses well-matched a symbiotic origin of metakaryotic organelles. Like most evolutionary theories with quite general aims, however, GT's too provided only partial causal analyses. It was also somewhat obsolete when born, as in hardly considering the substantial botanical knowledge on symbioses and wholly neglecting Schiaparelli's quite compatible theory of evolution through fixed types. This notwithstanding, GT had prestigeous followers in France, Germany and Italy till his treatment of heredity collided head-on with the results of extensive genetic mapping, giving the impression that his whole theory was wrong. Ironically, much at the same time the most plausible correct switch from haplont to diplont gender systems was being hinted by major empirical advances to be considered in due course and, if taken into account, it would have automatically corrected the only major deficiency in GT's theory. Most of its predictions were not affected by this mistake and kept having outstanding corroborations, as by Thomas Hunt Morgan on the effects of splitting early embryos (GT 1924) or, later on, by "epigenetic inheritance" turning out to be ubiquitous in sexually reproducing eukaryotes. In its light, standard multicellular developments are indeed akin to the evolution of symbioses, while GT's usage of the same term for biomolecules is somewhat improper, as shall become clear from posterior theories.

When GT died, in 1926 shortly after Grassi, Pierantoni was left as the major champion of his theory. He did not fall into the error on particulate inheritance, but failed to correct GT's position on the evolution of sex systems by which it is entailed. GT's and Pierantoni's elaborations were well accessible only to the very few who mastered zoology and microbiology, and had a good grasp of botany and quantitative reasoning. GT and Pierantoni wrote mostly in Italian and French, while knowledge on close microbial symbioses was scattered in heterogeneous literature also in Russian and German. Further, the whole field was being questioned because of ludicrous claims by amateurs, such as culturing mitochondria in artificial media. Then, also due to Buchner's position (see below), the major evolutionary roles of close symbioses were hardly considered outside Italy and Russia, and consensus on them was far from universal even among Russian botanists and Italian zoologists. Around-mid century also Pierantoni's treatments, including widely used biological textbooks, were becoming obsolete relative to advances in molecular biology, directly genotypical symbioses with viruses and viral-mediated gene transfers (e.g. Davoli 1943; l'Heritier 1948; Teissier 1952). Then even enthusiasts of the evolutionary roles of close symbioses might no longer explicitly rely for their syntheses on Giglio-Tos, or Pierantoni (e.g. Leonardi 1950). Meanwhile Buchner, who long survived Pierantoni, kept coming out with treatises on close microbial symbioses (e.g. 1965, the main one in the English) not covering or summarily dismissing many of them, and alluding to the symbiotic origin of metacaryotic organelles as the main example of "wrong paths to the study of symbiosis". Then, as English became the major vehicle of scientific communication, the Synthetic Theory could easily dismiss the evolutionary roles of symbioses other than as peculiar mechanisms justifying "the selection of alternative alleles" (Williams, 1965, as in 1972: 246). The dogma that metakaryotic organelles originated by "compartmentalization" from the nucleus was thus established, and even so cogent evidences of symbiote-to-host gene transfer as in teleostean luminescence would be regarded morbid imaginations, or Lysenkoist frauds. After close symbioses were once more generally accepted, monographs such as Smith and Douglas (1987) focused on scatters of "fashionable" cases. Most of the knowledge on close microbial symbioses by metazoans and protists up to mid-century is thus buried in Pierantoni's popularizations (e.g. 1923) and reviews, waiting for rediscovery.

Theories on Evolution and Symbiosis

Starting from Haeckel's Darwinism, symbiosis has been a central theme of neo-Lamarckian positions as by Giard, Mereschowskii and Creative Darwinism (Lysenkoism), presented as radical alternatives to Darwinism in the standard Western sense, or neo-Darwinism (e.g. Khakina 1979). No matter how illuminating some such interpretations have been and still are, they do not constitute full alternatives to neo-Darwinism and grossly fail to account for fast modes of change in higher eukaryotes. This should become clear by the end of this section, which compares the axiomatic structures and analytical back-ups of two major theoretical alternatives, through the special forms in which one of them applies to higher eukaryotes.

Today's Major Alternatives

Old Darwinian Theories (ODTs)

Let us call Old Darwinian Theories (ODTs) those arrived at by mid-century, in terms of a consensus among authors such as Wright, Schmalhausen, Cuénot and d'Ancona. The axiomatization of these theories by Botnariuc (1992), largely based on Zavadskij (1961), consists in ordering the phenomena of life and relevant features of non-living matter into three *hierarchies* which may consist in a number of distinct *levels* - i.e. the *somatic*, the *taxonomic* and the *organizational*. It will be convenient to start from the last one which better characterizes ODTs, and some of whose levels are not easily or "directly" observable.

A level of organization is defined to consist in an ensemble of biological systems that are equivalently organized and capable of independent life (in a sense such as individuals not being obligatory slaves), and the following ones are universal properties of life:

1. the *level of individuals*, regardless of complex links among them as through colonies;
2. the *level of populations of conspecifics* and higher such categories up to Linnean species;
3. the *level of communities* or biocoenoses, namely tight systems of relationships among very different forms, essential for their survival, and

4. the *biosphere* which connects these levels to non living matter through looser, less direct ties (regions where life is affected by artificial connections due to technological activities are distinguished as *noosphere*).

The coexistence of these levels of organization perhaps also with others, non universal ones, poses more or less acutely conflicting demands on any system - e.g. individual fertility in any one species being high enough to guarantee a prolonged survival of its local population, but not so high as to jeopardize the survival of the whole community. The responses to these demands thus tend to consist in compromises often biased in favor of more integrated systems, such as of families over individuals. In their earliest forms these responses may be purely individual, or "plastic" but somehow transmissible to the progeny through mechanisms primarily qualified by their persistence, penetrance and whether having all-or-none rather than incremental phenotypic effects - e.g. maternal inheritance, behavioral forms of transmission and epigenetic modifications which only endure one or very few generations. Since their inception with Lamarck, ODTs were mainly interested in how initially individual or plastic responses to sufficiently long lasting demands can give rise to "congenital", in a longer run even "irreversible" changes.

Also from the outset, one of ODTs' main purposes was to justify the taxonomic hierarchy historically, and as far as reasonable also causally, mainly by relying upon species or lower populational entities such as subspecies, biotypes and ecotypes, down to local populations and jordanions (often mistaken for "species"; see the glossary). These systemic entities are the only ones the taxonomic hierarchy shares with the organizational one while taxa higher than Linnean species, from genera up, are mere collections or additive entities. For easy identification and functional assessment taxonomy always relied on evident characters most of which retain useful functions, well aware that they might not reliably reflect genealogy - i.e. represent answers to similar demands by rather different organisms, such as through transformations induced by much the same microorganism (e.g. Scudo 1996). Genealogy, instead, is best revealed through great labors on as many "trifling characters" as feasible (Darwin, Origin Ch. XIV, cf. e.g. Cavalli-Sforza et al 1994). Having somehow achieved reasonably accurate phylogenies, ODTs attempt to interpret them causally through all sorts of somatic entities such as biomolecules (peptides, sugars, fatty and nucleic acids), cells and their components, tissues, organs or "fixed" motor patterns (Erbkoordinationen). The somatic hierarchy thus consists of an indefinite number of levels of integration, none of which is obviously capable of independent life under natural conditions in the sense above. ODTs further assume that no level of integration is universal in the ontogeny and phylogeny of living matter and, as we shall see in the last §, this assumption is expecially unpopular for nucleic acids.

Darwinism-Related Theories

Darwinism-related theories - in the main Western sense rather than in the Russian one - are characterized by assuming that evolution mainly or solely consists in answer to demands posed by "the environment", through gene substitutions by rare, “random” mutants due to fitness differences among individuals. Many such "theories" cannot be seriously discussed since they are not specified by suitable axiomatizations, such as the one Eldredge provided (1985) for the Synthetic Theory. This regards life as being organized in a number of hierarchical rows vaguely analogous to Zavadskii's only two of which, however, would be essential for evolutionary processes: The genealogical row consisting of various levels capable of reproducing, transmitting and expressing information and the ecological or economic one, comprising entities of relevance to energetic and material exchanges. The interactions between these two essential rows could only be assessed through individuals, the only real entity they share through gametes entering the genealogical row and somas making up the ecological one; other notions that might seem suited to assess these interactions would not be real, as for communities being mere "collections" of individuals. By New Darwinian Theories (NDTs) I shall intend those, now popular, that mainly differ from the Synthetic Theory by assuming that genotypic selection can operate in the framework of discrete phenotypic strategies. Many traits, notably behavioral, would thus move towards strategies that are evolutionary stable (ESS), by this intending a strategy which is the best answer against itself as well as against "feasible" alternatives when rare. Among such strategies, those selectively achieved through genotypic monomorphisms would be of special interest for their long term stability, being not invadible by mutants.

NDTs are usually axiomatized through the notion of a replicator, namely any entity whose properties can affect the probability of being copied and whose line of descent is in principle unlimited (cf. e.g. Hofbauer and Sigmund 1988). Since in theory any replicator could behave as a "unit of selection", at face value NDTs might seem to differ from ODTs mainly in terminology, perhaps also in emphasis, were it not that most replicators turn out to be quite unreliable. Genes are ideal replicators so long as they remain in a germ line, as they would nearly always result in identical copies. Other replicators might also be quite reliable, but they would only have scarce selective powers or, as for families, their selective effects would be much the same as those of more direct struggles among genes. Some NDT theorists might admit that communities or biocoenoses exist but they are not trustworthy replicators and thus often replace them with hypercycles, sorts of chain reactions originally introduced in molecular evolution. Then, faced by a number of dubious replicators among which only one is precise and selectively effective, most contemporary Darwinian theorists are forced to conclusions such as "ultimately, genes are the only replicators" (Maynard Smith 1991: 38). The original neo-Darwinist dichotomy between somas and genomes is thus reestablished, entailing severe struggles among genes even within the same organism with effects such as "Even with direct transmission ... there can be a conflict of interests between the genes of the host and the symbionts.... With indirect transmission the likelihood that a symbiont will evolve towards parasitism is far greater." (Ibid.: 35, my emphasis; indeed indirectly transmitted symbiotes commonly parasitize other hosts).

How ODTs Work for Metazoans and "High" Plants

Following authors such as Rosa (e.g. 1899), Giglio-Tos (e.g. 1911) and Schmalhausen (mainly 1946), ODTs regard the phenotypes of multicellular individuals as *reactions of their whole genomes to conditions of life.* As a rule, multicellular eukaryotes would at first react to any novel demand by some reversible change, often variable according to the intensity of the eliciting stimulus or its timing of action in development. If a novel demand would *uniformly* affect most or all individuals in a well isolated range(s) of a species, for a very long period of time, it might well select improvements upon the initially individual or "plastic" reaction that also turn it into being somehow "inheritable" (as in *organic selection* or, later on, *genetic assimilation*; cf. the logical-sociological scheme; Fig. 1). Far more often, however, a novel demand only affects part of the individuals of a species, with different incidences in populations and perhaps also with different intensities or timings in development. At first it thus results in mixtures of individuals variously changed, or not at all, *also depending upon their individual genotypes.* According to how the demand acts on which kind of traits, it might a priori seem that the most appropriate answer might consist in *any* degree of phenotypic specialization, since weakly multimodal phenotypic distributions are in fact common among closely related organisms for a variety of traits. However intermediates between the wide, unimodal spreads of phenotypic variability, usually called *reaction norms,* and sharp phenes with no intermediates, or *norms of reaction,* appear to be rare *within the same interbreeding population* - i.e. either such extreme seems to be clearly preferred even for traits which could vary in continuous forms. This preference could have a number of plausible, not mutually exclusive causes, whose relative weights might thus be hard to assess - e.g. a set of genes relevant for a trait having a single control mechanism or the challenge being intrinsically discontinuous. Whichever the cause, whole classes of traits end up changing just discontinuously as do "fixed" motor patterns (*Erbkoordinationen*), which develop in all or none fashion according to inducing stimuli, being likewise elicited by thresholds in other stimuli (Lorenz, e.g. 1965).

Then the choice between two reactions to unpredictable challenges would obviously be risky if occurring through a threshold for an eliciting stimulus which is (genotypically) fixed within any one species or population - e.g. a fixed threshold of switch to an alpine morph within a plant species would severely penalize individuals and populations at intermediate altitudes. As a rule, then, one expects the threshold(s) of choice among phenotypic reactions to conditions of life being genotypically variable to "spread the risk", with different advantages to different levels of organization and, possibly, also disadvantages to some of the levels (cf. below for gender morphs). Such arrangements are also expected since, as will be seen in the next section, any mode of selecting rare mutants in diplonts is more likely to result in stable polymorphisms than in fixations. On the other hand a direct determinism of norms by genotypic polymorphisms appears being very rare, if at all possible, for traits other than gender and colorations. Only after evolving into strict norms, reactions to conditions of life will substantially facilitate the invasion of novel habitats, where they can be further improved or modified. Then in most animals any two coexisting morphs tend to easily "segregate" into separate lineages, one of which is usually much closer than the other to the common ancestor (e.g. Rosa 1899). Through this segregation - fixing at different locations different alleles of the polymorphic loci controlling threshold choice (cf. Leonardi 1950 for Blanc) - given functions or expressions of a set of genes are becoming "irreversibly" specified. Fixing a single such polymorphic locus thus has much the same effect as a number of gene substitutions. This mode of divergence well justifies cases such as an animal species switching in mass to the morph of a closely related one when transplanted to a different habitat, or such a switch rarely occurring in apparently normal conditions (atavism). On the other hand flowering plants easily switch for their "standard" morphs, as between dry and water ones, single and double flowers.

To see how this reasoning applies, consider an animal hosting a bacterium which is moderately beneficial in some conditions of life and detrimental in others. Accordingly, the host would then benefit by either favoring its reproduction or preventing it as much as possible and - not surprisingly also for reasons hinted in the next section - such choices tend to occur in a threshold fashion. Once such a threshold choice is established, especially higher animal hosts tend to split into a lineage dependant on the symbiote and into another, specialized to different conditions, for which the bacterium is an occasional nuisance. In cross breeding animals reproductive isolation might directly result from this difference or even from differences of little or no adaptive significance, such as in karyotype. If the symbiote is obtained by reinfection from the wild at each generation, the bacterium would then benefit by specializing, as often through plasmid differences, into a lineage seeking this kind of host and another lineage avoiding such an host as far as possible. If suited, turning to direct transmission would benefit both the symbiote and the host, for whom it will be easier to incorporate in its genome key functions of the symbiote which might be ultimately discarded (cf. Pierantoni, e.g. 1923, for animal luminescence, and Scudo 1996). Such genomic acquisitions from bacteria are relatively common in lowish "mosaic" animals but ever less so in higher forms, up to apparently impossible in those with the most regulatory modes of development such as warm-blooded vertebrates. In the latter, directly genotypical symbioses with viruses still occur, but little is known on their relationships with the mobile elements which are highly repeated in their genomes (e.g. Ratner et al 1996). Noticing that the benefits derived by a host from a microbial symbiote might be hard to assess even if consistent, and physiological symbioses "on their way out" might persist while hardly beneficial, the long-term persistence in a standardized form of any close microbial interaction is the most practical and safest criterion to assess it as a symbiosis.

Other clear-cut examples of threshold modes of reaction are primitive gender forms in multicellular eukaryotes, such as the male versus female specialization of different branches in embryophytes or the sex switches in some metazoans and the "environmental" separation of the sexes in some others (e.g. Hartmann 1943; Bacci 1965). All diplont gender forms appear to have arisen, through such threshold reactions, by joining in the same haploid complement the two blocks of genes whose segregation previously specified haplont gender. In many bryophytes analogous switches to monoic gametophytes took place and they still do by genomic duplications, gametophites resulting by *apospory* from mature sporophytes when smashed. While this had been known for long, Von Wettstein and Straub (1942) managed to fully replicate in the laboratory the switch to monoicism by *Bryum* sp. - i.e. the monoic gametophytes so obtained, at first sexually sterile, were becoming fertile in a few generations of vegetative reproduction. At least in such cases, then, the switch away from segregational determinism of gametic or sporal gender seems to only entail a tuning of pre-existing signals, specifying that only either block of gender genes is activated at any one time or body position; simultaneous hermaphroditisms or, at the opposite, "environmental" separations of the sexes might then follow. On the other hand dimorphic loci often assume a partial control over the original gender morphs by favoring either, as in the temperature gender determinism of turtles or in the echiurid *Bonellia viridis*, part of whose larvae (*Spätmännchen*) never settle to the ground - i.e. they thus avoid developing into females even in the absence of mature female hosts. Such partial genotypic controls of diplont gender often evolved into stricter ones, up to the full controls by *a number of loosely linked or unlinked dimorphic loci* which are still relatively common among invertebrates (e.g. Bacci 1965, Scudo 1973). A progressively closer linkage among the heterozygous elements of such poly-factorial controls then gave rise to alternative gene blocks, up to sex chromosomes; loci outside the main controlling gene blocks remain evidently involved but, for long, in mysterious ways. Such sophisticated forms of control are now well understood, as for dimorphic genes specifying the gender-specific splicings in the primary transcripts of other genes, thus coding for different proteins in the sexes (e.g. *Drosophila*'s doublesex, cf. Marx 1955). From many such cases it is now evident that close linkage among the loci is not needed to specify whether or not to express a set of alleles, or in which of two alternative forms to express it.

ODTs assumed that collective switches in gene action were generally responsible for threshold morphs even before this became evident for gender so that, as pointed out above, many if not most multicellular adaptations would derive by easily "segregating" previously coexisting such morphs. These assumptions were mainly based on indirect suggestions such as the disruptive effects of inbreeding on development well known since Darwin, or the results of mathematical models such as mentioned in the next section. Once a threshold morph comes to characterize a "good species", its still substantially heterozygous genotypic determinism tends to be slowly changed to a functional monomorphism, by which the morph is manifested at earlier stages in ontogeny and usually modified later on in regulatory forms (cf. the glossary). All these data or expectations well agree with many genotypic polymorphisms being needed to buffer development against the disruptive effects of "normal" variations in long-established conditions of life. These polymorphisms are then expressed as phenotypic invariance so long as much the same conditions persist, but they will result in a spread of phenotypic variability if conditions change drastically - i.e. a hidden reserve of genotypic variability largely accounting for the higher hereditary variability in changed conditions that was evident since Darwin (Schmalhausen 1946). If two morphs were to coexist long enough, one might think that their heterozygous regulators would slowly achieve closer linkage for easier transmission, as in the highest forms of gender determinism. While no evidence appears available on this point, alternative blocks of closely linked genes might be unsuited for morphs other than gender and most of them, anyway, would not coexist long enough for such a linkage to evolve. In fact, particularly in higher animals, non-gender morphs soon tend to"segregate" into distinct lineages through changes in mating preferences (particularly easy when recognition is based on learning as in warm-blooded vertebrates, e.g. Scudo 1976).

The metazoans and, to a lesser extent, the "higher" plants are also characterized by occasional mass extinctions that might involve whole large taxa, and are sudden or suddenly terminated. After any such extinction major transformations among the "higher" surviving forms take place in relatively short time spans, and just minor ones occasionally follow till the next extinction. As closest analogy in protists, a few small or very small of their taxa suddenly disappear at the border of severe metazoan extinctions. Jointly with the phenogenetic "peculiarities" of animals and high plants outlined above, this basic difference from protists and prokaryotes is easily justified by associative properties first mainly inferred by a priori reasoning (e.g. d'Ancona 1954, cf. the next section). A young community replacing an extinct one tends to become enriched of different life-forms while still closely tied by strong competitive and predatory bonds, thus being also more fragile than poorer or longer established communities, less closely tied by such bonds - i.e. some perturbations are more likely to decimate or destroy part or all the characteristic forms of these richer, younger communities. Nevertheless, unless the perturbation is global, these younger communities are more apt to persist and spread than poorer or older ones since their higher quantities or turnovers of life make them more invasive in their patchy spatial distributions. Selection within the ephemeral local populations of such communities would just provide different, rough, largely haphazard sketches of genotypic adaptations; wider ranging, more permanent ones mostly derive from differential replacement and hybridization among these populations. This inter-population selection seems also possible in a single species regardless of its ties to others, but only under exceptional conditions and with different, usually lesser outcomes.

The better established a young community becomes by supplanting close competitors, the less its higher invasivity will matter and a higher local stability will rather be favored, through looser and more mutualistic interactions. The more stable a community thus becomes the less effective on it will be interpopulation selection and, unable to rapidly undergo further adaptive changes, its characteristic forms could be annihilated even by relatively mild global perturbations. As a simple analogy to these effects of complexity and specialization, an old fashioned mechanical typewriter tends to somehow work in most conditions, and most of its rare failures are simply remedied. At the opposite extreme, an advanced personal computer either writes well or not at all, and most of its failures are not simply remedied. The natural trend towards more mutualistic interactions in communities facilitates, even forces the establishment of all kinds of symbioses among component forms. Kostitzin's mathematical models (1934, 1937, cf. also Scudo & Ziegler 1978) are especially illuminating on this point as they suggest that appropriately mild forms of parasitism can easily switch to symbioses for two concurring reasons. On one hand different stable equilibria can easily arise within the same symbiotic or parasitic association, so that major changes may be triggered just by local populations switching by chance from one such equilibrium to another. On the other hand the dynamics of any parasitic association is radically changed by reliably changing the sign of the payoff to the host, so that even a minute change in absolute value easily results in mutualism, or symbiosis. This helps justifying why intermediates between erratic mutualistic interactions and standardized symbioses tend to be rare among organisms with substantial cellular differentiation (cf. above). In terms of all such mechanisms and modes of change - or of the massive induced changes their large genomes can easily afford - no wonder that multicellular organisms, "high" metazoans in particular, change for the most part jointly, perhaps radically, in rapid bursts following each mass extinction.

More Technical Comparisons Between Theories

ODTs and NDTs rely on analogous mathematical models save for ODTs not formalizing their game-strategical reasonings as above, and NDTs not relying upon topology. Even when the mathematical set up is identical, however, ODTs and NDTs *utilize it in very different ways to very different ends*. To display these differences let us informally consider some simple results of models, starting from Lotka-Volterra differential equations as originally proposed to represent, in special conditions, the struggle for existence among entities such as species or genes:

1. A predator and a prey will both survive in 1/2 of the parameter space by which they are represented, while only the prey might in the remaining 1/2 - i.e. these outcomes would be a priori equally likely if the parameters were chosen at random (e.g. Kostitzin 1937, Ch. IX).

2. When two species compete for common resources, four outcomes are a priori equally likely - i.e. both can stably coexist, or just either one does depending on initial numbers, or just one survives, each in 1/4 of the parameter space. In terms of competition between two haploid genotypic variants, starting from one of them

being rare for being selected against, by a random change in adaptive response (selective regime) this has 1/4 a priori probability of increasing towards fixation and the same of increasing to a stable coexistence with the other (becomimg even more rare with probability 1/2). In the special cases of "additive competitive effects", however, just one entity will generally survive, being in absolute the "more apt" (Ibid.).

3. Five outcomes are instead possible when two predators compete for the same prey - i.e. both predators coexist with the prey, either can depending upon initial numbers, just one can and neither does. It is seen at a glance that the last outcome has a priori probability 1/4 (versus 0 in point 2), while the others occur in substantial portions of the parameter space that cannot be easily specified - e.g. it is only evident that both predators coexist with the prey in less than 1/8 of this space. If neither predator is self-limited, however, they cannot stably coexist (Ibid.).

The selective effects of the struggle for existence within diploid, interbreeding species are also described by models like the above, but these are far too complex to be analyzed except in very special cases (e.g. Kostitzin 1937, cf. also Scudo & Zieger 1978). More usually, then, genotypic selection is approximated in terms of the frequencies of alleles at each locus and constant fitnesses of the genotypes; at variance from the above, this gives much the same results whether in terms of differential equations or of recursions. This approximation can be further simplified in terms of fitnesses relative to one genotype taken as a unit of measure, with no further loss of information for problems as in 6 or 7. Some simple results are as follows.

4. Two haploid alleles as in 2 cannot stably coexist by selection alone; starting from one being rare since unfavored, this has 1/2 a priori probability of becoming favored by a random change in selective regime, if so moving towards fixation.

5. At a diploid locus with two alleles and random mating, a random selective regime results into a stable polymorphism with a priori probability 1/3 or into an unstable one with the same probability. If one allele is rare at first, being at a disadvantage, it has again 1/2 a priori probability of becoming favored by a random change in a selective regime and, if so, it is twice more likely to move towards a stable polymorphism than towards fixation.

6. Fitness differences of comparable magnitudes have about twice stronger selective effects on the haplophase than on the diplophase (Scudo 1967).

7. For fitness differences of the same magnitude, the fixation of a rare allele is much faster in haplonts than in diplonts even with no dominance, while the speed of initial increase by a recessive allele and that of approach to fixation by a dominant one are close to zero (i. e. algebraic, Haldane 1924).

Through the frequency-fitness approximation, also random genotypic changes (drift) can be easily accounted for, but only in two extremes, wholly unrealistic cases for the dispersal and spatial distribution of a species. One consists in assuming that at each generation the individuals spread and mate, or somehow unite their gametes or spores, at random within their native population, unless moving to some other with given probabilities. Typical inferences are as follows.

8. In a species with substantial migration among the populations (of the order of a 1 % or more), chance has much the same effects as on an equally numerous species with wholly random dispersal within its whole range, i.e. these effects are substantial if the species consist in few individuals, while if numerous such as represented by thousands of reproducing individuals, drift can have substantial effects only in a very long run and if not contrasted by deterministic factors such as selection and mutation (e.g. Nagylaki 1992).

9. With random dispersal within each small population and low migration among them, the populations can easily differentiate at random while drift can be small or insignificant for the species as a whole; if migration is very low, *drift for the whole species will be much smaller than for an equally numerous, unsubdivided one.* Thus if gene frequencies change wholly at random, the rate of loss of alleles in a species tends to become algebraic as migration among the populations tends to zero, i.e. *as slow as that of an infinitely numerous species, no matter how few its populations are* (more than one), *or how small* (Karlin 1968, exp.: 152-156). Likewise, subdivision of a species into populations can much lower its *genotypic variance* when drift is deterministically contrasted, e.g. the variance of a species subdivided into two populations tends to 1/2 that of an equally numerous, unsubdivided one as migration among them tends to zero, to 1/3 if is subdivided into three, and so on (Scudo 1992).

10. The alternative to the above consists of assuming that an infinitely numerous population is continuously distributed in an open space, and its individuals disperse in the vicinities of their birthplaces according to some probability distribution. Even with very short-range dispersal, random drift will be small if contrasted by deterministic factors while, if not so contrasted, it might become substantial in the long run but only locally, remaining practically nihil for the species as a whole (Nagylaki, e.g. 1992).

Directly, then, these models tell precious little about real species, most of which are subdivided into local populations, *most of whose individuals breed in the vicinities of their native places* unless moving to a spot just across the nearest border. The only reasonable inference one can draw from these models is that, provided it is not contrasted by deterministic factors, random drift will be scarce for a species as a whole unless it is exceedingly small, or quite small and very close to random dispersal. These models, then, just quantify what is intuitively expected, namely that in order to be effective on a whole species drift ought to operate *at all places in the same direction* - i.e. to be strongly non-random.

ODTs tend to rely on the qualitative implications of concording results by different models on the same process since, by itself, any one model may not be very informative, and rather misleading in some respect. For instance point 4, predicting that a rare haploid mutant when becoming favored always moves towards fixation, sharply contrasts with 3 and thus suggests that the fitness approximation grossly underestimates the achievement of stable genotypic polymorphisms even just by ignoring self-limitation. Then point 7 predicts that a selective regime is not likely to approach the scarce tendencies to directly fix rare alleles in diplonts - quite likely scarcer than the 1/6 predicted by 5 - unless it lasts far longer (in generation time) than in a comparable haploid case. If generally indicative, these tendencies imply that cross-breeding diplonts respond to a change in selective regime primarily by altering the compositions of stable genotypic polymorphisms, to a lesser extent by selecting new ones only marginally by bringing initially rare alleles close to fixation through the same selective regime - e.g. quite likely 5 underestimates the fixation of an initially rare allele as twice more likely to occur through more selective regimes than through only one, even if this were to last indefinitely. Likewise points 8, 9 and 10 jointly imply that, if not much contrasted by deterministic factors, drift can be substantial for a whole species in the long run only if it consists of a single, smallish, well-mixed population or, if consisting of more than one, these are not long lasting, i.e. simply put, only so long as the species persists at the brink of extinction. Unless this were to occur most of the time, contrary to all evidence and common sense, drift per se could not explain the rates of base substitutions, especially high in third position (see below). On the other hand, initially rare mutants can easily proceed to fixation through interactions among substantial random changes within populations and both the intra- and the inter-population effects of more than one change in selective regime. As anticipated in the previous section, this far more strongly applies to diplonts than to haplonts and to the early stages of growth of communities whose local populations are not very long-lasting, and interpopulation selection synergetically operates on more of its characteristic forms, possibly on all. In Wright's expressive imagery (e.g. 1970) unless operating on very massive adaptive mutations, by acting deterministically on a species a novel selective regime could go hardly beyond trapping any one of its populations into the adaptive peak closest to its current state, even if only marginally more adaptive, i.e. only result in slow, modest and mostly local changes. On the other hand, if "helped" by massive drift within ephemeral populations and acting on not too rare alleles, the same regime would rapidly bring populations towards many different adaptive peaks. In sufficiently numerous trials, some populations will thus closely approach the highest possible peaks even if far away from their initial genotypic compositions. Especially in young, fragile communities these lucky local achievements can then spread to whole species by selective replacement and hybridization among their ephemeral populations, synergetically involving part or all characteristic forms. Most likely these differ also in their fragility and invasivity, thus making local collapses and reinvasions more likely than on a species "on its own". Incidentally, these forms of interdemic selection will also result in substantial rates of base substitutions by putting a premium, even if slight, on populations improving the efficiency of their individuals also chemistry-wise. On the other hand, relatively specific genotypic changes, such as "amplifications" of mobile elements, can be massively induced in diplonts by conditions of life (e.g. Ratner et al 1996) so that, if useful, they might approach fixation in a reasonable time just through mass selection.

NDTs differ from ODTs in quantification by mostly relying for each question on a single mathematical setup, regarded as best suited, with hardly any regard for the relative likelihoods and speeds of its possible outcomes. They thus deal with species dynamics mostly through special cases of Lotka-Volterra systems in continuous time, or minor variations thereof, and with genotypic change mostly through the frequency approximation in discrete time. As prime example of their scarce concern for the likelihoods and speeds of outcomes, NDTs oncur with traditional Western Darwinisms on adaptive changes being mostly due to the direct effects of substitutions by rare mutants, achieved by a single selective setup through small fitness differences. This choice stems from trying to avoid as much as possible Malthusian self-limitation, which NDTs generally accept for primary producers or basal elements of chains as the only way to prevent population explosions (e.g. High and Maynard Smith 1972, cf. Scudo 1997 for the origins of this position). Then even for a handful of species it is hard to coexist through adversary interactions (cf. e.g. point 1 with 3), and NDTs make it nearly impossible by neglecting the higher invasivity of such communities in their patchy distributions. That rich communities of this sort are so common (e.g. Cohen et al 1990) is thus for NDTs a hard theoretical challenge, about as hard as to explain why so many animals retain the male sex since their major argument, Müller's original one, broke down (Karlin, 1973). Further, to the limited extent that Müller's argument still holds, it does not square with prokaryotes having substantial recombination and relatively fast and easy substitutions by rare mutants, and yet having evolved at enormously slower rates per generation than multicellular eukaryotes (cf. the next §). NDTs eventually acknowledged that innumerable base changes (mostly in third position) cannot be justified as "one shot" selective substitutions, and attributed most of them to chance, alone though this hardly works unless "impossible" conditions are met (cf. again points 8, 9 and 10, recalling that for ODTs most base changes are still ultimately selective). These positions concur with NDTs ignoring alternative, threshold phenotypic responses to poorly predictable conditions of life, and rather accounting for phenotypic discontinuities through peculiar games of strategy (cf. again Scudo 1996b) or through "genetic assimilation" (cf. the previous section). Unless selection could get trapped into clear-cut choices by alternative responses, particularly for fixed motor patterns of animals, communities or biocoenoses could not even start building up. Ordinary communities being thus "impossible" for NDTs, no wonder that close microbial symbioses would be about as unlikely, the traditional ODT canon of assessing them through the long term persistence of standardized interactions would thus be absurd. Unless directly, rigorously proven otherwise, NDTs regard close microbial interactions as parasitic, and the repetitive elements of viral origin so common in the genomes of high eukaryotes would thus be "junk DNA". No wonder, then, that NDTs cannot rely on eukaryotes easily changing the numbers of copies and locations of genomic elements in answer to clues of external origin, with potentially meaningful morpho-functional consequences. By much downplaying or altogether discarding such "wholesale" modes of genomic change and selection, NDTs managed to maintain about as extreme "adaptivist" positions as those of older western Darwinisms.

Symbiogeneticists then react to NDTs much as did their predecessors, i.e. they agree with Darwinisms in not considering wholesale modes of selective change, but regard obvious failures what Darwinism's try to pass as intriguing theoretical challenges. Indeed, retail genotypic selection does not appear as having much to do with innovation, especially if "fast", not even for haplonts on which it is far more efficient than on diplonts (see below). Symbiogeneticists, instead, typically conclude that genotypic selection tout court would have hardly any power.

Origins of Prokaryotes: Two Alternative Scenarios

The origins of bacterial cells provide a strict test for theories as it demands choices between radical alternatives, matching very different tempos of changes. Very simple life-forms should have started sometime after the Earth approached its present mass and orbit around 4000 million years ago (Ma), while cosmic impacts might have been still too harsh for uninterrupted prokaryotic life. Stromatolites, the macroscopic manifestations of prokaryotic communities, are found since 3500 Ma and, if the earliest ones represent true prokaryotes, these would not have undergone any major change till about 1750 Ma (e. g. Schopf 1992). In spite of some good reasons to the contrary (Scudo 1996), the much larger cells which then appeared are regarded as nucleated from the outset. Earlier on oxygen would have been too scarce for them, and they would have hardly changed till evolving multicellularity a billion years later.

According to Darwinist-type reconstructions such as Eigen's (1987) or Maynard Smith and Szathmary's (1995, from here on abbreviated as MSS) a proteic origin of present-day life should not be seriously considered since proteins only clumsily produce *other* proteins, as for a handful of big enzymes being needed to synthetize the short peptide gramicidin S. Present-day life would have started with RNAs though problematically so, since nucleotides are not produced abiotically in solution (Miller, e.g. 1992). The dry-phase phosphorylation of nucleosides being too clumsy, MSS end up assuming a *direct dry phase origin of nucleotides.* RNAs could then evolve through hypercycles acting as replicators within compartments provided by lipidic membranes, though it is hard to see how these could have come about (see below). As to why RNAs then started coding for proteins, MSS propose they got "trapped into it" by regularly utilizing amino acids as *coenzymes of ribozymes*. Then it is even harder to justify how the triplet code would have *originated as universal* as MSS assume, since RNAs should have choosen among plenty of equally suitable coding possibilities by evolving *uniformly*, without branchings. Starting from a universal triplet code it is also hard to justifiy coding differences as posterior specializations, especially for viruses which are far more variable than bacteria in *all codes* as well as in the chemical composition of DNAs. Obviously from such beginnings the first cell (*progenote*) cannot easily come about, particularly if autotrophic as MSS assume. Nevertheless they are not concerned that all these feats should have happened in less, possibly much less than a half a billion years, and rather wonder about the subsequent, very long "prokaryotic stasis"thus being implied. For MSS further prokaryotic evolution is also problematic, as for the two *very different* membranes of the Gram negatives, since they virtually disregard symbiosis. For them the first explanation being considered starts from an *obcell* whose chromosome is attached *on the outside*, which would have flattened and then risen vase-like, finally enclosing its chromosome whithin the *same* double membrane. There is no need to go further with such "RNA Worlds" which mostly rely on chance, and only through miracles can satisfy constraints they regard as "sacred".

Since chemistry, physics and planetary science only vaguely hint at the conditions in which life arose (e. g. de Duve 1995), it seems wiser to assume that the molecules of life somehow originated in the order of increasing complexity of abiotic synthesis, namely peptides first, followed by sugars and fatty acids and nucleic acids last. If so it is most unlikely, for instance, that lipidic membranes have been the "intermediates" between biomolecules and cells on which many "RNA Wordls" rely. More generally, all the sacred constraints of such theories become irrelevant in largely deterministic theories of peptidic first origins, notably Cordón (1977), de Duve (1991, 1955) and Ratner et al (1996). Thus extant proteins should hardly be capable of producing other proteins since cellular coding has been in operation for so long, but this would have no implication as to what went on previously. Proteic functions being so loosely related to sequences, complex "communities" of simple peptides could have easily "reproduced" somewhat imprecisely, thus being able to quickly change much as do simple self-replicating molecules (e. g. Hong et al. 1992). That nucleotides are not produced in Miller-like conditions is likewise irrelevant since simple peptidic enzymes could phosphorylate nucleosides, also aboundantly produced in H_2-rich, reducing **media** (e.g. de Duve 1955). These media quite likely resulted from comets predominating over asteroids as the Earth's accretion was slowing down (the “late heavy bombardment”, e.g. Thomas et al 1996) Rather, the main problem for ODTs is how to bridge the enormous gap between "isolated" components of mostly peptidic communities and bacterial-like cells (not for NDTs, accounting even for high social behaviours through game struggles among automata). Giglio--Tos bridged this gap through small aggregates of simple, not better specified biomolecules, that would have directly resulted from physical forces (cf. ¤ 1). Cordón (1977, 1990) specifies such entities, the basibionts, as associations of α-poplypeptides akin to globular proteins which eventually became alive, with some spatial differentiation between enzymatic activities and energy reserves. He further states (Cordón 1977: 522, my translation) "At this stage in the origin of the first protoplasmatic individuals, one might say that their metabolism, phylogeny, ontogeny and reproduction (thus inheritance) were confounded " (so also were the basic organizational levels of ODTs). Although their descendants have been tightly bound to cells for so long, the chemical workings of basibionts could be plausibly inferred from the phylogeny of main metabolic pathways, reasoning on their steps much as on organs and functions in complex organisms.

One could object to Cordón's scheme that the assembly of peptides and selection on them would not have been precise enough to result in complex, organized communities and, later on, in basibionts. This objection overlooks substrates being able to direct the formation of efficient biocatalysts, as was recently proven for simple ones under prebiotic conditions (Kochavi *et al* 1996). Prior speculations on peptidic enzymes having also arisen in this way thus have a strong, even if somewhat indirect empirical suport. Further, self-assembly now seems to make sense as "ordinary" physics of "long distance" electromagnetic interactions, rather than as an extraordinary outcome of unlikely combinatorial feats (Preparata 1995). One might also objetct to theories of proteic first origins having assumed uniform chirality before coding just from a priori reasoning, namely expecting that minute differences between enantiomers would eventually tilt either way the balance of chiral reactions. This expectation now well agrees with the homogenizing effects of magnetic fields (cf. e.g. Bradley 1994), keeping in mind that most of the time the Eart's magnetic field was much stronger than now. Thus reassured about basibionts, let us turn to their evolution as according to Cordón. At first they reproduced by canalizing "environmental" biomolecules but, later on, they evolved ever more complex enzymatic activities, up to becoming autotrophs on energy from dissolved atmospheric gases. Some such basibionts organized into simple, flat colonies which, in turn, made room for other colonies to prosper as their saprophytes. These heterotrophs were more complex, vase-shaped, made up by a layer of basibionic individuals connected by descent and with a lipidic layer on the inside. To feed on ever larger chunks of decaying autotrophs their individual components were forced to differentiate, reaching cooperative, veritably cellular behaviours (Cordón, exp. 1990, Vol. 1); evidences for life much before 3.5OO Ma ago such as by Mojzis *et al* (1996) might well record these stages. Cordón's pubblished work stops at this point, suggesting that RNAs arose from nucleotides, common within such cells as coenzymes.

It is now evident that the coding machineries of cells originated from symbiotic combinations of different RNA entities (e.g.Ohnishi, 1990, Ratner *et al* 1996). It is also obvious that these combinations would have occurred far more easily within communities of peptidic protocells, somehow alive but not yet tightly closed, than starting from non-living membranes. The "uracil-(HMU)-only DNAs" of *Bacillus*' own viruses, just reduced RNAs, would still reflect the obvious first step of the process which eventually gave rise to the "thymidine only DNAs" of the high eukaryotes, leaving clear traces all the way up to low ones such as dinoflagellates. Being the closest extant representatives of DNA's first origins rules out viruses as "degenerated cells", leaving to choose whether they came about alongside the molecules of life versus being RNAs which "escaped the control of cells" (cf. Scudo 1996). It is then natural to ask whether parasitic behaviours by simple RNA entities, even if just partial and modest ones in part of their life cycles, might help solving such puzzles. Tentative scenarios to this end must also face the problem of mono- versus polyphyletic origins of nucleic acids, as for most processes involving symbioses usually posed in nonsensically extreme forms. If indeed the earliest DNAs were just reduced RNAs, and the process went on in a number of rather alike RNA species more of which left descendants, DNA strictly speaking would be polyphyletic. And yet it would have evolved much as if having a monophyletic origin and its "polyphylety" would be hard to sort out. Likewise that a number of *alike* RNAs independently "escaped the control of cells" hardly makes a difference from whether just one did. On the other hand partially independent origins of DNAs easily justify all sorts of coding differences among viroids, virusoids and viruses as mostly ancestral, some of which might have been retained also since they offer protection from bacterial restriction enzymes. Justifying *all* such differences as secondary adaptations out of strictly monophyletic origins -- as often assumed without any solid evidence -- would require, once more, most unlikely feats.

A plausible scenario along such lines could start from complex enough cells to harbour different nucleotides in sufficent numbers to easily polymerize, at first perhaps just for storage and transport of components. In some cell lines these Ur-RNAs would have evolved more specialized forms of "mutual instructing" (Ratner *et al* 1995) -- e. g. repair of proteic components or poorly specific supports for their production -- and some of them could be mildly parasitic on cell lines other than their "native" ones. Different lineages of such "viroids" then started building capsids from cellular amino acids to protect their chromosomes; thus shelterd they could reach unusual hosts, on which being “virulent”. The earliest codes would thus have *semi-independently* evolved while horizontal exchanges were still very easy, thus much diminishing initial coding diversity among these "Ur-viruses" and fully homogenizing possible chemical differences of their RNAs. Then some viruses would have profited by specializing a nucleic acid just for chromosomal replication, thus becoming more efficient "predators" of cellular nucleotides. Later on, part of them would have evolved reverse transcriptases, making it easier to alternate among different, still "open" host cells. Some such cells thus reacted to ever-increasing parasitic loads by developing specialized machineries of transport across membranes, eventually "sealing up". As suggested by de Duve (1991), symbiotic viruses probably started coding for cells by making "RNA copies" of their proteic components and retrotranscribing these primitive messengers. In this scenario the two main hypotheses on the origin of viruses are no longer alternative -- i. e. RNAs evolved alongside the molecules of life after having escaped the control of cells. Whether or not in this specific sequence, the nucleic machineries of extant prokaryotes most likely evolved as a *bona fide* biocoenotic process among many cell lineages and viral-like ones.

To the extent their components became coded natural selection would operate on cells genotypically, and its achievements would no longer be diluted or lost by approximate duplication. Although intrinsically much slower through this "irreversible" operation, natural selection would still have had major effects on cells so long as they did not fully close, since major symbioses could still readily occur. Thus the two very different membranes of Gram-negatives most liekly arose from an endocellular symbiosis, followed by chimeric fusion of the partners' cytoplasms and chromosomes. As cells sealed up, much reducing horizontal exchanges and making chimeras most unlikely, natural selection was left to operate mostly "on its own" through slowish, "one shot" substitutions by rare mutants (see „More Technical Comparisons Between Theories § 2). Prokaryotes could thus react to novel challenges mainly by obtaining an optional plasmid or an extra new gene, as alternative responses to poorly predictable differences in conditions of life. As a rule, however, any such response *would not result in further adaptive divergence* except through close symbioses. Further, barring ever less likely impacts so large as to evaporate or sterilize the whole oceanic mass, it is hard to envisage how communities of these hardy Ur-prokaryotes and their "viruses" could undergo mass extinctions akin to those at the root of major evolutionary changes in plants and animals. These communities would rather change piecemeal, as reactions to the chemical changes they were slowly inducing on the biosphere. Most likely, then, the earliest cells on record were not true prokaryotes but primitive stages in the long evolution of coding up to near-universal control of reproduction by a single chromosome, a process that might be easily mistaken as a stasis in the fossil record.

Acknowledgements. I am deeply indebted to Dr. T. Cordón, Professors S. Gimelfarb, P. Farinella and V. Sgaramella for providing me with some key informations, to H. E. A. Schenk and P. Nardon for useful suggestions to the text, to my wife Katherina, Dr. O. Fiorani and Mrs. S. Zanoli for their editorial help.

References

Bacci G (1965) Sex determination, Pergamon, Oxford
Botnariuc N (1992) Evolutionismul în impas? (English summary: Evolutionism in impass?). Acad. Române, Bucharest
Bradley D (1994) A new twist in the tale of nature's asymmetry. Science 264: 908
Buchner P (1965) Endosymbiosis of animals with plant microorganisms. Interscience, New York
Cavalli-Sforza LL, Menozzi P, Piazza A (1994) The history and geography of human genes. Univ. Press, Princeton
Cohen JL, Briand F, Newman CM (1990) Community food webs. Springer, Berlin
Cordón F (1977) La alimentación, base de la biologia evolucionista. Alfaguara, Madrid
Cordón F (199O) Tratado evolucionista de biologia. Parte segunda. 2 Vols, Aguilar, Madrid
D'Ancona U (1954) The struggle for existence. Brill, Leiden
Davoli R (1943) I virus filtrabili nelle malattie dell'uomo. Vallecchi, Firenze
De Duve C (1991) Blueprint for a Cell. Patterson, Burlington
De Duve (1995) Vital dust, the origin and evolution of life on Earth. Basic Books, New York
Del Giudice E (1995) Quantum electrodynamics' coherence in matter. World Scient, Singapore

Eigen M (1987) Stufen zum Leben. Piper, München
Eldrege N (1985) Unfinished synthesis. Univ Press, Oxford
Giglio-Tos E (1900) Les problemes de la Vie. Part. I: La substance vivante. Published by the Author, Tourin.; 1903 - Part. II : L'ontogénèse et ses problèmes. Ibid.; 1905 La fécondation et l'hérédité.; 1910 La variation et l'origine des espèces. Ibid., Cagliari.
Giglio-Tos E (1911) La via nuova della biologia. Ricci, Firenze
Giglio-Tos E (1915) A proposito delle mie leggi sull'ibridismo. Published by the Author, Cagliari
Giglio-Tos E (1924) Alcune mie curiose previsioni verificate dall'embriologia sperimentale. Rend Accad Lincei 33: 451-455.
Grassi GB (1925) L'interpretazione di Giglio-Tos dei fenomeni fondamentali della vita. Cappella, Ciriè
Hartmann M (1943) Die Sexualität. Fischer, Jena (also 1956 ed)
High J, Maynard-Smith J (1972) Can there be more predators than preys? Theor Pop Biol 3: 290-299
Hofbauer J, Sigmund K (1988) The theory of evolution and dynamical systems. Univ Press, Cambridge
Hong J-I, Feng Q, Rotello V, Rebek J Jr (1992) Competition, cooperation and mutation: Improving a synthetic replicator by light irradiation. Science 255: 848-850
Karlin S (1969) Equilibrium behavior of population genetics models with non-random mating. Gordon & Breach, New York
Karlin S (1973) Sex and infinity: a mathematical analysis of the advantages and disadvantages of genetic recombination. In: Bartlett, Hiorns (eds) Population dynamics. Academic Press, New York, pp 155-194
Khakina LN (1979) Problema simbiogeneza: istorik-kritichesky ocerk issledovani otchestrennykh botanikov. Acad. Nauk, Leningrad (in English: Margulis L, Mc Men Amin M (eds) Concepts of symbiogenesis - A historical and critical study of the research of Russian botanists, Yale Univ Press, New Haven 1992)
Kochavi E, Bar-Nun A, Flemimger B (1996) Substrate-directed formation of small biocatalysts under prebiotic conditions. 50 in ISSOL 1996, Paris
Kostitzin VA (1934) Symbiose, Parasitisme et évolution. Hermann, Paris
Kostitzin VA (1937) Biologie mathématique. Colin, Paris
Leonardi P (1950) L'evoluzione dei viventi. Morcelliana, Brescia
L'Heritier P (1948) Sensitivity to CO_2 in *Drosophila* - a review. Heredity 2: 325-348
Marx J (1995) Tracing how the sexes develop. Science 269: 1822-1824
Lorenz K (1965) The evolution and modification of behavior. Un Press, Chicago
Maynard-Smith J (1991) A Darwinian view of symbiosis. In: Margulis L, Fester R (eds) Symbiosis as a source of evolutionary innovation. MIT Press, Cambridge, pp 26-39
Maynard-Smith J, Szathmary E (1995) The major transformations in evolution. Freeman, Oxford
Miller SL (1992) The prebiotic synthesis of organic compounds as a step toward the origin of life. In: Schopf JW (ed) Major events in the history of life. Jones & Bartlett, Boston, pp 1-28
Mojzis S, McKeegan KD, Harrison TM, Nutman ,AP, Arrhenius G (1996) New evidence for life on Earth by 3870 Ma. c4.13 in ISSOL 1996 Paris
Nagylaki T (1992) Introduction to theoretical population genetics. Springer, Berlin
Ohnishi K (1990) Cell machinery as a symbiotic complex of contemporary RNA organisms including tRNAs, rRNAs and mRNAs: Evolution from RNA individuals to cell individuals by cooperative symbiosis. In: Nardon P et al (eds) Endocytobiology IV. INRA, Paris, pp 593-600

Pierantoni U (1923) Gli animali luminosi. Sonzogno, Milan
Pierantoni U (1940) Nozioni di biologia. UTET, Tourin
Ratner VA, Zharkikh AA, Kolchanov N, Rodin SN, Solovyov VV, Antonov AS (1996) Molecular evolution. Springer, Berlin
Rosa D (1899) La riduzione progressiva della variabilità e i suoi rapporti coll'estinzione e coll'origine delle specie. Clausen, Tourin
Sapp J (1994) Symbiosis and disciplinary demarcations: The boundaries of the organism. Symbiosis 17: 91-115
Schmalhausen II (1946) Faktorii evolutsii; teoria stabilizirushego otbora. Akad Nauk, Moskva (Factors of evolution: The theory of stabilizing selection, also in abridged English edition, Blackiston, Philadelphia, 1949)
Schopf JW (1992) The oldest fossils and what they mean. In: Schopf JW (ed) Major events in the history of life. Jones & Bartlett, Boston, p 29- 64
Scudo FM (1967) Selection on both haplo and diplophase. Genetics 56: 693-704
Scudo FM (1973) The evolution of sexual dimorphism in plants and animals. Boll Zool 40: 1-28
Scudo FM (1976) "Imprinting", speciation and avoidance of inbreeding. In: Novak V (ed) Evolutionary biology. Acad Sci, Praha, pp 375-392
Scudo FM (1992a) Vito Volterra, "ecology" and the quantification of "Darwinism". In: Amaldi E et al (eds) Convegno internazionale in memoria di Vito Volterra. Lincei, Rome, pp 313-333
Scudo FM (1992b) A Darwinian theory of molecular evolution. In: Ratner VA, Kolchanov NA (eds) Modelling and computer methods in molecular biology and genetics. Nova Sci, New York, pp 369-378
Scudo FM (1996) Symbiosis, the origins of major life forms and systematics: A review with speculations. In: Ghiselin, MT Pinna G (eds) Systematic biology as an historical science. Mem Soc Ital Sci Nat, Milan, pp 95-108
Scudo FM (1997) Lotka, Volterra, d'Ancona and Kostitzin: Some puzzles in the mathematical theory of the struggle for existence. To appear in Atti Ist Lomb
Scudo FM, Ziegler JR (eds) (1978) The golden age of theoretical ecology: 1923-1940. Springer, Berlin
Smith DC, Douglas AE (1987) The biology of symbiosis. Arnold, London
Teissier G (1952) Dynamique des populations et taxinomie. Ann Soc Roy Zool Belg 83: 23-44
Thomas PJ, Chyba CF, McKay CP (eds) (1996) Comets and the origin and evolution of life. Springer, Berlin
Von Wettstein F, Straub J (1942) Experimentelle Untersuchungen zum Artbildungsproblem. III: Weitere Beobachtungen an polyploiden Bryum-Sippen. Vererbungs 80: 272-279
Williams GC (1963) Adaptation and natural selection (4th print). Univ Press, Princeton
Wright S (1970) Random drift and the shifting balance theory of evolution. In: Kojma K (ed) Mathematical topics in population genetics. Springer, Heidelberg, pp 1-31
Wright S (1982) Character change, speciation and the higher taxa. Evolution 36: 427-443
Zavadskij KM (1961) Ucenie o vide. Nauka, Leningrad
Zook D (1994) Integrating symbiosis into mainstream science: Penetrating the curriculum membrane. Symbiosis 17: 117-126

Part II

Symbiotic Systems

Adaptation, Signal Transduction, Taxonomy and Evolution

2.1 Molecular Approach to Taxonomy of Endocytobionts

Progress in the Studies of Endosymbiotic Algae From Larger Foraminifera

J. J. Lee[1], J. Morales[1], J. Chai[1], C. Wray[2] and, R. Röttger[3]
[1]Department of Biology, City College of City of New York, New York, New York, USA
[2]Department of Biology, University of California, Los Angeles, USA
[3]Institute for General Microbiology, University Kiel, Germany

Key words: Larger foraminifera, algal endosymbionts, *Cyclopeus carpenterii*, *Amphora bigibba*, *Sorites marginalis*, surface antigens of endosymbiotic diatoms.

Summary: The endocytobiotic diatoms isolated from a previously unexamined host, *Cyclopeus carpenterii*, turned out to be quite unusual. They belong to a new genus related to *Nitzschia*, but which has a canopium attached to the keel. *Amphora bigibba* was also isolated from specimens of the same host. Progress is also being made in clarifing the description of the endosymbiotic *Nitzschia frustulum* symbiotica variants found in Caribbean hosts. The sequence of ssrRNA of the endosymbiotic dinoflagellate from the larger foraminifera, S*orites marginalis*, has been completed. It is not identical to any previously sequenced endosymbiotic dinoflagellate.

Progress is being made in the identification of surface antigens particular to endosymbiotic diatoms. In Western blots with polysera we have identified a 104 kDa protein band found on the surfaces of five species of endosymbiotic diatoms but not on non-symbiotic species. In blocking experiments we were able to show that the antiserum-coated endosymbionts were digested at much higher rates than controls.

Introduction

It is now fairly widely known that the larger foraminifera in today's shallow, well-illuminated tropical and semi-tropical seas are the hosts for a diversity of endosymbiotic algal types (unicellular red algae, dinoflagellates, diatoms, and chlorophytes; reviewed by Lee and Anderson 1991). Although we have been studying symbionts in foraminifera for 30 years (Lee et al. 1965), and those in larger foraminifera for 25 years (Lee and Zucker 1969), it is dispiriting to acknowledge how slow progress has been made toward answering some of the most basic questions about this ecologically and evolutionary significant phenomenon. Apologies having been given, we take the opportunity of this Endocytobiology Colloquium to report progress on a number of ongoing projects in our laboratory.

Research over the past 16 years has confirmed speculations that diatoms are endosymbionts in four families of larger foraminifera. Analysis of the endosymbionts from more than 2000 hosts has shown that the relationships between host species and diatom endosymbionts are not finical (reviewed in Lee 1994). Six species, *Nitzschia frustulum* var. *symbiotica*, *N. panduriformis* var. *continua*, *N. laevis*, *Fragilaria shiloi*, *Amphora roettgeri*, *A. erezii*, are involved in over 75% of the associations, but any one of several dozen diatom species can establish themselves in most of the host species examined. Collaborative efforts between our group and others recently provided collection opportunities of new hosts (*Cyclopeus carpenterii*, *Amphistegina gibbosa*, and *Heterostegina antillarum*) from new locations (Sesoko Jima, Japan and Conch Reef, Florida).

One of the most interesting aspects of each new examination of endosymbionts from different habitats has been the discovery of new and unusual diatoms (e.g., Lee and Reimer 1983; Reimer and Lee 1984; Lee and Xenophontos 1989). The unusual forms, perhaps, are not so strange when one considers the fact that frustule development is suppressed when the symbionts are within their hosts and that the symbionts may be passed frustuleless from one generation to the next during asexual reproduction. Since searches of the habitats where the hosts are found revealed that the species found as endosymbionts are rare, or absent (Lee et al. 1989, 1992) it is not unreasonable to expect that the endosymbiotic habitat might also be a reservoir of undiscovered taxa.

The identity of the endosymbionts from Caribbean hosts was really unknown before we began the present study. An unknown species of *Nitzschia*, described in a report of an unspecified number of isolations of diatoms from *Amphistegina gibbosa* from La Parguera, Puerto Rico (Hallock et al. 1986), piqued curiosity. It had characteristics which did not quite fit those of *N. frustulum* var. *symbiotica*, *N. laevis* or several other related taxa. This, in itself, is not surprising since the *N. "lanceolateae"*, includes many taxa with overlapping characters (e.g., outline, size, striation, fibula density; reviewed by Lange-Bertalot 1980). The resumption of field work by Dr. Pamela Hallock, University of South Florida, on larger foraminifera from Conch Reef, Florida Keys, gave us the opportunity to do more detailed examination of the endosymbiotic diatoms of hosts (*A. gibbosa* and *Heterostegina antillarum*) from this habitat. Through the cooperation between Dr. Rudolf Röttger, University of Kiel (Germany) and Dr. Johann Hohenegger, University of Vienna (Austria), we obtained specimens of *Cyclopeus carpenterii* from Sesoko Jima. This is a larger foraminifera which had not been examined previously, but which was expected, because of its relatives, to be a host for endosymbiotic diatoms. The diatoms we isolated in culture from all three hosts were sufficiently different from known taxa to merit more focused morphological study. This report details the results of our continued study of these taxa.

Materials and Methods

Diatom Isolation. Two hosts, *Amphistegina gibbosa* and *Heterostegina antillarum*, were collected by Dr. Pamela Hallock on Conch Reef on 30 April and 13 December 1993. Specimens of *Cyclopeus carpenterii* were sent to us by Dr. Rudolf Röttger in February 1995. They were from cultures maintained by him from collections originally made by Dr. Johann Hohenegger. The foraminifera were washed and resuspended in filtered seawater. They were then transferred to insulated containers and sent by express mail to our laboratory at City College of New York where the isolations were carried out. The methods for isolation were the same as previously published (Lee et al. 1989, 1991, 1992).

The endosymbionts from each foraminifera were aseptically isolated and transferred to Erdschreiber medium in a test tube. The tubes were incubated in front of a bank of fluorescent lights, (14 h light/10 h dark) in a temperature controlled room (25 °C) for approximately 10-14 days; and aliquots were removed from each tube to be prepared and examined in a scanning electron microscope (SEM). Because microscopic observations showed that the endosymbionts were unusual, aliquots of each isolate were aseptically transferred to fresh media. They were serially transferred at monthly intervals and examined again to study the stability of the characters we found in this study.

All stubs were prepared from 1 ml aliquots from primary isolation cultures and successive serial transfers from them. The sample was oxidized with 0.1 ml 30% H_2O_2 and then passed through a Nucleopore R (5μm) filter. The filter was dried in an oven at 60 °C overnight, mounted on a stub with double-stick tape, and coated with 10 nm of Au/Pd mixture in a Polaron sputter coater. We found it necessary to produce fragments in order to understand the internal structure of the frustule of the new genus. To accomplish this, we used a Pasteur pipette to chop oxidized frustules of a centrifuged dense culture in the bottom of a test tube. Stubs were examined on a digital SEM (Zeiss DSM 950) and photographed on Kodak T-Max 100 film. The film was developed with T-Max developer and printed on Kodak Polymax RC paper. Measurements were facilitated by the electronic measurement and scan rotation features built into the SEM.

Dinoflagellate Isolation. The hosts, *Sorites marginalis*, (Fig. 1) were collected with the aid of SCUBA by Dr. Pamela Hallock at Conch Key, Florida. They were sent in insulated containers to the City College of New York where the isolations of the symbionts were accomplished. Ten hosts were transferred to sterile seawater in the wells of sterilized spot plates placed on the stage of a dissecting microscope equipped with both above stage fiber optic cold light and below stage dark field illumination. There they were aseptically and carefully brushed with ethyl alcohol sterilized sable artist brushes (#0000). After ten aseptic changes of seawater and brushes, the specimens were examined at relatively high magnification (60x), using both types of illumination, to see if all surfaces of the test had been freed of colonizing contaminants. The hosts were then transferred to seawater containing 1 mM EDTA and observed under the dissecting microscope as they were being decalcified. When the process was completed, the decalcifying medium was withdrawn and replaced with sterile normal seawater. The hosts were then dissected with sterile glass needles which were used to tear open the organic layers of the chamberlet walls and release the symbionts. Individual and small groups of symbionts were removed from the wells with the aid of 10 μl capillary pipettes and transferred to ISM medium (Lee et al. 1980) which contained GeO_2 (0.1 μM) to inhibit the growth of any contaminating diatoms which might have survived the washing process.

Molecular Systematics Methodology. One culture of *Symbiodinium* from *Sorites marginalis* was harvested by centrifugation. DNA isolation followed the techniques of Rowan and Powers (1992). PCR reactions were carried out in a Perkin-Elmer 4000 thermal cycler programmed with a standard cycle of 1 min at 94 °C, 1.25 min at 42 °C, and 2 min at 72 °C for 35-45 cycles. The product was run at 130 V for 40 min on a 1% agarose gel in 1x TBE buffer and stained with ethidium bromide. PCR and sequencing oligonucleotide primers were designed after Medlin et al. (1988). Positive amplifications of the complete nuclear small-subunit (18S) rDNA were cloned using the components of the TA cloning kit from Invitrogen. Cloned DNA was sequenced as described by Toneguzzo et al. (1988). The divergent regions of the 18S rDNA were sequenced for multiple clones to investigate sequence heterogeneity; no nucleotide differences were apparent in the regions examined. Complete clones were sequenced for both strands. DNA sequence was read directly into the computer using Macvector 4.11 (Eastman Kodak) and verified by the program's computer voice-assisted read back. DNA sequences were aligned using MALIGN (version 3.1) (Wheeler and Gladstein 1992). Complete dinoflagellate 18S rDNA sequences taken from the NCBI database (release 6.0) and incorporated into the alignments and phylogenetic reconstructions were: *Cryptothecodinium cohnii* (M34847), *Prorocentrum micans* (M14649), *Symbiodinium microadriaticum* (Rowan and Powers 1992), *Symbiodinium pilosum* (Rowan and Powers 1992) and *Symbiodinium* sp. (Rowan and Powers 1992). Phylogenetic reconstructions were carried out using PAUP (Swofford 1991). A single most parsimonious tree was found after exhaustive search using PAUP.

Diatom Working Library. In order to compare symbiotic with nonsymbiotic diatoms, a diatom working library was needed. According to the previous report (Lee et al. 1992), the prevalent endosymbiotic diatom species are: *Nitzschia frustulum var. symbiotica* (25.9%), *Amphora erezii* (15.4%), *A. roettgeri* (14.6%), *F. shiloi* (12.0%), *Cocconeis andersonii* (7.6%), *Nitzschia laevis* (5.9%), *N. panduriformis* (4.8%), *N. valdestriata* (3.6%), *A. tenerrima* (1.9%), *Navicula muscatinei* (1.2%), and *Navicula hanseniana* (1.1%). Among them, *F. shiloi, Cocconeis andersonii, Nitzschia laevis, N. panduriformis, A. tenerrima, A. roettgeri, Amphora* sp. (*halamphora*), *Navicula muscatini*, and *Navicula hanseniana* have been isolated from their host foraminifera *Amphistegina lobifera* and cloned on agar plates. *F. shiloi, C. andersonii, N. laevis, A. tenerrima, A.* sp.

(halamphora), A. roettgeri, and *Navicula hanseniana* grow well in 2-L flasks of Erdschreiber medium.

For comparison, six species of free-living diatoms, *Amphora tenerrima* variety I (AR1) and *A. tenerrima* variety II (EN), *Nitzschia laevis* variety (X72) and *A. luciae* variety II (XJA1), and *Navicula* sp. (A201) and *Navicula viminoides* (XD32), were isolated from the IOLR mariculture facility in Elat on the Red Sea. They belong to three genera which have symbiotic relatives.

Protein Extraction. Cells from six symbiotic species and five nonsymbiotic species were harvested by centrifugation and washed with sterile seawater. Cell pellets were resuspended in lysis buffer A (50 mM Tris, pH 7.0 / 10 mM NaCl / 100 μM Phenylmethylsulfonyl Fluoride (PMSF) / 1 μM leupeptin / 25% glycerol) and sonicated at full power with the aid of a Branson model 4 x 30 s with a 1-min interval on ice. The number of sonication cycles was determined by microscope examination. For some species, four runs are enough to break frustules, but for others it may take up to seven runs to break 60% of the cells.

Homogenates were centrifuged at 150000 x g for 1 h. The soluble proteins in the supernatants were collected and adjusted to 5 x sample buffer and used as references. The pellets containing both frustules and membranes were resuspended in 2 x sample buffer without bromophenol blue (2% SDS / 100 mM dithiothreitol / 60 mM Tris, pH 6.8 / 25 % Glycerol) and kept in a refrigerator overnight. Proteins were extracted from both frustules and membranes and collected by centrifugation at 150000 x g for 1 h. The concentration of both soluble and membranous proteins was determined by Bicinchoninic Acid Assay (Bollag and Edelstein 1991). Samples were denatured in boiling water for 5 min and separated on 9 or 10% acrylamide gel.

Western Blot Identification. After electrophoresis, proteins were transferred from gels to nitrocellulose membranes. One copy was stained with Amido black to verify the transfer. The other copy was blocked with Blotto/Tween blocking solution for 30 min and then washed with PBS 3 x 10 min. Polyclonal antibodies were raised in rabbits respectively against *F. shiloi, N. panduriformis, N. f. symbiotica,* and *A. tenerrima*. Each of them was incubated with the blot for 2 h. Because the rabbits developed different titers of antibodies in their sera it was necessary to dilute them differentially. The polyserum anti-*F. shiloi* was diluted by 1:50, while anti-*A. tenerrima* by 1:30, anti-*Nitzschia frustulum* var. *symbiotica* by 1:10, and anti-*Nitzschia panduriformis* by 1:4. After incubation, the membrane was washed 3 x 10 min with PBS. Goat anti-rabbit IgG conjugated with horseradish peroxidase (HRP) or protein A-HRP in 1:1000 dilution was used as the second antibody to react with the blot for 2 h. The blot was then washed for 3 x 10 min with PBS. Diaminobenzidine (DAB) was used as substrate to produce brown color reaction.

Proteins from eight symbiotic species including *F. shiloi, C. andersonii, N. laevis, A. tenerrima, A. roettgeri, Amphora* sp. *(halamphora), N. muscatinei, N. hanseniana* and four nonsymbiotic species including *Nitzschia laevis* variety (x 72), *Amphora luciae* variety II (XJA1), *Navicula viminoides* (XD32), and *Amphora tenerrima* variety I (AR1) were separated on the 9 or 10% acrylamide gel in equal amounts.

Affinity Purification of Antibodies. After western blotting, two bands, the most common band shared by symbionts 104 kDa-polypeptide from the blot probed with anti-*F. shiloi* polyserum and the the most popular band in both symbionts and nonsymbionts 66 kDa-polypeptide from the blot probed with anti-*A. tenerrima,* were excised from the nitrocellulose paper and cut into small pieces and transferred to a microcentrifuge tube. Each fragment was eluted with 200 µl elution buffer (5 mM glycine-HCl, pH 2.3 / 500 mM $NaCl_2$ / 0.5 % Tween20 / 100 µg/ml of BSA) for 5 x 60 seconds. The elutes were combined and neutralized immediately by the addition of Na_2HPO_4 to a final concentration of 50 mM. To test whether the elute is pure rabbit antibody against the single polypeptide which was excised, the same protein samples were separated on another SDS-PAGE and then transferred onto the nitrocellulose paper. Individual lane was sliced from the nitrocellulose membrane and incubated with the elute for 2 h. The original antiserum against *F. shiloi* was used as control. The second antibody treatment was same as before. After staining, only one band was seen on the the slice of nitrocellulose membrane which was treated with the elute. The control slices had same number of bands as usual. It means that the elute has pure antibody which is specific for one diatom polypeptide. These two elutes were used to block the special surface antigen on each species of diatom.

Functional Blocking Experiment. Before performing this experiment, four species of symbiotic diatoms, *N. laevis, A. roettgeri, F. shiloi, Navicula* sp. (W) and three species of nonsymbiotic diatoms, *Navicula* sp. (A201), *Navicula viminoides* (XD32), and *Nitzschia laevis* variety (X72), were labeled with 10 µCi radioisotope C^{14} for each 10 ml culture. The cells of each species were washed five times. 0.1 ml supernatant was removed to detect radioactivity (which is called background activity) and the pellet was resuspended in 1.4 ml of seawater. 0.1 ml of each species was removed to count cells and another 0.1 ml was removed to detect radioactivity (which is called sample activity). We estimated one-cell radioactivity by subtracting background from the sample and then averaged by the cell count. 0.3 ml of each species was incubated with the elutes from the antiserum or the antiserum itself and the last 0.3 ml was used as control. Each fraction of diatoms was provided to eight individuals of *Amphistegina lobifera.* Two of the forams were removed to detect radioactivity which was then averaged by 2 and divided by one-cell activity (this is the estimated number of diatom cells ingested by each foram). The rest of the forams were removed to the Erlenmyer flasks and incubated with 0.1 ml KOH and 0.1 ml seawater. A piece of millipore filter paper was placed in each flask to trap the CO_2 produced. Every two forams from each flask were put into a scintillation vial to detect radioactivity (which was used to estimate the number of diatom cells each foram kept inside). The filter paper was removed from each flask to detect radioactivity (this was used to estimate the number of diatom cells each foram digested).

Results

Diatoms

New Diatom Genus. A new diatom genus was isolated from all of the specimens of *Cyclopeus carpenterii* examined. Two specimens of *Amphistegina gibbosa* from Conch Key also contained a similar organism. Light microscope observations suggested that all the isolates of the new genus were identical. In valve view, the new diatom is an elongated oval, with an average length in fourth serial transfer cultures of 9.19 μm (0.40) and width of 3.80 μm (0.49). The valve face consists of three elements: (1) a contoured, striated face plate; (2) a medial raised raphe-sternum with a longitudinal uninterrupted exterior raphe slit; (3) a longitudinally slightly curved conopeum (Figs. 2 and 4). Normally the conopeum arches from the sternum to rest on the most raised part of the valve plate. The form of the three elements is such that, together, they form a longitudinal subconopeal grove between them (Fig. 4) The grove is best seen when the conopeum is partially removed. There are 60 striae/10 μm, with 8 punctae/μm, in the valve plate and jacket. We did not find much variation in either the striae or the punctae.

Although their proportions are visible in cingulate view (Fig. 4), one cannot appreciate, without disruptive treatment, that two lateral chambers form a large part of the internal space of the valve. The chamber is formed by two silicon sheets, the external valve face and a basal sheet. There is a copular canal formed just beneath the junction of these. The margin of the ventro-lateral external valve face sheet and the basal sheet form the medial surface of the canal. Copulae and a valve jacket form the external surface. This canal is best seen in partially destroyed frustules (Fig. 7). There is a single pore which opens between each lateral valve chamber and the canal beneath it. On the exterior, the canal seems to be open to the surface in common with the opening to the lateral valve chamber. There are five to eight fibulae which arch around the canal (Fig. 7). The fibulae seem to support the thin siliceous sheets since they seem very slightly depressed between them (Fig. 7).

The raphe-sternum runs from pole to pole along the center of the valve (Figs. 2, 4 and 7). It divides the interior of the valve into two lateral chambers. The raphe slit is usually central and continuous, but occasionally abnormalities are observed (e.g., Fig 2). Terminal fissures vary from slightly bent (Fig. 2) to strongly hooked. The helicoglossae are slightly bulbous, hollow and perforate both on the exterior and internally. The internal terminal fissure is sometimes quite branched and coiled. Except for the terminal ends, the external face of the sternum is heavily silicified and solid. There are three sets of six to eight fibulae in this organism's raphe-sternum system. One set of fibulac is at the level where the conopium is attached to the sternum; the second is at the level of the tubular interior; and the third connects the external face elements to the tubular and medially perforated internal valve face.

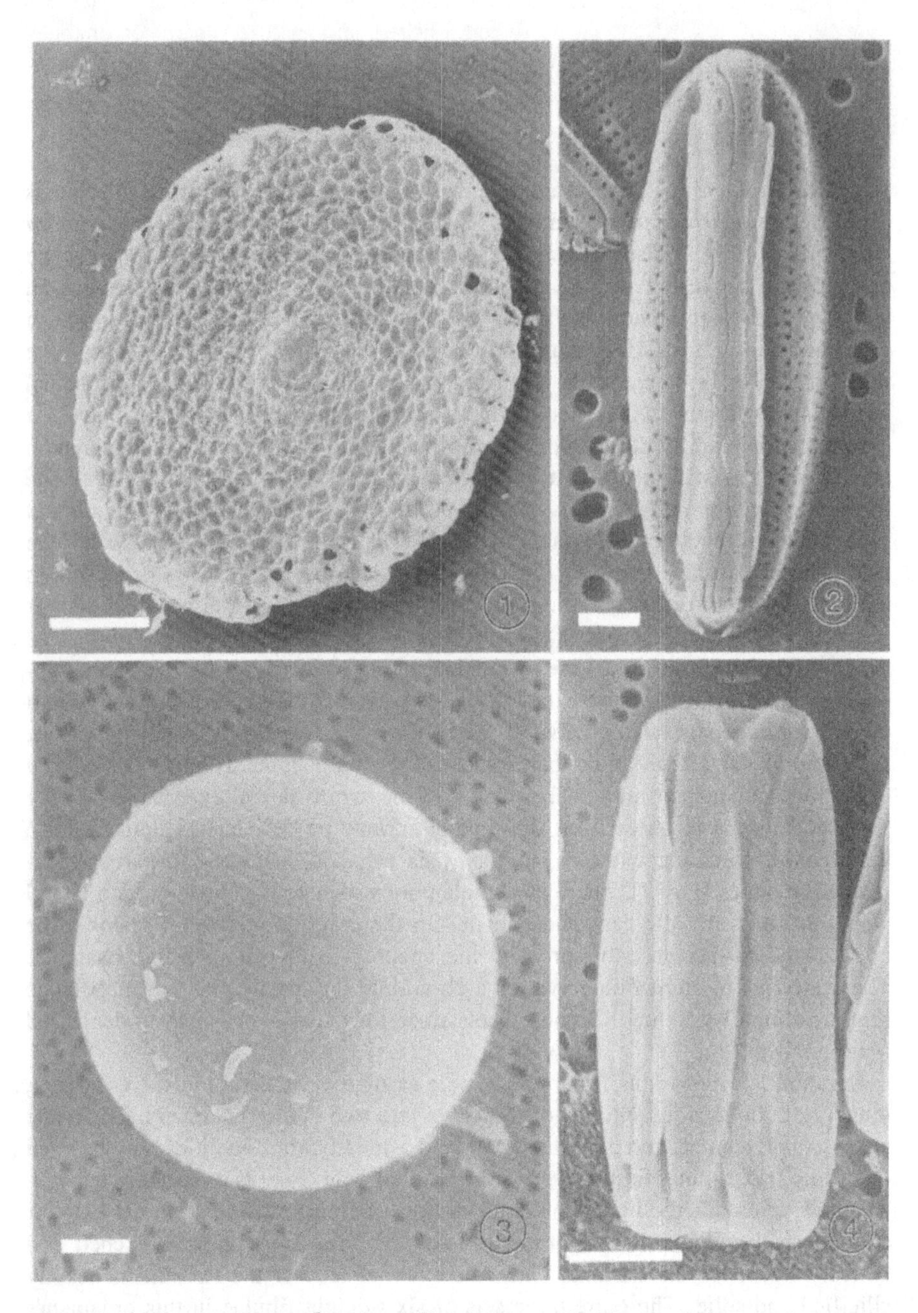

Fig. 1-4. 1 *Sorites marginalis*, from Conch Key, Florida, a foraminiferal host for endosymbiotic dinoflagellates. Scale = 200 µm. **2** Valve view of new endosymbiont genus. Scale = 1 µm. **3** Vegetative stage of endosymbiotic dinoflagellate isolated from *Sorites marginalis*, from Conch Key, Florida. Scale = 2 µm. **4** Singular view of new endosymbiont genus. Scale = 1 µm

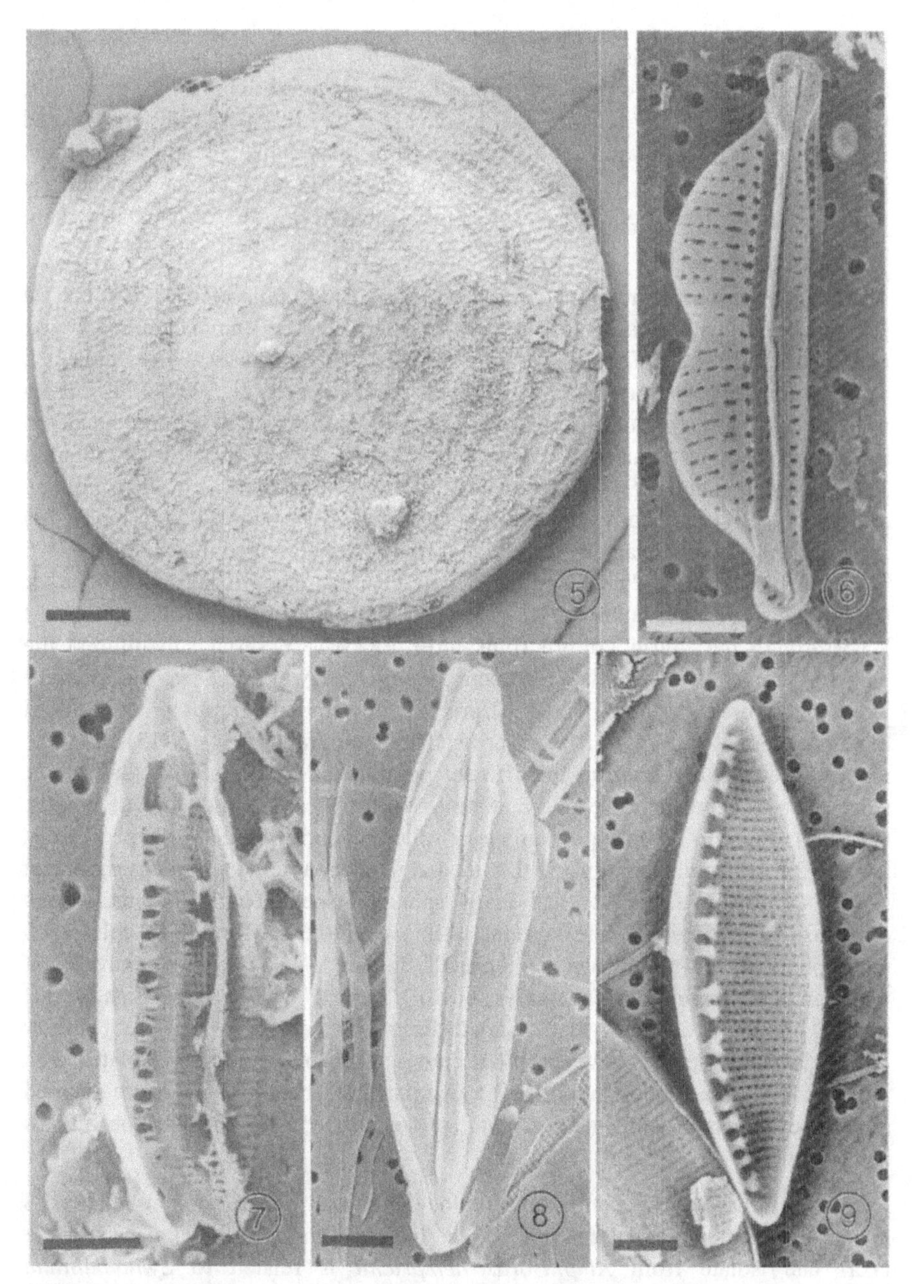

Fig. 5-9. 5 *Cyclopeus carpenterii* host of new endosymbiont genus. Scale = 5 μm. **6** *Amphora bigibba*, an endosymbiotic diatom isolated from *Cyclopeus carpenterii*. Scale = 1 mm. **7** A partially digested frustule of a new endosymbiont genus isolated from *Cyclopeus carpenterii*. Scale = 2 μm. **8** Valve view of a new endosymbiont genus isolated from *Amphistegina gibbosa* from Conch Key, Florida. Scale = 2 μm. **9** A variant of *Nitzschia frustulum* var. *symbiotica* isolated from *Amphistegina gibbosa* from Conch Key, Florida. Scale = 2 μm

Since transfers of the isolates from *Amphistegina gibbosa* did not grow, and we used the remainder of the primary culture to prepare microscope slides and stubs, we had little opportunity to study the detailed structure of the diatoms we isolated from that host. The diatoms we isolated from specimens 35 and 36 seemed identical (Fig. 8). We believe that there is a reasonable resemblance between the forms isolated from both hosts (cf. Fig. 2 with Fig. 8). Perhaps, they are congeneric.

***Nitzschia frustulum* var. *symbiotica* Complex.** The populations of diatoms we isolated from most specimens of the two Caribbean hosts, *Amphistegina gibbosa* and *Heterostegina antillarum*, were quite different from those we had collected earlier from other hosts at Indo-Pacific field stations (Lee et al. 1989, 1992). Even in primary isolation cultures, many diatoms had valve structure, often considered as abnormalities in cultured diatoms (e.g., Cholnoky-Pfannkuche 1971; Schmid 1979; Estes and Dute 1994). The range of variation we found was in the following characters: (1) striae count; (2)keel punctae count; (3) position of the raphe; (4) form of the raphe; (5) shape of the cell; (6) fibular structure and distribution.

To see whether these "abnormalities" were characteristic of the populations, or only found in primary cultures, we examined the organisms in serial cultures. At least through the fourth culture (plus three PRIMARY serial subcultures) the cultures seemed to faithfully represent the picture we found in the primary isolation cultures. Figures 10 and 12 are low power photographs of cultures of diatoms isolated from two different hosts. The photographs (Figs. 9-13) show the ranges of character variation in different populations of symbionts. Culture 27 (Fig. 11) was representative of populations with low striae and punctae counts. Culture 1 (Fig. 10) had a high number of spherical and irregularly-shaped forms. Culture 3 had many specimens with a medial displaced raphe and many spherical forms. Culture 23 (Fig. 12) had many spherical specimens with circular and otherwise "abnormal" raphae systems. Almost all the specimens in culture 40 were spherical or oval with very "abnormal" fibulae. If we look at the extremes of variation in morphological characteristics of "normal" valves (e.g., Figs. 9, 11, and 13) one might conclude that we are dealing with a number of different closely related lancelate species. The problem comes when one examines isolates with intermediate characters or "abnormal" valve patterns. It is clear that more detailed analysis of the small lancelate *Nitzschia* sp. we isolated from the Caribbean hosts is in order.

Identity of the *Sorites* Symbiont. Phylogenetic analysis of zooxanthellae 18S rDNA sequences indicates that none of the foraminiferal dinoflagellates are sister taxa. The isolate from *Amphisorus hemprichii* is related to *Symbiodinium microadriaticum* (isolated from the upside-down jellyfish *Cassiopeia*) and the isolate from *Marginopora kudakajimensis* is sister to an isolate from an Hawaiian anemone. The 18S rDNA sequence from the *Sorites marginalis* isolate is shown in Fig. 14. Initial sequence comparisons suggest that the dinoflagellate isolated from

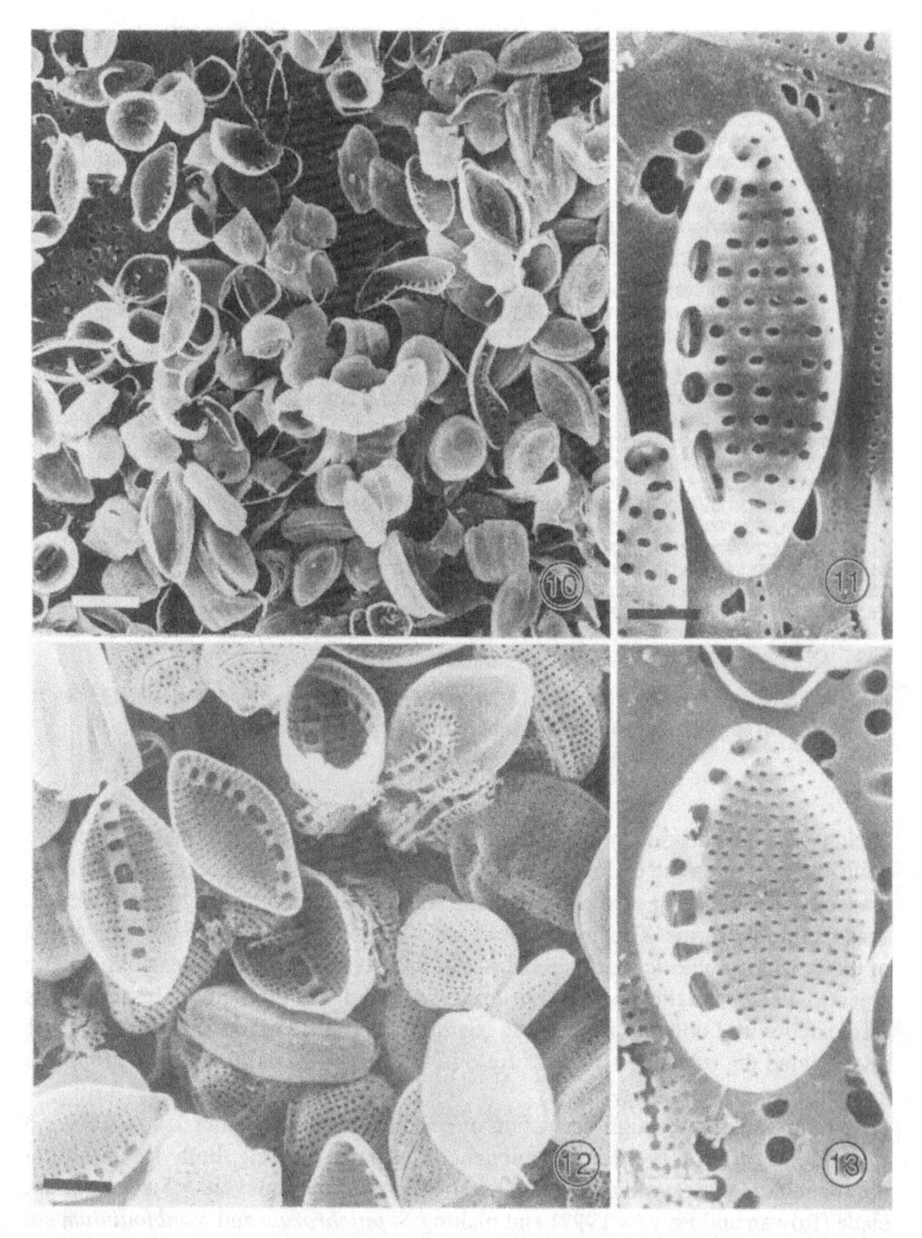

Fig. 10-13. 10 A variant (Culture 1) of *Nitzschia frustulum* var. *symbiotica* isolated from *Amphistegina gibbosa* from Conch Key, Florida. Scale = 5 µm. **11** A variant (Culture 27) of *Nitzschia frustulum* var. *symbiotica* isolated from *Amphistegina gibbosa* from Conch Key, Florida. Scale = 1 µm. **12** A variant (Culture 23) of *Nitzschia frustulum* var. *symbiotica* isolated from *Amphistegina gibbosa* from Conch Key, Florida. Scale = 5 µm. **13** A variant (Culture 40) of *Nitzschia frustulum* var. *symbiotica* isolated from *Amphistegina gibbosa* from Conch Key, Florida. Scale = 1 µm

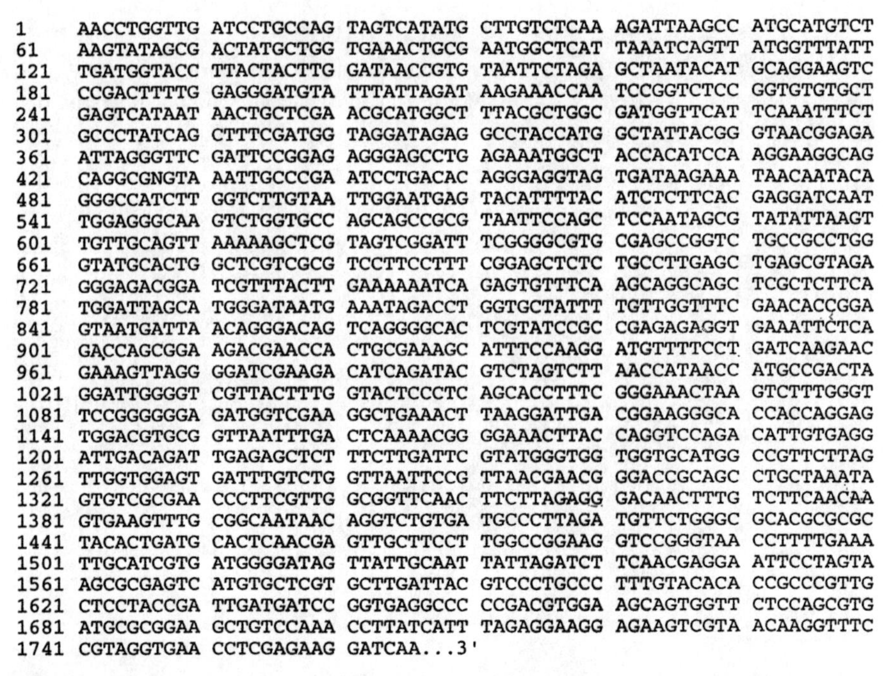

```
1    AACCTGGTTG ATCCTGCCAG TAGTCATATG CTTGTCTCAA AGATTAAGCC ATGCATGTCT
61   AAGTATAGCG ACTATGCTGG TGAAACTGCG AATGGCTCAT TAAATCAGTT ATGGTTTATT
121  TGATGGTACC TTACTACTTG GATAACCGTG TAATTCTAGA GCTAATACAT GCAGGAAGTC
181  CCGACTTTTG GAGGGATGTA TTTATTAGAT AAGAAACCAA TCCGGTCTCC GGTGTGTGCT
241  GAGTCATAAT AACTGCTCGA ACGCATGGCT TTACGCTGGC GATGGTTCAT TCAAATTTCT
301  GCCCTATCAG CTTTCGATGG TAGGATAGAG GCCTACCATG GCTATTACGG GTAACGGAGA
361  ATTAGGGTTC GATTCCGGAG AGGGAGCCTG AGAAATGGCT ACCACATCCA AGGAAGGCAG
421  CAGGCGNGTA AATTGCCCGA ATCCTGACAC AGGGAGGTAG TGATAAGAAA TAACAATACA
481  GGGCCATCTT GGTCTTGTAA TTGGAATGAG TACATTTTAC ATCTCTTCAC GAGGATCAAT
541  TGGAGGGCAA GTCTGGTGCC AGCAGCCGCG TAATTCCAGC TCCAATAGCG TATATTAAGT
601  TGTTGCAGTT AAAAAGCTCG TAGTCGGATT TCGGGGCGTG CGAGCCGGTC TGCCGCCTGG
661  GTATGCACTG GCTCGTCGCG TCCTTCCTTT CGGAGCTCTC TGCCTTGAGC TGAGCGTAGA
721  GGGAGACGGA TCGTTTACTT GAAAAAATCA GAGTGTTTCA AGCAGGCAGC TCGCTCTTCA
781  TGGATTAGCA TGGGATAATG AAATAGACCT GGTGCTATTT TGTTGGTTTC GAACACCGGA
841  GTAATGATTA ACAGGGACAG TCAGGGGCAC TCGTATCCGC CGAGAGAGGT GAAATTCTCA
901  GACCAGCGGA AGACGAACCA CTGCGAAAGC ATTTCCAAGG ATGTTTTCCT GATCAAGAAC
961  GAAAGTTAGG GGATCGAAGA CATCAGATAC GTCTAGTCTT AACCATAACC ATGCCGACTA
1021 GGATTGGGGT CGTTACTTTG GTACTCCCTC AGCACCTTTC GGGAAACTAA GTCTTTGGGT
1081 TCCGGGGGGA GATGGTCGAA GGCTGAAACT TAAGGATTGA CGGAAGGGCA CCACCAGGAG
1141 TGGACGTGCG GTTAATTTGA CTCAAAACGG GGAAACTTAC CAGGTCCAGA CATTGTGAGG
1201 ATTGACAGAT TGAGAGCTCT TTCTTGATTC GTATGGGTGG TGGTGCATGG CCGTTCTTAG
1261 TTGGTGGAGT GATTTGTCTG GTTAATTCCG TTAACGAACG GGACCGCAGC CTGCTAAATA
1321 GTGTCGCGAA CCCTTCGTTG GCGGTTCAAC TTCTTAGAGG GACAACTTTG TCTTCAACAA
1381 GTGAAGTTTG CGGCAATAAC AGGTCTGTGA TGCCCTTAGA TGTTCTGGGC GCACGCGCGC
1441 TACACTGATG CACTCAACGA GTTGCTTCCT TGGCCGGAAG GTCCGGGTAA CCTTTTGAAA
1501 TTGCATCGTG ATGGGGATAG TTATTGCAAT TATTAGATCT TCAACGAGGA ATTCCTAGTA
1561 AGCGCGAGTC ATGTGCTCGT GCTTGATTAC GTCCCTGCCC TTTGTACACA CCGCCCGTTG
1621 CTCCTACCGA TTGATGATCC GGTGAGGCCC CCGACGTGGA AGCAGTGGTT CTCCAGCGTG
1681 ATGCGCGGAA GCTGTCCAAA CCTTATCATT TAGAGGAAGG AGAAGTCGTA ACAAGGTTTC
1741 CGTAGGTGAA CCTCGAGAAG GATCAA...3'
```

Fig. 14. Small ribosomal subunit RNA DNA-sequence of the symbiotic algae *Symbiodinium* sp. isolated from the Caribbean host *Sorites marginalis*

Sorites lies outside the *Symbiodinium* clade. This may be reasonable since we have not yet gotten the endosymbiont to produce a motile stage in culture. We are presently manipulating media components and cultural conditions to see if we can produce motile forms. The nonmotile stage of the *Sorites* isolate (Fig. 3) is typical of many endosymbiotic dinoflagellates, but it is so simple that it does not give any clues to its systematic position. In order to avoid the problematic assumptions associated with bootstrap resampling (Sanderson 1989; Hillis and Bull 1993), we investigated the robustness of our phylogenetic hypothesis for the *Marginopora kudakajimensis* and *Amphisorus*
hemprichii isolates through derivation of Bremer support (Bremer 1988; Källersjö et al. 1992) There is strong phylogenetic support placing both foraminiferal dinoflagellate endosymbiotic sequences within a monophyletic *Symbiodinium* clade (Rowan and Powers 1992) and making *S. pulchrorum* and *Symbiodinium* sp. (Kudaka Jima) sister taxa. Strict consensus of the 245 phylogenetic hypotheses at least 16 steps longer than our most parsimonious tree still resolve the *Symbiodinium* clade and the early branching of *S. pulchrorum* and *Symbiodinium* sp. (Kudaka Jima) (Lee et al. 1995). Much more work (e.g., karyotype, morphological and fine structural examination of the motile stage, more extensive 18S rDNA molecular comparisons) remains before we can place the symbiont

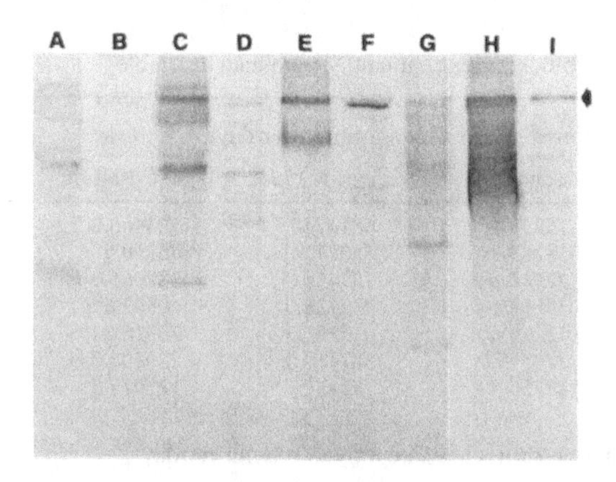

Fig. 15. Immunoblot, probed with anti-*F. shiloi* polyserum, of surface proteins from nonsymbiotic diatom: **A** *A. tenerrima* variety I, and symbiotic diatoms: **B** *N. hanseniana*, **C** *C. andersonii*, **D** *A. roettgeri*, **E** *A. tenerrima*, **F** *A.* sp. (*halamphora*), **G** *N. muscatinei*, **H** *F. shiloi*, **I** *N. laevis*. Arrow indicating the common protein band of 104 kDa molecular weight exclusively shared by all symbionts

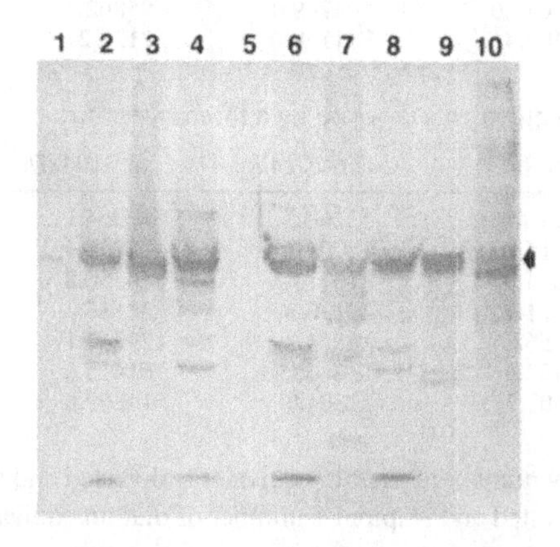

Fig. 16. Immunoblot, probed with anti-*A. tenerrima* polyserum, of surface proteins from nonsymbiotic diatoms: **2** *A. tenerrima* variety I, **3** *A. luciae* variety II, **4** *N. laevis* variety, **5** *N. viminoides*, and symbiotic diatoms: **6** *F. shiloi*, **7** *N. laevis*, **8** *A. roettgeri*, **9** *C. andersonii*, **10** *A. tenerrima*. **1** Immunobolt probed with anti-66 kDa antibody elute. Arrow indicating the 66kDa-polypeptide band shared by both symbiotic and nonsymbiotic diatom species

from *Sorites marginalis* in the systematic constellation of endosymbiotic dinoflagellates.

Western Blots. The results showed that these four polysera recognized many different proteins on the surfaces of the diatoms. The blot with the antiserum raised against *F. shiloi* (Fig. 15) recognized a polypeptide band of about 104 kDa molecular weight. It was only found on the symbiotic species we tested. None of

Table 1. Result of blocking experiment (* no data available)

A. Number of diatoms ingested by a foraminifer/day

Diatom	Control	anti-*F. shiloi*	anti-104kDa	anti-66kDa
N. laevis	15459.1	48147.0	183460.0	26269.5
A. roettgeri	18183.8	55877.4	45260.5	22145.4
XD32	27045.6	107456.0	222065.0	100723.0
A201	32117.4	232738.0	40803.6	51817.4
F. shiloi	130745.0	61732.3	70316.8	*
Navicula sp.	220519.0	387410.0	220697.0	*
X72	82756.9	77245.5	121919.0	*

B. Number of diatom equivalents kept by a foraminifer/day

Diatom	Control	anti-*F. shiloi*	anti-104kDa	anti-66kDa
N. laevis	9279.14	6409.9	7551.0	5495.8
A. roettgeri	7293.4	6481.42	7169.08	7053.3
XD32	7739.3	3865.3	4561.8	3398.8
A201	10637.3	8467.7	7175.5	4957.6
F. shiloi	68423.2	18936.3	30287.6	*
Navicula sp.	150647.0	104719.0	95862.3	*
X72	10400.4	7711.9	8126.2	*

C. Number of diatom equivalents digested by a foraminifer/day

Diatom	Control	anti-*F. shiloi*	anti-104kDa	anti-66kDa
N. laevis	6024.1	12268.3	10964.1	9554.21
A. roettgeri	10915.2	12009.2	15755.4	8644.33
XD32	19054.9	31071.4	16356.8	16372.7
A201	20821.1	29130.3	22113.3	12152.1
F. shiloi	28856.8	16874.3	29267.1	*
Navicula sp.	50957.5	112462.0	61416.0	*
X72	40601.3	11801.0	14809.6	*

D. Digestion rate: daily number of diatom equivalents digested and respired / (number of diatom equivalents digested and respired + number of diatom equivalents captured)

Diatom	Control	anti-*F. shiloi*	anti-104kDa	anti-66kDa
N. laevis0	0.393649	0.6563	0.592371	0.631406
A. roettgeri	0.599454	0.649812	0.687274	0.551323
XD32	0.711157	0.793553	0.795591	0.829759
A201	0.661861	0.774801	0.75284	0.710811
F. shiloi	0.296637	0.47121	0.491433	*
Navicula sp.	0.252759	0.517828	0.390492	*
X72	0.796077	0.604963	0.649765	*

the nonsymbiotic species had this protein. This could be one of the symbiotic markers we are looking for. The blot with anti-*Nitzschia frustulum* var. *symbiotica* antiserum showed that a band of high molecular weight (113 kDa) was shared by four symbiotic species, *F. shiloi, Cocconeis andersonii, Amphora* sp. (*halampho-*

ra), and *Amphora tenerrima*, and two nonsymbionts, *Navicula* sp. and *Navicula vimnoides*. A band of 66 kDa on the blot with anti-*Nitzschia panduriformis* antiserum was found in four symbiotic species, *F. shiloi, Cocconeis andersonii, Nitzschia laevis,* and *Amphora* sp. (*halamphora*), and three nonsymbionts, *Navicula* sp., *Navicula viminoides*, and *Amphora tenerrima* variety I. This might be a common protein shared by both symbiotic and nonsymbiotic diatoms. This band was also found on the blot with anti-*Amphora tenerrima* antiserum (Fig. 16). Because of denaturation, some proteins may not be recognized by antisera which were prepared with native diatom frustules.

Affinity Purification of Antibodies. After staining, only one band was seen on the the slice of nitrocellulose membrane which was treated with the elute. The control slices had same number of bands as usual. This means that the elute has pure antibody which is specific for one diatom polypeptide. These two elutes were used to block the special surface antigen on each species of diatom.

Blocking Experiment. The results (Table 1) seem to show that either original antiserum or purified antibodies improve both ingestion and digestion rates of the host species. The original antiserum worked a little better than individual elutes. The purified anti-104 kDa antibody elute had stronger action than the anti-66 kDa antibody possibly because the 104 kDa polypeptide is exclusively shared by symbionts and the 66 kDa polypeptide is common to both symbionts and nonsymbionts. It seems that several proteins work together to play the role during recognition. Each of them makes some contribution to the process.

More comparisons will be done in order to find more proteins which are only shared by symbionts but not nonsymbionts. Further blocking experiments will be done with both individual immuno-affinity purified antibodies, and with combinations of them, to test whether several proteins might work together in the recognition process.

References

Bollag DM, Edelstein SJ (1991) In: Protein Methods. Wiley-Liss, New York, pp 60-62
Bremer K (1988) Evolution 42: 795-803
Hallock P, Forward LB, Hansen HJ J (1986) Foram Res 16: 224-231
Hillis DM, Bull JJ (1993) Syst Biol 42: 182-192
Källersjö M, Farris JS, Kluge AG, Bult C (1992) Cladistics 8: 275-287
Lange-Bertalot H (1980) Int J Diat Res 3: 41-77
Lee JJ (1994) Proc 11th Int Diat Symp p 21-36
Lee JJ, Anderson OR (1991) In: Biology of Foraminifera. Lee JJ, Anderson OR (eds) p 157-220
Lee JJ, Faber WW, Lee RE (1991) Symbiosis 10: 47-51
Lee JJ, Faber WW, Nathanson B, Rottger R, Nishihira M, Kruger R (1992) Symbiosis 14: 265-281
Lee JJ, Freudenthal H, Kossoy V, Be A (1965) J Protozool 12: 531-542

Lee JJ, McEnery ME, ter Kuile B, Erez J, Röttger R, Rockwill RF, Faber WW Jr, Lagziel A (1989) Micropaleontol 35: 353-366
Lee JJ, Reimer CW (1983) In: "Proceedings of the 7th International Diatom Symposium, Philadelphia, August 22-27, 1982". Mann DG (ed) pp 327-343
Lee JJ, Xenophontos X (1989) Diatom Res 4: 69-77
Lee JJ, Zucker W (1969) J Protozool 16: 71-81
Medlin L, Elwood HJ, Stickel S, Sogin ML (1988) Gene 71: 491-499
Reimer CW, Lee JJ (1984) Proc Aca Nat Sci Phil 136: 194-199
Rowan R, Powers DA (1992) Proc Natl Acad Sci 89: 3639-3643
Sanderson MJ (1989) Cladistics 5: 113-129
Schmid A-MM (1979) Protoplasma 99: 99-115
Swofford DL (1991) PAUP, Phylogenetic Analysis Using Parsimony (software). Illionois Nat His Surv, Champaign, Illionois
Toneguzzo F, Glynn S, Hayday A (1988) Biotechniques 6: 460-469
Wheeler WC, Gladstein D (1992) MALIGN (alignment software). Amer Mus Nat Hist

Phylogeny Reconstruction Based on Molecular Property Patterns

W.Schmidt
Max-Delbrück-Center for Molecular Medicine, Department of Bioinformatics, Robert-Rössle-Str. 10, 13122 Berlin, Germany

Summary: An essentially new method to relate taxa of any kind by means of a predefined set of dichotomous properties (i.e., either present or absent) will be presented. The main feature of the analysis is the usage of a sophisticated distance measure to describe the pairwise dissimilarities quantitatively in dependence on the individual properties. The distance measure implies in a natural way the derivation of a tree structure by successive joining of taxa with minimum distance (neighbor joining). The distances are based on the jointly compatible properties and, consequently, they are referred to as the consensus taxa with respect to the properties. Further, the branch lengths of the resulting tree will be calculated simultaneously to the branching structure and is no longer the result of a second procedure. Moreover, the distances can be interpreted by means of a (nonprobabilistic) information concept and negative or other noninterpretable distances never occur. Generally, the proposed tree reconstruction method is not based on stochastic or other mathematical models of the underlying evolutionary processes and can be interpreted best in terms of discrete information theory. The proposed method is highly flexible and can be applied to data of various types. In the present chapter this is shown by reanalyzing published data and comparison of the trees derived with the property pattern method and with other established methods: amino acid sequences, restriction sites data, gene frequencies, and sensitivities of bacterial protein synthesis to different antibiotics. In all cases we found that the tree derived with the property pattern method is in good accordance with the biological expectation.

Introduction

A dichotomous tree is the most common way to represent phylogenetic relationships and usually it can be derived from a matrix of distances between all possible pairs of taxa. The inner nodes of trees are interpreted as indicating the fission events having taken place in the evolutionary history of the taxa under investigation. Consequently, the branch lengths of the tree should be closely related to the time periods from one branching event to the next one and to the present time. On the other hand, the final nodes of the tree represent today's taxa and, consequently, the internal nodes can be identified with (hypothetical) ancestral taxa. For instance, if the (present-time) taxa are nucleotide or amino acid sequences then the ancestral taxa should also be sequences of the corresponding type defined by the condition that the total number of substitutions to realize the observed sequences has to be minimal. Of course, the latter condition is a sound principle from the biological point of view. Nevertheless, it is artificial to some degree because it cannot be assumed that the evolutionary process is strictly governed by this principle. The more that problem will occur if other than sequence data have to be analyzed because then no "elementary" evolutionary events (as mutations, deletions, insertions in case of sequences) will be known to formulate reasonable hypotheses.

A special problem connected with the interpretation of phylogenetic trees is the identification of the branches with the time scale. Of course, one of the basic intentions of phylogenetic analyses is to gain insights about the real (mostly probable) time course of the evolution process and, doubtless, the branch lengths in the "correct" tree are highly correlated with the corresponding times, but that identification (after a suitable transformation, if necessary) requires additional hypotheses, for instance, the assumption of homogeneity of mutation rates and their independence on the time (constancy). However, assumptions of this type are questionable and, at least in some cases, far from the biological reality. Thus, it is highly desirable to have tree reconstruction methods which avoid as far as possible restricting mathematical and biologically unjustified assumptions. In that case, the interpretation of the branch lengths and their translation on the real-time axis will be a separate problem and here, depending on the data, mathematical models can be useful to specify and to quantify the underlying hypotheses.

As mentioned above, the presently used tree reconstruction algorithms are based on distance measures quantifying the relationships (dissimilarities) between the taxa under investigation. Then, usually, the tree calculation consists of two separate procedures: determination of the tree topology (branching structure) and determination of the branch lengths. In case of an additive distance measure, for instance, the branch lengths can be estimated by least-squares methods from the condition that the total length of the way within the tree connecting any two of the taxa should be (approximately) equal to the distance between these taxa. However, here also negative branch lengths can occur which obviously cannot be interpreted biologically. To avoid this one can introduce the additional condition of nonnegativity (Klotz and Blanken 1981). but this seems to be highly artificial. The deeper reason for that problem seems to be that the additivity criterion is generalized in an unjustified way. Indeed, additivity should include only such branch sequences connecting present-time taxa with any of their ancestors and not any other ways. However, of course, the distances to the ancestors are not known and, consequently, the available data are insufficient to solve the problem without introducing additional estimation principles (maximum parsimony, minimum total length, etc.). As already noted, such more or less artificial estimation principles can be the reason for incalculable biases of the estimated tree structure.

We propose an essentially new method to reconstruct phylogenetic trees avoiding the above-mentioned difficulties completely. It is based on a special structure of the initial data describing the taxa by means of a predefined totality of discriminating properties which are either present or absent for any given taxon. Thus, the initial data can be represented by means of a matrix with elements 0 and 1 indicating the absence and the presence of a property (represented by the rows) for a given taxon (the columns of the matrix). At first glance, this data structure seems to be a heavy restriction of the applicability of the method, but we will show that also continuous data can be described by means of appropriately defined property patterns.

The distance between any two taxa has to be evaluated on the basis of those properties which are present for only one of both. The simplest way to do that is to define the distance simply as the total number of properties in which the taxa are different (Hamming distance). However, this measure has (at least) two important disadvantages. Firstly, it is completely unspecific, i.e., all properties contribute equally to the measure and, for instance, there is no difference between highly informative and redundant and dependent properties. Secondly, it is problematic to use it for phylogenetic studies because the calculation of distances to ancestral taxa would presuppose that the properties of the ancestors are completely known. Obviously that assumption is unacceptable without any additional principle, e.g., to postulate minimum total number of property changes. The proposed distance measure is somewhat more complicated, but it is based strictly upon the common properties of the considered taxa whereby the distribution of these properties within the totality of all taxa is taken into consideration. Indeed, this leads to a different evaluation of the properties. In principle, the tree reconstruction algorithm is predetermined by the distance measure and, as a byproduct of that algorithm, the occurrence of noninterpretable (e.g., negative) branch lengths is impossible. The branch length itself can be interpreted as information loss if going back from any taxon to any one of its ancestors whereby the information is the natural measure of the degree of identifiability of the taxon from its (known) properties.

Mathematical Background and Algorithm

We give only a short summary of the mathematical framework. A more detailed representation can be found in the paper by Schmidt and Müller (1995; in press).

Let S be the set of taxa (species, populations, gene or protein sequences, etc.) and let P be the set of properties. We assume that each property p is dichotomous, meaning that for each taxon p is either present or absent. Hence, each taxon X defines a subset P_X of S consisting of those properties which are present on X. The basic mathematical entities on which the phylogenetic analysis will be based are the partitions of the set S here denoted by the letter u and indexed if necessary. A partition u of S is defined as a subdivision of S into some nonempty and jointly distinct classes. The totality of partitions of S forms a mathematical structure which is denoted as lattice (cf. Birkhoff 1948). The binary operations in that lattice we denote as addition and multiplication, respectively. Additionally, in that structure we have a partial ordering. Here, the relation $u_1 < u_2$ is equivalent to the condition that each class of u_1 is completely contained (subset) in a unique class of u_2 (dominance relation). With respect to this ordering, the product and the sum of partitions u_1 and u_2 are defined qualitatively: $u_1 \cdot u_2$ is the largest partition which is dominated by both factors and, equivalently, $u_1 + u_2$ is the smallest partition which dominates both of the summands.

With these definitions, for any taxon X from the corresponding property set P_x we derive the associated partition u_x: Each class of u_x consists of elements of S which are identical with respect to the properties of X, meaning that properties which are not present on X will not be considered. By definition, each property p defines a dichotomous partition u_p (split, cf. Bandelt and Dress 1992) of S consisting of exactly two classes containing the elements having and having not that property. As a basic statement we note that all of the splits derived from the properties of a taxon X dominate the associated partition u_x. The opposite statement is not true, generally: depending on the internal structure of the property set, it is possible that some of the property-associated splits dominate the partition u_x associated with a taxon X not having that property. Thus, we denote properties of this type to be compatible with the taxon X. Consequently, the set P_x of the properties of X is a (not necessarily proper) subset of the set c_x of the X-compatible properties.

Now, the generalization of the notion "compatibility" is straightforward: The property p is said to be compatible with a given partition u if $u < u_p$ and the corresponding set of properties is denoted by C(u). Then, we have the central theorem (Schmidt and Müller 1995: theorem 2) justifying the property pattern method as tree reconstruction method: If u_1 and u_2 are arbitrary partitions of S then the compatibility set of the sum partition is given by the equation

$$C(u_1 + u_2) = C(u_1) * C(u_2),$$

("*" = set intersection).

From that equation we conclude that the sum partition $u_1 + u_2$ can be interpreted as consensus partition of u_1 and u_2 with respect to the property set. With respect to the phylogenetic tree, the sum partitions represent the corresponding ancestral taxa which are represented by the partitions $u_1$and u_2, respectively. We remark that this does not include the precise description of the ancestral taxa by means of their complete property set (present or only compatible). The only assumption which is underlying here is that the jointly compatible properties of any two taxa are also compatible with their common ancestor. This seems to be justified from the biological point of view.

The tree calculation is based on a specific distance measure applying within the totality of partitions of the set S. The precise definitions are given elsewhere

(Schmidt and Müller 1995). Here we note that the distance between partitions $u_1$and u_2 is equal to the sum of the distances of u_1 and u_2 to the consensus partition $u_1 + u_2$, and these individual distances allow for a very natural interpretation in terms of information theory. Thus, the distance is defined as the total information loss with respect to identifiability if going back from taxa to their common ancestor.

With these preparations, the tree calculation is based on a straightforward algorithm according to the neighbor-joining principle (Saitou and Nei 1987). The first step is to calculate the associated partitions to represent the present-time taxa for the subsequent analysis. Then, as the second step, calculate the pairwise distances. In the following step, determine the minimum distance pair of taxa and replace the members of that pair by their "consensus" taxon defined as the corresponding sum partition. Note that this step can be ambiguous, possibly as a consequence of a too small or too uninformative property set. After that step, modify the distance matrix by calculating the distances of all other taxa to this newly defined taxon. The mentioned steps must be repeated until there is only one taxon. Obviously, the described algorithm calculates both the branching structure of the tree and the branch lengths. Indeed, the branch lengths are the distances to the consensus partition in the corresponding step of the procedure. As already mentioned, the length of any branch is the information difference between the taxon and its immediate ancestor which are connected by that branch. Thus, as a byproduct, there is never a problem concerning the interpretability of these branch lengths.

Applications

As noted in the preceding section, the proposed property pattern method is based on dichotomous character data and data of that type are frequently used in phylogenetic studies. Moreover, also more complex data can be described in that way by introducing a number of formal properties. For instance, one can proceed as follows. Let p be a property with arbitrary values (e.g., real numbers). Then, if the number of taxa is k, obviously only the corresponding k values $f_1, \ldots, f_k$ of the given property are of interest for the phylogenetic analysis. However, these values can be represented by binary coding using $r = [{}^2\log(k)] + 1$ digits ([..] = integral part). Now, we define that the property p_i $(1 \le i \le r)$ is present if and only if digit i in that coding is 1. Note that the number r can be reduced if not all values f_i are different. We remark also that the indicated way is not unique but from the general theory (Schmidt and Müller 1995) it follows that the distance measure is essentially independent of that coding. Consequently, the fact that the availability of binary character data is a prerequisite of the property pattern method is no essential restriction. In the following we show how the method can be applied to data of different types.

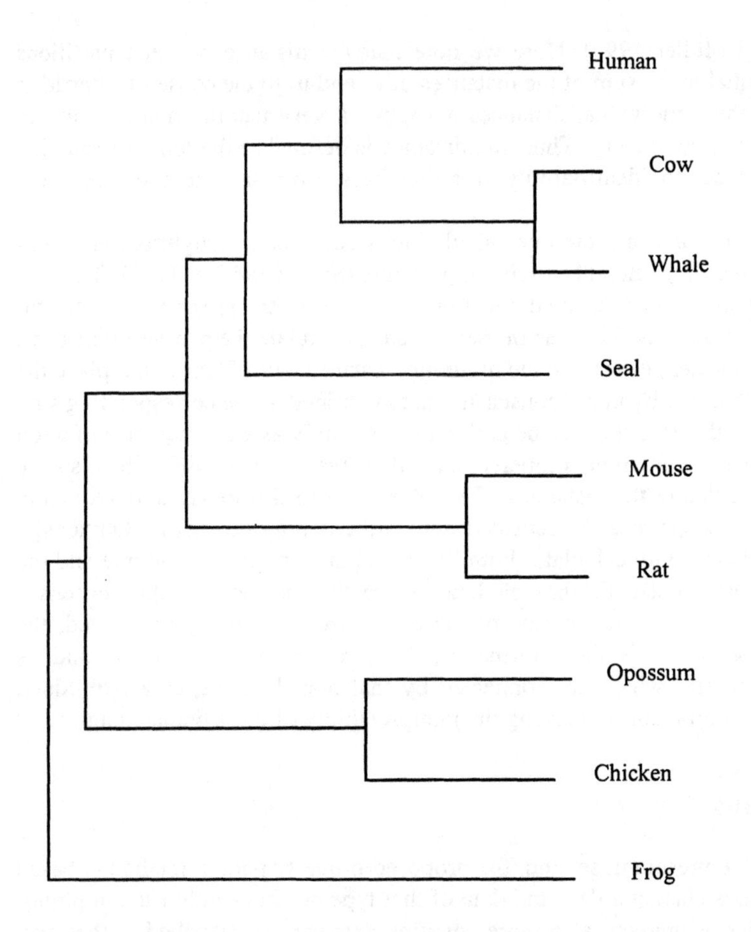

Fig. 1. Property pattern tree of mtDNA-encoded cytochrome b (length of the amino acid sequence: 376) from 9 taxa. Positions with gaps and regions where the alignment was ambiguous were excluded from the analyses. The ML tree derived from 13 proteins including cyto-chrome b is shown in Cao et al. (1994). The data sources were also cited there

Amino Acid Sequences

In the following example we used four (hydrophobicity, charge, small, aromatic) of the 11 steric and biophysical amino acid properties proposed by Taylor (1986), and Zvelebil et al. (1987). Then the phylogenetic relationships among Primates (human), Artiodactyla (cow), Cetacea (whale), Carnivora (seal), and Rodentia (mouse and rat) were estimated from the inferred amino acid sequences (cytochrome b, total length 376) of the mitochondrial genomes with the property pattern method. Additionally, the sequences from Marsupialia (opossum), Aves (chicken), and Amphibia (frog) were taken as an outgroup. More details concerning these data can be found in the paper of Cao et al. (1994). The property

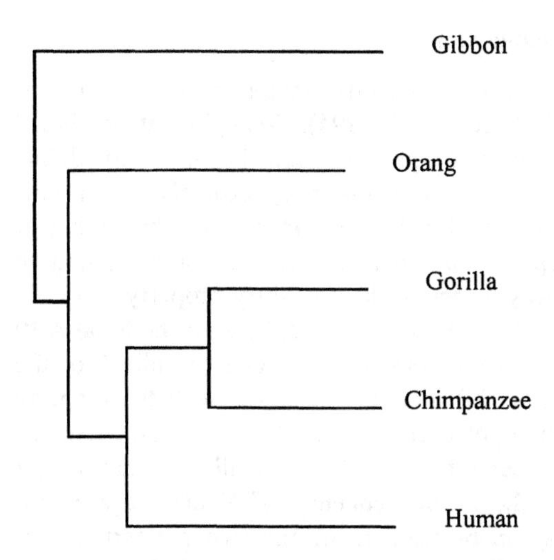

Fig. 2. Phylogenetic tree derived from the restriction sites data (Table 1 in Felsenstein 1992) for five hominoid species with the property pattern method. The ML tree of Felsenstein and the property pattern tree are essentially identical

table of the nine taxa has been constructed as follows: Every amino acid defines four steric and biophysical property values (0 or 1) and the totality of these four-tupels form the complete property table. Thus, we have $4 \cdot 376 = 1504$ properties and, consequently, the property table has 1504 lines and 9 columns. From that table the tree can be derived with our computer program PROPTREE. The result is shown in Fig. 1. Obviously, the tree structure is in good coincidence with the biological expectations. Nevertheless, there are some minor differences to the ML (maximum likelihood) tree in the paper of Cao et al. (1994), which is based on a totality of 13 amino acid sequences.

Phylogenies From Restriction Sites

We reanalyzed restriction sites data for five hominoid species (Human, Chimpanzee, Gorilla, Orang, Gibbon) with the property pattern method. The data of Ferris et al. (1981) (represented in Table 1 of Felsenstein 1992) can be analyzed by maximum likelihood using the PHYLIP program package, and the result is shown in Fig. 1 of Felsenstein's paper. Because the data are dichotomous ("+" is the presence of a site, "-" its absence) the data can be used immediately by taking "+" as 1 and "-" as 0. The resulting tree obtained with the property pattern method is shown in Fig. 2, and obviously it is nearly identical (also with respect to the relative branch lengths) to the ML tree obtained by Felsenstein. We note that the computational amount of the property pattern method was significantly smaller (approximately 10 min on a PC).

Phylogenetic Analysis of Gene Frequencies

Gene frequencies are classical parameters to study phylogenetic relationships between human populations (Cavalli-Sforza et al. 1994). To apply distance-based tree reconstruction methods the gene frequencies can be used to define appropriate distance measures (e.g., Mahalanobis distance) possibly after a suitable transformation of the raw data. As already mentioned at the beginning of "Applications" also data of this type can be analyzed with the property pattern method, but we used a simplified way to derive the necessary property table. To show this we considered the allele frequencies (A, A1, A2, B, 0) of the usual AB0 blood group system. The data were taken from the table in Appendix 1 of the monograph of Cavalli-Sforza et al. (1994). For the reanalysis with the property pattern method we chose 36 human populations (Fig. 3). To define the necessary properties we proceeded as follows: Let X be one of the five alleles. Then within the 36 populations we determined the mean frequency of X and its standard deviation. These two parameters can be used to define two properties. By definition, for a given population the first property is present if its X frequency lies within the 1-sigma range around the mean; in the opposite case the property is assumed to be absent. For the second property we define that it is present if the X frequency is not larger than the mean, and absent in the other case. Consequently, because we have 5 alleles, we obtain a totality of 10 properties describing the 36 populations. Thus, the property pattern method can be applied on that property table. The result obtained with the program PROPTREE is shown in Fig. 3. As can be seen by comparison the resulting tree is highly similar to the trees in Cavalli-Sforza et al. (1994), Figs. 2.3.2.A and 2.3.2.B, which are based on 120 allele frequencies from different gene systems including the AB0 gene using two different distances measures.

Phylogenies of Bacterial Orders From Sensitivities to Antibiotics

Recently, Sanz et al. (1994) have published the results of a phylogenetic analysis of data describing the sensitivity of the cell-free protein synthesis of some members of the archaeal order Sulfolobales to 40 antibiotics with different specificities. The sensitivity patterns were compared with those of Sulfolobus solfataricus and other archaeal, bacterial, and eukaryotic systems. The sensitivity results have been used to estimate the phylogenetic relationships among the members of the order Sulfolobales, and the evolutionary significance of the results has been analyzed in the context of the phylogenetic position of this group of extreme thermophilic microorganisms. A total of 40 antibiotics consisting of three subgroups (only bacterial-targeted antibiotics, only eukaryotic-targeted antibiotics, antibiotics affecting both bacteria and eukarya) have been used for the analysis. The intensity of the protein synthesis inhibition has been classified according to four levels (++,+,+-,-) defined in Table 1 of the paper of Sanz et al. (1994). To apply the property pattern method to these data, the four levels of reaction have to be described by properties. This can be done in several ways.

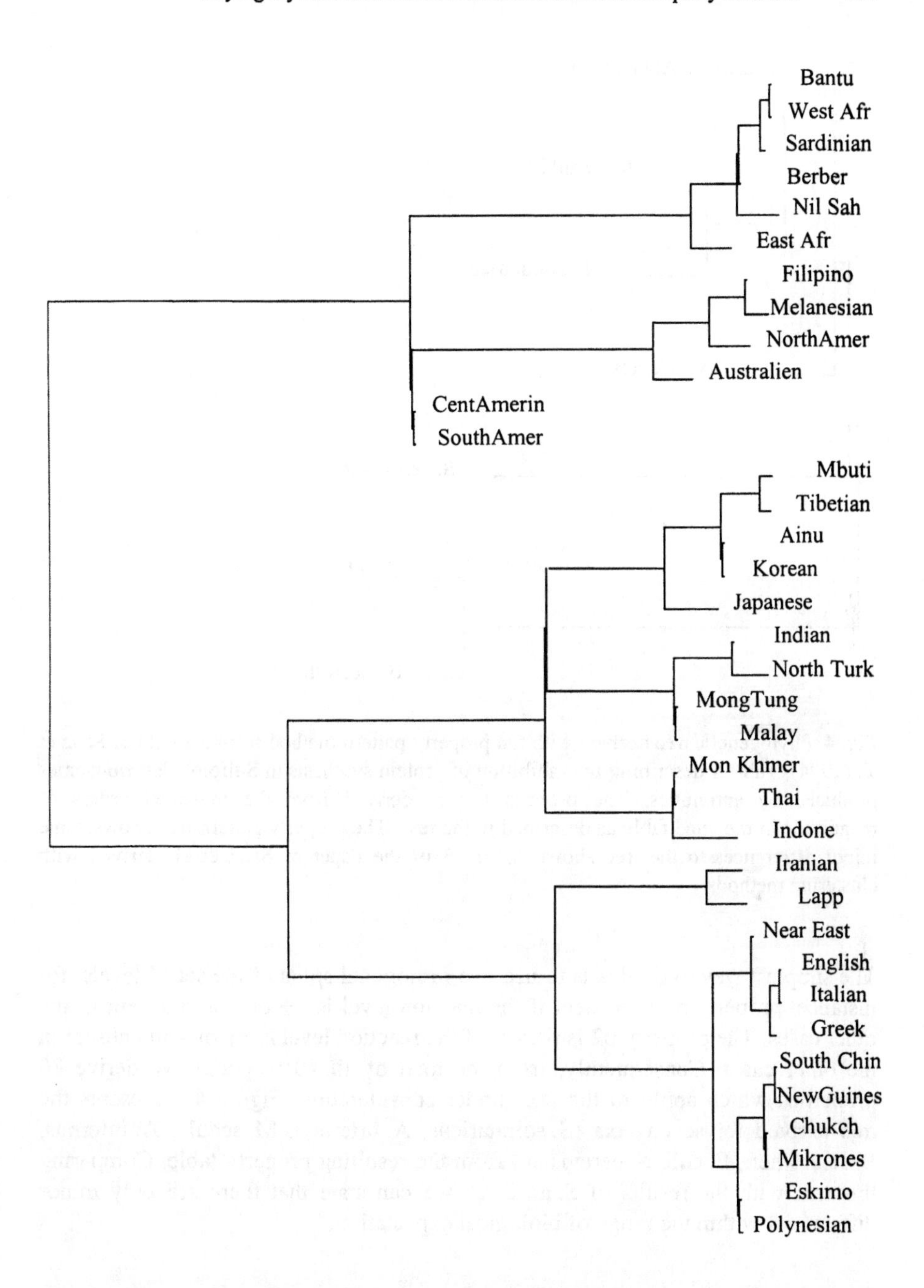

Fig. 3. Property pattern tree of 36 human populations derived from the allele frequencies of the AB0 blood group system. The required properties were derived from the statistical distributions of these frequencies (see text). The tree structure shows high similarity to the trees derived from many genes (including AB0) with other methods on the basis of Nei's distance (Nei 1987). For comparison of the results see Cavalli-Sforza et al. (1994)

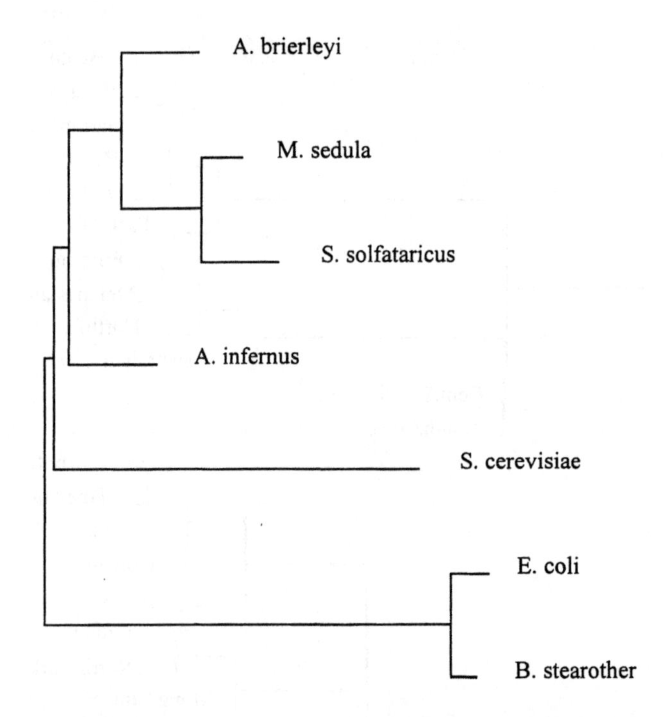

Fig. 4. Phylogenetic tree derived with the property pattern method from the data of Sanz et al. (1994), Table 1 describing the inhibition of protein synthesis in Sulfolobales' ribosomes produced by antibiotics. The properties were derived from the indicated orders of magnitude in the cited table as described in the text. The property pattern tree shows some minor differences to the tree shown in Fig. 3 of the paper of Sanz et al. derived with clustering methods

The simplest way to do this is to use two orthogonal splits of the set of levels: for instance, property p1 is present if the reaction level is ++ or + and absent in the other cases. The property p2 is present if the reaction level is ++ or - and absent in the other cases. Consequently, from the total of all 40 reactions we derive 80 properties which apply to the taxa under consideration. Figure 4 represents the tree derived for seven taxa (S. solfataricus, A. brierleyi, M. sedula, A. infernus, B. stearother., E. coli, S. cerevisiae) from the resulting property table. Comparing this tree with the results of Sanz et al. we can state that there are only minor differences within the range of biological expectation.

Summary and Conclusions

The proposed tree reconstruction method differs from other presently favored methods in some essential points. Probably the main point is that the method is not based on detailed mathematical models of the underlying evolutionary processes and also not on global estimation principles and hypotheses, e.g., minimum total distances, maximum parsimony or minimum evolutionary effort (substitutions). Indeed, additional hypotheses of that kind may not only be the reason that the tree reconstruction becomes an extremely complex estimation problem but, moreover, they can lead to biased estimations. It is well-known that evolutionary processes can be more or less far from optimality with respect to these criteria. Thus, avoiding additional hypotheses in phylogenetic tree reconstructions is highly desirable. The proposed method is based on the description of the investigated taxa by means of patterns of dichotomous properties (i.e., properties which are either present or absent). Data of this type allow for the derivation of a distance measure which is highly appropriate for tree reconstructions for several reasons: Firstly, the measure itself includes the basic step of the tree calculation (neighbor joining) by defining the common reference taxon (ancestor) of a given pair of taxa as the consensus with respect to the properties and, simultaneously, the distances of the members of that pair to the common ancestor. As a consequence of this feature, the determination of the tree topology and the evaluation of the branch lengths are no longer distinct steps of the tree reconstruction. In particular, uninterpretable (e.g., negative) branch lengths can never occur.

We remark that the restriction of the method to patterns of dichotomous properties is not essential. Indeed, we have shown that data of any other type describing the taxa can be used to define properties representing the original data completely.

The distance measure itself defining both the branching structure and the branch lengths of the tree can be interpreted as an information measure. Indeed, a partition is an incomplete description of the taxa, defining only classes of taxa but not the taxa itself. In particular, in that sense the unit partition (putting all taxa into one class) has no resolution power. It follows that the mean size of the classes of a given partition is strictly related to the information content of that partition and, on the other hand, it is easy to see that the introduced distance measure is immediately derived from that mean. Especially it can be shown that the distance of a taxon to any of its ancestors is precisely the information defect between these taxa. Thus, the information measure defines the scale of the tree and its translation into the time scale has to be based on adequate mathematical models.

References

Bandelt H-J, Dress AWM (1992) Split Decomposition: A New and Useful Approach to Phylogenetic Analysis of Distance Data. Molecul Phylogenetics and Evolution, Vol 1/3: 242-252

Birkhoff G (1948) Lattice Theory. 2nd Ed, New York

Cao Y, Adachi J, Janke A, Pääbo S, Hasegawa M (1994) Phylogenetic Relationships Among Eutherian Orders Estimated from Inferred Sequences of Mitochondrial Proteins: Instability of a Tree Based on a Single Gene. J Mol Evol 39/5: 519-527

Cavalli-Sforza LL, Menozzi P, Piazza A (1994) The History and Geography of Human Genes. Princeton University Press, Princeton, New Jersey

Felsenstein J (1992) Phylogenies from Restriction Sites: A Maximum- Likelihood Approach. Evolution 46/1: 159-173

Ferris SD, Wilson AC, Brown WM (1981) Evolutionary tree for apes and humans based on cleavage maps of mitochondrial DNA. Proc Natl Acad Sci 78: 2432-2436

Nei M (1987) Molecular Evolutionary Genetics. New York, Columbia University press

Saitou N, Nei M (1987) The neighbor-joining method: A new method for reconstructing phylogenetic trees. Mol Biol Evol 4: 406-425

Sanz JL, Huber G, Huber H, Amils R (1994) Using Protein Synthesis Inhibitors to Establish the Phylogenetic Relationships of the Sulfolobales Order. J Mol Evol 39: 528-532

Schmidt W (1995) Phylogeny reconstruction of protein sequences based on amino acid properties. J Mol Evol, in press

Schmidt W, Müller E-Ch (1995) A distance measure based on binary character data and its application to phylogeny reconstruction. Bull Math Biol, in press

Taylor WR (1986) Identification of Protein Sequence Homology by Consensus Template Alignment. J Mol Biol 188: 233-258

Zvelebil MJ (1987) Prediction of Protein Secondary Structure and Active Sites Using the Alignment of Homologous Sequences. J Mol Biol 195: 957-961

2.2 Endocytobionts in Protists and Invertebrates

Symbiosis and Macromolecules

K. W. Jeon
Department of Biochemistry, University of Tennessee, Knoxville, Tennessee 37996, USA

Key words: symbiosis, macromolecules, amoeba, endosymbionts.

Summary: Several symbiont- and host-produced macromolecules are involved in the host-symbiont interactions in amoeba/X-bacteria symbiosis. These molecules appear to play some roles in the avoidance of digestion of symbionts by the host, maintenance of the structural integrity of symbiosomes, prevention of lysosomal fusion with symbiosomes, and physiological interactions between host and symbionts. However, the precise roles of these molecules are not known and further studies are needed for clarifying such roles. The amoeba-bacteria symbiosis is a good model to study roles of symbiont- and host-produced macromolecules because new symbiosis can be easily established and cellular character changes including a host's dependence on symbionts can be experimentally reproduced under laboratory conditions.

Introduction

Various symbiont- and host-produced macromolecules are known to be involved in several steps during the establishment and maintenance of symbiotic relationships (Fig. 1). In endosymbiosis, host cells usually provide a suitable "shelter" and supply important material needs for endosymbionts. Meanwhile, endosymbionts must overcome many difficulties in order to survive inside a host cell, and hence the cytoplasm of a host cell is considered to be a hostile environment for endosymbionts, including intracellular parasites (Moulder 1974; Choi et al. 1991; Morioka and Ishikawa 1992). The mechanisms whereby endosymbionts avoid digestion by their hosts and continue to multiply are of primary importance in understanding the stability of host-symbiont relationships (Joiner et al. 1990; Hall and Joiner 1991; Parham 1992). Many of the studies on intracellular symbiosis have been concentrated on the four steps involved in the establishment of a stable symbiosis, viz., initial recognition, entry of symbionts into host cells, symbionts' avoidance of digestion or ejection by the host, and sustained multiplication of both partners. However, there are still many questions unanswered about the mechanisms for each of the steps.

Once inside the host cell, symbionts must first avoid destruction by the host. Some endosymbionts are able to prevent lysosomal fusion with symbiosome membranes (Ahn and Jeon 1979, 1982; Hart and Young 1979; Pfeifer 1987; Frëhel and Rastogi 1989; Holtzman 1989; Goren and Mor 1990; Ishibashi and Arai 1990; Berger and Isberg 1993; Sadosky et al. 1993; Britton et al. 1994). Recent studies show that symbionts may also modulate acidification of symbiosomes and hence indirectly prevent lysosome-symbiosome fusion (Eissenberg et al. 1993; Sturgill-Koszycki et al. 1994). Some soluble products of bacteria also inhibit such fusion (Frenchick et al. 1985), while free Ca^{2+} is needed for lysosomal fusion (Jaconi et al. 1990), and low temperature delays the process (Haylett and Thilo 1991). The Mip protein potentiates intracellular infection of macrophages (Engleberg et al. 1989)] and protozoa by *Legionella* (Cianciotto and Fields 1992), and genes have been identified that control phagosome-lysosome fusion and organelle recruitment (Berger and Isberg 1993; Hacker et al. 1993, Sadosky et al. 1993). Mip-like protein was also found in *Chlamydia* (Lundermose et al. 1993) and a DNA fragment has been identified that is associated with entry and survival of *Mycobacterium tuberculosis* (Arruda et al. 1993).

Here, we will consider several symbiont- and host-produced macromolecules found in amoeba/X-bacteria symbiosis as a model system. In the symbiosis between the xD strain of *Amoeba proteus* and X-bacteria, the new symbionts enclosed in symbiosomes may be integrated in the host within a short period of time, bringing about cellular character changes and causing the hosts to become dependent on symbionts (Jeon 1994, 1995).

Symbiont-Produced Molecules

Several symbiont-derived macromolecules have been identified that are found not only inside symbionts but also in the host cytoplasm and on symbiosome membranes. If protein synthesis by symbionts is selectively inhibited by treating them with antibiotics, the host amoebae die (Kim and Jeon 1986), suggesting that some of the proteins play an essential role(s) in maintenance of symbiotic relationships.

The S29x Protein. Among symbiont-derived proteins, the 29-kDa protein (S29x) is most prominent (Ahn and Jeon 1983; Park and Jeon 1988, 1990; Pak and Jeon 1994). Since S29x is a major bacterial protein, we suspect that it plays an important role in the survival of X-bacteria inside amoebae. In our studies, we first established S29x to be X-bacterial in origin by following its appearance and disappearance with the presence and absence of X-bacteria in xD amoebae (Kim and Jeon 1986). Then, we cloned and sequenced X-bacteria's *s29x* gene (Pak and Jeon 1994). The gene has an open reading frame of 774 nucleotides and encodes

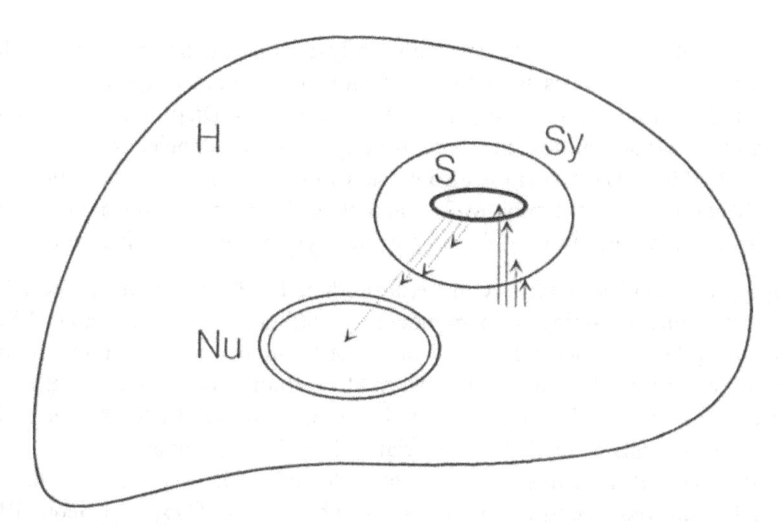

Fig. 1. Sources and destination of symbiont-produced (dotted-line arrows) and host-produced (solid-line arrows) macromolecules in an xD amoeba containing symbiosomes. H Host; Nu Nucleus; S Symbiont; Sy Symbiosome. The symbiont and organelles are not drawn to scale

258 amino acids. The native S29x protein has M_r of about 87000, a trimer of 29-kDa polypeptide (Kim and Jeon 1986).

Next, we examined the production and transport of S29x. One of the interesting properties of S29x is its ready transport into the amoeba cytoplasm across the symbiosome membrane (SM), but the mechanism for its transport across the two membrane systems is not known. The same protein synthesized in *E. coli* transformed with *s29x* is also exported to the medium (Pak and Jeon 1994). Results of pulse-labeling studies using ^{32}P-methionine in isolated X-bacteria and transformed *E. coli* showed that newly synthesized S29x starts to be exported into the medium within 5 min (J.W. Pak and K.W. Jeon, unpublished data). However, S29x does not have a characteristic signal sequence for a secreted protein. The hydropathy pattern of S29x shows more hydrophilic amino acids with no noticeable hydrophobic domains. The molecular masses of S29x determined by SDS-PAGE are the same for both S29x obtained from bacteria (X-bacteria living inside xD amoebae or transformed *E. coli* grown in vitro) and S29x isolated from the amoeba cytoplasm or the culture medium, respectively. This indicated that S29x was not altered during transport across membranes (Pak and Jeon 1994). Thus, it appears that S29x is transported across bacterial and symbiosome membranes by a mechanism not involving the signal peptide.

A 96-kDa Symbiosome-Membrane Protein. The fact that symbiosome membranes are initially derived from the plasma membrane via phagolysosomal membranes but do not fuse with lysosomes suggests that they contain some components that may actively interfere with the lysosomal fusion as a "fusion-blocking" factor. This contrasts with routine fusion of ordinary phagosomes with lysosomes. One of the unique antigens on the SM is a 96-kDa protein as determined by indirect immunofluorescence microscopy and immunoelectron microscopy (Ahn et al. 1990). The protein is also present in the symbiotic bacteria but not on any cell component of symbiont-free amoebae. Thus, the 96-kDa protein may play a role in the prevention of fusion between symbiosomes and lysosomes.

Lipopolysaccharides (LPS). X-bacteria-produced LPS is found on the SM as shown by immunostaining with monoclonal antibodies (mAbs) against LPS of X-bacteria (Choi and Jeon 1992). When fixed and permeabilized xD amoebae were immunostained with anti-LPS mAbs, fluorescence was confined to symbiosomes. Isolated X-bacteria were also immunostained by the mAbs, and the mAbs also recognized purified X-bacterial LPS on immunoblots.

If a purified anti-LPS mAb was injected into individual xD amoebae, the anti-LPS mAb showed specific staining of symbiosomes (Choi and Jeon 1992), indicating that X-bacterial LPS were exposed on the cytoplasmic side of symbiosome membranes. Results of double-staining experiments, in which the LPS on symbiosome membranes and a lysosomal-membrane protein (LMP) were identified with specific mAbs, showed that an injected anti-LPS mAb abolished the fusion-avoiding property of symbiosomes and caused them to fuse with lysosomes (Kim et al. 1994). Thus, symbiosomes coated with an injected anti-LPS mAb fused with lysosomes tagged with Texas Red-conjugated streptavidin, while symbiosomes in noninjected xD amoebae or those injected with unrelated mAbs did not fuse with lysosomes. However, it is not yet known how the LPS are transported and added to the symbiosome membrane in xD amoebae.

Heat-Shock Proteins (HSP) of X-Bacteria. X-bacteria contain much 67-kDa protein (GroELx), its gene having a high degree of identity with those of other bacterial *groEL* genes, including that of *Legionella pneumophila* and complementing mutations of *groESL* in *E. coli* (Ahn et al. 1991, 1994). We obtained mAbs against the protein, which cross-reacted with GroEL from various bacterial strains and mitochondria of eukaryotic cells (Choi et al. 1991). We cloned and sequenced the *groEx* operon (Ahn et al. 1994) The *groELx* gene appeared to be controlled by a novel and potent second promoter (*P2*) in addition to the heat-shock consensus promoter (*P1*). The *groELx* gene was expressed in a temperature-dependent manner in transformed *E. coli*, producing more GroELx at 37 °C than at 24 °C (Ahn et al. 1991).

The view that the inside of a host cell is a hostile environment to intracellular parasites was first proposed by Moulder (Moulder 1974) and later further elaborated (Moulder 1985). Initially, the view was based on the perception that host cells presented barriers to intracellular parasites for their initial entry, their subsequent survival, and multiplication. Moulder considered the intracellular environment to be comparable to common abiotic stress conditions. However, this view cannot be verified with X-bacteria since there are no free-living forms of X-bacteria available for comparison.

Since the inside of a living cell is a potentially hostile environment, it is not surprising that endosymbionts contain large amounts of stress proteins, as do intracellular parasites and semi-autonomous cell organelles. We assume that the stress proteins in endosymbionts play some role as molecular chaperones, such as enhancing transport and stabilizing imported proteins (Kakeda and Ishikawa 1991; Lindquist 1992; Ohtaka et al. 1992; Ohtaka and Ishikawa 1993). Mitochondria and chloroplasts also contain stress proteins that are related to those of prokaryotic cells, and protein translocation and folding in mitochondria are known to be mediated by HSP (Koll et al. 1992). This is consistent with the view that these organelles arose by serial endosymbiosis.

It is assumed that the roles of GroE analogs present in X-bacteria are similar to those found in other cells (Abshire and Niedhardt 1993) and organelles and that these proteins provide some protective benefits to symbionts as well as stabilizing macromolecules imported from host cell cytoplasm. Thus, GroEx could be involved in the transport and stabilization of such imported proteins. The *groE* gene is thought to play an important role for bacteria, since *groE* mutants are defective in cell growth and the gene products are essential for bacteria at all temperatures (Ellis and Hemmingsen 1989).

Host-Produced Molecules

A few host-derived proteins have been found inside symbiosomes and on symbionts as well as on the SM. While their roles are not clearly known, it seems that they are involved in the maintenance of symbiotic relationships as summarized below.

Spectrin. Spectrin with two subtypes of 225 and 220 kDa is present on the SM as well as on other membranes (Choi and Jeon 1989). Spectrin accounts for about 5% of the whole-cell protein. Although it is not unique to symbiosome membranes and, hence, may not be the only "fusion-preventing" factor, it is worth noting that the plasma membrane and nuclear envelope that have attached spectrin do not fuse with lysosomes. Thus, it is possible that spectrin may play a role in the prevention of lysosomal fusion with symbiosomes in addition to its possible role in maintaining the structure of symbiosomes.

Actin. Amoeba's actin (43 kDa) is selectively accumulated in symbiosomes and attached to the surface of X-bacteria (Kim and Jeon 1987) as well as on SM (Oh and Jeon 1995). However, it is not known how the protein is transported into symbiosomes or accumulated by X-bacteria. The role of actin inside the symbiosome is not clear.

Amoeba's Myosin. Amoeba's myosin co-localizes with actin on symbiosome membranes (Oh and Jeon 1995) as well as on the cytoplasmic side of the plasmalemma as detected with mAbs against amoeba myosin. It appears that amoeba's cytoskeletal proteins are involved in the formation and maintenance of the symbiosome. The time when myosin is first added to symbiosome membranes during the biogenesis of symbiosomes has not been determined.

Lethal Factor. xD amoebae contain a new protein with M_r above 200000, which exerts a lethal effect when injected into normal D amoebae (Jeon and Lorch 1979). This protein is synthesized by xD amoebae as a result of harboring symbionts. An active lethal factor against a D amoeba can be detected in newly infected D amoebae as early as 2 weeks after induced infection with X-bacteria. The gene encoding the protein has not been identified yet.

Loss of D-Amoeba Proteins. When D amoebae are infected with X-bacteria, amoebae lose several proteins, one of which is SAM synthetase-like 45-kDa protein (T.A. Ahn and K.W. Jeon, unpublished data). The partial nucleotide sequence has a 55 - 70% identity with SAM synthetase genes from *E. coli*, yeast, and rat. It appears that xD amoebae no longer produce the 45-kDa protein after they are infected with X-bacteria. Thus, the host cells may become dependent on symbionts because they lose one or more gene products of their own as a result of symbiosis.

Effects of Symbiosis on the Host

The presence of X-bacteria causes various physiological changes in host amoebae (Jeon and Jeon 1976; Ahn and Jeon 1979; Jeon 1992, 1995). They include: (1) Accelerated initial growth during the initial phase of experimental infection; (2) increased sensitivity to starvation; (3) increased sensitivity to over-feeding; (4) newly acquired temperature sensitivity; (5) increased sensitivity to overcrowding; (6) altered nucleocytoplasmic compatibility; (7) newly acquired nuclear lethal effect; (8) appearance of symbiont-synthesized proteins in the host cytoplasm; and (9) specificity of symbiotic relationship.

Symbionts and host amoebae are mutually dependent for survival. X-bacteria do not grow *in vitro*, and xD amoebae lose viability if X-bacteria are selectively removed from them. However, aposymbiotic amoebae may be rescued by reintroducing X-bacteria within a day after all X-bacteria have been removed (Lorch and Jeon 1980).

Acknowledgments. I thank Dr. C. Alex Shivers for his critical reading of the manuscript. Our own research work was supported in parts by grants from the National Science Foundation and the Faculty Research Fund of the University of Tennessee.

References

Abshire KZ, Niedhardt FC (1993) J Bacteriol 175: 3734-3743
Ahn GS, Choi EY, Jeon KW (1990) Endocyt Cell Res 7: 45-50
Ahn TI, Jeon KW (1979) J Cell Physiol 98: 49-58
Ahn TI, Jeon KW (1982) Exp Cell Res 137: 253-268
Ahn TI, Jeon KW (1983) J Protozool 30: 713-715
Ahn TI, Leeu HK, Kwak IH, Jeon KW (1991) Endocyt Cell Res 8: 33-44
Ahn TI, Lim ST, Leeu HK, Jeon KW (1994) Gene 128: 43-49
Arruda S, Bomfim G, Knights R, Huima-Byron T, Riley LW (1993) Science 261: 1454-1457
Berger KH, Isberg RR (1993) Mol Microbiol 7: 7-19
Britton WJ, Roche PW, Winter N (1994) Trends Microbiol 2: 284-288
Choi EY, Ahn GS, Jeon KW (1991) BioSystems 25: 205-212
Choi EY, Jeon KW (1989) Exp Cell Res 185: 154-165
Choi EY, Jeon KW (1992) J Protozool 39: 205-210
Cianciotto NP, Fields BS (1992) Proc Natl Acad Sci USA 89: 5188-5191
Eissenberg LG, Goldman WE, Schlesinger PH (1993) J Exp Med 177: 1605-1611
Ellis RJ, Hemmingsen SM (1989) Trends Biol Sci 14: 339-342
Engleberg NC, Carter C, Weber DR, Cianciotto NP, Eisenstein BI (1989) Infect Immun 57: 1263-1270
Frenchick PJ, Markham RJ, Cochrane AH (1985) Amer J Vet Res 46: 332-335
Frëhel C, Rastogi N (1989) Acta Leprol (Geneve) 7 Suppl 1: 173-174
Goren MB, Mor N (1990) Biochem Cell Biol 68: 24-32
Hacker J, Otto M, Wintermeyer E, Ludwig B, Fischer G (1993) Int J Med Microbiol Virol 278: 348-358
Hall BF, Joiner KA (1991) Immunol Today 12: A22-27
Hart PD, Young MR (1979) Exp Cell Res 118: 365-375
Haylett T, Thilo L (1991) J Biol Chem 266: 8322-8327
Holtzman E (1989) Lysosomes, Springer-Verlag, New York 439 pp
Ishibashi Y, Arai T (1990) FEMS Microbiol Immunol 2: 89-96
Jaconi ME, Lew DP, Carpentier JL, Magnusson KE, Sjögren M, Stendahl O (1990) J Cell Biol 110: 1555-1564
Jeon KW (1992)J Protozool 39: 199-204
Jeon KW (1994) Molec Cells 4: 259-265
Jeon KW (1995) Trends Cell Biol 5: 137-140
Jeon KW, Jeon MS (1976) J Cell Physiol 89: 337-347
Jeon KW, Lorch IJ (1979) Int Rev Cytol Suppl 9: 45-62
Joiner KA, Fuhrman SA, Miettinen HM, Kasper LH, Mellman I (1990) Science 249: 641-646
Kakeda K, Ishikawa H (1991) J Biochem 110: 583-587
Kim HB, Jeon KW (1986) Endocyt Cell Res 3: 299-309
Kim HB, Jeon KW (1987) Endocyt Cell Res 4: 151-166
Kim KJ, Na YE, Jeon KW (1994) Infect Immun 62: 65-71
Koll H, Guiard B, Rassow J, Ostermann J, Horich AL, Neupert W, Hartl F-U (1992) Cell 68: 1163-1175
Lindquist S (1992) Curr Opin Genet Dev 2: 748-755
Lorch IJ, Jeon KW (1980) J Protozool 27: 423-426
Lundermose AG, Rouch DA, Penn CW, Pearce JH (1993) J Bacteriol 175: 3669-3671

Morioka M, Ishikawa H (1992) J Biochem 111: 431-435
Moulder JW (1974) J Infect Dis 130: 300-306
Moulder JW (1985) Microbiol Rev 49: 298-337
Oh SW, Jeon KW (1996) J Eukaryot Microbiol 43: 7A
Ohtaka C, Ishikawa H (1993) J Mol Evol 36: 121-126
Ohtaka C, Nakamura H, Ishikawa H (1992) J Bacteriol 174: 1869-1874
Pak JW, Jeon KW (1994) Mol Biol Cell Suppl 5: 375a
Parham P (1992) Nature 356: 291-292
Park MS, Jeon KW (1988) Endocyt Cell Res 5: 215-224
Park MS, Jeon KW (1990) Endocyt Cell Res 7: 37-44
Pfeifer U (1987) In: Lysosomes: Their Role in Protein Breakdown. H Glaumann, FJ Phillips (eds). Academic Press, New York, p 3-59
Sadosky AB, Wiater LA, Shuman HA (1993) Infect Immun 61: 5361-5373
Steinhoff U, Golecki JR, Kazda J, Kaufmann SH (1989) Infect Immun 57: 1008-1010
Sturgill-Koszycki S, Schlesinger PH, Charkraborty P, Haddix PL, Collins HL, Fok AK, Allen RD, Gluck SL, Heuser J, Russell DG (1994) Science 263: 678-681

Acidification in Digestive Vacuoles is an Early Event Required for *Holospora* Infection of *Paramecium* Nucleus

M. Fujishima and M. Kawai
Biological Institute, Faculty of Science, Yamaguchi University, Yamaguchi 753, Japan.

Key words: *Paramecium caudatum*, digestive vacuole, proton pump, H^+-ATPase, concanamycin, *Holospors obtusa* (infection).

Summary: Specific inhibitors of vacuolar-type H^+-ATPase, concanamycin A and concanamycin B, inhibited acidification of lysosomes and digestive vacuoles of the ciliate *Paramecium caudatum*. *Holospora obtusa* engulfed in the host digestive vacuoles could not escape there and could not infect the host macronucleus in the presence of these antibiotics. Indirect immunofluorescence microscopy with polyclonal antibodies specific for 57 kDa subunit of V-ATPase of chromafin granule showed that antigens were present in the digestive vacuole membranes. These results show that V-ATPase is present in the digestive vacuole membrane and maintains acidic pH inside the vacuole, and that acidification of the digestive vacuole is a required event for bacteria escape from the digestive vacuole and for infection into the host macronucleus.

Introduction

Gram-negative bacterium *Holospora obtusa* is a macronucleus-specific symbiont of the ciliate *Paramecium caudatum*. When the infectious forms are added to paramecia, they are soon engulfed into the host digestive vacuoles, escape there, appear in the host cytoplasm and infect the host macronucleus within 15 min after mixing. *Paramecium* digestive vacuoles can be classified into four developmental stages, according to Allen and Staehelin (1981), of the University of Hawaii. DV-I is a spherical vacuole. As soon as the vacuoles are pinched off from the host buccal cavity, they are soon fused with acidosomes, so the pH inside the vacuole decreases from 7 to 3 within about 5 minutes. At the same time, the DV-I membrane pinches off and moves into the cytoplasm, so this membrane replacement decreases the volume of the vacuole, and the vacuole differentiates into a DV-II vacuole. After that, the DV-II vacuole fuses with primary lysosomes, increases its volume, and differentiates into a DV-III vacuole. In this DV-III , infectious forms of *H. obtusa* are partially digested. The DV-III vacuole then decreases its acid-phosphatase activity and becomes a DV-IV vacuole . This DV-IV vacuole fuses with the host cytoploct, and from there holosporas are ejected from the cell. During this process, we found that bacteria can escape from the host digestive vacuoles only when they are in the DV-I stage, but they cannot escape from DV-II or later stages of the vacuoles.

Therefore, the majority of the bacteria fail to escape from the DV-I vacuole, and they are eventually digested and ejected from the host cell. DV-I vacuoles differentiate to DV-II within 10 min, so only a small number of the bacteria can appear in the host cytoplasm. Compared with the isolated infectious forms, bacteria in the cytoplasm decrease periplasmic region and look dark under a phase-contrast microscope and decrease buoyant density before they infect the host macronucleus with this special tip.

After the infection, the bacteria form constrictions to differentiate the reproductive short forms within 2 days after the infection. So the question arising is, why and how infectious form escapes only from DV-I vacuoles. There are two possibilities, namely, (1) *Holospora* cannot escape from the acidified digestive vacuole, and (2) acidification of DV-I vacuole and membrane replacement during DV-I to DV-II transformation is required for escaping from the digestive vacuoles. In order to answer the question, in the present study, we examined effects of concanamycin A and B on bacterial escaping from the host digestive vacuole. These antibiotics are known as specific inhibitors of vacuolar-type H^+-ATPase (V-ATPase), and inhibit acidification of lysosomes in other cell types. Furthermore, we examined whether pH inside the vacuoles becomes acidic by these antibiotics using an acridine orange. We also examined whether the host digestive vacuole membrane contains V-ATPase by indirect immunofluorescence microscopy using polyclonal antibody specific for 57 kDa subunit of chromafin granules.

Materials and Methods

Cells and Culture. Cells used in this study were *H. obtusa*-bearing and nonbearing strain FK-1 of *P. caudatum* (syngen unknown, mating type E). Culture medium and culture methods were described in our previous paper (Dohra et al. 1994). Cells were grown at 25 °C.

Isolation of *H. obtusa*. Infectious forms of *H. obtusa* were isolated from host homogenates by Percoll density gradient centrifugation (Fujishima et al. 1990) and kept at 4 °C until use.

Staining with Acridine Orange and Treatment with Concanamycin. Acridine orange was dissolved in deionized water at concentration of 5 mM, and kept at 4 °C as a stock solution. This dye, at a final concentration of 2 μM, was added to *Paramecium* cells for 10 min at 25 °C. Then, the cells, washed twice with Dryl's solution (Dryl 1959) and collected by centrifugation with a hand-operated centrifuge, were suspended in Dryl's solution with or without the presence of concanamycin A or concanamycin B. The living cells were then examined under a fluorescence microscope to determine whether the *Paramecium* digestive vacuoles had incorporated protons.

Concanamycin A and concanamycin B were dissolved in 100% methanol at concentration of 11.5 mM and stocked at -30 °C. These reagents were added to *Paramecium* cells at final concentrations of 0.1, 1 and 2.5 μM.

Immunofluorescence Microscopy. *Paramecium* cells were dried on glass slides and fixed with cold ethanol (-20 °C) for 15 min, washed with phosphate buffered saline (PBS) for 10 min, and then reacted with rabbit polyclonal antibody raised against the 57 kDa subunit of the vacuolar type H^+-ATPase that originated from bovine chromafin granules, for 1 h at room temperature. The cells were then washed three times with PBS for 10 min each, reacted with fluorescein isothiocyanate (FITC)-conjugated goat anti-rabbit antibody for 1 h at room temperature, washed three times with PBS and observed under a fluorescence microscope (Olympus, BH2-RFL).

Results and Discussion

Concanamycin A and B are known as strong inhibitors of the V-ATPase of the digestive vacuole membrane, lysosome membrane and so on, whereas other types of ATPase, e.g. F-ATPase of the mitochondria and P-ATPase of the plasma membrane, are not affected by these antibiotics. These antibiotics inhibit the pumping function of V-ATPase, and cause neutralization of the content of the vacuoles. Therefore, if the acidification of the host digestive vacuole is a required event for bacterial escape from the digestive vacuole, we can expect that these antibiotics prevent the bacterial escape from the host digestive vacuole.

When living *Paramecium* cells were stained with 2 μM acridine orange, their digestive vacuoles and lysosomes showed yellow or brilliant vermilion fluorescence. This indicates that pH inside the vacuoles is acidic. However, when the cells were treated with 1 or 2.5 μM of concanamycin A or concanamycin B, fluorescence of the brilliant vermilion in the digestive vacuoles and primary lysosomes is completely inhibited. Instead, green fluorescence appeared. This indicates that pH inside these vacuoles became neutral. Thus, acidification in digestive vacuoles and in primary lysosomes is inhibited by V-ATPase inhibitors.

In order to examine effect of inhibition of the acidification of the host digestive vacuoles, *Paramecium* cells and isolated infectious forms of *H. obtusa* were mixed for 2 min. During this, bacteria were engulfed into the host digestive vacuole DV-I. Then, the cells were treated with concanamycin A or with concanamycin B, and it was observed whether the bacteria could escape from the host DV-I vacuole or not, at 30 and 60 min after the treatment.

When the cells were treated with concanamycin A, bacteria in the host digestive vacuoles could not escape from there to appear in the host cytoplasm. The bacteria remained in DV-II or more later stages of the vacuole, but no bacteria appeared in the host cytoplasm and in the macronucleus. The same result was obtained when the cells were treated with concanamycin B.

In concentrations of 1 and 2.5 μM, both inhibitors completely prevented the bacterial escape from the host digestive vacuoles and the inhibitors of these concentrations also prevented the bacterial infection into the macronucleus. However, 0.1 μM of the inhibitors could not prevent the bacterial escape and the infection into the macronucleus. Thus, our results strongly suggest that acidification of the host digestive vacuole is an indispensable phenomenon for the bacterial escape from the host digestive vacuoles, and the results also show that bacterial escape from the host digestive vacuole is also necessary for infection into the host macronucleus.

Our results strongly suggest a possibility that proton-pomp is present in the *Paramecium* digestive vacuole membrane and its pumping function maintains acidic pH inside the vacuole. However, nothing is known about the localization of the vacuolar-type proton-pump in *Paramecium.* In order to prove the existence of the V-ATPase in digestive vacuole membrane of *Paramecium,* we carried out indirect immunofluorescence microscopy using polyclonal antibodies against 57 kDa subunits of the V-ATPase of chromafin granule.

V-ATPase can be divided into two parts, the one is hydrophilic part consisted of 9 subunits, and the other is hydrophobic part consisted of 7 subunits. Among these subunits, 57 kDa subunits are involved in hydrophilic part (Moriyama and Nelson 1989). Indirect immunofluorescence microscopy showed that FITC-fluorescence appears in spherical digestive vacuoles. Fluorescence was also observed in the cytoplasm, probably acidosomes and lysosomes. Thus, our results strongly suggest that V-ATPase is present in the spherical digestive vacuole membrane of *P. caudatum.*

Acknowledgments. This study was supported by grants to M. Fujishima from Japan Ministry of Education, Science and Culture, Grant-in-Aid for Scientific Research (05454028) and International Scientific Research Program (Joint Research) (07044202).

References

Allen RD, Staehelin LA (1981) J Cell Biol 89: 9-20
Dohra H, Fujishima M, Fok A, Allen, RD (1994) J Euk Microbiol 41: 503-510
Dryl S (1959) J Protozool 6 (Suppl) 25
Fujishima M, Nagahara K, Kojima Y (1990) Zool Sci 7: 849-860
Moriyama Y, Nelson, N (1989) Biochem Biophys Acta 980: 241-247

Monoclonal Antibody Specific for Activated Form of *Holospora obtusa*, a Macronucleus-Specific Bacterium of *Paramecium caudatum*

M. Kawai and M. Fujishima
Biological Institute, Faculty of Science, Yamaguchi University, Yamaguchi, 753, Japan

Key words: *Paramecium caudatum*, *Holospora obtusa* (activated form), monoclonal antibody.

Summary: Two monoclonal antibodies (mAb IR-4-2 and mAb AF-1) were developed by injecting sonicated debris of the activated forms of *Holospora obtusa* into mice. Indirect immunofluorescence microscopy showed that the antigen of mAb IR-4-2 was located inside the cell and that ther amount increases in the activated form. On the other hand, the antigen of mAb AF-1 appears only in the activated form, and is present inside the cell. The immunoblot showed that the molecular weights of the antigen of mAb AF-1 are 58.8 and 22.9 kDa. These results show that new antigens appear when the infectious forms transform the activated forms in early infection. We found that the AF-1 antigens were inducible *in vitro* if the infectious forms were treated with a buffer of low pH (pH 3-4) for 15-20 min. This strongly suggests that a trigger for the appearance of the AF-1 antigens in activated forms is low pH in the host digestive vacuoles.

Introduction

The Gram-negative bacterium *Holospora obtusa* is a macronucleus-specific symbiont of the ciliate *Paramecium caudatum*. This macronucleus-specific endonuclear symbiont grows by binary fissions of the reproductive short forms (1.5-2 µm in length) when the host cell is growing, but ceases growth and differentiates into the infectious long form (13 µm in length) when the host cell starves (Görtz 1983; Görtz et al. 1989; Fujishima et al. 1990a, b). This infectious form can be released from its host after a preceding cell division (Wiemann 1990). The infectious form consists of three distinct regions: a periplasm, a cytoplasm and a recognition tip at one end of the bacterium (Görtz et al. 1989; Görtz and Wiemann 1989; Fujishima et al. 1990a). When the isolated infectious forms are mixed with paramecia, they soon are engulfed into the digestive vacuoles. *Paramecium* digestive vacuoles can be classified into four developmental stages, according to Allen and Staehelin (1981) of the University of Hawaii. Bacteria can escape from the host digestive vacuoles only when they are in the DV-I stage, but they cannot escape from DV-II or later stages of the vacuoles. Unlike the isolated infectious forms, bacterial cells in the cytoplasm look dark under a phase-contrast microscope. The bacteria then infect the macronucleus with their recognition tip first. The infected bacteria form constrictions to differentiate the reproductive short forms within 2 days after the infection.

In order to investigate the bacterial changes that are required for the infection, we attempted to develop monoclonal antibodies that were specific for the bacteria moving from the host digestive vacuoles to the host macronucleus.

Materials and Methods

Stocks and Culture. Stocks used in this experiment are strain GT703S16 (syngen 2, mating type O) and FK-1 (syngen unknown, mating type E) of the ciliate *P. caudatum*. *H. obtusa*-bearing *Paramecium* cells and the aposymbiotic cells were cultivated in modified lettuce juice medium (KH_2PO_4 was used instead of $NaH_2PO_4 \cdot 2\ H_2O$) at 25 °C, as decribed in our previous paper (Fujishima et al. 1990).

Isolation of *H. obtusa*. Reproductive short forms of buoyant density 1.09 g/ml were isolated by 40% (v/v) Percoll density gradient centrifugation from homogenates of the isolated host macronuclei as described in our previous papers (Fujishima et al. 1990a, b). For isolation of the infectious forms with a buoyant density of 1.16 g/ml, homogenates of infectious form-bearing host cells were directly subjected to 70% (v/v) Percoll density gradient centrifugation (Fujishima et al. 1990a, b). For isolation of the activated forms, the isolated infectious forms and paramecia were mixed at densities of 1.2×10^8 bacteria/ml and 1×10^4 paramecia/ml for 1 h at 25 °C, homogenized and centrifuged in 40% (v/v) Percoll density gradient centrifugation. After the centrifugation, a band consisting of the reproductive forms, the infectious forms and the activated forms was collected, washed with 10 mM Na,K-phosphate buffer, pH 6.5 and stored at -85 °C prior to use.

Production of Monoclonal Antibodies. Monoclonal antibodies (mAbs) were obtained by routine procedures according to Galfre and Milstein (1981). The activated forms (2×10^7 cells) were sonicated in 125 μM of a Na,K-phosphate buffer for 1 min on ice. The cell debris was then mixed with an equal volume of Freund's complete adjuvant and injected intraperitoneally into 6-week-old BALB/c mice. The first injection was followed by two booster injections at 2-week intervals with the same antigen but emulsified in Freund's incomplete adjuvant. The final injection was given 4 days before the cell fusion. Excised spleen cells were fused with mouse myeloma cell line NS-1, using polyethylene glycol 4000 (E. Merck) and placed in 96-well plates. Culture fluids were then collected and hybridoma cells producing antibodies were screened by indirect immunofluorescence staining, and the cells were cloned by limiting dilution (Coffino et al. 1972).

Indirect immunofluorescence staining was done as follows. Isolated bacteria suspended in 10 mM phosphate buffer, pH 6.5, were air-dried on cover glasses, fixed in 100% ethanol for 10 min at -20 °C, and then the cover glasses were washed twice with phosphate-buffered saline (PBS; 137 mM NaCl, 2.68 mM KCl, 8.1 mM $NaHPO_4 \cdot 12\, H_2O$, 1.47 mM KH_2PO_4, pH 7.2) for 15 min each. The cover glasses were incubated in a monoclonal antibody-containing culture supernatant for 1 h at room temperature. The cover glasses were washed three times with PBS for 15 min each and incubated in fluorescein isothiocyanate (FITC)-conjugated anti-mouse IgG diluted 100-fold with PBS for 1 h at room temperature. Observations were carried out with fluorescent optics (Olympus, BH2-RFL) at a magnification of 400 x.

Electrophoresis and Immunoblotting. Cell pellets of *H. obtusa* were boiled with Laemmli's lysis buffer (Laemmli 1970), and the lysate was used for sodium dodecyl sulfate-polyacrylamide gel electrophoresis (SDS-PAGE). In each lane, lysates of 1 x 10^6 cells for the infectious forms, 2 x 10^7 cells for the reproductive forms and 1 x 10^6 cells for the activated forms of *H. obtusa* were cast. These numbers of bacteria showed almost the same protein content. The gels were then stained with a Silver-stain kit (Wako Pure Chemical).

After SDS-PAGE, proteins of the gel were transferred to an Immobilon-P membrane (Millipore) for immunoblotting. The membrane was then incubated with mAbs-containing culture supernatant containing 5% (w/v) skim milk as the primary antibodies for 1 night at 25 °C, then incubated in biotinylated anti-mouse IgG as the secondary antibodies, and finally with an avidin GH-biotinylated alkaline phosphatase H complex using a Vectastain ABC-AP kit (Vector Laboratories, Inc.). Molecular weights were measured with a prestain marker kit (Bio-Rad) and a biotinated marker kit (Bio-Rad).

Treatment with Low pH. Isolated infectious forms were suspended in 60 mM glycine-HCl (pH 3) for 10, 15, 20, 30, 40, 50 and 60 min and then the bacteria were observed by indirect immunofluorescence microscopy with mAb AF-1.

Results and Discussion

Indirect Immunofluorescence Microscopy

MAb IR-4-2 stains all bacteria of life cycle but, slightly sonicated bacteria well, and the activated form shows especially strong fluorescence. Thus, the antigens are inside the cell and their amount increases in the activated form. MAb AF-1 is the activated form-specific monoclonal antibody This antibody does not react with the reproductive form and infectious form, but only reacts with activated forms.

This indicates that the antigen appears only in the activated form, and is present inside the cell, because the fluorescence in sonicated cell is strong.

Immunoblotting

By immunoblotting, it was found that the molecular weights of the AF-1 antigen were 58.8 kDa and 22.9 kDa proteins. Both antigens were deteted only in the activated forms and could not be detected in the reproductive and infectious forms.

Treatment with Low pH

We found that an acidic pH induces the activated form-specific antigen AF-1 in the infectious long forms. In the control, isolated infectious forms were suspended in pH 6.5 buffer for 60 min at 25 °C. However, no AF-1 antigens appear in the infectious form. On the other hand, AF-1 antigens appeared at pH 4, but only temporarily. The bacteria were suspended in pH 4 buffer for 10, 20 and 60 min and then observed. No AF-1 antigens appear at 10 min, but the antigens appear at 20 min, and later the fluorescence decreases. Thus, our results show that acidic pH induces the appearance of the activated form-specific AF-1 antigen. This strongly suggests that, in the host cell, the AF-1 antigen is induced by the acidic pH in the host digestive vacuoles. So far, many mAbs had been developed against *H. obtusa* as shown in Table 1. However, mAb AF-1 is the first monoclonal antibody to be developed that is specific to the activated form, though its function in the infection process is unknown.

Acknowledgements. This study was supported by grants to M. Fujishima from Japan Ministry of Education, Science and Culture, Grant-in-Aid for Scientific Research (05454028) and International Scientific Research Program (Joint Research) (07044202).

References

Allen RD, Staehelin LA (1981) J Cell Biol 89: 9-20
Coffino P, Baumal R, Laskov RSM (1972) J Cell Phisiol 79: 429-440
Fujishima M, Nagahara K, Kojima Y (1990a) Zool Sci 7: 849-860
Fujishima M, Sawabe H, Iwatsuki K (1990b) J Protozool 37: 123-128
Fujishima M, Nagahara K, Kojima Y, Sayma Y (1991) Europ J Protistol 27: 119-126
Galfre G, Milstein C (1981) Methods Enzymol 73: 3-47
Görtz H-D (1983) Intern Rev Cytol, suppl 14: 145-176
Görtz H-D, Ahlers N, Robenek H (1989) J Gen Microbiol 135: 3079- 3085
Görtz H-D, Wiemann M (1989) Europ J Protistol 24: 101-109
Laemmli UK (1970) Nature 277: 680-685
Wiemann M (1990) J Protozool 36: 176-179

Interactions of Host Paramecia With Infectious *Holospora* Endocytobionts

E. Baier and H.-D. Görtz
Biological Institute, University of Stuttgart, Pfaffenwaldring 57, 70550 Stuttgart, Germany

Key words: *Paramecium caudatum*, *Holospora obtusa*, macronuclear envelope, infection, protein interaction.

Summary: *Holospora obtusa* invades the nuclei of *Paramecium caudatum*. This chapter describes the investigation of the interactions of bacterial surface structures with components of the nuclear envelope of the host. Digoxigenin-labeled bacterial surface proteins were found to bind to blotted nuclear envelope proteins, indicating a possible interaction in the invasion process.

Introduction

Within the infection cycle of *Holospora obtusa* in its paramecia hosts there is a step where the infectious forms of the bacteria get into the macronucleus of the paramecia and where bacterial surface structures and host nuclear envelope components must interact (Görtz et al. 1987). To find out which proteins are involved in this interaction, we isolated periplasmic proteins from the infectious form of *Holospora obtusa* and macronuclear envelope proteins from *Paramecium caudatum.* Binding reactions were screened on blotted paramecia proteins, with bacterial proteins labeled with digoxigenin-3-O-methylcarbonyl-ε-aminocapron acid N-hydroxy-succinimide ester (digoxigenin), anti-digoxigenin antibody, and the carbazole reaction.

Materials and Methods

Infected and uninfected *Paramecium caudatum* strains were cultured as described by Görtz and Dieckmann (1980). The infectious form of *Holospora obtusa* was isolated from the host cells as described by Schmidt et al. (1987) and Görtz et al. (1988). Proteins were extracted from the isolated bacteria by alkaline lysis as described by Schmidt et al. (1987) and labeled with digoxigenin-3-O-methylcarbonyl-ε-aminocapron acid N-hydroxy-succinimide ester as described by Boehringer Mannheim. Macronuclei (Fig. 1), their nuclear envelopes (NE) and the

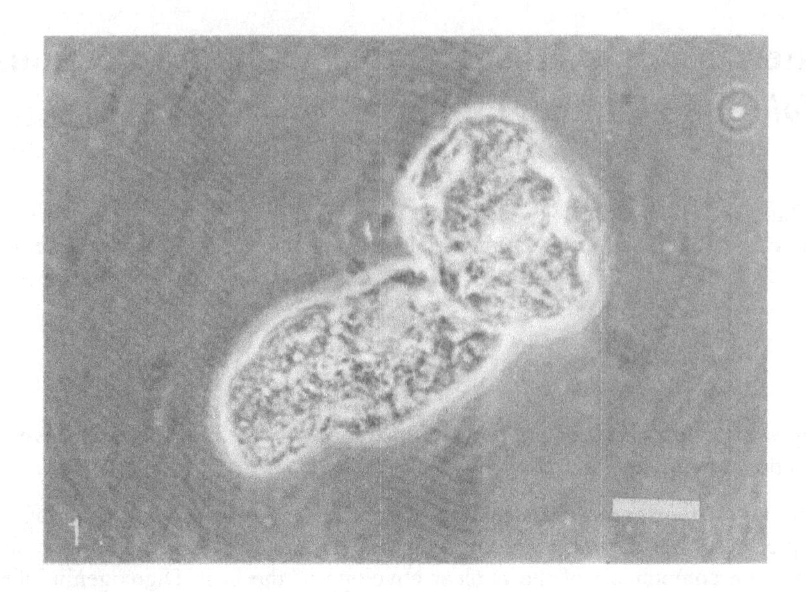

Fig. 1. Isolated macronuclei of *Paramecium caudatum*. Bar = 8 μm

NE proteins of uninfected *Paramecium caudatum* were isolated as described by Freiburg (1985) and Kaufmann et al. (1983) with the modification of using a lower detergent concentration.

For SDS-PAGE, nuclear envelope proteins were separated according to Laemmli (1970) using a 15% separation gel and a 5% stacking gel. NE proteins were lysed with 5.5 μl 1N NaOH, neutralized with 1N HCl, stabilized with 1 μl 7% PMSF, diluted with 110 μl lysis buffer (9 M urea, 2% w/v Triton X-100, 5% ß-mercaptoethanol) and centrifuged for 5 min at 15000 x g. The proteins were in the supernatant and electrophoresed. Protein gels were blotted on PVDF membrane (Fig. 2a) and the binding reaction was carried out as follows: the membrane was blocked overnight at 4 °C (blocking buffer: 0.1% Tween 20, 10% slim milk powder in PBS), washed three times for 10 min at room temperature (washing and dilution buffer: 0.1% Tween 20 in PBS) and incubated with the labeled bacterial proteins for 1 h (dilution factor 1:250). After three times of washing, the anti-digoxigenin antibody labeled with horseradish peroxidase (Boehringer Mannheim) was incubated for 2.5 h (dilution factor 1:250). After three washing steps the membrane was incubated with the substrate (5 ml carbazole solution [1% w/v in acetone], 50 μl H_2O_2, 50 ml 0.1 M sodium acetate, pH 5.5, 46 ml H_2O_{dd}] for 5-10 min until the staining reaction was ready. Then the membrane was washed with water and dried.

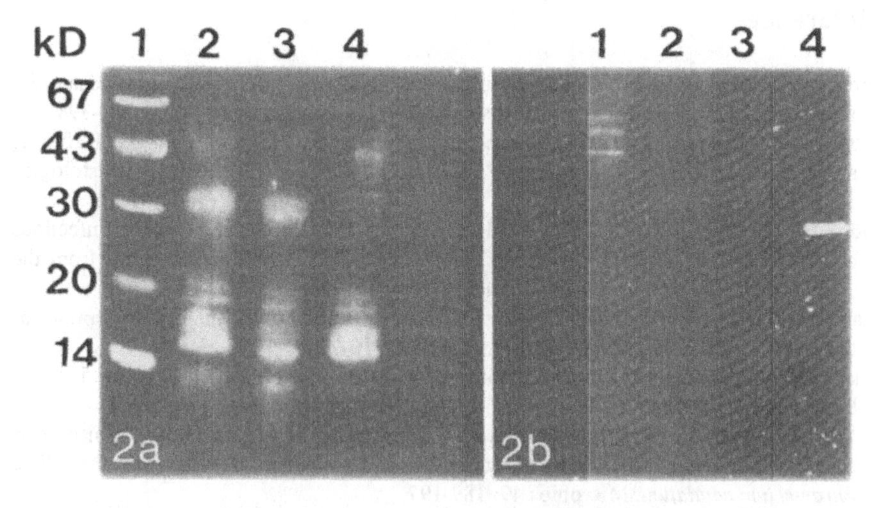

Fig. 2a. Blot stained with amido black. *Lane 1* Marker; *lanes 2-4* decreasing concentrations of NE proteins. **b** Immunoblot; interactions between NE proteins and *Holospora* proteins labeled with digoxigenin; anti-digoxygenin antibody labeled with horseradish peroxidase, carbazole reaction. *Lanes 1-3* Decreasing concentrations of NE proteins interacting with bacterial proteins; *lane 4* Marker

Results and Discussion

The binding reaction resulted in bands of different molecular weight on the blot. At least three distinct bands were located at molecular weights between 43 kDa and 67 kDa (Fig. 2b), two weak bands between 30 kDa and 43 kDa, and one at about 30 kDa. This is the first observation of a binding reaction of individual proteins of the infectious *Holospora* bacteria to proteins of the nuclear envelope of the host cell. These proteins of both organisms may be candidates for having a function in the invasion process. Although the binding of bacterial proteins to the blotted NE proteins is reproducible, it is not clear whether these interactions are really specifically involved in the invasion process or perhaps unspecific reactions. Alternatively, the binding will be analyzed using the reciprocal procedure: blotted bacterial proteins will be incubated with labeled NE proteins.

To further identify the nuclear proteins binding the bacterial proteins, a different lysis procedure is now being developed to get the NE proteins in solution for the IEF. The reaction will be controlled also on the cytological level.

References

Freiburg M (1985) Isolation and characterization of macronuclei of Paramecium caudatum infected with the macronucleus-specific bacterium Holospora obtusa. J Cell Sci 73: 389-398

Görtz H-D, Dieckmann J (1980) Life cycle and infectivity of *Holospora elegans* Haffkine, a micronucleus-specific symbiont of *Paramecium caudatum* (Ehrenberg). Protistologica 16: 591-603

Görtz H-D, Freiburg M, Wiemann M (1988) Polypeptide differences between infectious and reproductive forms of *Holospora obtusa*, an endonucleobiotic bacterium from the macronucleus of *Paramecium caudatum*. Endocyt & Cell Res 5: 233-244

Kaufmann SH, Gibson W, Shaper JH (1983) Characterization of the major polypeptides of the rat liver nuclear envelope. J Biol Chem 258: 2710-2719

Laemmli UK (1970) Cleavage of structural proteins during the assembly of the head of bacteriophage T4. Nature 227: 680-685

Schmidt HJ, Freiburg M, Görtz H-D (1987) Comparison of the infectious form of two bacterial endonucleobionts, *Holospora elegans* and *Holospora obtusa*, from the ciliate *Paramecium caudatum*. Microbios 49: 189-197

Appearance of Viable Bacteria in *Acanthamoeba royreba* After Amoebic Exposure to Megarad Doses of Gamma Radiation

A. A. Vass[1], R. P. Mackowski[2], and R. L. Tyndall[1,2]
[1]Oak Ridge National Laboratory, Building 9207, MS 8077, Oak Ridge, TN 31781-8077, USA
[2]Department of Ecology and Evolutionary Biology, University of Tennessee, Knoxville, TN 37996, USA

Key words: Amoebae, derivars, irradiation, *Legionella*, DNA

Summary: We previously demonstrated that *Legionella* infection of *Acanthamoebae royreba* could result not only in the expected amplification of *Legionella*, but also in the appearance of bacteria other than *Legionella*. Extensive controls in these experiments were always negative for the presence of such bacterial "derivars." In related studies on the radiation resistance of *Acanthamoebae*, cultures of *A. royreba* were shown to be highly resistant to gamma as well as alpha and UV irradiation. Cloned, axenic cultures of *A. royreba* were placed in antibiotic-free medium in sealed tubes and exposed to 1 Mrad of gamma radiation. The amoebae were not killed by this dosage. Of particular interest was the observation that bacterial growth occurred after several months, in some of the unopened, irradiated amoebae cultures stored at room temperature. Bacterial outgrowths were not seen in control amoebae populations which were not irradiated and incubated for the same length of time. In each experiment, the radiation-induced bacterial derivars differed morphologically from each other, some required specialized medium for growth and were resistant to radiation as was the „parent" amoebae.

This result appears related to the appearance of bacterial derivars from *Acanthamoebae* occurring after *Legionella* infection except the induction is caused by irradiation. That the results of the *Legionella* and irradiation experiments are not caused by contamination is indicated by the controls in each experiment as well as by the similarity of some unusual properties of many of the bacterial derivars which include pigmentation, crystal formation, biodispersant production and lack of reactivity with many carbohydrates. A comparison of the DNA fingerprints of the amoebae and bacteria shows that the majority of the DNA bands in the bacterial profile are not discernible in the amoebic DNA fingerprints.

One explanation of these and previous results is that bacterial endosymbionts in *Acanthamoebae*, although not detected by light or electron microscopy, are nevertheless rescuable by infection with *Legionella*, gamma irradiation, or possible other environmental stressors.

Introduction

Several genera of bacteria have been known to survive high doses of gamma radiation (Keller and Maxcy 1984). Most notable is *Micrococcus* which was shown to survive gamma radiation dosages as high as 1.2 Mrad (megarads). Most of the organisms which have been tested at these high dosages have been prokaryotes, although simple eukaryotes, such as *Tetrahymena*, have survived after exposure to 1 Mrad.

Free-living amoebae are complex eukaryotes and as such have true nuclei and divide by binary fission preceded by DNA synthesis, replication and mitosis. They also possess mitochondria, ribosomes, Golgi structures, endoplasmic reticulum and microtubules.

We have shown that free-living amoebae isolated from environmental sources often harbor bacteria (Tyndall et al. 1991). Such bacteria, unlike typical non-amoebae-associated bacteria, often produce crystals, pigments and biodispersants (Dietz et al. 1994). Additionally, free-living amoebae can be infected by and amplify *Legionella* (Tyndall and Dominigue 1982). We have also shown that infection of free-living amoebae with *Legionella* may result in the appearance of bacterial derivars other than *Legionella* (Tyndall et al. 1993).

In view of these observations, this study was carried out:

1. To determine if differences exist between different genera of free-living amoebae with regard to radiation resistance after exposure to varying radiation types and dosages
2. To determine whether or not free-living amoebae can protect sequestered bacteria from the effects of radiation
3. To determine if *Legionella* and other bacteria associated with amoebae are more resistant to irradiation than unassociated bacteria; and ultimately
4. to characterize bacterial derivars arising from radiation of free-living amoebae.

Material and Methods

Amoebae/Bacterial Growth. Bacterial derivars, bacteria other than *Legionella* which appeared after a 2 week co-infection of axenic *Acanthamoebae* at 37 °C with single colony purified *Legionella,* as well as *Legionella* per se, were cultured on Buffered Charcoal Yeast Extract (BCYE) medium at 37 °C. Intra-amoebic bacteria from environmental free-living amoebae, i.e., Oak Ridge Consortium 46C (Tyndall et al. 1991), were isolated and cultured on Trypticase Soy Agar (TSA) and modified peptone medium (MPM). All other bacteria were grown on TSA. *Acanthamoebae* and *Hartmannella* were grown in 712 medium (Cote 1984) at 37 °C and 23 °C, respectively. *Naegleria* was grown in CGVS medium at 37 °C. Various *A. royreba* clones as well as uncloned *A. royreba* were used in some experiments (Tyndall and Vass 1993). All amoebae were irradiated in the trophozoite stage so that the cyst form would not impart additional protection. In order to obtain amoebae which sequestered bacteria, bacterial cultures were fed to the amoebae 24 h prior to the beginning of each experiment, and the amoebae were stained with acridine orange confirming the presence of intra-amoebic bacteria.

Gamma Irradiation. A ^{60}Co irradiator (Model 109) emitting 0.30145 MR/h was used as the source of gamma radiation. Preparatory to irradiation, all organisms were diluted to a concentration of 10^6 cells/ml in 712 medium and placed in glass test tubes with Teflon-lined caps. After irradiation and prior to plating, a serial dilution was performed on each culture. Survival was compared with unirradiated controls and measured 5 days after plating the amoebae in microtiter plates containing 0.3 ml 712 medium. In addition, amoebae were placed on non-nutrient agar plates (NNA) spread with a lawn of *Escherichia coli* to determine whether or not they were able to ingest bacteria after irradiation.

Alpha/UV Irradiation. ^{251}Am was used as the alpha source which produced a dose rate of 0.95 Gy/min. The bottom of a 60 x 15 mm petri dish was cut off, and a mylar sheet stretched and cemented over the bottom. To obtain a monolayer, 10^6 organisms in 50µl of medium were spread over the center of this sheet. The mylar sheet was then placed directly over the alpha source. Controls (mylar sheets prepared in the same manner, inoculated with organisms, but not irradiated) were set up at the same time. After irradiation, the mylar sheets were washed twice with 2 ml 712 medium containing 0.05 mg gentamicin/ml. A serial dilution was then performed on each culture. Survival was compared with unirradiated controls and measured 5 days after plating the amoebae in microtiter plates containing 0.3 ml of 712 medium. In addition, amoebae were placed on NNA plates spread with a lawn of *E. coli* to determine whether or not they were able to ingest bacteria after irradiation.

The UV irradiator consisted of a UV light source emitting 254 nm wavelength light (15 W) with a dose rate of 109.1 mW/s. This was housed in a wooden box with the light source 21 cm above the base of the box. Glass petri plates were used for this experiment. After irradiation, the petri dishes containing the amoebae and mammalian cell cultures were titrated and plated.

Extraction of Amoebic/Derivar DNA. Amoebic/derivar DNA could not be extracted by simply boiling. Instead, 0.5 ml of 0.2 N NaOH was added to bacterial and amoebic pellets and allowed to stand at room temperature for 1 h. The material was then neutralized with glacial acetic acid.

DNA Amplification Fingerprinting (DAF). DNA amplification was performed following the procedure of Caetano-Anolles et al. (Caetano-Anolles et al. 1992).

Results

Table 1 shows that *Acanthamoebae* and *Hartmannella* were able to survive gamma radiation dosages up to 2 Mrad with *A. royreba* being the most resistant

Table 1. Ability of select amoebae to survive gamma radiation

	Radiation Dosage (megarads)					
Test organism	**0**	**0.3**	**1.0**	**1.22**	**2.0**	**5.0**
Hartmannella veriformis	100[a]	100	100	100	10	nt
Naegleria sp.	100	100	0[b]	nt	nt	nt
Acanthamoeba royreba	100	100	100	90	10	0
Acanthamoeba royreba (clone)	100	100	100	80	10	0
Acanthamoeba castellani	100	nt	20	nt	nt	nt
Acanthamoeba astronyxis	100	nt	10	nt	nt	nt
Acanthamoeba sp. (environmental isolate)	100	100	50	nt	nt	nt
Acanthamoeba royreba infected with *Legionella pneumophila*	50	nt	nt	80	nt	nt

nt, not tested.
[a]Approximate percent survival.
[b]Amoebae physically survived the radiation dosage, but were unable to replicate.

Table 2. Ability of select eukaryotes to survive 5.3 MeV alpha particles[a]

Test organism	**Dosage time (h)**[b]	**Survival (%)**
Acanthamoeba royreba	3.0	100
Acanthamoeba astronyxis	3.0	50-60
Naegleria lovanensis	3.0	40-65
Acanthamoeba lenticulata	3.0	95-100
Hartmannella veriformis	3.0	60-80
Normal primary rat tracheal cells	0.083	30
Mink lung cells	0.083	30

[a]Nuclear size average cross section 6.2 μm^2.
[b]Maximum times tested.

Acanthamoebae tested. *Naegleria* was much more sensitive to gamma radiation, only being able to withstand 0.3 Mrad. Initial infection of *A. royreba* with *Legionella* was detrimental to the amoebae, but the deleterious effects of the radiation treatment on the *Legionella* allowed the amoebae to flourish.

Table 2 illustrates that all the amoebae tested were able to survive a minimum of 3 h of 5.3 MeV alpha radiation. The average cross-sectional area of the amoebic nuclei was 6.2 μm^2. Additional increases in the dosage times were not possible due to a drying out of the material placed on the mylar membrane. As with gamma radiation, *A. royreba* was the most resistant amoebae tested.

Table 3. Ability of select eukaryotes to survive UV irradiation[a]

Test organism	Dosage time (min)[b]	Survival (%)
Mouse mammary carcinoma cell line (LT-1)	0.5	0.5
Acanthamoeba royreba	0.5	100
	1.0	100
	2.0	~50
	3.0	~3
	4.0	<1
	5.0	0
Hartmannella veriformis	0.5	100
	2.0	50
	3.0	~20
	4.0	~10
	5.0	0

[a]Wavelength 254 nm.

Table 3 compares the ability of select eukaryotes to withstand the killing effects of UV irradiation. For this experiment, the two genera of amoebae which were most resistant to gamma and alpha radiation were selected as the test organisms and compared with a mammalian cell culture line (LT-1). As seen in Table 3, 95.5% of the LT-1 cells were killed after only a 30 s exposure. A 2-minute exposure was required to kill 50% of *Acanthamoebae* and *Hartmannella* cultures, with survival of a portion of the populations occurring even after 4 min. None of the amoebae were able to survive an exposure of 5 min.

Table 4 shows that bacteria, isolated from environmental sources which were not associated with amoebae, were sensitive to gamma radiation, with more than 99% of the populations being killed with only 0.3 Mrad. Many of the bacteria which were isolated from amoebae, i.e., consortium 46C, were able to withstand 0.3 Mrad with a 50% survival rate. Bacterial derivars and amoebae-associated bacterial derivars were by far the most resistant bacteria tested with a significant percentage of the population being able to survive 1 Mrad. As seen in Table 1, these were nevertheless not as resistant to gamma radiation as the parent amoebae. Bacterium 15 isolated from consortium 46C, when sequestered within amoebae, was protected to a certain extent by the amoebae and able to survive 0.3 Mrad when compared with the same bacteria in an unprotected state outside the amoebae which were killed by this dosage.

Surprisingly, like bacterial derivars arising after *Legionella* infection, the appearance of bacterial derivars was also seen after axenic cloned cultures of *A. royreba* were irradiated with 1 Mrad gamma radiation. No bacterial outgrowths were seen in control amoebae populations which were not irradiated. In each

Table 4. Ability of select bacteria and bacterial derivars to survive gamma radiation

	Radiation Dosage (megarads)		
Test organism	**0**	**0.3**	**1.0**
Environmental bacteria:			
Escherichia coli	100[a]	1	nt
Pseudomonas maltophilia	100	0	nt
Aeromonas hydrophilia	100	0	nt
Klebsiella oxytoca	100	0	nt
Flavobacterium sp.	100	1	nt
Legionella pneumophila	100	nt	0
Amoebae-associated bacteria:			
#1, 46C[b]	100	50	nt
#2, 46C	100	50	nt
#3, 46C	100	50	nt
#4, 46C	100	50	nt
#5, 46C	100	50	nt
#9, 46C	100	20	nt
#15, 46C	100	0	nt
#15, 46C[b]	100	+[c]	nt
Bacterial derivars:[d]			
#1	100	50	10
#2	100	50	10
#3	100	80	50
#4	100	80	50

nt, not tested.
[a]Approximate percent survival.
[b]Bacterium #15 irradiated while sequestered inside *A. royreba.*
[c]Percent survival could not be determined due to bacterial association with amoebae.
[d]Bacteria other than *Legionella* which appeared after the co-infection of *Acanthamoebae* with *Legionella*.

experiment, the radiation-induced bacterial derivars differed morphologically, and they shared unusual properties which included pigmentation, crystal formation, biodispersant production, and lack of reactivity with many carbohydrates, and some require specialized media for growth (Table 5). Subsequent DNA fingerprinting of bacterial derivars revealed bands which differed from those seen in amoebae (Fig. 1). Some derivars have similar DNA profiles while others appeared unrelated. Various clones of *A. royreba* had identical DNA profiles.

Table 5. Comparison of Legionella and radiation derivars vs. extra-amoebic bacteria

Observation	*Legionella* derivar	Radiation derivar	extra-amoebic bacteria
Production of biogenic crystals	+	+	-
Production of biodispersants	+	+	rare
Production of both crystals and biodispersants	+	+	-
Pigmented	+	+	rare
Ability to ferment common sugars[a]	-	-	+
Degradation of complex polymers	+	+	rare

[a]Sugars tested include: glucose, mannitol, inositol, sorbitol, rhamnose, sucrose, xylose, lactose, melibiose, amygdalin, and arabinose.

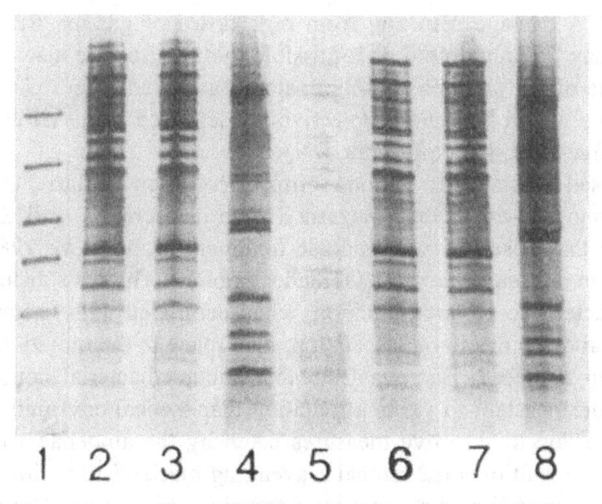

Fig. 1. *Lanes 2, 3, 6, and 7* show DNA profiles of *A. royreba* clones 4, 5, 7, and 11, respectively. *Lanes 4, 5, and 8* show DNA profiles of bacterial derivars isolated from *A. royreba* clones 5, 7, and 11. The biomarker profile is shown in *lane 1*.

Discussion

Two unusual observations resulted from this study. First, the marked resistance of some free-living amoebae to various forms of radiation was evident, and second, the appearance of bacterial derivars in some clones of gamma irradiated *A. royreba* was surprising.

Killing of eukaryotic cells by gamma radiation is generally mediated through DNA damage by oxygen radicals, while death by alpha or UV radiation generally results from direct DNA damage or formation of thymidine dimers, respectively. Considering that mammalian cells can only tolerate 800-1000 rads of gamma irradiation, the ability of *Acanthamoebae* to tolerate 1 million rad is surprising. Similar, but somewhat less dramatic, was the resistance of *Acanthamoebae* to alpha and UV radiation when compared with mammalian cells.

The ability of free-living amoebae per se to survive high doses of gamma, UV and alpha radiation is poorly understood. It is possible that they either possess (alone or in tandem) one of several known repair mechanisms or contain extremely potent radical scavengers which serve to protect the amoebic, as well as the bacterial, DNA from damage. Alternatively, the amoebae may possess an as yet unknown protective/repair mechanism. Since bacteria which are sequestered inside amoebae are also more resistant to gamma radiation, it is possible that the same mechanism(s) which protect the amoebae also allow for the survival of intra-amoebic bacteria when subjected to large doses of radiation. For instance, we have observed that *Acanthamoebae* contain high levels of catalase (unpublished results) which is known to protect against hydrogen peroxide, one of the major causes of DNA damage resulting from high doses of gamma irradiation. Since *A. royreba* may be polyploidic, it is possible that this feature also allows for survival of some of the amoebic genetic material after radiation exposure. It is unlikely that this would have any protective influence on bacterial endosymbionts, L-forms, or free-floating cytoplasmic DNA.

The increased resistance to various forms of radiation (gamma, UV, and alpha) seen with amoebae-associated bacteria (bacterial derivars and *Legionella*) is analogous to that observed with biocide treatment (King et al. 1988) where the amoebae serve as a sanctuary for the endosymbiont. The resistance of amoebae-associated bacteria to gamma radiation, when compared with bacteria which did not have an amoebic origin or association, may relate to the radiation resistance of the associated amoebae. Bacteria from the amoebic/bacterial consortium (46C) were also more resistant to gamma radiation than typical environmental isolates. This could be due to selective measures taken by the amoebae when engulfing bacteria, or the result of some radical scavenging mechanism within the amoebae, or it could be due to their previous intra-amoebic association. The intra-amoebic bacteria and all of the culturable derivars share certain commonalities which, in combination, are unusual in bacteria which have had no amoebic association and may impart additional protection against the types of radiation insults studied herein. These characteristics include being highly pigmented, nonfermentative, and catalase positive. In addition, many of these intra-amoebic bacteria produce biogenic crystals, biodispersants, degrade complex polymers, and lack recognizable enzymatic capabilities, as shown in Table 5.

Somewhat of a mystery is the fact that radiation-induced derivars also have increased resistance to gamma radiation, similar to that seen in the amoebae from which they were isolated. As noted previously with *Acanthamoebae*, bacterial derivars and bacteria isolated from amoebae also contain high levels of catalase. These radiation induced derivars also share the characteristics of other intra-amoebic bacteria. These derivars were detectable only after months of room temperature incubation of the irradiated cloned amoebae in sealed test tubes. Controls consisting of unirradiated amoebae never yielded bacteria, and none were seen in the unirradiated controls using various staining procedures such as acridine orange, DAPI, Gram, Giemsa, and acid fast. One or more of these staining procedures will reveal the presence of endosymbionts as well as membrane deficient forms, i.e., L-forms. All the radiation derivars are morphologically different and present similar yet different DNA profiles. All the DNA profiles of *A. royreba* are very similar, if not identical, and show few similarities with the radiation-induced bacterial derivars. While the possibility of the expression of a latent endosymbiont activated by radiation insult has not been totally ruled out, many observations tend to indicate that this has not occurred. Firstly, no bacteria were ever isolated from control tubes, nor were they ever seen in the amoebae by various staining procedures. Secondly, all the derivars differ morphologically and in their DNA profiles. Thirdly, the derivars share common characteristics with the *Legionella* derivars and those of other known intra-amoebic bacteria, and these characteristics appear to be common to amoebae-associated bacteria. Finally, the radiation-induced derivars are much more resistant to gamma radiation than either intra-amoebic *Legionella* or environmental bacteria with no amoebic association.

Various possible scenarios may explain the above and previous observations:

1. Although not noticeable in these studies, nonculturable endosymbionts have been seen by other investigators in *Acanthamoebae* cultures, and infection of *Acanthamoebae* by *Legionella* or irradiation may rescue bacterial endosymbionts from the amoebae.
2. Genetic recombination of the bacterial genome from the *Legionella* inoculum with genetic information from incomplete bacterial or virus-like particles potentially present in *Acanthamoebae* may result in bacteria that were not originally present.
3. A combination of amoebic genetic information with bacterial DNA of the inoculum may result in the formation of newly constituted bacteria.

Whatever the mechanisms, the results to date indicate that some protozoan populations such as *Acanthamoebae* may provide a mechanism for rapid genetic rescue or recombination, resulting in previously undetected or unculturable microbes.

As seen in DNA fingerprints of irradiation-induced bacterial derivars and *A. royreba*, only a few DNA bands in the bacterial derivars were analogous to the bands from *A. royreba*. Major DNA bands found in the bacterial derivars which did not appear in the amoebae can now be exploited to produce genetic probes. These molecular probes prepared against the newly derived bacteria and reacted with amoebic DNA should indicate which of the previously mentioned scenarios are operative.

Acknowledgments: We would like to thank Peter Gresshoff, University of Tennessee, for assistance in interpretation of DNA fingerprint gels and Peggy

Terzhagi-Howe, Oak Ridge National Laboratory, for data on mammalian cell sensitivity to alpha radiation.

References

Caetano-Anolles G, Bassam BJ, Gresshof PM (1992) Mol Gen Genetics. 235: 157-165

Cote R (ed) (1984) ATCC Media Handbook, First Edition. American Type Culture Collection, Rockville, MD

Dietz AJ, Vass AA, Mackowski RP, Tyndall RL (1994) Abstracts of Annual Meeting, American Society for Microbiology. p 130

Keller LC, Maxcy RB (1984) Appl Environ Microbiol 47(5): 915-918

King CH, Shotts Jr EB, Wooley RE, Porter KG (1988) Appl Environ Microbiol 54: 3023-3033

Tyndall RL, Dominigue EL (1982) Appl Environ Microbiol 44: 954-959

Tyndall RL, Ironside KS, Little CD, Katz DS, Kennedy JR (1991) Appl Biochem Biotechnol 28/29: 917-925

Tyndall RL, Vass AA (1993) Endocytobiology V, Fifth International Colloquium on Endocytobiology and Symbiosis, pp 515-522. Tübingen University Press

Tyndall RL, Vass AA, Fliermans CB (1993) *Legionella*, Current Status and Emerging Perspectives. In: Barbaree JM, Breiman RF, Dufour AP (eds) American Society for Microbiology, ASM Press, Washington DC, pp 142-145

Endosymbiosis of *Sogatodes orizicola* (Muir; Insecta) With Yeast-Like Symbionts

E. Kreil, H. Tauchert, G. Hoheisel, and S. Richter
Department of Microbiology and Genetics, Institute of Zoology, Faculty of Biosciences, Pharmacy and Psychology, University of Leipzig, Talstrasse 33, 04103 Leipzig, Germany

Key words: *Sogatodes orizicola* (Muir), insects, endosymbiosis, TEM-identification, preparation of endosymbionts (YLS).

Summary: We demonstrate that yeast-like symbionts, which are apparently endosymbionts, are present and multiply in the fat body of the insect *Sogatodes orizicola*. Transfer mechanisms explain the essential presence for host organisms. Ultra-structural studies established the eukaryotic nature of these yeast-like symbionts. A method for their efficient isolation is described, and initial results showing the transfer of radiolabelled phosphate from the host to the symbiont are presented.

Introduction

Endosymbiosis of insects with microorganisms is a well-known phenomenon, especially since the intensive investigations of Paul Buchner (1965). In this chapter, we investigate an example in the plant sap-sucking insect *Sogatodes orizicola,* a member of the family Delphacidae in the order Homoptera. This insect has considerable economic impact as a pest, causing extensive damage to the rice plant, upon which it feeds. Accordingly, various attempts have been made to control its reproduction, for example by infection with symbionts (Fröhlich 1989) . For this, and for deeper theoretical understanding, it is necessary to understand the exact nature of the symbionts, as well as the mechanism of their transfer between generations.

Former investigations of Lienig (1983, 1993) and Fröhlich (1989) showed that *Sogatodes orizicola* develops through five larval stages, the duration of which are temperature-dependent. In younger larval stages, symbionts are diffusely distributed in the fat body. From the fourth larval stage onwards, however, they concentrate in syncytial ranges near the ovarioles. From morphological investigations, Lienig (1993a,b) and Fröhlich (1989) conclude that these symbionts infect primary oocytes in much the same way as described for *Laodelphax striatellus* (Fall)

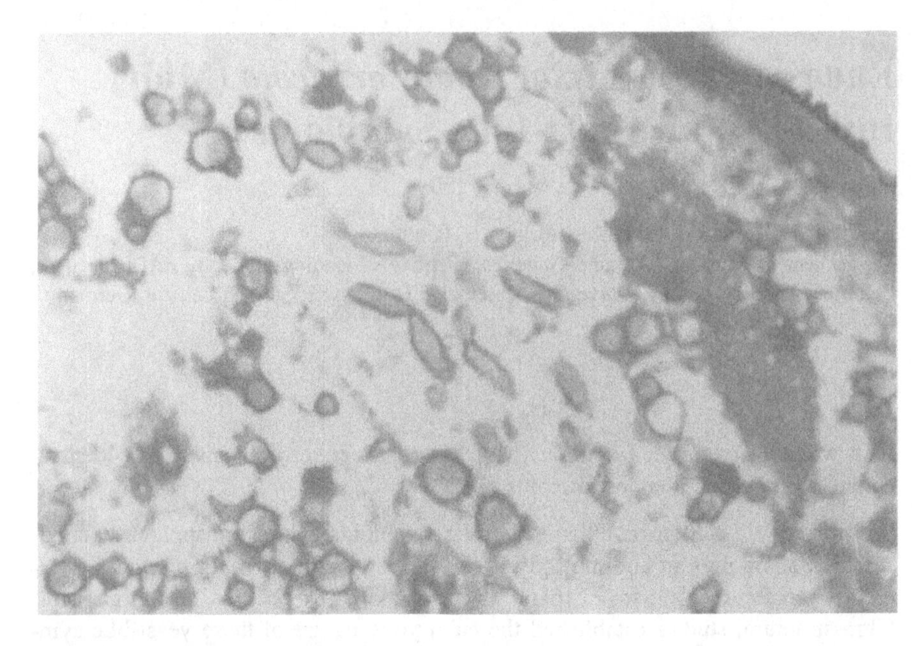

Fig. 1. Micrograph of *Sogatodes orizicola* (female, abdomen), overview (100x and 400x original)

by Noda and Kowahara (1993). Lienig and Fröhlich also described the formation and growth of a symbiont ball at one pole of the egg cell by replication of symbionts in accordance with the observations of *Euscelidius variegatus* by Schwemmler (1991). During embryogenesis, these symbionts infect cells of the fat body and replicate there.

The main symbionts of *Sogatodes orizicola* appear yeast-like in form and shape. However, their true nature is still unknown due to the lack of ultrastructural evidence for substructures characteristic of eukaryotic cells.

Material and Methods

Sogatodes orizicola was cultivated on rice plants as *in vitro* cultures according to Lienig (1993a,b) and Fröhlich (1989). Adult insects were collected and prepared for morphological investigations and the isolation of symbionts by the standard techniques described in method books. Difficulties arose due to the high fat content in the abdomen. Therefore, a lot of care was necessary during fixation with glutaraldehyde and osmium tetroxide, the Durcupan embedding procedure and especially during cutting.

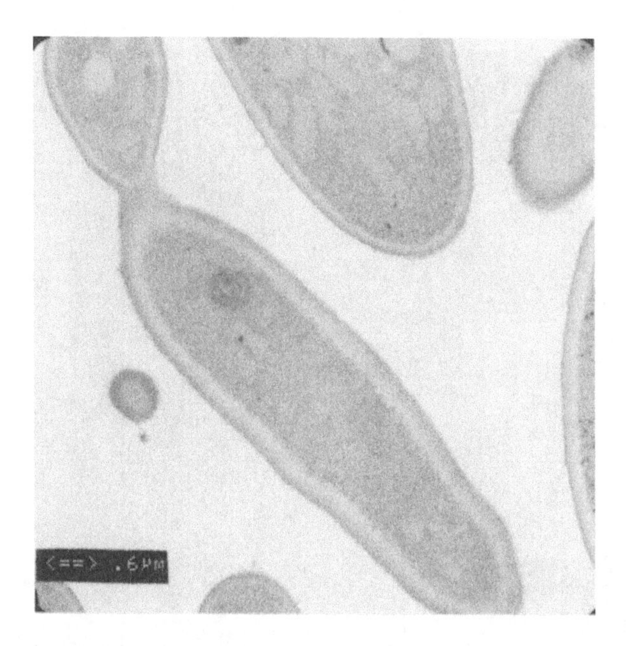

Fig. 2. Transmission electron micrograph of a symbiont cell in abdomen of *Sogatodes orizicola* with budding

Results

Adult male and female insects were first investigated by light microscopy. In a transverse section through the abdomen of an adult female (Fig. 1), symbionts can be seen in the fat body. At this phase of development and symbiosis, they are diffusely distributed and oval in shape, measuring approximately 11.5 µm in length and 5 µm in diameter. In several individuals, replication by budding seems to be visible. This assumption was confirmed by transmission electron microscopy.

Figure 2 shows an ultrathin (0.1 µm). section through a bud ligating from the mother cell. Characteristic (yeast-like) cell wall structures are visible in both these compartments. At a 20000-fold magnification (Fig. 3), all of the substructures characteristic for eukaryotic cells are revealed. For example, a nucleus with double membrane, rough endocytoplasmic reticulum, mitochondria and ribosomes. The supposition that besides the main symbiont, another existed, we could not confirm morphologically until now.

Due to the heat lability of endocytobiotes, exposure to temperatures in excess of 32 °C for several days leads to the death of the host when adult and the prevention of adult metamorphosis in larvae.

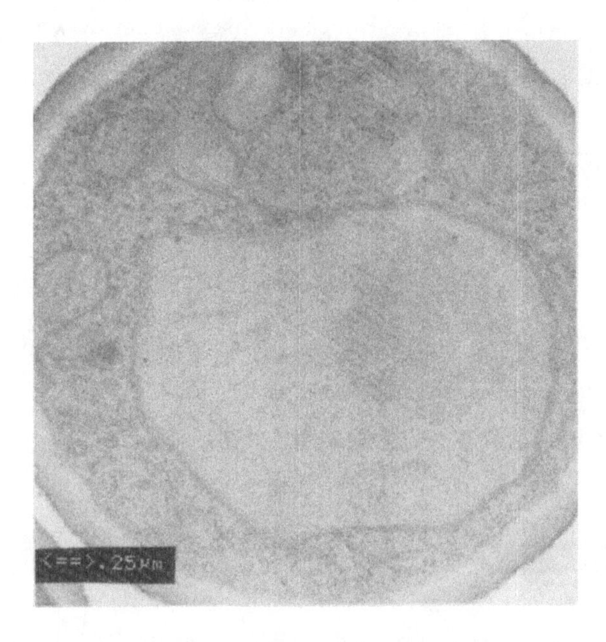

Fig. 3. Cellular substructures of a symbiont cell from the abdomen of a female *Sogatodes orizicola*

By adding radiolabeled phosphate (40 Mbg ^{32}P) to the nutrition fluid of the rice plants and subsequent autoradiography, we could follow substrate transfer from the rice plant to the insects and to the yeast-like symbionts (Fig. 4). Transmission of labeled phosphate to the symbionts could only be estimated after their isolation. In accordance with the recommendations of Noda and Omura (1992), we used a modified separation method with centrifugation in a Percoll gradient. In this way, we were able to isolate a very homogeneous fraction of symbiontic yeast-like cells in which ^{32}P-radioactivity was also detectable.

Discussion

The preliminary investigation of Lienig (1993a,b) and Fröhlich (1989) is in general agreement with the results presented here. We could show that *Sogatodes orizicola* lives in endocytobiosis with yeast-like symbionts, demonstrate the asexual replication of the symbionts, and discovered that temperature exceeding 32 °C are lethal for the adult insects and precludes adult development in larvae. The latter result is quite likely due to the reduction or loss of symbionts. Ultrastructural investigations established that these symbionts are yeast-like eukaryotic cells.

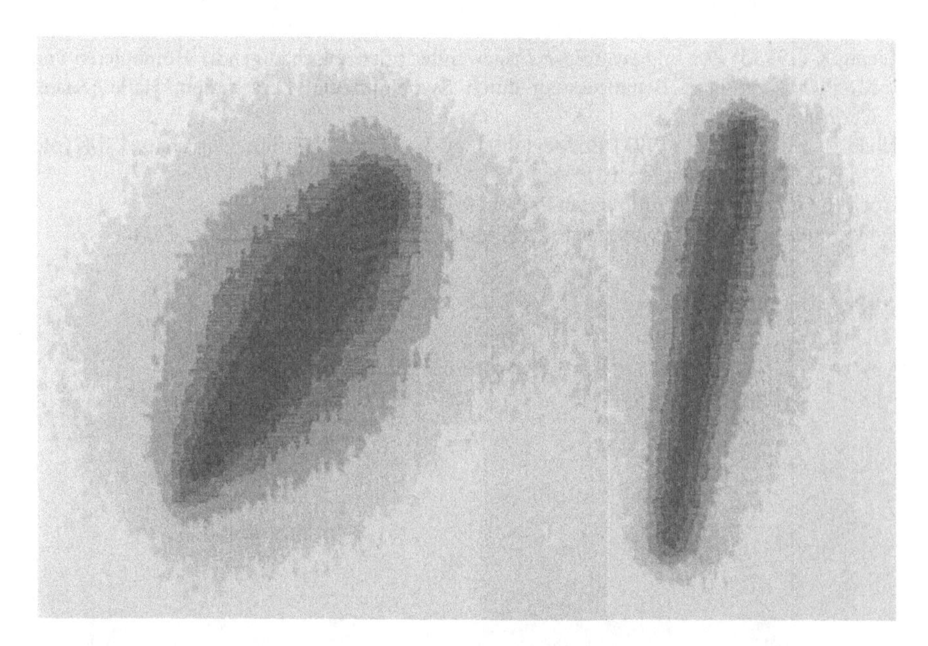

Fig. 4. Autoradiography of rice leaves and of one animal (*Sogatodes orizicola*) after administration of ^{32}P to the nutrition fluid of the rice plant

Endocytobiosis with yeast-like symbionts has been reported in only very few cases up until now (Fröhlich 1989, Schwemmler 1991), and it is still not very well understood. The first steps towards unraveling the complexity of host-symbiont interactions have now been taken. Symbionts being in the extracellular stage can be isolated by preparation and centrifugation on a Percoll density gradient, and initial experiments showing the incorporation and transfer of radiolabeled phosphate have been achieved. Further studies on the metabolism of symbionts and on their genome structure are in progress.

Acknowledgments. Support by the Deutsche Forschungsgemeinschaft is gratefully acknowledged.

References

Buchner P (1965) Endosymbiosis of animals with plant microorganisms. John Wile & Sons, New York

Fröhlich G (1989) Beitr Ent 39: 393-412

Harrison CP, Douglas AE, Dixon AFG (1989) J Invertebr Pathol 53: 427-428

Ishikawa H (1982) Comp Biochem Physiol 72 B: 239-247

Lienig K (1993a) Beitr Ent 43: 445-449

Lienig K (1993b) Zur Endosymbiose ausgewählter pflanzenschädigender Homopteren und Möglichkeiten ihrer Beeinflussung durch Symbiontizide, Dissertation, Halle (Saale) Germany
Noda H, Kowahara N (1993) Endocytobiology V, 151-153 Philips University Uji-Kyoto (Japan), Tübingen University Press
Noda H, Omura T (1992) J Inverteb Pathol 59: 104-105
Schwemmler W (1991) Symbiogenese als Motor der Evolution. Paul Parey, Berlin

A Chaperonin-Like in the Principal Endocytobiotes of the Weevil *Sitophilus*

H. Charles, A. Heddi and P. Nardon
Laboratoire de Biologie Appliquée INSA-406, UA-INRA 203, SDI-CNRS 5128, 20 Avenue A. Einstein, 69621 Villeurbanne Cedex, France

Key words: chaperonin, endocytobiosis, symbiosis, *Sitophilus*, weevil.

Summary: Chaperonins are ubiquitous proteins found in all prokaryotic and eukaryotic cells. They are expressed at high level in several parasitic bacteria and in at least two types of endocytobiosis: in amoebae and in aphids.

In this study, we have investigated the presence of a chaperonin-like, named symbionin, in the principal endocytobiotes of the weevil *Sitophilus*. This protein of 60 kDa (pHi = 6.1) shares a high immunological homology with the groEL chaperonin of *E. coli* and is present in the three symbiotic species of *Sitophilus* (*S. oryzae*, *S. granarius* and *S. zeamais*). In isolated endocytobiotes, the symbionin represents 40% in quantity of the total expressed proteins. These results show the important role of symbionin in the maintenance of the symbiocosme.

Introduction

Chaperonins are ubiquitous proteins which ensure that the folding of polypeptides and their assembly into oligomeric structure occur correctly (Hendrick and Hartl 1993). They are often described in parasitic bacteria because of their high immunogenecity (Shinnick 1991). However, the biochemical function of the chaperonins during the invasion of the host by the bacteria is not known. Considering that endocytobiotes could have evolved from parasitic bacteria (Paillot 1933), it is very interesting to know whether symbiotic organisms still express such a protein or not. It must be pointed out that a chaperonin is expressed at high level in some of the oldest known descendants of endocytobiotes such as mitochondria and chloroplasts (Hemmingsen et al. 1988; Gupta et al. 1989).

In the case of endocytobiosis, a chaperonin-like protein has already been described in two models: the pea aphid *Acyrthosiphon pisum* (Ishikawa 1984) and the amoebae *Amoebae proteus* (Jeon 1983). In these two models, the genes encoding the chaperonin are located downstream of an *E. coli* groES-like gene. Both genes (symL and groEx) are heat shock regulated (Ohtaka et al. 1992; Ahn et al. 1994).

In this study, we have investigated the presence of a chaperonin-like protein in our endocytobiosis model: the weevil *Sitophilus*. This coleoptera (Curculionidae) harbors symbiotic Gram-negative bacteria located in the larval bacteriome and transmitted to the offspring by maternal inheritance (Nardon 1971). The host and the bacteria share very intimate biochemical and genetic interactions. For instance, symbiotes supply the host with vitamins and interfere therefore with its metabolism (Gasnier-Fauchet and Nardon 1987; Wicker 1983). As physiological consequences, we noticed that the symbiotic bacteria increase the fertility and the development rate of the host insect (Nardon and Grenier 1988). Moreover, symbiotic bacteria are completely dependent on the host genome since they cannot divide outside the host cell. They are also involved in an unidirectional cytoplasmic incompatibility phenomenon (Nardon and Grenier 1991 1993).

Four species of *Sitophilus* were studied: *S. oryzae, S. granarius S. zeamais* and *S. linearis*. The weevil chaperonin, called symbionin - in reference to the pea aphid symbionin (Ishikawa 1984) - shows a high immunological homology with the GroEL protein of *E. coli*. Analysis of various biochemical properties (MW, pHi and amino acid composition) leads to classification of this protein within the Hsp60 family. Immunoblotting experiments show that symbionin is overexpressed and accumulated inside the endocytobiotes. Taken together, these observations and those from the literature point out the principal role that symbionin plays in the maintenance of symbiotic equilibrium inside the symbiocosme.

Material and Methods

Insects. *S. oryzae*, *S. granarius* and *S. zeamais* were reared on wheat at 27.5 °C and 75% relative humidity. The fourth instar larvae were taken from inside the grain 21 days after egg laying (28 days for the aposymbiotic strain) and were then homogenized for electrophoresis or fixed for immunohistochemistry.

Insect Medium (MSM). Tris-HCl (20 mM) pH 7.2; Mannitol (250 mM); EDTA (1 mM); $MgCl_2$ (5 mM); KH_2PO_4 (10 mM); KCl (20 mM); GTP (0.5 mM); ATP (2 mM); ADP (2 mM); Pyruvate (5 mM); malate (5 mM); amino acid pool (0,5 mM); BSA (0.4% w/v).

Immunohistochemistry. Fourth instar larvae were fixed in alcoholic Bouin's solution and embedded in paraffin. Sections were mounted on poly-L-lysin coated microscope slides, treated 30 min with 2% H_2O_2 in methanol and probed with anti-symbionin antiserum using Vectastain Elite ABC Kit (Vector). The immunostained sections were then counterstained with 0.1% toluidine blue. For the control, preimmune serum was used instead of anti symbionin antiserum.

Bidimensional Electrophoresis. Isolated endocytobiotes (see isolation method in Heddi 1991) of fourth instar *Sitophilus oryzae* larvae were homogenized (1 ml/bacteriome) in MSM saturated with phenyl-thiourea (PTU). The sample was then desalted by ultrafiltration and resolubilized in solution A: urea (9.5 M); Triton X-100 (16% v/v); dithiothreitol (5% w/v); ampholytes 5/7 (4% v/v); ampholytes 3/10 (2% v/v). The proteins were then isoelectrofocuzed in capillary tubes of 0.5 mm and finally separated by SDS-PAGE.

Immunoblotting. The proteins extracted from endocytobiotes of fourth instar *Sitophilus oryzae* larvae were separated by SDS-PAGE, transferred to a nitrocellulose membrane and probed with anti-GroEL antiserum using Vectastain Elite ABC Kit (Vector).

***In vivo* Protein Synthesis and Western Blot Analysis.** Isolated symbiotic bacteria were incubated for 1 h at 27 °C in sterile MSM medium with ^{35}S methionine (Sp. Act.: 1000 Ci/mmol, 50 µCi/ml). They were centrifuged at 8000 g for 10 min and crushed in solution B : Tris-HCl (1.5 M) pH 6.8; dithiothreitol (1,5% w/v); SDS (2% w/v); glycerol (11% v/v); bromophenol blue (75 mg/ml). The neosynthesized labeled proteins were then separated on 10-20% polyacrylamide gel and detected by fluorography. The intensities of protein bands were scanned with the image processing and analysis program *NIH Image* (National Institutes of Health, USA).

Amino Acid Composition. After SDS-PAGE electrophoresis, the symbionin band was transferred to an Immobilon membrane and hydrolysed in 6N HCl containing 1% thioglycolic acid at 120 °C for 24 h. Amino acids in the hydrolysate were resolved through an autoanalyzer Beckman 6 300.

Results

Immunohistochemistry of Cross Sections

The use of a polyclonal antibody against the chaperonin GroEL of *E. coli* (Hara et al. 1990) allowed us to reveal the presence of symbionin in all symbiotic strains of *Sitophilus* (*S. oryzae, S. granarius* and *S. zeamais*). The specific coloration of the antibody was found in the larval bacteriome and in the apical bacteriome of the adult female ovaries (Fig. 1A, B). The other tissues were free of symbionin. Figure 1C shows an enlargement of a *S. granarius* ovocyte with some specifically stained endocytobiotes.

This prokaryotic protein was not found in the aposymbiotic species *S. linearis* nor in the heat-treated aposymbiotic strain of *S. oryzae*.

Biochemical Characterization

Whole proteins of isolated endocytobiotes were extracted and separated on a bidimensional electrophoresis gel (Fig. 2). A molecular weight of 57 kDa and a pHi of about 6.1 were found for the *Sitophilus oryzae* symbionin. Immunoblotting experiments (Fig. 3) also confirm the presence of symbionin inside the endocytobiotes. Finally, the amino acid composition of the protein was determined and compared with those of three other chaperonins: GroEL of *E. coli*, symbionin of

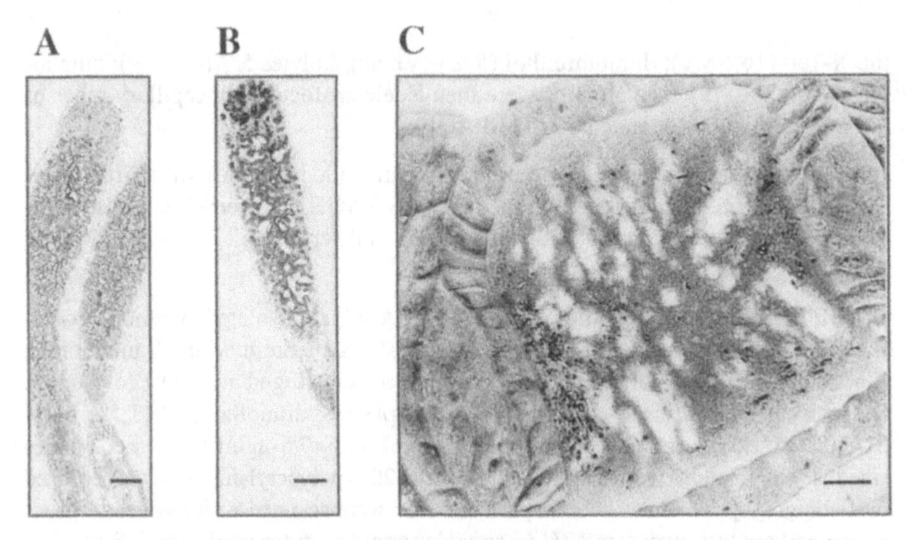

Fig. 1. Immunohistochemistry of *Sitophilus granarius* cross sections. **A** and **B** Adult ovaries and their apical bacteriome; bars = 100 µm. For the control **A**, preimmune serum was used instead of anti-symbionin antiserum. **B** The presence of symbionin was visualized in the dark inside the bacteriome **B**. **C** An enlarged image of an ovocyte with immunostained endocytobiotes at the posterior pole; bar = 10 µm)

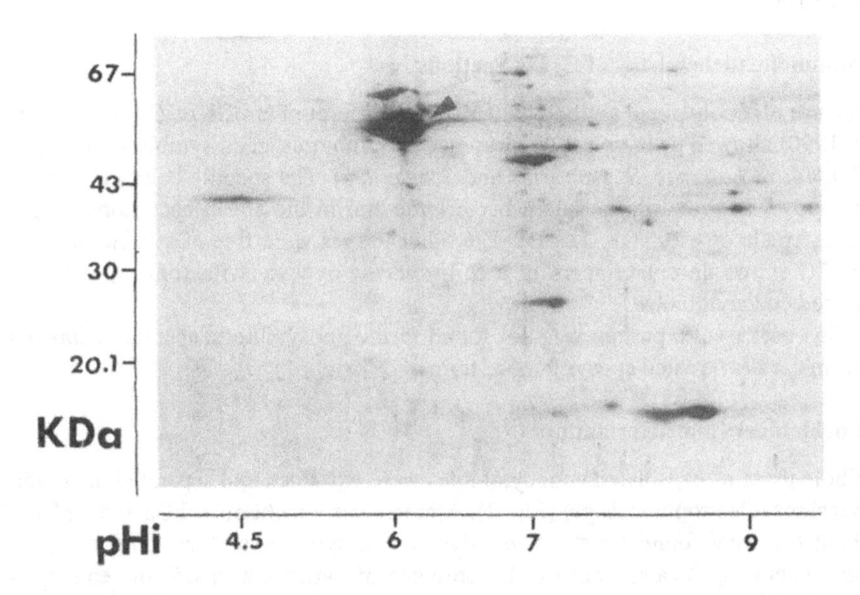

Fig. 2. Bidimensional electrophoresis. Proteins of isolated endocytobiotes of fourth instar *Sitophilus oryzae* larvae were separated on bidimensional electrophoresis gel. The symbionin is visualized on the gel (see arrow head)

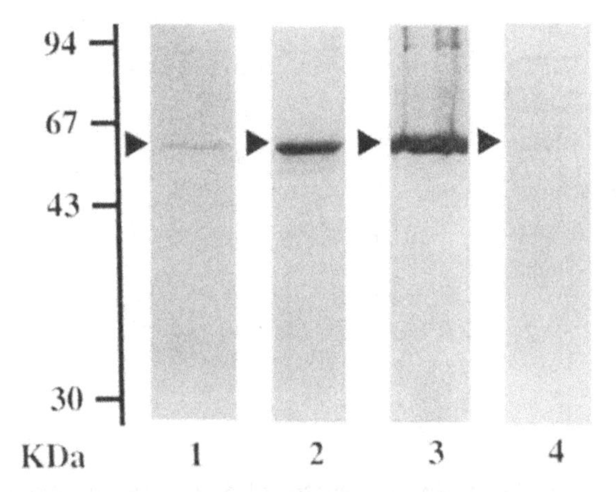

Fig. 3. Immunoblotting. The extracted proteins were separated by SDS-PAGE, transferred to a nitrocellulose membrane and probed with anti-GroEL antiserum. *1* Whole proteins of the symbiotic strain of *S. oryzae*; *2* proteins of isolated bacteriocytes; *3* proteins of isolated endocytobiotes; *4* whole proteins of the aposymbiotic strain of *S. oryzae*.

Acyrthosiphon pisum endocytobiotes and GroEx of *Amoebae proteus* endocytobiotes (Table 1).

In vitro Neosynthesized Proteins

The isolated endocytobiotes cannot divide outside their host, but they can survive for more than 1 month in MSM medium. Their viability was assessed under the microscope by the movement of the bacteria (Fig. 4). From this analysis it was found that a large part of the bacterial metabolism *in vitro* is focused on the production of symbionin. The band corresponding to symbionin represents about 40% of total neosynthesized proteins but the purpose of this important energetic cost for the bacteria is still not understood.

Discussion

All the studied symbiotic species of *Sitophilus* harbor endocytobiotes expressing symbionin at a high level. This protein shares the biochemical characteristics of the Hsp60 protein family. Its molecular weight is 57 kDa and its pHi is about 6.1. Moreover, its global amino acid composition is very close to that of other chaperonins.

As symbionin is expressed at a high level inside the symbiotic bacteria, we can thus assume that it has an important function in the symbiotic relationships. The most interesting point is to understand why this protein is produced by the bacte

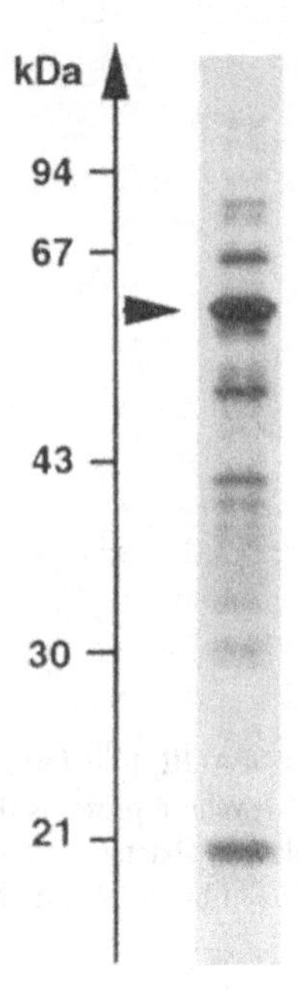

Fig. 4. Endocytobiotes of *S. oryzae*. Electrophoresis of *in vitro* neosynthesized proteins. Symbiotic bacteria were, directly after isolation, incubated for 1 h at 27 °C in sterile MSM medium with ^{35}S methionine. After electrophoresis, the intensities of protein bands were scanned with the *NIH-image* program

Table 1. Amino acid compositions (% mol) of *S. oryzae* symbionin. Comparison between the amino acid composition of *S. oryzae* symbionin and three other chaperonins: GroEL (*E. coli*), symbionin (*Acyrtosiphon pisum* endocytobiotes) and GroEx (*Amoeba proteus* endocytobiotes). The amino acid composition of the three chaperonins were found in Swiss prot database

	S. oryza symbionin	*E. coli* GroEL	*A. pisum* symbionin	*A. proteus* GroEx
Ala	8,7	13,7	10,8	12,36
Arg	3,9	4	6,8	4,91
Asx	11,2	11,8	11,2	9,82
Glx	10,9	9,8	8,8	11,27
Gly	12	10,4	10,5	10,55
His	1,7	0,2	0,7	0,73
Ile	5,5	6	6,7	6,73
Leu	8,7	7,5	9,3	6,55
Lys	7,2	7,3	8,3	7,09
Met	2	4	4,3	5,27
Phe	3,2	1,3	1,5	1,82
Pro	5,15	2,6	1,5	2,00
Ser	6,2	3,1	5,1	4,55
Thr	4,5	6	3,5	5,82
Tyr	2,4	1,3	1,9	0,91
Val	6,9	10,6	9,3	9,09

ria. The presence of such a protein in parasitic bacteria as well as in three models of endocytobiosis (the weevil, the pea aphid and amoebae) corroborate with a basic function, conserved during evolution, in the maintenance of the symbiotic equilibrium.

Finally, we have formulated four hypotheses on the possible function of the symbionin in the weevil: (1) the symbionin could be a storage protein, the host could then use it by lysing its own endocytobiotes; (2) the symbionin is expressed to fold and assemble the endocytobiote's own proteins; (3) the symbionin could fold and assemble some host proteins imported from the host cytosol, these proteins could be used by the bacterial machinery or re-exported in the host cytosol afterwards; (4) the protein syntheses could be induced by the host, and the significant energetic cost for the bacteria could inhibit its growth potential. By this means, the host could control the bacterial division and preserve itself from invasion by the bacteria.

Currently, the two first hypotheses do not seem very probable (Charles et al. 1995), but it is still not an obvious choice between the two last hypotheses. In one case, the protein seems to be beneficial to the bacteria, and in the other case to the host insect. We are investigating the biochemical mechanism of the symbionin in one way and the expression and the regulation of its encoding gene in another way. These experiments will surely help us to better understand this fascinating model.

Acknowledgments. We would like to thank Dr. Ishikawa for the kind gift of anti-GroEL antibody and for useful discussions.

References

Ahn TI, Lim ST, Leeu HK, Lee JE, Jeon KW (1994) A novel strong promoter of the *groEx* operon of symbiotic bacteria in *Amoeba proteus*. Gene 128: 43-49

Charles H, Ishikawa H, Nardon, P (1995) Presence of a protein specific of endocytobiosis (symbionin) in the weevil *Sitophilus*. C R Acad Sci Paris, Life sciences 318: 35-41

Gasnier-Fauchet F, Nardon P (1987) Comparison of sarcosine and methionine sulfoxide levels in symbiotic larvae of two sibling species, *Sitophilus oryzae* L. and *S. zeamais* mots. (Coleoptera : Curculionidae) Insect Biochem 17: 17-20

Gupta RS, Picketts DJ, Ahmad S (1989) A novel ubiquitous protein chaperonin supports the endosymbiotic origin of mitochondrion and plant chloroplast. Biochem biophys Res Comm 163: 780-787

Hara E, Fukatsu T, Ishikawa H (1990) Characterization of symbionin with anti-symbionin antiserum. Insect Biochem 20: 429-436

Heddi A, Lefèbvre F, Nardon P (1991) The influence of symbiosis on the respiratory control ratio (RCR) and the ADP/O ratio in the adult weevil *Sitophilus oryzae*. (Coleoptera, Curculionidae) Endocyt & Cell Res 8: 61-73

Hemmingsen SM, Woolford C, Vies SM, Tilly K, Dennis DT, Georgopoulos CP, Hendrix RW, Ellis RJ (1988) Homologous plant and bacterial proteins chaperone oligomeric protein assembly. Nature 333: 330-334

Hendrick JP, Hartl FU (1993) Molecular chaperone functions of heat-shock proteins. Ann Rev Biochem 62: 349-384

Ishikawa H (1984) Characterization of the protein species synthesized *in vivo* and *in vitro* by an aphid endosymbiont. Insect Biochem 14: 417-425

Jeon KW (1983) integration of bacterial endosymbionts in Amoeba. Int R Cytol Supplement 14: 29-47

Nardon P (1971) Contribution à l'étude des symbiotes ovariens de *Sitophilus sasakii* : localisation, histochimie et ultrastructure chez la femelle adulte. C R Acad Sci Paris 272D: 2975-2978

Nardon P, Grenier AM (1988) Genetical and biochemical interactions between the host and its endocytobiotes in the weevils *Sitophilus* (Coleoptera, Curculionidae) and other related species. In: Scannerinni S, Smith DC, Bonfante-Fasolo P, Gianinazzi-Pearson V ed. Cell to Cell Signals in Plant, Animal and Microbial Symbiosis, NATO ASI Series H17. Berlin, Springer Verlag, 255-270

Nardon P, Grenier AM (1991) Serial endosymbiosis theory and weevil evolution: the role of symbiosis. In: Symbiosis as a source of evolutionary innovation. L Margulis and R Fester, 153-169

Nardon P, Grenier AM (1993) Symbiose et evolution. Ann Soc Entomol Fr (NS) 29: 113-140

Ohtaka C, Nakamura H, Ishikawa H (1992) Structures of chaperonins from an intracellular symbiont and their functional expression in *Escherichia-Coli* GroE mutants. J Bacteriol 174: 1869-1874

Paillot A (1933) L'infection chez les insectes - Immunité et symbiose. Trévoux. G Patissier

Shinnick TM (1991) Heat shock proteins as antigens of bacterial and parasitic pathogens. In: Current topics in microbiology and immunology. Heatshock proteins and immune response. SHE Kaufman (ed). Berlin, Springer-Verlag, 145-160

Wicker C (1983) Differential vitamin and choline requirements of symbiotic and aposymbiotic *S. oryzae* (Coleoptera : Curculionidae). Comp Biochem Physiol 76A: 177-182

2.3 Symbiotic Plant Microbe/Fungus Interactions

Host Signals Dictating Growth Direction, Morphogenesis and Differentiation in Arbuscular Mycorrhizal Symbionts

M. Giovannetti
Dipartimento di Chimica e Biotecnologie Agrarie, Centro di Studio per la Microbiologia del Suolo, Università di Pisa, Via del Borghetto 80, 56124 Pisa, Italy

Key words: symbiosis, arbuscular mycorrhizas, morphogenesis, infection structures, host signals, *Glomus mossae, Ocimum basilicum, Pisum sativum*

Summary: Arbuscular mycorrhizal (AM) fungi are biotrophic organisms living in the roots of host plants. They produce spores in the soil which are capable of germination in the absence of host roots. Thus, fungal hyphae have evolved an efficient system for locating their specific hosts, which have assured their survival for 400 million years.

AM symbiosis is established through a cascade of recognition events leading the two partners, host plant and fungal symbiont, to complete morphological and physiological integration. Our studies have shown that host-derived signals are the early cues to which AM fungi respond, by changing the direction of hyphal elongation and initiating a different branching pattern, prior to the formation of infection structures. Such morphogenetic events have been described in detail, using a simple experimental model system, capable of enhancing the phenomenon.

Spatio-temporal developmental studies allowed time-sequencing of the infection process, showing that host-derived signals act as triggers of cellular commitment to the symbiotic status in AM fungi. Many challenges remain concerning the nature of host-derived signals; the problem of which molecules are involved in fungal morphogenesis, and the question as to whether the signaling system functions by a morphogenetic gradient or by yes/no answers.

Introduction

Microorganisms play an important role in the functioning and biodiversity of plant ecosystems: they are essential in biogeochemical cycles and particularly in the degradation of organic matter, which makes mineral nutrients available for plant nutrition. Thus, the events occurring in the soil and roots are of primary importance; among these, the most common phenomenon observed in the root systems of different plant species is mycorrhizal symbiosis. In nature, mycorrhizas represent the absorbing root system of about 90% land plants (Harley and Smith

1987; Read 1991). In diverse land plant ecosystems, different symbiotic associations between soil fungi and plants have evolved, originating different mycorrhizal types. Notwithstanding their diversity, all mycorrhizas have in common the fact that mineral nutrition is operated mainly by fungal hyphae, which explore the soil, take up and transfer nutrients to host plants. Each mycorrhizal type has become prominent in a defined biome; for example ericoid mycorrhizas predominate on acidic soils, at high latitudes and altitudes, whereas arbuscular mycorrhizas predominate on mineral soils, at low latitudes. It has been shown that the hyphal network occurring around roots can infect different plant species living in the same habitat and transfer photosynthates from one species to another (Grime et al. 1987; Martins 1993): these findings contributed to the understanding of the main role played by mycorrhizas in the maintenance of biodiversity, particularly in low fertility soils (Giovannetti and Gianinazzi-Pearson 1994).

The real meaning of mycelial networks colonizing the soil has been recently studied in *Armillaria bulbosa*: it has been shown that a unique individual, clone 1, explored with its hyphae about 15 ha of forest soil, reaching the approximate weight of 10 000 kg (Smith, Bruhn and Anderson 1992). Such outstanding extension of fungal hyphae is possible because single hyphae are capable of fusion by anastomosis. Moreover, hyphae are capable of self/nonself discrimination, so that only genetically compatible hyphae can fuse and originate myceliar networks in the soil (Brasier 1992). The same mechanism is at work in those plant ecosystems, for example in Mediterranean vegetation, where all different mycorrhizal types coexist: the hyphae of fungi forming different mycorrhizas never fuse together and many hyphal networks, formed by genetically different fungi meet, intersect and explore the soil for nutrients.

Another striking phenomenon is that these individual mycelial networks of mycorrhizal fungi are able to recognize unequivocally, and discriminate without mistakes, the roots of their hosts from those of the many surrounding nonhosts. Recently, the existence of unequivocal mechanisms of plant/fungus recognition has been shown in arbuscular mycorrhizas by using a simple experimental model system (Giovannetti et al. 1993a, b; Giovannetti et al. 1994). The same system was used in the present work for spatio-temporal studies of recognition responses occurring in AM symbionts in the presence of host-derived signals.

Materials and Methods

Plant and Fungal Material. The following host plant species were used: *Ocimum basilicum* L.; *Pisum sativum* L. The AM fungus *Glomus mosseae* (Kent isolate), maintained in *Medicago sativa* pot cultures was used.

Experimental Design. Sterile seeds were germinated in sterile sand. Six days after germination, the root system of each seedling was sandwiched between two 47 mm diameter millipore membranes (0.45 μm diameter pores). Another membrane, containing ten germinated sporocarps of *G. mosseae*, was placed over the millipore sandwich containing the roots, using clips to hold the membranes together. Alternatively, the root system of each seedling was sandwiched between membranes, one of which contained ten germinated sporocarps of *G. mosseae*. Each sandwiched root system was placed into a 10-cm diameter pot, and then buried with sterile, acid-washed quartz grit (2-5 mm diameter). The experiments were performed under controlled conditions (20-24 °C, 16-h photoperiod, 100 μE $m^{-2}s^{-1}$, 60% RH,) and pots were watered daily.

Sequential harvests were carried out over periods from 0 to 120 h; hyphal growth and/or appressorium formation were monitored by opening the sandwiches and staining the mycelium with 0.05% trypan blue in lactic acid. Arbuscule formation was monitored after clearing and staining the roots (Phillips and Hayman 1970). The roots were examined under a dissecting microscope, and the pieces showing appressoria were selected and mounted on slides for observations under a Reichert-Jung Polyvar light microscope equipped with epifluorescence and Nomarski interference contrast optics. Five replicate plants were used at each harvest.

Results

On the basis of the present spatio-temporal studies and of our previous work (Giovannetti et al. 1993b; Giovannetti and Citernesi 1993), the time sequencing of host root recognition, infection and colonization process is represented in Fig. 1.

The differential growth pattern of *G. mosseae* hyphae was observed 24 h after the beginning of the interaction between the symbiont and the host plants *O. basilicum* and *P. sativum*. As early as 36 h after the beginning of the interaction, infection structures were formed. Appressoria were shown to increase in number with time, to reach values of 20 ± 3/plant in *O. basilicum* after 72 h. The first arbuscules were observed 72 h after the beginning of the interaction between *G. mosseae* and the hosts *O. basilicum* and *P. sativum*, whereas in the *G. mosseae*-*H. annuus* interaction, arbuscules were observed as early as 42 h.

It is interesting to compare the previous results with those obtained with *P. sativum*, in both the susceptible cultivar Frisson (myc^{+}) and the resistant mutant P2 (myc^{-}) (Duc et al. 1989). In Frisson, the infection process proceeded at the same rate as in the other host plants; on the contrary, in the myc^{-} mutant P2, though the differential growth pattern of hyphae and appressorium formation occurred in the same time range, the number of appressoria increased, because intraradical fungal penetration was hindered and the fungus continued growing and producing appressoria along the roots (Gollotte et al. 1993). Arbuscule formation was monitored in the susceptible cultivar Frisson: the first arbuscule was observed after 72 h.

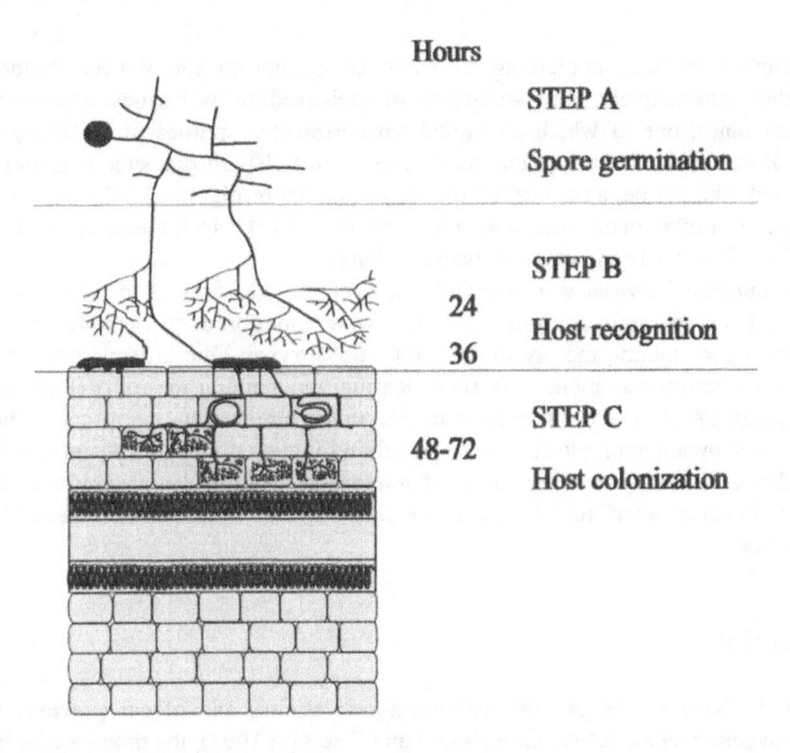

Fig. 1. The spatio-temporal development of arbuscular mycorrhizal symbiosis. **Step A** A germinated spore shows a linear growth pattern, consisting of branches extending in all directions, functional for soil exploration and for an efficient exploitation of resources. **Step B** As early as 24 h after the perception of host-derived signals, a differential hyphal growth pattern is expressed, functional to the location of infection sites. As early as 36 h after the beginning of symbionts interaction, dramatic morphogenetical changes occur in hyphal tips, leading to the formation of appressoria. **Step C** After root penetration, intercellular hyphae colonize the root, producing intracellular branched structures, the arbuscules, as early as 42-72 h after the beginning of the interaction

The present spatio-temporal studies confirm that host-derived signals capable of eliciting fungal responses occur only along host roots. In the interaction *G. mosseae-O. basilicum* the elicited area was 569 μm wide (ranging from 504 to 605 μm), whereas root diameter measured 282 ± 18 μm: thus, the signals emitted by host roots were able to spread in the millipore membrane no more than 143 μm from the source.

Moreover, these studies show that host-derived signals are capable of eliciting sharp threshold responses in AM fungal hyphae. In fact, the departure from the normal growth pattern to the elicited one was never observed to be gradual. On the contrary, when hyphae reached the area of the millipore membrane overlying roots, they dramatically altered their growth pattern. This fungal behavior strongly suggests also chemotropism (Koske 1982; Gemma and Koske 1988), though this remains to be shown, because by using our experimental system it was not possible to monitor alterations in single hyphae, which occur well before 24 h. Other systems should be developed, able to monitor hyphal growth from 0 to 24 h.

Discussion

The present work shows the spatio-temporal developmental sequence of events leading to the establishment of arbuscular mycorrhizal symbiosis. It also confirms that host-derived signals are the early cues to which AM fungi respond, by changing the direction of hyphal elongation and initiating a different branching pattern, prior to the formation of appressoria. Moreover, the experimental model system was validated, by showing that the sequential infection events are predictable in space and time.

Previous works have shown that plant hosts of ectomycorrhizae, arbutoid and ericoid mycorrhizae, as well as non-mycorrhizal plants, were unable to elicit any morphogenetical response, either differential growth pattern or appressoria (Giovannetti et al. 1993b; Giovannetti et al. 1994). Such results showed that the signals eliciting fungal responses by AM fungi occur only in plant species able to establish a symbiosis. The host plants used here confirmed the occurrence of host-derived signals triggering recognition responses in AM fungi as early as 24 h and 36 h respectively for the differentiation of growth pattern and appressorium formation, and as early as 72 h for the completion of the symbiotic interaction, the formation of arbuscules (Giovannetti and Citernesi 1993).

These studies suggest that host-derived signals are short-range ones: in fact, fungal hyphae growing a few microns apart from the source of the signals did not respond to them. This was confirmed by the vigorous responses occurring in the hyphae growing just over the source, which suggested the possibility of threshold responses.

Till now, it has not been possible to identify host-derived signals triggering such responses. Some authors have suggested flavonoids and phenolic compounds as possible candidates (Gianinazzi-Pearson et al. 1989; Nair et al. 1991; Siqueira et al. 1991; Tsai and Phillips 1991; Phillips and Tsai 1992).

In conclusion, the development of AM fungi can be envisaged as a programmed sequence of phenotypic changes, resulting from differential gene expression, in space and time, elicited by different host signals (Gianinazzi-Pearson and Gianinazzi 1989; Giovannetti et al. 1994; Bonfante and Perotto 1995). Such signals act as triggers of commitment to the symbiotic status in AM fungi, because without them the fungi both stop growing in 10-21 days and are unable of developing infection structures (Hepper 1983; Giovannetti et al. 1993b; Giovannetti et al. 1994).

Further characterization of the phenomenon, by developmental studies of the first 12-24 h of symbionts' interaction, will provide clues to understand the life strategies of AM fungi and might pave the way to studies on survival of these individuals in nature.

Acknowledgments: I am indebted to Dr. Cristiana Sbrana for her help in preparing the manuscript, and to Dr. C. Logi for his skillful help in preparing Fig 1.

References

Bonfante P, Perotto S (1995) Strategies of arbuscular mycorrhizal fungi when infecting host plants. The New Phytologist 130: 3-21

Brasier C (1992) A champion thallus. Nature 356: 382-383

Duc G, Trouvelot A, Gianinazzi-Pearson V, Gianinazzi S (1989) First report of non-mycorrhizal plant mutants (Myc-) obtained in pea (*Pisum sativum* L.) and fababean (*Vicia faba* L.). Plant Science 60: 215-222

Gemma JN, Koske RE (1988) Pre-infection interactions between roots and the mycorrhizal fungus *Gigaspora gigantea*: chemotropism of germ tubes and root growth response. Transactions of the British Mycological Society 91: 123-132

Gianinazzi-Pearson V, Branzanti B, Gianinazzi S (1989) In vitro enhancement of spore germination and early hyphal growth of a vesicular-arbuscular mycorrhizal fungus by host root exudates and plant flavonoids. Symbiosis 7: 243-255

Gianinazzi-Pearson V, Gianinazzi S (1989) Cellular and genetical aspects of interactions between host and fungal symbionts in mycorrhizae. Genome 31: 336-341

Giovannetti M, Avio L, Sbrana C, Citernesi AS (1993a) Factors affecting appressorium development in the vesicular-arbuscular mycorrhizal fungus *Glomus mosseae* (Nicol. & Gerd.) Gerd. & Trappe. The New Phytologist 123: 114-122

Giovannetti M, Citernesi AS (1993) Time-course of appressorium formation on host plants by arbuscular mycorrhizal fungi. Mycological Research 97: 1140-1142

Giovannetti M, Gianinazzi-Pearson V (1994) Biodiversity in arbuscular mycorrhizal fungi. Mycological Research 98: 705-715

Giovannetti M, Sbrana C, Avio L, Citernesi AS, Logi C (1993b) Differential hyphal morphogenesis in arbuscular mycorrhizal fungi during pre-infection stages. The New Phytologist 125: 587-594

Giovannetti M, Sbrana C, Logi C (1994) Early processes involved in host recognition by arbuscular mycorrhizal fungi. The New Phytologist 127: 703-709

Gollotte A, Gianinazzi-Pearson V, Giovannetti M, Sbrana C, Avio L, Gianinazzi S (1993) Cellular localization and cytochemical probing of resistance reactions to arbuscular mycorrhizal fungi in a "locus a" myc- mutant of *Pisum sativum* L..Planta 191: 112-122

Grime JP, Mackey JML, Hillier SH, Read DJ (1987) Floristic diversity in a model system using experimental microcosms. Nature 328: 420-422

Harley JL, Smith SE (1983) Mycorrhizal symbiosis. Academic Press, London.

Hepper CM (1983) Limited independent growth of a vesicular-arbuscular mycorrhizal fungus in vitro. The New Phytologist 93: 537-542

Koske RE (1982) Evidence for a volatile actractant from plant roots affecting germ tubes of a VA mycorrhizal fungus. Transactions of the British Mycological Society 79: 305-310

Martins MA (1993) The role of the external mycelium of arbuscular mycorrhizal fungi in the carbon transfer process between plants. Mycological Research 97: 807-810

Mosse B (1988) Some studies relating to "independent" growth of vesicular-arbuscular endophytes. Canadian Journal of Botany 66: 2533-2540

Nair MG, Safir GR, Siqueira JO (1991) Isolation and identification of vesicular-arbuscular mycorrhiza-stimulatory compounds from clover (*Trifolium repens*) roots. Applied and Environmental Microbiology 57: 434-439

Phillips JM, Hayman DS (1970) Improved procedures for clearing and staining parasitic and vesicular-arbuscular mycorrhizal fungi for rapid assessment of infection. Transactions of the British Mycological Society 55: 158-161

Phillips DA, Tsai SM (1992) Flavonoids as plant signals to rhizosphere microbes. Mycorrhiza 1: 55-58

Read DJ (1991) Mycorrhizas in ecosystems. Experientia 47: 376-391

Siqueira JO, Safir GR, Nair MG (1991) Stimulation of vesicular-arbuscular mycorrhiza formation and growth of white clover by flavonoid compounds. The New Phytologist 118: 87-93

Smith ML, Bruhn JN, Anderson JB (1992) The fungus *Armillaria bulbosa* is among the largest and oldest living organisms. Nature 356: 428-431

Tsai SM, Phillips DA (1991) Flavonoids released naturally from alfalfa promote development of symbiotic *Glomus* spores in vitro. Applied and Environmental Microbiology 57: 1485-1488

Role of Fungal Wall Components in Interactions Between Endomycorrhizal Symbionts

A. Gollotte[1], C. Cordier, M. C. Lemoine and V. Gianinazzi-Pearson
Laboratoire de Phytoparasitologie, INRA/CNRS, Centre de Microbiologie du Sol et de l'Environnement, INRA, BV1540, 21034 Dijon Cédex, France.
[1]Present address: Scottish Agricultural College, Land Resources Department, Doig Scott Building, Craibstone Estate, Aberdeen AB29TQ, UK

Key Words: Endomycorrhiza, elicitors; ß-1,3 glucans, chitin, plant defense reactions, symbiosis-related genes.

Summary: Although endomycorrhizal symbioses represent the most widespread of plant-microbe interactions, the mechanisms leading to reciprocal functional compatibility are still unknown. In order to better understand why plant defense reactions are only weakly and locally induced in endomycorrhizas, it is interesting to determine whether endomycorrhizal fungi possess elicitors of plant defense reactions. In fact, walls of both ericoid and arbuscular mycorrhizal fungi contain chitin, β-1,3 glucans and/or chitosan and glycoproteins which could potentially induce plant defense reactions. However, important wall modifications occur in these endomycorrhizal fungi during the root colonization process and therefore they could have strategies of self-camouflage so that they would not be recognized by the plant as non-self. Alternatively, plant defense reactions could be induced but then repressed. Evidence for the existence of such phenomena comes from studies of Myc^{-1} Nod^{-} pea mutants which do not form arbuscular mycorrhizas nor nodules with rhizobial bacteria. Myc^{-1} mutants are characterized by the elicitation of strong plant defense reactions in epidermal and hypodermal cells in contact with fungal appressoria. Moreover, the fact that these mutants are resistant to both symbioses means that some plant genes are specifically involved in symbiosis establishment. Therefore, one role of symbiosis-related (SR) genes could be to repress a strong expression of plant defense reactions during endomycorrhization. Expression of these SR genes could somehow interact with fungal compatibility factors by mechanisms which are discussed.

Introduction

Although mutualistic endomycorrhizal symbioses are frequently ignored by plant physiologists and ecologists, they represent the most widespread of plant-microbe interactions, as they are constantly present in the large majority of plant species. They probably play an important role in most terrestrial ecosystems as endomycorrhizal fungi are involved in nutrient cycling and transfer of soil elements to their host plants, in exchange for carbon compounds (Allen 1991). They also confer plants with better growth as well as greater resistance towards biotic and abiotic stresses (Gianinazzi and Gianinazzi-Pearson 1988). The genetic and molecular factors responsible for these highly compatible interactions are not really understood. In this review, we will consider the mechanisms which could be involved in enabling this general plant susceptibility towards endomycorrhizal fungi in ericoid and arbuscular mycorrhizas.

Endomycorrhizal fungi can colonize the entire root system except apices and the central cylinder. Moreover, intracellular fungal structures can occupy a large volume of the host cell, up to 40% in the case of arbuscules (Alexander et al. 1988), without inducing any signs of plant cell damage, but rather a phenomenon of cell differentiation. These features raise the question as to why the fungus is not recognized as a pathogen and how this highly sophisticated mutualistic interaction can take place.

Incompatible interactions between plants and pathogenic fungi are characterized by the induction of active defense reactions at the site of penetration, with the formation of a physical barrier containing callose, phenolic compounds, proteins and silicon (Heath 1980). This is associated with the neosynthesis of antimicrobial compounds, phytoalexins, and of "pathogenesis-related" PR proteins including chitinases and β-1,3 glucanases (Graham and Graham 1991). This results in a hypersensitive reaction leading to localized cell death which limits fungal development and causes plant resistance (Freytag et al. 1994). "Elicitors" of these defense and hypersensitive reactions can be proteins, lipophilic substances like ergosterol or result from fragmentation of either plant or fungal cell wall polysaccharides (Hahn et al. 1989, Boller 1995). Chitinous compounds and β-glucans can be released during spore germination, as well as from hyphae by plant hydrolytic enzymes like chitinases and β-1,3 glucanases (Waldmüller et al. 1992, Yoshikawa et al. 1993). Moreover, some fungi secrete pectolytic enzymes leading to plant tissue maceration and release of oligogalacturonides with an elicitor activity (Yoshikawa et al. 1993). In fact, defense reactions are also induced in compatible interactions but later (Collinge et al. 1994). Endomycorrhizas indeed resemble compatible interactions with biotrophic fungi, with similarities at the cellular level, including a rejuvenation of host cells and the creation of an interface adapted to nutrient transfer to the fungus (Bonfante-Fasolo and Perotto 1990). However, in the case of endomycorrhizas, the interface is adapted to reciprocal nutrient exchanges between the two partners (Smith and Smith 1990).

It is therefore interesting, in the light of these observations, and in order to better understand the mechanisms leading to the widespread compatibility between endomycorrhizal partners, to consider whether elicitors are generated during endomycorrhizal symbiosis establishment and whether plant defense reactions are induced.

Ericoid Mycorrhizal Fungi

The ericoid type of endomycorrhizas is formed by a small group of plant species belonging to the Ericaceae and a few ascomycetous fungi such as *Hymenoscyphus ericae* or *Oidiodendron griseum* (Perotto et al. 1995). Ericoid mycorrhizal fungi colonize the one or two layers surrounding the central cylinder of the fine hair roots. In these cells, they form coils which are surrounded by wall material deposited by the host and a membrane continuous with the plant plasmamembrane. This interface is considered as a key structure for reciprocal nutrient exchanges between endomycorrhizal partners.

The wall of *H. ericae* contains chitin, β-1,4 glucans and β-1,3 glucans (Perotto et al. 1995; Figs. 3, 4). The same components have been detected in *O. griseum* although ultrastructural wall morphology is different (Perotto et al. 1995). In pure culture, the surface of the cell wall of *H. ericae* is covered by a disorganized fibrillar material. When hyphae grow in the presence of a host plant, this fibrillar sheath becomes more abundant and organized, and enters in close contact with host wall components (Fig. 1), suggesting that it may be involved in recognition phenomena (Bonfante-Fasolo et al. 1984). In fact, isolates which produce an abundant fibrillar sheath have a greater ability to infect roots (Gianinazzi-Pearson and Bonfante-Fasolo 1986). However, fibrils disappear from the surface of intracellular hyphae (Fig. 1). Fibrils also develop when the fungus enters in contact with the root surface of non host plants (Fig. 2), but they later subsist on hyphae which develop parasitically inside the root and can even colonize the central cylinder (Figs. 3,4).

Cytochemical and immunocytochemical analyses have indicated the presence of β-1,4 or β-1,6 glucans and proteins in the fibrils, but their actual composition is not really known. The use of lectins has indicated the presence of sugar residues like mannose and glucose, suggesting in particular the presence of glycoproteins (Bonfante-Fasolo et al. 1987). It is known that ericoid mycorrhizal fungi excrete a wide range of enzymes like acid phosphatase, carboxyproteinase, chitinase, polygalacturonase, β-1,4 glucanase and β-1,3 glucanase which may be important for their saprophytic activity (Straker and Mitchell 1986; Leake and Read 1991, Peretto et al. 1993, Varma and Bonfante 1994). Polygalacturonase has indeed been immunolocalized in the fibrillar sheath of extraradical hyphae but it is almost absent from intracellular hyphae in host roots, whereas it is still present in intraradical hyphae in non host plants (Perotto et al. 1995). An acid phosphatase, a mannoprotein, has also been revealed in both the fungal wall and the extracellular fibrils by immunocytochemistry. However, although its presence is indicated by immunocytochemistry in the fungal wall of intracellular hyphae, it is not active. In contrast, in non host plants, acid phosphatase activity remains in intracellular hyphae. This indicates that host cells somehow inhibit acid phosphatase activity of intraradical hyphae (Lemoine et al. 1992). Moreover, an invertase, another mannoprotein, comigrates with one acid phosphatase isoform (Straker et al. 1992).

Recently, it has been shown that invertase from *Saccharomyces cerevisiae* can act as an elicitor of plant defense reactions. However, the presence of 9 to 11 mannosyl residues is required on the glycoprotein for elicitation activity. Plant mannoproteins contain 8 mannosyl residues whereas fungi possess a ninth residue in a specific position which may contribute to plants not recognizing these glycoproteins as self proteins (Basse et al. 1992). Therefore, these mannoproteins could not only have physiological activities but also be involved in recognition phenomena with host and non host roots via their glycosidic chains. We speculate that in Ericaceae, host cells could cut some mannosyl residues of fungal acid phosphatase glycoprotein, resulting in a loss of enzyme activity and in the non-elicitation of plant defense responses.

Arbuscular Mycorrhizal Fungi

The most common type of endomycorrhizas, however, is certainly arbuscular mycorrhizas which are characterized by a lack of host specificity as they are formed by members of at least 80% of plant families (Newman and Reddell 1987). The fungi are classified in the Zygomycetes, order Glomales and comprise about 130 species (Morton 1990). Therefore, arbuscular mycorrhizal fungi (AMF), which appeared about 400 million years ago (Simon et al. 1993; Remy et al. 1994), must have developed a highly sophisticated system of recognition and compatibility with their host plants which is responsible for the establishment of the mutualistic symbiosis. During their colonization process, AMF form, on the root surface, appressoria which are essential for further penetration (Giovannetti et al. 1993), and then abundantly develop in the cortical parenchyma. At this level, they can enter cells in which they highly ramify into specialized haustoria, the arbuscules. In fact, a very sophisticated interface develops in infected cells. As is the case in ericoid mycorrhizas, the fungus is surrounded by a wall-like material, or matrix, deposited by the host cell and the periarbuscular membrane derived from the host plasmamembrane. Furthermore, as the fungus branches, important modifications occur at the interface, comprising a thinning out of both fungal wall and of the host matrix material.

Wall Structure and Modifications During Root Colonization

Zygomycetes mainly contain chitin and chitosan, but no β-1,3 glucans in their walls (Bartnicki-Garcia 1968). Early cytological studies of spores and extraradical hyphae of AMF have shown that their walls contain a high quantity of chitin and proteins and minor components like chitosan and alkali soluble polysaccharides with glucuronic acid, glucose, rhamnose and mannose residues (Bonfante-Fasolo and Grippiolo 1984). The presence of chitin in AMF has since been confirmed in several species either by cytochemical or chemical analyses

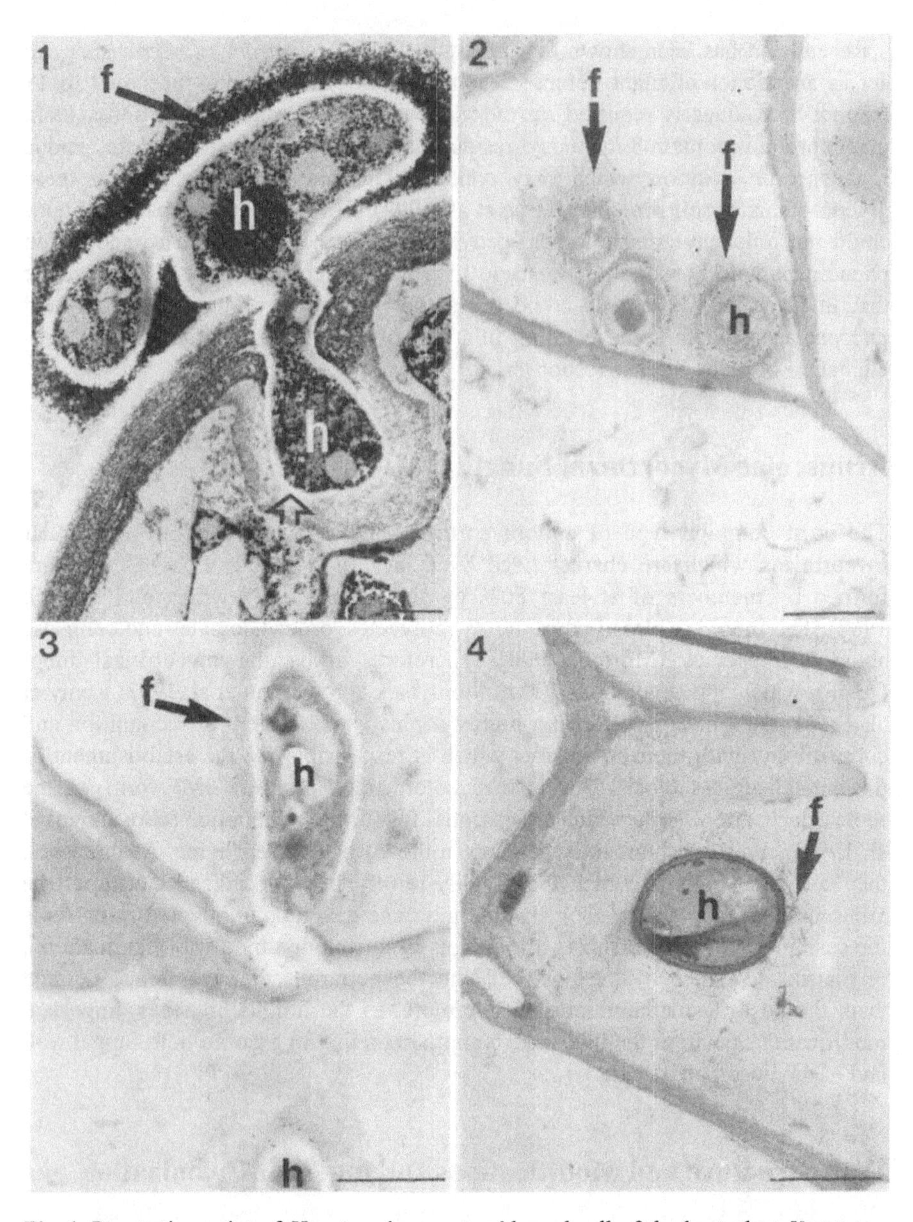

Fig. 1. Penetration point of *H. ericae* in a root epidermal cell of the host-plant *Vaccinium corymbosum*. An organized fibrillar material (f) is observed around external hyphae (h), which disappears from the surface of intraradicular hyphae (h) (open arrow). Bar = 1 µm

Figs. 2-4. Infection pattern of *H. ericae* in roots of a non-host plant, pea cv. Frisson after PATAg reaction for β-1,4 and β-1,6 glucans (**4**) and immunolabelling with β-1,3 glucan antibodies (**2, 3**). External hyphae (h) are surrounded by a fibrillar material (f) (**2**), maintained around intracellular hyphae in cortical cells (**3**) and in the central cylinder (**4**) of pea. Bars = 1 µm

(Frey et al. 1994; Grandmaison et al. 1988; Sbrana et al. 1995; Lemoine et al. 1995). More recently, it has been shown that species of Glomineae unexpectedly contain β-1,3 glucans in their walls whereas these glucans have not been detected in species of Gigasporineae (Gianinazzi-Pearson et al. 1994a).

However, the wall composition of AMF is not static and depends on the stage in the fungal life cycle. In fact, a gradual thinning of walls and simplification of components occurs from extraradical to intraradical hyphae (Bonfante-Fasolo and Grippiolo 1982). Chitin can be detected in all ontogenic stages of the Glomineae and the Gigasporineae but is in a fibrillar state in spores and is amorphous in arbuscules (Grandmaison et al. 1988; Bonfante-Fasolo 1988; Bonfante-Fasolo et al. 1990a; Lemoine et al. 1995). Moreover, there is a progressive disappearance in wall ß-1,3 glucan detection during root colonization (Lemoine et al. 1995). These observations, first made for *G. mosseae* and *A. laevis* colonizing pea and tobacco have been confirmed for *G. mosseae* developing in tomato roots (Figs. 5-8). Extraradical hyphal walls contain a large quantity of β-1,3 glucans (Fig. 5). In intercellular hyphae, immunolabelling is less abundant (Fig. 6) and no β-1,3 glucans can be detected in the fungus at its penetration point into cortical parenchyma cells (Fig. 7) nor in arbuscules (Fig. 8).

In contrast to what happens in these mycorrhizal associations, no modifications in wall thickness or composition have been observed in other fungi infecting pea roots, like *Chalara elegans*, *Aphanomyces euteiches* (Fig. 10), *Rhizoctonia* sp. or *H. ericae* (Figs. 3, 4; Gollotte et al. 1996a). Similarly, the walls of *Phytophthora nicotianae* var. *parasitica* do not appear to be affected during fungal colonization in tomato roots (Fig. 11; Cordier et al., unpublished observations).

According to the steady-state model for apical fungal wall synthesis (Wessels et al. 1990; Gooday 1995), arbuscules could be considered as apices (Bonfante-Fasolo 1988; Fig. 12). In this model, chitin and β-1,3 glucan chains are synthesized at the apex. As the hypha elongates, homologous chains of chitin and β-1,3 glucans form hydrogen bonds. At the same time, β-1,3 glucans become ramified with β-1,6 chains and covalent linkages develop between chitin fibrils, β-glucans and proteins, leading to a progressive cell wall rigidification. In arbuscules, which have an amorphous structure, individual β-glucan chains could be synthesized but solubilized during sample preparation, leading to their non-detection using immunocytochemical techniques (Lemoine et al. 1995). In contrast, development of hydrogen and covalent bonds would occur in walls of spores, extraradical and intercellular hyphae, leading to their fibrillar aspect (Fig. 12). As such, the arbuscule wall would not be a rigid structure and could be influenced by changes in osmotic pressure in the dedifferentiating host cell, so conferring plasticity and inducing hyphal branching.

However, other phenomena could also be involved in the development of the highly branched pattern of arbuscules which differs from the less ramified structure of haustoria formed by biotrophic pathogenic fungi. It has been suggested that hyphal ramification in filamentous fungi is due to localized wall softening by enzymes like chitinases and β-1,3 glucanases (Gooday 1995).

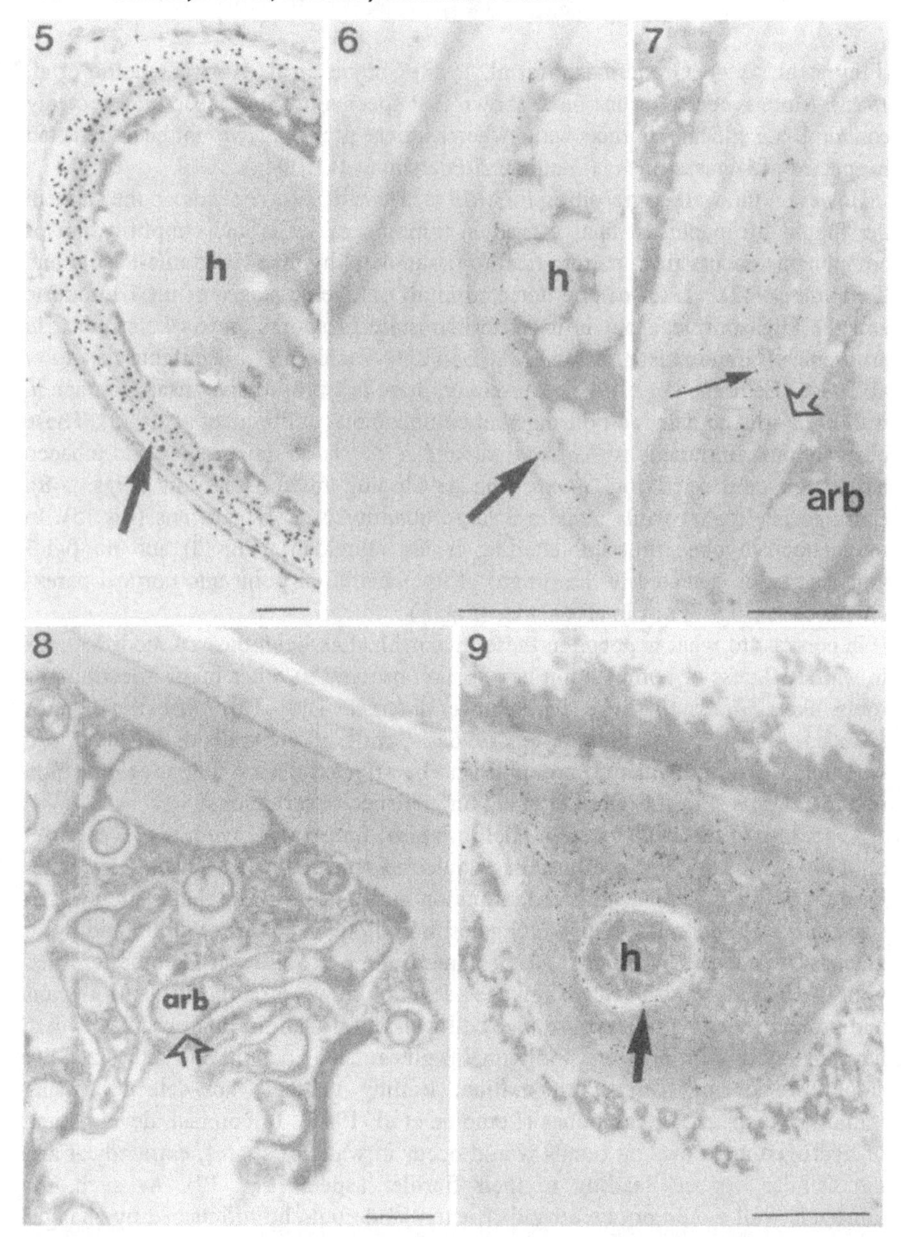

Figs. 5-8. Ultrastructural localization of β-1,3 glucans in the cell wall of *G. mosseae* colonizing tomato roots. Immunogold labelling is strong in walls of external hyphae (h) (**5**), weaker over an intercellular hyphal (h) wall (**6**) in the cortical parenchyma and absent from large (**7**) and thin (**8**) arbuscule branches (arb) (open arrow). Scattered β-1,3 glucans are localized in host wall material at the penetration point of *G. mosseae* into a cortical cell (small arrow) (7). Bars = 1 μm

Fig. 9. Ultrastructural localization of β-1,3 glucans in the cell wall of the pathogenic fungus *C. elegans* (h) and in wall appositions formed by colonized pea roots. Bar = 1 μm

There is as yet no evidence for the presence of such enzymes in AMF though they must exist. Nevertheless, the fact that arbuscules only develop in plant cortical parenchyma and that their morphology depends on the host plant could mean that branching may result from the action of plant hydrolytic enzymes. Weak activation of defense-related chitinase and β-1,3 glucanase genes has recently been shown to occur specifically in arbuscule containing host cells (Lambais and Mehdy 1995; Blee and Anderson 1996). A new chitinase isoform has indeed been shown to appear in mycorrhizal roots as compared to non mycorrhizal roots (Dumas-Gaudot et al. 1994). This isoform is different from isoforms induced by pathogenic fungi provoking either a hypersensitive response or having a compatible interaction with pea roots (Dumas-Gaudot et al. 1994; Dassi et al. 1996). Its role still has to be determined but it could be instrumental in changing fungal wall morphology.

Fungal wall modifications are probably important not only for nutrient exchanges but also for recognition phenomena between the two symbiotic partners.

Why is There no Strong Plant Defense Reaction in Arbuscular Mycorrhizas?

Although AMF walls contain chitin, chitosan and/or β-1,3 glucans, which are potential elicitors, it is now well accepted that only very localized and transient defense reactions are induced by AMF in their hosts. These remain at a low level and uncoordinated as compared to plant-pathogen interactions (Gianinazzi-Pearson et al. 1996). For example, β-1,3 glucans (callose) can only be detected in the material deposited by the plant around hyphae at the point of penetration into cortical parenchyma cells (Figs. 5-8), and this is an extremely limited reaction as compared to the extensive callose deposition interactions with pathogenic fungi (see Fig. 9).

This could be explained by the fact that there is a lack of generation of active elicitors of plant defense reactions during symbiosis establishment. It has been shown that both length and structure of elicitors are important for inducing defense responses (Hahn et al. 1989). Therefore, it may be possible that polysaccharide fragments are liberated from AMF during their development inside the root but without having the structure required for elicitation. Moreover, the fungal wall may be protected from active plant hydrolytic enzymes. It has indeed been shown that chitinases and β-1,3 glucanases, able to inhibit hyphal growth of pathogenic fungi by attacking their apices, have no effects on the ectomycorrhizal fungus *Heleboma crustuliniforme* nor on *H. ericae* (Mauch et al. 1988; Arlorio et al. 1991, 1992). Furthermore, although *H. crustuliniforme* is able to release some elicitors in pure culture, they are not active because they seem to be cleaved by plant chitinases and β-1,3 glucanases (Salzer et al. 1996). As mentioned above, in arbuscular mycorrhizas, plant chitinase genes may be locally activated in arbuscule-containing parenchyma cells, although immunolocalization of chitinase

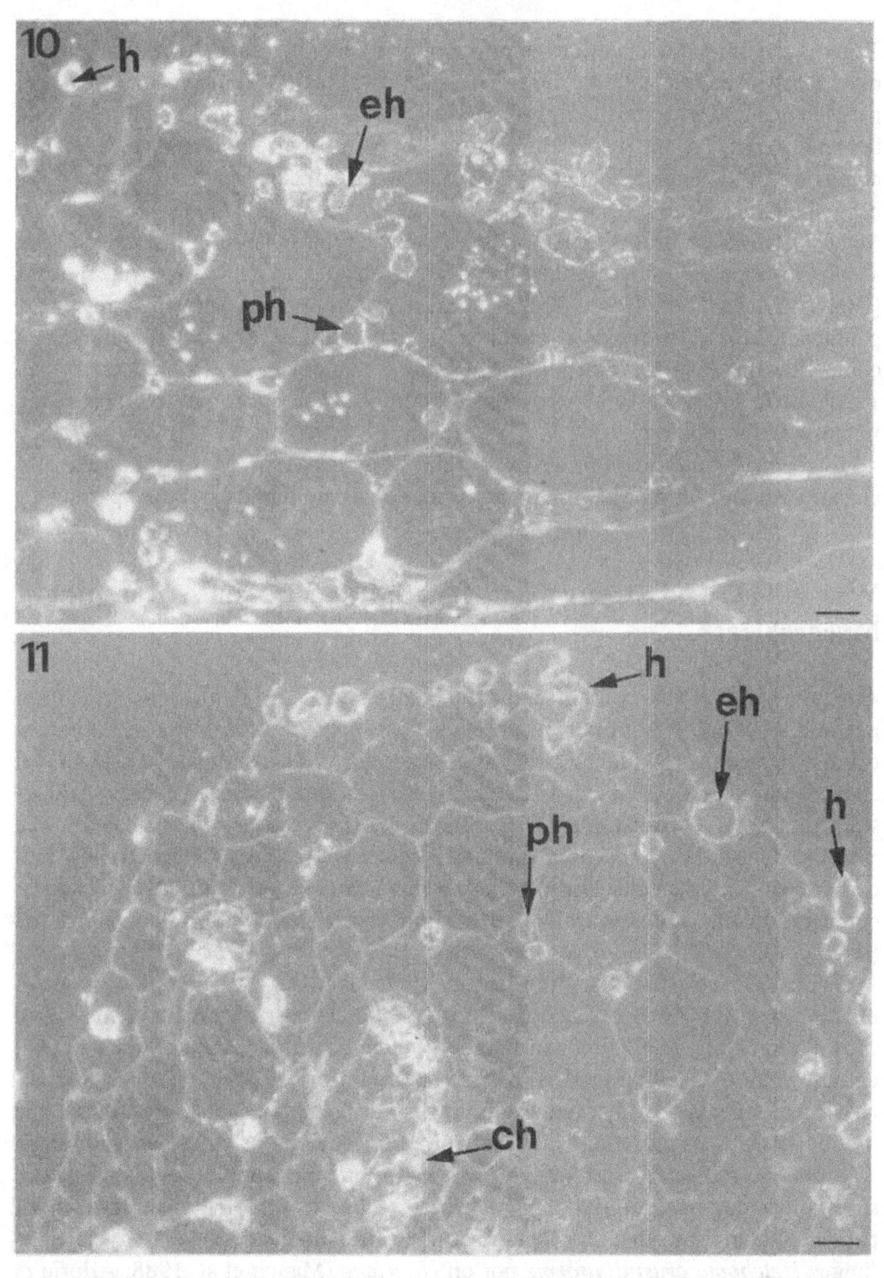

Figs. 10-11. Thin sections of pea (**10**) and tomato (**11**) roots colonized by *A. euteiches* and *P. nicotianae* var. *parasitica* respectively, after immunogold labelling with β-1,3 glucan antibodies and silver enhancement. Observations in epipolarised light show the same immunogold detection of β-1,3 glucans in fungal walls, as the fungi develop in the rhi - zosphere (h) and in epidermal (eh), cortical parenchyma (ph) and central cylinder (ch) root tissues. Bars = 10 μm

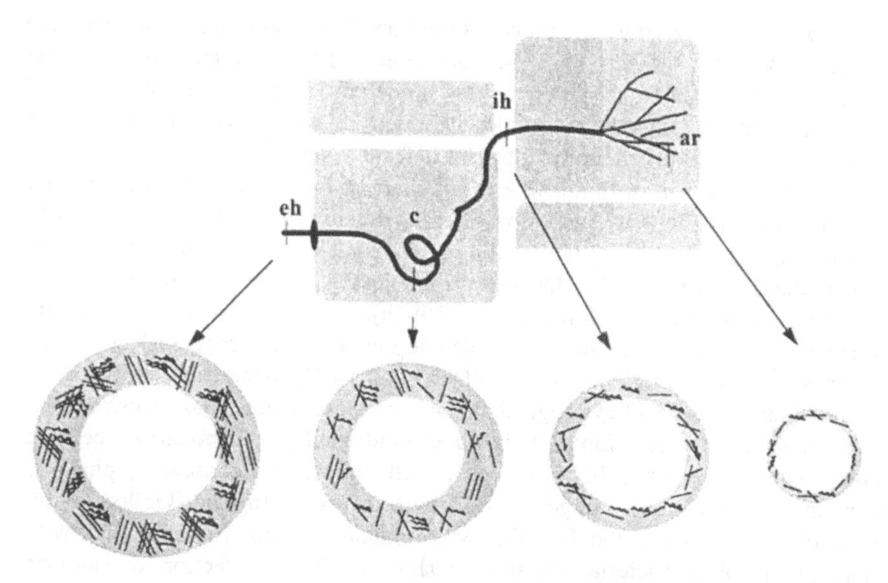

Fig. 12. Model of suggested wall structure of arbuscular mycorrhizal fungi in a host root, adapted from the steady-state theory of apical wall growth (Wessels et al. 1990; Gooday 1995). In arbuscules (ar), the wall may be formed by nascent chains of chitin (straight lines) and β-1,3 glucans (wavy lines), giving a plastic structure. Progressively, in intercellular hyphae (ih) and coils (c), covalent linkages between chitin and glucan chains could develop, as well as hydrogen bonding between homologous chains, resulting in a more rigid structure. In the mature walls of external hyphae (eh), all these arrangements may be present at the same time

indicates its presence in intercellular spaces and in vacuoles but not in direct contact with the fungus (Spanu et al. 1989; Blee and Anderson 1996). Therefore, mycorrhizal fungi could have a strategy of self camouflage during host root colonization. Biotrophic pathogenic fungi seem to have developed a similar strategy with important modifications of their walls in different stages of infection (Mendgen and Deising 1993).

Proteins are also potential elicitors. Not much is known about protein synthesis in AMF which do not have saprophytic activity like ericoid mycorrhizal fungi. Total extractable polygalacturonase activity is at the same level in non-infected and infected leek roots, but a new isoform has been shown to be produced in arbuscular mycorrhizas and it has indeed been immunolocalized within the interfacial material of arbuscules (Peretto et al. 1995). This enzyme is thought to be of fungal origin and its activity may make pectins in the interface available to the fungus. In fact, the presence of unesterified pectins in the interface depends on the host plant species. For example, in leek and in tomato, pectins are present in the interfacial material all around arbuscules (Bonfante-Fasolo et al. 1990b, Cordier et al., unpublished results). In contrast, in clover, maize and pea, pectins can only be detected around large arbuscule branches but they disappear around fine arbuscule branches (Bonfante and Perotto 1993; Gollotte et al. 1994). We may speculate that some host plants have the potential to somehow repress activity of fungal polygalacturonase. Oligogalacturonides liberated by such activity may not have the necessary length for elicitation of plant defense responses.

Another explanation of this low level of plant defense reactions in arbuscular mycorrhizas could be that expression of plant defense genes is elicited by AMF but then repressed either by plant or fungal factors. The study of the behaviour of Myc^- mutants resistant to AMF has given some interesting information concerning possible mechanisms (Gollotte et al. 1996b). Most of these mutants are characterized by blocking of fungal development on the root surface at the appressorium level (Myc^{-1} phenotype) (Duc et al. 1989, Gianinazzi-Pearson et al. 1991). Ultrastructural and cytological analyses of these mutants have shown that a typical wall defense reaction is induced in epidermal and hypodermal cells in contact with appressoria. In these cells, wall appositions containing phenolic compounds, callose and pathogenesis-related proteins are formed (Gollotte et al. 1993, 1995). Moreover, the fact that Myc^- mutants cannot form nodules with symbiotic rhizobial bacteria whilst they show unaltered infection phenotypes towards a number of pathogens indicates that the mutation has affected genes specifically involved in symbioses (Gianinazzi-Pearson et al. 1994b). These results suggest that AMF can generate elicitors of plant defense reactions but that in the compatible interactions with roots, these responses are repressed directly or indirectly by the expression of plant symbiosis-related (SR) genes (Gollotte et al. 1993).

The mechanisms leading to repression of plant defense reactions during arbuscular mycorrhizal development are still unknown but several hypotheses can be formulated. Like biotrophic pathogenic fungi (Briggs and Johal 1994), AMF could have compatibility factors which are somehow activated by the expression of plant SR genes. Such compatibility factors, induced by products of SR genes, could affect the synthesis or activity of molecules encoded by plant defense genes. For example, it has been discovered that in compatible interactions, the pathogen *Phytophthora sojae* f. sp. *glycinea* secretes a protein that inhibits soybean β-1,3 glucanase activity and which could release elicitors from fungal walls, so that a resistance reaction does not occur (Ham et al. 1995). If AMF possess the capacity of synthesizing such proteins, this could explain why β-1,3 glucanases and chitinases are induced during early stages of AMF colonization and then repressed (Spanu et al. 1989, Volpin et al. 1994). This suppression only occurs at low soil phosphorus concentration and increased chitinase and β-1,3 glucanase activities occur at high phosphorus concentrations which could explain the localized activation of corresponding genes in arbuscule-containing cells where phosphorus is being released to the host cell (Lambais and Mehdy 1995, Blee and Anderson 1996).

Alternatively, compatibility factors could act as "suppressors" of plant defense reactions by recognizing a specific receptor encoded by SR genes, and thus interfere with the signal transduction pathway of elicitation. Evidence for the existence of such phenomena comes from a number of other plant-microbe interactions. *Phytophthora* species possess water soluble glucans and glycoproteins which have a suppressive activity towards plant defense reactions (Storti et al. 1988, Sanchez et al. 1993). However, their structure and mode of action are unknown. The nature of suppressors of defense reactions has been described in the pea pathogen *Mycosphaerella pinodes*. They are small glycopeptides which are able to affect both P-type ATPase activity and transmembrane signaling, leading to a delay in plant defense responses, finally responsible for compatibility (Shiraishi 1994).

Compatibility factors are also known in the Legume-*Rhizobium* symbiosis; they are the bacterial Nod factors necessary for symbiosis establishment. Not only their structure is essential for host specificity but they are also able to induce plant defense reactions. They are lipo-oligosaccharides containing an oligochitin chain capable of inducing chitinase activity. Mellor and Collinge (1995) hypothesized that some host plant molecules could temporarily bind to Nod factors and delay activation of chitinase whereas in non host plants, elicitation immediately occurs and Nod factors are quickly destroyed by the enzyme. Furthermore, when the nodule is fully developed, host cells must synthesize molecules which repress synthesis of Nod factors in bacteroids which would elicit defense reactions and lead to premature nodule abortion. Similar phenomena could occur in arbuscular mycorrhizas. It has indeed been recently shown that Nod factors are able to increase root colonization by mycorrhizal fungi and it has been suggested that some chitin chains released from the fungal wall may have a role in recognition phenomena with host roots (Xie et al. 1995).

Conclusions

Symbiosis establishment must result from a complex series of events involving recognition between the partners and limitation of elicitation of plant defense reactions. Initially during evolution, recognition phenomena may have occurred between widely occurring molecules, which may explain their maintenance with the appearance of new plant species so that arbuscular mycorrhizas now represent the most widespread of plant-microbe interactions. Subsequently, more complex recognition mechanisms must have developed which have led to the appearance of associations with narrower host specificity as in ericoid mycorrhizas, ectomycorrhizas, the Legume-*Rhizobium* symbiosis and plant-pathogen interactions.

We have suggested the existence of several mechanisms responsible for the low activation of plant defense responses in endomycorrhizas. In fact, different processes must occur at different stages of symbiosis establishment and functioning but plant SR genes may somehow be involved in modifying fungal metabolism which could be responsible for elicitation.

However, we are far from really understanding these processes. Future research will have to concentrate on identifying, firstly how fungal wall metabolism is modified by the host plant during root colonization, and secondly the molecular interactions between fungal compatibility factors and expression products of symbiosis-related genes.

Acknowledgments. The authors are grateful to Dr L. Harrier for critically reading the manuscript. This research program was financially supported by the Burgundy Regional Council (contrat de plan no 3060A) and a european AIR-Project (no 3-CT94-0809).

References

Alexander T, Meier R, Toth R, Weber HC (1988) Dynamics of arbuscule development and degeneration in mycorrhizas of *Triticum aestivum* L. and *Avena sativa* L. with reference to *Zea mays* L. New Phytol 110: 363-370

Allen MF (1991) The Ecology of Mycorrhizae. Cambridge University Press, Cambridge

Arlorio M, Ludwig A, Boller T, Mischiati P, Bonfante P (1991) Effects of chitinase and β-1,3-glucanase from pea on the growth of saprophytic, pathogenic and mycorrhizal fungi. Giorn Bot Ital 125: 956-958

Arlorio M, Ludwig A, Boller T, Bonfante P (1992) Inhibition of fungal growth by plant chitinases and β-1,3-glucanases. A morphological study. Protoplasma 171: 34-43

Bartnicki-Garcia S (1968) Cell wall, chemistry, morphogenesis, and taxonomy of fungi. Annu Rev Microbiol 22: 87-108

Basse CW, Bock K, Boller T (1992) Elicitors and suppressors of the defense response in tomato cells. Purification and characterisation of glycopeptide elicitors and glycan suppressors generated by enzymatic cleavage of yeast invertase. J Biol Chem 267: 10258-10265

Blee KA, Anderson AJ (1996) Defense-related transcript accumulation in *Phaseolus vulgaris* L. colonized by the arbuscular mycorrhizal fungus *Glomus intraradices* Schenck and Smith. Plant Physiol 110: 675-688

Boller T (1995) Chemoperception of microbial signals in plant cells. Annu Rev Plant Physiol Plant Mol Biol 46: 189-214

Bonfante-Fasolo P, Grippiolo R (1982) Ultrastructural and cytochemical changes in the wall of a vesicular-arbuscular mycorrhizal fungus during symbiosis. Can J Bot 60: 2303-2312

Bonfante-Fasolo P, Gianinazzi-Pearson V, Martinengo L (1984) Ultrastructural aspects of endomycorrhiza in the Ericaceae. IV. Comparison of infection by *Pezizella ericae* in host and non-host plants. New Phytol 98: 329-333

Bonfante-Fasolo P, Grippiolo R (1984) Cytochemical and biochemical observations on the cell wall of the spore of *Glomus epigaeum*. Protoplasma 123: 140-151

Bonfante P, Perotto S, Testa B, Faccio A (1987) Ultrastructural localization of cell surface sugar residues in ericoid mycorrhizal fungi by gold-labelled lectins. Protoplasma 137: 25-35

Bonfante-Fasolo P (1988) The role of the cell wall as a signal in mycorrhizal associations. In: S Scannerini, D Smith, P Bonfante-Fasolo, V Gianinazzi-Pearson (eds) Cell to Cell Signals in Plant, Animal and Microbial Symbiosis. Springer-Verlag, Berlin, Heidelberg pp 219-235

Bonfante-Fasolo P, Perotto S (1990) Mycorrhizal and pathogenic fungi: do they share any features? In K Mengden, DE Lesemann (eds) Electron Microscopy Applied in Plant Pathology. Springer-Verlag, Berlin, pp 265-275

Bonfante-Fasolo P, Faccio A, Perotto S, Schubert A (1990a) Correlation between chitin distribution and cell wall morphology in the mycorrhizal fungus *Glomus versiforme*. Mycol Res 94: 157-165

Bonfante-Fasolo P, Vian B., Perotto S, Faccio A, Knox JP (1990b) Cellulose and pectin localization in roots of mycorrhizal *Allium porrum* labelling continuity between host cell wall and interfacial material. Planta 180: 537-547

Bonfante P, Perotto S (1993) Affinity techniques to investigate the plant-microbe interface in mycorrhizal symbiosis associations. Trends Microbiol Ecol 243-246

Briggs SP, Johal GS (1994) Genetic patterns of plant-host parasite interactions. Trends Genet 10: 12-16

Collinge DB, Gregersen PL, Thordal-Christensen H (1994) The induction of gene expression in response to pathogenic microbes. In: AS Basra (ed) Mechanisms of Plant Growth and Improved Productivity: Modern approaches and perspectives. Marcel Dekker, New York, pp 391-433

Dassi B, Dumas-Gaudot E, Asselin A, Richard C, Gianinazzi S (1996) Chitinase and β-1,3 glucanase isoforms expressed in pea roots inoculated with arbuscular mycorrhizal and pathogenic fungi. Eur J Plant Pathol 102: 105-108

Duc G, Trouvelot A, Gianinazzi-Pearson V, Gianinazzi S (1989) First report of non-mycorrhizal plant mutants (myc⁻) obtained in pea (*Pisum sativum* L.) and fababean (*Vicia faba* L.). Plant Sci 60: 215-222

Dumas-Gaudot E, Asselin A, Gianinazzi-Pearson V, Gollotte A, Gianinazzi S (1994) Chitinase isoforms in roots of various pea genotypes infected with arbuscular mycorrhizal fungi. Plant Sci 99: 27-37

Frey B, Vilarino A, Schüepp H, Arines J (1994) Chitin and ergosterol content of extraradical and intraradical mycelium of the vesicular-arbuscular mycorrhizal fungus *Glomus intraradices*. Soil Biol Biochem 26: 711-717

Freytag S, Arabatzis N, Halbrock K, Schmelzer E (1994) Reversible cytoplasmic rearrangements precede wall apposition, hypersensitive cell death and defense-related gene activation in potato/*Phytophthora infestans* interactions. Planta 194: 123-135

Gianinazzi S, Gianinazzi-Pearson V (1988) Mycorrhizae: a plant's health insurance. Chimicaoggi 8: 56-58

Gianinazzi-Pearson V, Bonfante-Fasolo P (1986) Variability in wall structure and mycorrhizal behaviour of ericoid fungal isolates. In: V Gianinazzi-Pearson, S Gianinazzi (eds) Physiological and Genetical Aspects of Mycorrhizae. INRA Publications, Paris, pp 563-567

Gianinazzi-Pearson V, Gianinazzi S, Guillemin JP, Trouvelot A, Duc G (1991) Genetic and cellular analysis of the resistance to vesicular-arbuscular (VA) mycorrhizal fungi in pea mutants. In: H Hennecke, DPS Verma (eds) Advances in Molecular Genetics of Plant-Microbe Interactions. Kluwer Academic Publishers, Boston, London, pp 336-342

Gianinazzi-Pearson V, Lemoine MC, Arnould C, Gollotte A, Morton JB (1994a) Localization of ß-1,3 glucans in spore and hyphal walls of fungi in the Glomales. Mycologia 86: 477-484

Gianinazzi-Pearson V, Gollotte A, Franken P, Gianinazzi S (1994b) Gene expression and molecular modifications associated with plant responses to infection by arbuscular mycorrhizal fungi. In: M Daniels (ed) Advances in Molecular Genetics of Plant-Microbe Interactions. Kluwer Academic Publishers, Boston, London, pp 179-186

Gianinazzi-Pearson V, Dumas-Gaudot E, Gollotte A, Tahiri-Alaoui A, Gianinazzi S (1996) Cellular and molecular defense-related root responses to invasion colonization by arbuscular mycorrhizal fungi. New Phytol 135: 45-57

Giovannetti M., Avio L, Sbrana C, Citernesi AS (1993) Factors affecting appressorium development in the vesicular-arbuscular mycorrhizal fungus *Glomus mosseae* (Nicol. & Gerd.) Gerd. & Trappe. New Phytol 123: 114-122

Gollotte A, Gianinazzi-Pearson V, Giovannetti M, Sbrana C, Avio L, Gianinazzi S (1993) Cellular localization and cytochemical probing of resistance reactions to arbuscular mycorrhizal fungi in a 'locus a' mutant of *Pisum sativum* (L.). Planta 191: 112-122

Gollotte A, Gianinazzi-Pearson V, Gianinazzi S (1995) Etude immunocytochimique des interfaces plante-champignon endomycorhizien à arbuscules chez des pois isogéniques myc$^+$ ou résistant à l'endomycorhization (myc$^-$). Acta Bot Gallica 141: 449-454

Gollotte A, Lemoine MC, Gianinazzi-Pearson V (1996a) Morphofunctional integration and cellular compatibility between endomycorrhizal symbionts. In: KG Mukerji (ed) Concepts in Mycorrhizal Research. Kluwer Academic Publishers, Dordrecht, pp 91-111

Gollotte A, Gianinazzi-Pearson V, Dumas-Gaudot E, Giovannetti M, Lherminier J, Berta G, Gianinazzi S (1996b) Plant mutants as models for studying the cellular and molecular bases of plant-fungal interactions in arbuscular mycorrhiza. In: C Azcon-Aguilar, JM Barea (eds) Mycorrhizas in integrated systems from genes to plant development. European Commission,EUR 16728, Luxembourg, pp 189-194

Gooday G (1995) The dynamics of hyphal growth. Mycol Res 99: 385-394

Graham TL, Graham MY (1991) Cellular coordination of molecular responses in plant defense. Mol Plant-Microbe Interact 4: 415-422

Grandmaison J, Benhamou N, Furlan V, Visser SA (1988) Ultrastructural localization of N-acetylglucosamine residues in the cell wall of *Gigaspora margarita* throughout its life-cycle. Biol Cell 63: 89-100

Hahn MG, Bucheli P, Cervone F, Doares SH, O'Neil RA, Darvill A, Albersheim P (1989) Roles of cell wall constituents in plant-pathogen interactions. In: T Kosuge, ED Nerar (eds) Plant-Microbe Interactions. Molecular and Genetic Perspectives. pp 131-181

Ham KS, Albersheim P, Darvill AG (1995) Generation of β-glucan elicitors by plant enzymes and inhibition of the enzymes by a fungal protein. Can J Bot 73: S1100-S1103

Heath MC (1980) Reactions of nonsuscepts to fungal pathogens. Annu Rev Phytopathol 18: 211-236

Lambais MR, Mehdy MC (1995) Differential expression of defense-related genes in arbuscular mycorrhiza. Can J Bot 73: S533-S540

Leake JR, Read DJ (1991) Experiments with ericoid mycorrhiza. Methods Microbiol 23: 435-459

Lemoine MC, Gianinazzi-Pearson V, Gianinazzi S, Straker CJ (1992) Occurrence and expression of acid phosphatase of *Hymenoscyphus ericae* (Read) Korf and Kernan, in isolation or associated with plant roots. Mycorrhiza 1: 137-146

Lemoine MC, Gollotte A, Gianinazzi-Pearson V, Gianinazzi S (1995) β(1-3) glucan localization in walls of the endomycorrhizal fungi *Glomus mosseae* and *Acaulospora laevis* during colonization of host roots. New Phytol 129: 97-105

Mauch F, Hadwiger LA, Boller T (1988) Antifungal hydrolases in pea tissue. I. Purification and characterization of two chitinases and two β-1,3-glucanases differentially regulated during development and in response to fungal infection. Plant Physiol 87: 325-333

Mellor RB, Collinge DB (1995) A simple model based on known plant defense reactions is sufficient to explain most aspects of nodulation. J Exp Bot 46: 1-18

Mendgen K, Deising H (1993) Infection structures of fungal plant pathogens. A cytological and physiological evaluation. New Phytol 124: 193-213

Morton JB (1990) Species and clones of arbuscular mycorrhizal fungi (Glomales, Zygomycetes): their role in macro- and microevolutionary processes. Mycotaxon 37: 493-515

Newman EI, Reddell P (1987) The distribution of mycorrhizas among families of vascular plants. New Phytol 106: 745-751

Peretto R, Bettini V, Bonfante P (1993) Evidence of two polygalacturonases produced by a mycorrhizal ericoid fungus during its saprophytic growth. FEMS Microbiol Lett 114: 85-92

Peretto R, Bettini V, Favaron F, Alghisi P, Bonfante P (1995) Polygalacturonase activity and location in arbuscular mycorrhizal roots of *Allium porrum* L. Mycorrhiza 5: 157-163

Perotto S, Peretto R, Faccio A, Schubert A, Varma A, Bonfante P (1995) Ericoid mycorrhizal fungi: cellular and molecular bases of their interactions with the host plant. Can J Bot 73: S557-S568

Remy W, Taylor TN, Hass H, Kerp H (1994) Four hundred-million-year-old vesicular arbuscular mycorrhizae. Proc Natl Acad Sci USA 91: 11841-11843

Salzer P, Hebe G., Reith A, Zitterell-Haid B, Stransky H, Gaschler K, Hager A (1996) Rapid reactions of spruce cells to elicitors released from the ectomycorrhizal fungus *Heleboma crustuliniforme*, and inactivation of these elicitors by extracellular spruce cell enzymes. Planta 198: 118-126

Sanchez LM, Doke N, Kawakita K (1993) Elicitor-induced chemiluminescence in cell suspension cultures of tomato, sweet pepper and tobacco plants and its inhibition by suppressors from *Phytophthora* spp. Plant Sci 88: 141-148

Sbrana C, Avio L, Giovannetti M (1995) The occurrence of calcofluor and lectin binding polysaccharides in the outer wall of arbuscular mycorrhizal fungi spores. Mycol Res 10: 1249-1252

Shirashi T, Yamada T, Saitoh K, Kato T, Toyoda K, Yoshioka H, Kim HM, Ichinose Y, Tahara M, Oku H (1994) Suppressors: determinants of specificity produced by plant pathogens. Plant Cell Physiol 35: 1107-1119

Simon L, Bousquet J, Levesque RC, Lalonde M (1993) Origin and diversification of endomycorrhizal fungi and coincidence with vascular land plants. Nature 363: 67-69

Smith SE, Smith FA (1990) Structure and function of the interfaces in biotrophic symbioses as they relate to nutrient transport. New Phytol 114: 1-38

Spanu P, Boller T, Ludwig A, Wiemken A, Faccio A, Bonfante-Fasolo P (1989) Chitinase in roots of mycorrhizal *Allium porrum*: regulation and localization. Planta 177: 447-455

Storti E, Pelucchini D, Tegli S, Scala A (1988) A potential defense mechanism of tomato against the late blight disease is suppressed by germinating sporangia derived substances from *Phytophthora infestans*. J Phytopathol 121: 275-282

Straker CJ, Mitchell DT (1986) The activity and characterization of acid phosphatases in endomycorrhizal fungi of the Ericaceae. New Phytol 104: 243-256

Straker CJ, Schnippenkoetter WH, Lemoine MC (1992) Analysis of acid invertase and comparison with acid phosphatase in the ericoid mycorrhizal fungus *Hymenoscyphus ericae* (Read) Korf and Kernan. Mycorrhiza 2: 63-67

Varma AK, Bonfante P (1994) Utilization of cell-wall related carbohydrates by ericoid mycorrhizal endophytes. Symbiosis 16: 301-313

Volpin H, Elkind Y, Okon Y, Kapulnik Y (1994) A vesicular arbuscular mycorrhizal fungus (*Glomus intraradix*) induces a defense response in alfalfa roots. Plant Physiol 104: 683-689

Waldmüller T, Cosio EG, Grisebach H, Ebel J (1992) Release of highly elicitor-active glucans by germinating zoospores of *Phytophthora megasperma* f. sp. *glycinea*. Planta 188: 498-505

Wessels JGH, Mol PC, Sietsma JH, Vermeulen CA (1990) Wall structure, wall growth, and fungal cell morphogenesis. In: PJ Kuhn, AP Trinci, MJ Jung, MW Goosey, LG Copping (eds) Biochemistry of Cell Walls and Membranes in Fungi. Springer-Verlag, Berlin, Heidelberg, New York, London, Paris, Tokyo, Hong-Kong, pp 81-93

Xie ZP, Staehelin C, Vierheilig H, Wiemken A, Jabbouri S, Broughton WJ, Vogeli-Lange R, Boller T (1995) Rhizobial nodulation factors stimulate mycorrhizal colonization of nodulating and nonnondulating soybeans. Plant Physiol 108: 1519-1525

Yoshikawa M, Yamaoka N, Takeuchi Y (1993) Elicitors: their significance and primary modes of action in the induction of plant defense reactions. Plant Cell Physiol 34: 1163-1173

Control of Elicitor-Induced Reactions in Spruce Cells by Auxin and by Enzymatic Elicitor Degradation

P. Salzer, R. Mensen, G. Hebe, K. Gaschler and A. Hager
Botanisches Institut der Universität Tübingen, Auf der Morgenstelle 1, 72076 Tübingen, Germany

Key words: Ectomycorrhiza, defense reactions, elicitors, auxin, elicitor inactivation.

Summary: Elicitors from the ectomycorrhizal fungi *Hebeloma crustuliniforme* (Bull.) Quel., *Amanita muscaria* (L.) Pers., *Suillus variegatus* (Sw.: Fr.) O. K. and the spruce pathogenic fungus *Heterobasidion annosum* (Fr.) Bref. induced an extracellular alkalinization response in suspension-cultured cells of *Picea abies* (L.) Karst. within 4 to 6 min. In addition, phosphorylation of a 63-kDa protein and dephosphorylation of a 65-kDa protein were induced in spruce cells by 4 min after elicitor application. Both the alkalinization and the protein phosphorylation were inhibited by nanomolar concentrations of the protein kinase inhibitor staurosporine. Furthermore, the elicitors induced the synthesis of chitinase and extracellular ß-1,3-glucanase as well as the accumulation of peroxidase in the walls of spruce cells. On the other hand, auxins prevented the induction of peroxidase accumulation in spruce cell walls. This inhibiting effect was found for elicitors from *Hebeloma crustuliniforme*, *Suillus variegatus* and *Heterobasidion annosum*. Moreover, enzymes which were secreted by the spruce cells decreased the effectiveness of the elicitors from both the ectomycorrhizal and the pathogenic fungus. These findings indicate that no fundamental difference exists in the reactions which are induced in host cells by elicitors from the ectomycorrhizal fungi *Hebeloma crustuliniforme*, *Amanita muscaria*, and *Suillus variegatus* and the pathogenic fungus *Heterobasidion annosum*. We assume that the production of auxin by the invading fungus and the inactivation of fungal elicitors by extracellular enzymes of the plant itself could be inductive to root colonization and ectomycorrhiza formation.

Introduction

Most forest trees of the northern hemisphere are in symbiotic association with ectomycorrhiza-forming fungi. They belong to the most advanced orders of Basidiomycetes (Boletales, Russulales and Telephorales) and Ascomycetes (Pezizales). In contrast, *Heterobasidion annosum* (Basidiomycetes, Russulales) is strongly pathogenic to spruce and fir (Heneen et al. 1994, Kottke et al. 1996). Ectomycorrhizas are characterized by the differentiation of a hyphal mantle and a Hartig net. The dense fungal network of the mantle, which completely encloses the lateral roots takes over the function of the rhizodermis. Emanating hyphae which thoroughly penetrate the soil improve the mineral supply of the ectomycorrhizal roots. In the Hartig net which is formed in the root cortex by synenchymatous hyphae growing in the walls of the host cells (Blasius et al. 1986) the mutualistic exchange of nutrients occurs (Harley and Smith 1983, Kottke and Oberwinkler 1986, Kottke et al. 1996). The fungus supplies the plant with minerals and water and in return receives carbohydrates which are delivered by the plant.

The intensive penetration of the plant tissue by the fungus raises some general questions: (1) do the plant cells recognize the invading fungus from the beginning as an ectomycorrhizal partner?; (2) by which mechanisms are plant defense reactions attenuated or avoided during ectomycorrhiza formation? Induction of strong plant defense reactions surely would prevent the outcome of a balanced symbiotic ectomycorrhizal association (Salzer et al. 1996a). To answer these questions, we studied the effects which were induced in spruce cells by elicitors released from various ectomycorrhizal fungi and the pathogenic fungus *Heterobasidion annosum*. To get some insight into mechanisms which might attenuate induction of these elicitor-induced plant defense reactions we studied the influence of hormones and host cell-secreted enzymes on elicitor-induced defense reactions. To simplify matters we worked with suspension-cultured cells which were obtained from radicles of *Picea abies* seedlings.

Materials and Methods

Chemicals. Staurosporine was purchased from Boehringer (Mannheim, Germany), purified crab shell chitin, ß-glucuronidase type H-2 (crude solution from Helix pomatia), laminarin, N-acetylglucosamine, guaiacol and polyvinylpolypyrrolidone (PVPP) were from Sigma (Deisenhofen, Germany). CM-curdlan-RBB was obtained from Blue Substrates (Göttingen, Germany). Tetramethylethylenediamine and ammonium peroxydisulfate were from Biorad (Richmond, California, USA) and Protogel (30% acrylamide, 0.8% bisacrylamide stock solution) from Hölzel (Manville, New Jersey, USA). [^{32}P]Pi and Hyperfilm MP were purchased from Amersham Buchler (Braunschweig, Germany). The other chemicals used were p.a. grade from Merck (Darmstadt, Germany). All media were prepared with bidistilled water.

Cultures. Suspension cultures of *Picea abies* (L.) Karst. were grown in 4X-Gamborg's medium (Gamborg et al. 1968) and were subcultured weekly by transferring 20 ml of cell suspension into 60 ml of fresh medium. The cultures were shaken in 200 ml flasks on a rotary shaker at 100 rpm at 24 to 26 °C in the dark. Cultures with high concentrations of auxins contained 2 $\mu g\ ml^{-1}$ 2,4-dichlorophenoxyacetic acid (2,4-D), 0.5 μg ml^{-1} naphthylacetic acid (NAA) and 0.5 μg ml^{-1} indoleacetic acid (IAA). To obtain suspension cultures with low concentrations of auxins, 20 ml of a 7-day-old suspension culture originally containing high auxin concentrations was transferred to 60 ml fresh culture medium without auxins.

Amanita muscaria (L.) Pers. was isolated from fruit bodies growing underneath *Picea abies* trees. *Suillus variegatus* (Sw.: Fr.) O. K. was isolated from a fruitbody growing under *Pinus mugo* Turra (Sirrenberg et al. 1995). *Hebeloma crustuliniforme* (Bull. ex Fries) Quel. (strain Tü 704) has been isolated from fruit bodies on *Picea abies* and was kindly supplied by Dr. I. Kottke (Tübingen, Germany). *Heterobasidion annosum* (Fr.) Bref. (HA 34) strains were a kind gift from Dr. A. Honold (Tübingen). Suspension cultures of the fungi were grown in MMNc medium as described by Sirrenberg et al. (1995).

Isolation of Elicitors. Elicitors from suspension-cultured fungi were prepared from 3- to 4-week-old cultures. One gram of hyphae was consecutively washed in 40 ml bidistilled water, in 40 ml 0.1 M and in 40 ml 0.5 M potassium phosphate (KPi) buffer (pH 7.2). Then, the hyphae were homogenized in 0.5 M KPi buffer, using an Ultraturrax T25 (Janke and Kunkel, Staufen i. Br., Germany) 4 x 30 s (24000 rpm), and sonified 4 x 20 s (70 Watt, Sonifier B12, Branson, Danbury, Connecticut, USA). The homogenate was centrifuged (5 min, 4000 x g) and the pellet was subsequently washed 8 times in 0.5 M, then 8 times in 0.1 M KPi buffer, 8 times in bidistilled water, and was finally air-dried. To release soluble elicitors 3 mg dry weight (DW) of the walls were shaken for 30 min at room temperature in a mineral solution containing 10 mM $NaNO_3$, 1 mM KCl, 1 mM Na_2SO_4, 1 mM $Mg(NO_3)_2$, and 1 mM $CaCl_2$. The insoluble fragments were sedimented by centrifugation (5 min, 4000 x g) and the elicitors in the supernatant (soluble elicitors) were used for experiments. To release putative elicitors from suspension-cultured spruce cells the same procedure was followed as for fungal hyphae. For experiments performed under sterile conditions the dried cell wall fragments were autoclaved in bidistilled water.

Induction of Extracellular Alkalinization. The experiments were performed with spruce cells 5 to 9 d after subculture as previously described by Salzer et al. (1996b). Briefly: spruce cells were washed in mineral solution (10 mM $NaNO_3$, 1 mM KCl, 1 mM Na_2SO_4, 1 mM $Mg(NO_3)_2$, and 1 mM $CaCl_2$), then 0.5 g of spruce cells were incubated in 20 ml of this solution and the pH of the medium was continuously measured by pH electrodes. To induce extracellular alkalinization, 400 μl of soluble elicitors were added to the spruce cells.

In Vivo Phosphorylation. Spruce cells were washed with mineral solution and 0.2 g cells were incubated in 1.5 ml mineral solution supplemented with 10 mM Mes, pH 5.0. Twenty minutes after application of [^{32}P]Pi (780 kBq), 30 μl of the soluble elicitors (dissolved in water) were added. To control samples, an equivalent amount of water was added. The phosphorylation reaction was stopped within 10 s at the indicated times, ranging from 0 to 20 min, by adding 300 μl of 30%

(w/w) trichloroacetic acid and transferring the cells into liquid N_2 according to the method described by Felix et al. (1991). After three thaw and freeze cycles the samples were stored on ice for 30 min. Then the samples were centrifuged (15000 x g, 5 min). The pellet was washed three times with 80% acetone. Subsequently, the proteins were solubilized in a buffer containing 62.5 mM Tris-HCl, pH 8.0, 10% (w/v) glycerol and 2% (w/v) SDS by heating to 90 °C for 10 min. The protein concentration of the samples was determined with the DC-protein assay (Biorad, Richmond CA, USA) following the manufacturer's instructions. Twenty min before subjecting the samples to SDS-PAGE, dithiothreitol was added to a final concentration of 1 mM. SDS-PAGE was performed according to Laemmli (1970) using a 5% acrylamide stacking gel and a 10% separation gel with 2 μg proteins per lane. The gels were dried and the phosphorylated proteins were detected by autoradiography with a Hyperfilm MP.

Induction of ß-1,3-Glucanase, Chitinase and Peroxidase. Autoclaved elicitors were aseptically added to the spruce suspension cultures. To control cells an equivalent amount of autoclaved, bidistilled water was added. In the following days, samples were taken. Endo-acting ß-1,3-glucanase activity in the culture medium of the spruce cells was measured using CM-Curdlan-RBB as described by Wirth and Wolf (1992). Chitinase was isolated from spruce cells by homogenizing the cells in 50 mM citrate-NaOH buffer, pH 5.0 with mortar and pestle. The homogenate was centrifuged (20 min, 38000 x g) and the chitinase activity in the supernatant was measured as described by Sauter and Hager (1989). Ionically wall-bound peroxidase activity was determined after extraction of the enzyme from cell walls. To this end, the spruce cells were homogenized with mortar and pestle in a mixture of acetone (3 parts) and water (1 part), containing 0.5 g PVPP per gram cell fresh weight. The homogenate was vacuum-filtrated and dried under vacuum at 4 °C. Soluble peroxidase was extracted from the dried homogenate with 50 mM Tris-HCl, pH 7.2 for 1 h at 4 °C. Soluble peroxidase was separated from the cell wall-associated peroxidase by centrifugation (5 min, 4000 x g). Then, cell walls were washed with 1% (v/v) Triton X-100, followed by 3 washes with 50 mM Tris-HCl buffer (pH 7.2). Finally, ionically cell wall-bound peroxidase was released from the cell walls with 1 M NaCl (3 h at 4 °C). After removing the cell walls by centrifugation (5 min, 4000 x g) released peroxidase activity was measured in 50 mM sodium phosphate buffer, pH 4.6, containing 4 mM guaiacol and 2 mM H_2O_2. As a measure of peroxidase activity the slope of the continuously registrated absorbance at 450 nm was taken.

Isolation and Partial Purification of Extracellular Spruce Enzymes. Enzymes which were constitutively secreted by spruce cells into the culture medium were precipitated with ammonia sulfate (70% saturation) and were subsequently chromatographed on a Sephadex G-75 gelfiltration column as previously described by Salzer et al. (1996b). The fractions with the highest chitinase and glucanase activities were used for the enzymatic treatment of the elicitors. Prior to elicitor treatment, the enzymes were dialyzed against the mineral solution (10 mM

$NaNO_3$, 1 mM KCl, 1 mM Na_2SO_4, 1 mM $Mg(NO_3)_2$, 1 mM $CaCl_2$) and the activity of chitinase and ß-1,3-glucanase was determined with colloidal chitin and laminarin as substrates at 37 °C.

Treatment of Soluble Elicitors With Extracellular Spruce Enzymes. To study the elicitor inactivation by extracellular spruce enzymes, soluble elicitors (400 µl) obtained from fungal cell walls were incubated with spruce cell enzymes (50 µl) containing chitinase (20 - 60 pkat) and ß-1,3-glucanase activity (40 - 80 pkat) for 90 min at 37°C. Other enzymatic activities were not determined. Before adding to the spruce cells the mixtures (400 µl soluble elicitors with 50 µl enzymes) were boiled for 5 min. In control samples in which the elicitors and an equivalent amount of spruce enzymes were separately incubated for 90 min at 37 °C, 400 µl soluble elicitors and 50 µl spruce enzymes were separately boiled, then mixed and finally added to the spruce cells.

Results and Discussion

Rapid Elicitor-Induced Reactions in Spruce Cells

Cell walls of the ectomycorrhizal fungi *Hebeloma crustuliniforme*, *Amanita muscaria*, *Suillus variegatus* as well as walls from the spruce pathogenic fungus *Heterobasidion annosum* released elicitors which induced an extracellular alkalinization response in spruce cells (Fig. 1). Only gradual differences in the effectiveness of the elicitors in inducing this reaction were observed. For *Hebeloma crustuliniforme* it was recently shown that elicitors capable of inducing extracellular alkalinization and H_2O_2 synthesis were released from the living fungus without being influenced by the plant (Salzer et al. 1996b). Such elicitor release which occurs independently from host plants was also demonstrated for pathogenic fungi. For example, germinating zoospores of *Phytophthora megasperma* released glucan elicitors from their cell walls (Waldmüller et al. 1992). These findings indicate that elicitors act as primary signals irrespective of the symbiotic or pathogenic character of the fungus.

In addition to extracellular alkalinization, fungal cell wall elicitors induced a complex of rapid reactions in spruce cells (Salzer et al. 1996b). Release of Cl^- was the most rapid elicitor-stimulated response occurring in spruce cells. Using ion-selective electrodes an increase in the extracellular Cl^- concentration was detectable less than 1 min after elicitor application. Efflux of K^+ was induced with a lag of 1 to 3 min after addition of the elicitors. Within the same time span elicitor-induced uptake of $^{45}Ca^{2+}$ commenced, with highest uptake rates between 3 to 8 min after elicitation (Salzer et al. 1996b). Moreover, synthesis of H_2O_2 is induced in spruce cells by elicitors of the ectomycorrhizal fungus *Hebeloma crustuliniforme* (Schwacke and Hager 1992). Such rapid reactions were also induced by elicitors from pathogenic fungi in cells of tobacco, carrot and parsley. In this case

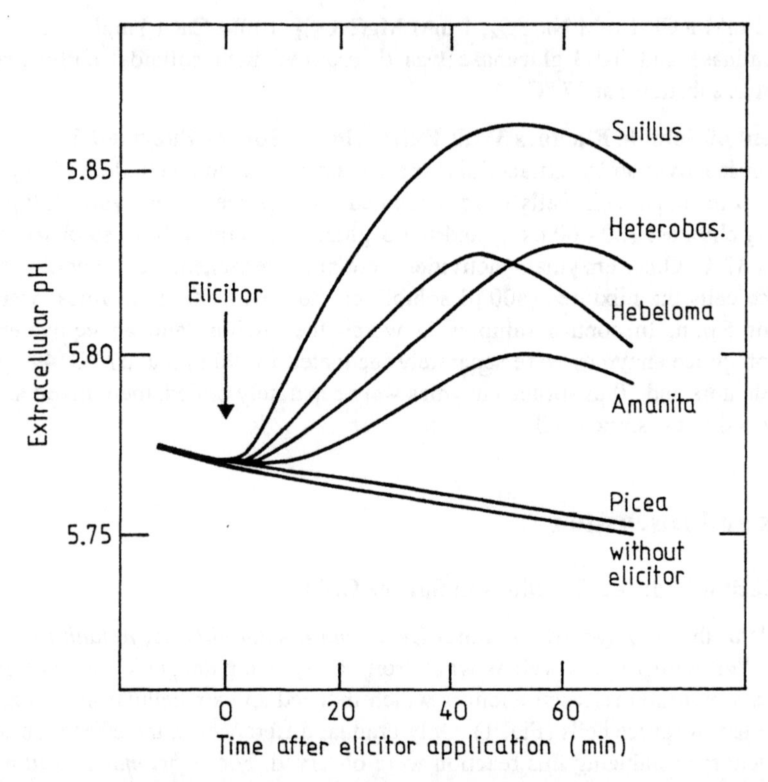

Fig. 1. Extracellular alkalinization response of suspension-cultured spruce cells induced by elicitors from the ectomycorrhizal fungi *Suillus variegatus*, *Hebeloma crustuliniforme*, *Amanita muscaria* and the pathogenic fungus *Heterobasidion annosum*. Soluble elicitors were released from the fungal cell walls to mineral solution; 400 µl of the elicitor-containing solution was added to the spruce cells. In control experiments, 400 µl of the mineral solution without elicitors was added. No pH effect could be induced in the *Picea abies* cells by putative elicitors prepared from spruce cell walls

they are assumed to be the initial reactions of the hypersensitive response (Atkinson et al. 1990, Bach et al. 1993, Nürnberger et al. 1994).

Signal Transduction of Rapid Elicitor-Induced Reactions in Spruce Cells

In spruce cells elicitor-induced phosphorylation of a 63-kDa protein and dephosphorylation of a 65-kDa protein were detectable already 4 min after elicitor application. This elicitor-induced protein phosphorylation was inhibited by nanomolar concentrations of the protein kinase inhibitor staurosporine (Fig. 2). The previous findings that the elicitor-induced extracellular alkalinization and the H_2O_2-synthesis in spruce cells were also inhibited by nanomolar concentrations of staurosporine (Salzer et al. 1996b, Schwacke and Hager 1992) and the promptness

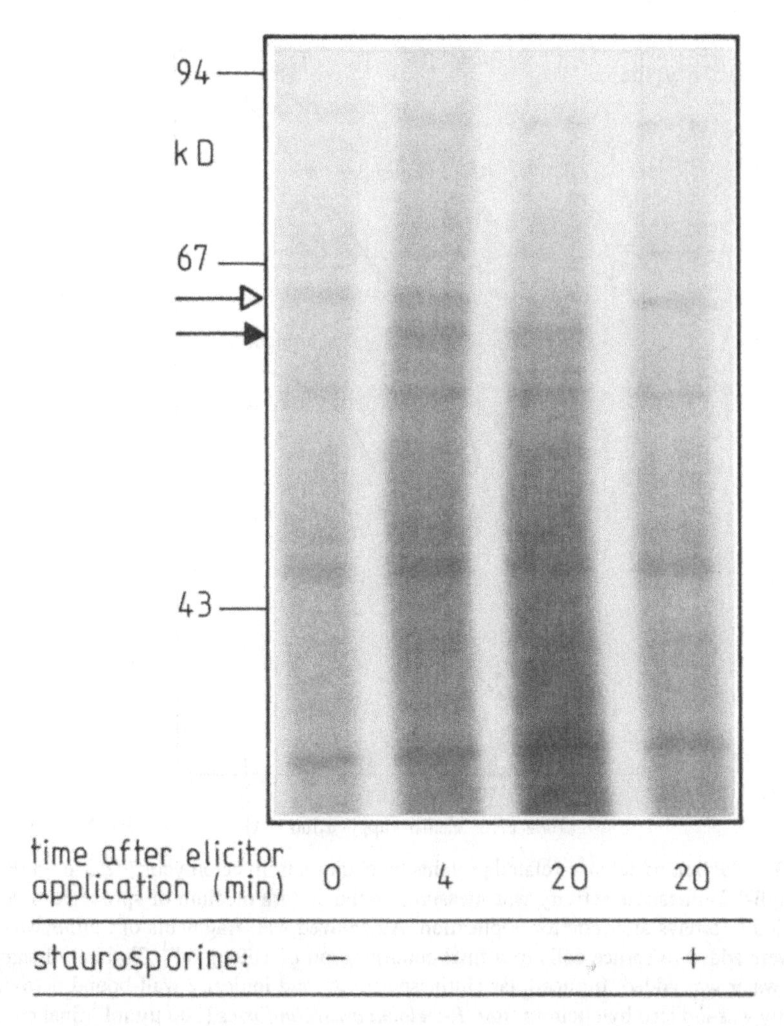

Fig. 2. Elicitor-induced rapid protein phosphorylation is sensitive to the protein kinase inhibitor staurosporine. Soluble elicitors from *Hebeloma crustuliniforme* induced the phosphorylation of a 63-kDa protein after 4 min. The phosphorylation was complete after 20 min. Staurosporine (500 nM) inhibited the elicitor-induced changes in the pattern of the phosphorylated proteins when applied 2 min prior to the elicitor. ^{32}Pi was added to the cells 20 min prior to the elicitors. The phosphorylation was stopped 0, 4 or 20 min after elicitor application. After extraction the proteins were separated by SDS-PAGE and autoradiographed. Addition of water without elicitors resulted in a phosphorylation pattern after 0, 4, and 20 min similar to the pattern shown for 0 min after elicitor application

of the phosphorylation strongly indicate that the phosphorylated 63-kDa protein is directly involved in the signal transduction cascade leading to the above mentioned reactions.

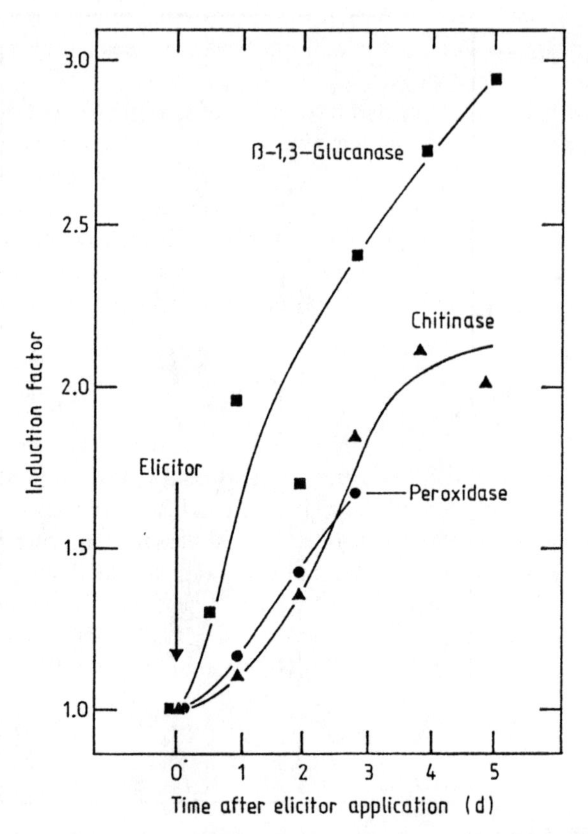

Fig. 3. Induction of defense-related proteins by elicitors from ectomycorrhizal fungi. Endo-acting ß-1,3-glucanase activity was measured in the culture medium of spruce cells during a period of 5 days after elicitor application. Autoclaved wall fragments of *Suillus variegatus* were added to spruce cells to a final concentration of 100 µg ml^{-1}. To control samples 1 ml water was added. Intracellular chitinase activity and ionically wall-bound peroxidase activity was induced by elicitors from *Hebeloma crustuliniforme* (100 µg ml^{-1} final concentration). Induction factors are calculated from the enzymatic activities measured in elicitor-treated spruce cells related to the activity in the control cells at the corresponding time

The transient character of the elicitor-induced Ca^{2+} influx in *Picea abies* cells and the finding that removal of extracellular Ca^{2+} by the Ca^{2+} chelator EGTA delayed the extracellular alkalinization and inhibited the H_2O_2 synthesis in spruce cells suggest that an increase in the cytoplasmic Ca^{2+} concentration is also part of the signal transduction cascade which is activated by elicitors from ectomycorrhizal fungi (Salzer et al. 1996b, Schwacke and Hager 1992). Influx of Ca^{2+} and phosphorylation of proteins are also elements in signal transduction cascades activated by elicitors of pathogenic fungi in various plant cells (Choi and Bostock 1994, Dietrich et al. 1990, Tavernier et al. 1995). Thus, the finding of Ca^{2+} as

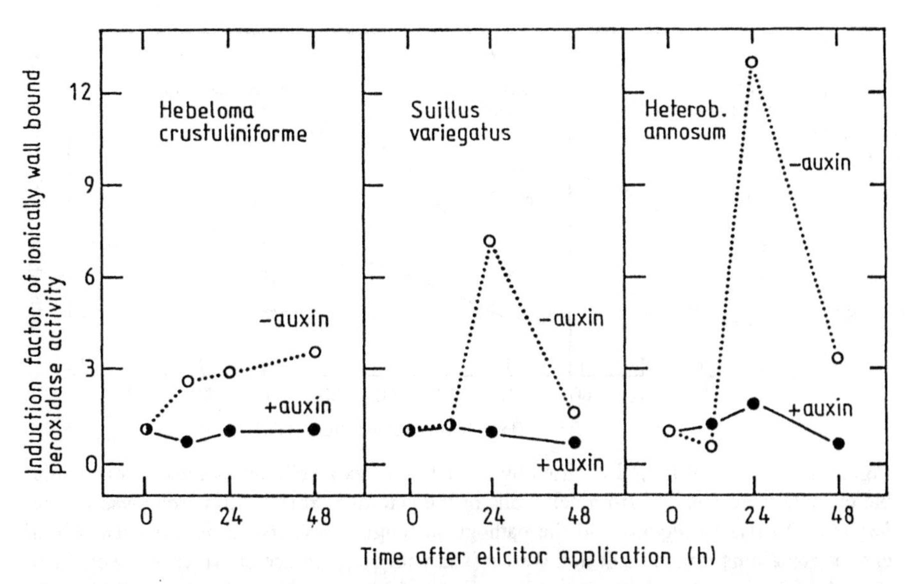

Fig. 4. Suppression of elicitor-induced peroxidase accumulation in spruce cells by auxins. Induction of ionically cell wall-bound peroxidase activity by autoclaved fungal cell wall fragments (final concentration 100 μg ml^{-1}) was measured over a period of 2 days in spruce cell cultures containing high amounts of auxins (2 μg 2,4-D ml^{-1}, 0.5 μg AN ml^{-1} and 0.5 μg IAA ml^{-1}) and in cultures without auxins. The induction factors were calculated from the corresponding control samples where water was added to the spruce cells instead of elicitors

second messenger and the involvement of protein phosphorylations further underline the similarities in the reactions induced by elicitors from ectomycorrhizal and pathogenic fungi.

Induction of Defense-Related Proteins in Spruce Cells

Elicitors of ectomycorrhizal fungi also induced typical defense-related proteins in spruce cells. Increased synthesis of endo-acting ß-1,3-glucanase and chitinase were induced in less than 24 h (Fig. 3). In plant pathogen interactions the synergistic effect of these hydrolytic enzymes often prevents the penetration of the plant tissue by fungal pathogens (Mauch et al. 1988). Incorporation of peroxidase in the spruce cell wall is a further effect induced by elicitors from ectomycorrhizal fungi (Fig. 3). Peroxidases catalyse the formation of physical barriers by cross-linking of extensin and by formation of lignin (Brownleader et al. 1995, Harkin and Obst 1973). Interestingly, induction of physical barriers by the vesicular arbuscular (VA) mycorrhiza-forming fungus *Glomus mosseae* prevented colonization of pea mutants which were unable to undergo mycorrhization (Golotte et al. 1993).

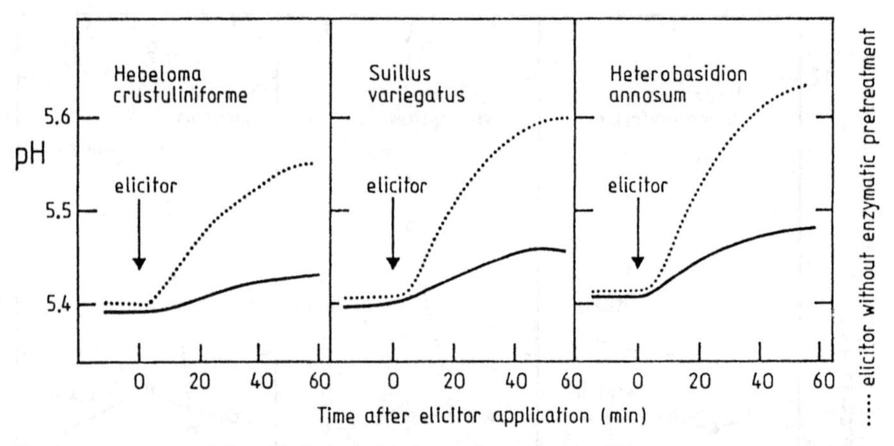

Fig. 5. Inactivation of fungal elicitors by the action of extracellular spruce enzymes. Soluble elicitors were released from cell walls of the ectomycorrhizal fungi *Hebeloma crustuliniforme*, *Suillus variegatus* and the pathogenic fungus *Heterobasidion annosum*. 400 µl elicitor-containing mineral solution were mixed with 50 µl spruce enzymes and were incubated for 90 min at 37 °C. In control experiments 50 µl enzymes and 400 µl soluble elicitors were separately incubated for 90 min at 37 °C. The diagram shows the time courses of the extracellular alkalinization of spruce cells induced by 450 µl of the mixtures of elicitors treated with enzymes and elicitors incubated without enzymes

All these similarities of the reactions induced by elicitors from ectomycorrhizal and pathogenic fungi indicate that the ability of the ectomycorrhizal partners to limit these inducible reactions is a key event which decides whether a compatible interaction can be accomplished or not.

Attenuation of Elicitor-Induced Reactions in Spruce Cells

Our studies on spruce cells gave hints that at least two mechanisms decreased the intensity of inducible reactions. (1) High concentrations of auxins in the culture medium of the spruce cells reduced the amount of elicitor-induced peroxidase accumulation in the walls of the spruce cells (Fig. 4). Suppression of specific chitinase isoforms in spruce cells by auxins was also demonstrated (Salzer et al. 1996a). Synthesis of auxin is widespread among ectomycorrhizal fungi (Gogala 1991). Therefore, it should be expected that fungi capable of synthesizing auxin should have advantages in colonizing the root tissues due to the hormonal suppression of plant defense responses. However, this advantage would not be restricted to ectomycorrhiza-forming fungi, because the effectiveness of elicitors from the pathogenic fungus *Heterobasidion annosum* was also reduced by auxins (Fig. 4).

Enzymatic inactivation of elicitors by the action of extracellular plant enzymes is a further strategy to avoid elicitor-induced reactions in the host cells (Salzer et al. 1996b). Again, such an enzymatic inactivation would be a mechanism which does not discriminate between elicitors of ectomycorrhizal and pathogenic fungi

(Fig. 5). In fact, soluble elicitors from the ectomycorrhizal fungi *Hebeloma crustuliniforme* and *Suillus variegatus* as well as elicitors from the pathogenic fungus *Heterobasidion annosum* were inactivated by the enzymes secreted by spruce cells. Irrespective of the unspecificity of hormonal suppression of plant defense responses and of inactivation of fungal elicitors by extracellular enzymes of the host plant, both mechanisms could effectively contribute to a successful interaction of ectomycorrhizal fungi and their host roots. However, the risk that such mechanisms were exploited by pathogenic fungi is lowered by the fact that the hyphal mantle of the ectomycorrhiza exerts a protective function against such pathogens (Farquhar and Peterson 1991). In addition, the surplus of ectomycorrhizal fungi colonizing the forest soil should successfully compete with a minority of root pathogenic fungi.

Acknowledgments: We thank Dr. A. Honold, Dr. I. Kottke and Dr. A. Sirrenberg (Tübingen, Germany) for fungal strains of *Amanita muscaria*, *Suillus variegatus*, *Hebeloma crustuliniforme* and *Heterobasidion annosum*. This work was supported by the Deutsche Forschungsgemeinschaft. Ruth Mensen and Katja Gaschler were financed by the Graduiertenkolleg "Organismische Interaktionen in Waldökosystemen" and Gerhard Hebe by a scholarship of the Landesgraduiertenförderungsgesetz.

References

Atkinson MM, Keppler LD, Orlandi EW, Baker CJ, Mischke CF (1990) Involvement of plasma membrane calcium influx in bacterial induction of the K^+/H^+ and hypersensitive responses in tobacco. Plant Physiol 92: 215-221

Bach M, Schnitzler JP, Seitz HU (1993) Elicitor-induced changes in Ca^{2+} influx, K^+efflux, and 4-hydroxybenzoic acid synthesis in protoplasts of *Daucus carota* L. Plant Physiol 103: 407-412

Blasius D, Feil W, Kottke I, Oberwinkler F (1986) Hartig net structure and formation in fully ensheathed ectomycorrhizas. Nordic J Bot 6: 837-842

Brownleader MD, Ahmed N, Traevan M, Chaplin MF, Dey PM (1995) Purification and partial characterization of tomato extensin peroxidase. Plant Physiol 109: 1115-1123

Choi D, Bostock RM (1994) Involvement of de novo protein synthesis, protein kinase, extracellular Ca^{2+}, and lipoxygenase in arachidonic acid induction of 3-hydroxy-3-methylglutaryl coenzyme A reductase genes and isoprenoid accumulation in potato (*Solanum tuberosum* L.). Plant Physiol. 104: 1237-1244

Dietrich A: Mayer JE, Hahlbrock K (1990) Fungal elicitor triggers rapid, transient, and specific protein phosphorylation in parsley cell suspension cultures. J Biol Chem 165: 6360-6368

Farquhar ML, Peterson RL (1991) Later events in suppression of Fusarium root rot of red pine seedlings by the ectomycorrhizal fungus Paxillus involutus. Can J Bot 69: 1372-1383

Felix G, Grosskopf DG, Regenass M, Boller T (1991) Rapid changes of protein phosphorylation are involved in signal transduction of the elicitor signal in plant cells. Proc Natl Acad Sci USA 88: 8831-8834

Gamborg OL, Miller RA, Ojima K (1968) Nutrient requirements of suspension cultures of soybean root cells. Exp Cell Res 50: 151-158

Gogala N (1991) Regulation of mycorrhizal infection by hormonal factors produced by hosts and fungi. Experientia 47: 331-340

Gollotte A, Gianninazzi-Pearson V, Giovanetti M, Sbrana C, Avio L, Gianninazzi S (1993) Cellular localization and cytochemical probing of resistance reactions to arbuscular mycorrhizal fungi in a 'locus a' myc- mutant of Pisum sativum L.. Planta 191: 112-122

Harkin J.M., Obst J.R. (1973) Lignification in Trees: Indication of exclusive peroxidase participation. Science 189: 286-287

Harley JL, Smith SE (1983) Mycorrhizal symbiosis. Academic Press, London

Heneen WK, Gustafsson M, Karlsson G, Brismar K (1994) Interactions between Norway spruce (*Picea abies*) and *Heterobasidion annosum*. I. Infection of nonsuberized and young suberized roots. Can J Bot 72: 872-883

Kottke I, Oberwinkler F (1986) Mycorrhiza of forest trees - structure and function. Trees 1: 1-24

Kottke I, Münzenberger B, Heyser W, Oberwinkler F (1996) Structural approach to function in ectomycorrhizas. In: Rennenberg H, Eschrich W, Ziegler H, (eds) Trees - Contributions to modern tree physiology. SPB Academic Publ, The Hague, in press

Laemmli UK (1970) Cleávage of structural proteins during the assembly of the head of bacteriophage T4. Nature 227: 680-685

Mauch F, Mauch-Mani B, Boller T (1988) Antifungal hydrolases in pea tissue. II. Inhibition of fungal growth by combinations of chitinase and ß-1,3-glucanase. Plant Physiol 88: 936-942

Nürnberger T, Nennstiel D, Jabs T, Sacks WR, Hahlbrock K, Scheel D (1994) High affinity binding of a fungal oligopeptide elicitor to parsley plasma membranes triggers multiple defence responses. Cell 78: 449-460

Salzer P, Münzenberger B, Schwacke R, Kottke I, Hager A (1996a) Signalling in ectomycorrhizal fungus root interactions. In: Trees - Contributions to modern tree physiology. Rennenberg H, Eschrich W, Ziegler H eds, SPB Academic Publishing, The Hague, in press

Salzer P, Hebe G, Reith A, Zitterell-Haid B, Stransky H, Gaschler K, Hager A (1996b) Rapid reactions of spruce cells to elicitors released from the ectomycorrhizal fungus *Hebeloma crustuliniforme*, and inactivation of these elicitors by extracellular spruce cell enzymes. Planta 198: 118-126

Sauter M, Hager A (1989) The mycorrhizal fungus *Amanita muscaria* induces chitinase activity in roots and in suspension-cultured cells of its host Picea abies. Planta 179: 61-66

Schwacke R, Hager A (1992) Fungal elicitors induce a transient release of active oxygen species from cultured spruce cells depending on Ca^{2+} and protein kinase activity. Planta 187: 136-141

Sirrenberg A, Salzer P, Hager A (1995) Induction of mycorrhiza-like structures and defence reactions in dual cultures of spruce callus and ectomycorrhizal fungi. New Phytol. 130: 149-156

Tavernier E, Wendehenne D, Blein JP, Pugin A (1995) Involvement of free Calcium in action of Cryptogein, a proteinaceous elicitor of hypersensitive reaction in tobacco cells. Plant Physiol 109: 1025-1031

Waldmüller T, Cosio EG, Grisebach H Ebel J (1992) Release of highly elicitor-active glucans by germinating zoospores of *Phytophthora megasperma* f. sp. *glycinea*. Planta 188: 498-505

Wirth SJ, Wolf GA (1992) Microplate colorimetric assay for endo-acting cellulase, xylanase, chitinase, 1,3-ß-glucanase and amylase extracted from forest soil horizons. Soil Bio Biochem 24: 511-519

A Novel IS Element is Present in Repeated Copies Among the Nodulation Genes of *Rhizobium 'hedysari'*

F. Meneghetti, S. Alberghini, E. Tola, A. Giacomini, F.J. Ollero, A. Squartini and M.P. Nuti
Dipartimento di Biotecnologie Agrarie, Università di Padova, Strada Romea 16, 35020 Legnaro (Padova), Italy

Key words: *Hedysarum coronarium*, IS element, nodulation genes, *Rhizobium 'hedysari'*

Summary: A genetic analysis of the regions required for symbiotic host plant nodulation by *Rhizobium 'hedysari'* led to the discovery of a 0.8 kb DNA element which presents the features of an insertion element and has no homology at DNA level with known sequences. The element, named ISRh1, appears to be specific for this *Rhizobium* species, and is present in all the strains tested in different numbers of copies. The majority of such copies are located on plasmids and the region of the nodulation and host specificity determinants seems to represent a hotspot for the insertion.

The complexity of hybridization patterns of strains from different isolation sites within the Mediterranean basin, is proposed as the basis for strain fingerprinting, and for evolutionary studies to track the spread of *Rhizobium 'hedysari'* in relation to that of its host legume *Hedysarum coronarium*.

Introduction

The nitrogen-fixing symbiosis between sulla (*Hedysarum coronarium* L.) and its highly host-specific bacterial partner *Rhizobium 'hedysari'* has recently received a increasing attention. This perennial forage crop is appreciated in several countries facing the Mediterranean for its agronomical properties, among which tolerance to drought, salinity, and alkaline soil reaction feature. These characteristics, along with its high productivity and good forage quality, make this plant a suitable choice for marginal soil cultivation, including pliocenic clays and semi-arid lands (Sarno et al. 1978; Lupi et al. 1988). This member of the tribe *Hedysareae* is distributed in Spain, Algeria and southern Italy, and has been then introduced for cultivation in Israel (Gurfel et al. 1982). The physiology and genetics of the symbiotic interaction have been studied at molecular level (Cabrera et al. 1979; Casella et al. 1984; Espuny et al. 1987; Ollero et al. 1989; Ollero et al. 1993; Squartini et al. 1993). Genetic determinants involved in symbiotic interactions are carried on large plasmids (Ollero et al. 1989, 1993). The taxonomic status of the bacteria as possible new species is presently under investigation.

Insertion elements are not new to *Rhizobiaceae* (Ruvkun et al. 1982; Dusha et al. 1987; Wheatcroft and Watson 1988a, b; Wheatcroft and Laberge 1991; Soto et al. 1992) although in most cases those described are not species-specific. Their potential as tools in strain characterization has been demonstrated (Wheatcroft and Watson 1988a; Kosier et al. 1993; Rice et al. 1994).

The present work, describing the finding of a novel rhizobial insertion element, stemmed from a project aimed at the determination of nucleotide sequence of the nodulation and host-specificity gene region of *Rhizobium 'hedysari'* strain IS123.

Materials and Methods

Bacterial Strains, Plasmids, and Culture Conditions. Strains and plasmids used are listed in Table 1. Rhizobia were grown in TY (Beringer 1974) or BIII (Dazzo 1984) media at 28 °C. *E. coli* was grown in LB at 37 °C. Antibiotics were used at the following concentrations: ampicillin, 50 µg/ml, kanamycin, 30 µg/ml.

DNA Manipulations and Sequencing. Molecular biology standard techniques, including cloning and restriction mapping were performed as described by Sambrook et al. (1989). For DNA hybridizations, digoxygenin labeled probes were obtained by a DIG-nonradioactive DNA labeling and hybridization kit (Boehringer Mannheim. Inc., Germany). Both labeling and hybridization were carried out as described by the manufacturer. Double stranded DNA fragments were cloned into pUC19 and sequenced by the di-deoxy nucleotide chain termination method (Sanger et al. 1977) using fluorescent di-deoxy nucleotides and an Applied Biosystems 373A DNA sequencer. Sequences were analyzed using the IG Suite Molecular Biology Software System (Intelligenetics Inc), rel. 5.4.

Results and Discussion

Cloning and Sequencing of the DNA Regions Involved in the Nodulation Process in *R. 'hedysari'*

Starting from plasmid pR2, a smaller R-prime derivative of the large natural symbiotic plasmid of strain IS123 (Ollero et al. 1993), a clone bank was obtained by cloning partially *Bgl*II-cleaved pR2 DNA in the unique *Bam*HI site of vector pRL497. Hybridization experiments with heterologous gene probes for the common (*nodDABC*) and host-specific (*nodFE*) nodulation genes from *Rhizobium leguminosarum* bv. *viciae* strain 1003 (Squartini et al. 1988) and bv. *trifolii* respectively led to the isolation of two clones, pBPZ14 and pBPZ18. From other clones which contain DNA fragments from both plasmids, it was possible to determine that the fragments contained in pBPZ14 and pBPZ18 lie adjacent in the *R. 'hedysari'* sym-plasmid. A genetic map of this region is shown in Fig. 1.

Table 1. Bacterial strains and plasmids

Strain or plasmid	Relevant traits	Source or reference
Rhizobium 'hedysari' IS123	Wild type, isolated from southern Spain	F.J. Ollero/F. Temprano
Rhizobium 'hedysari' CC1335	Wild type, isolated from southern Spain	J. Brockwell
Rhizobium 'hedysari' HCNT1	Wild type, isolated from Volterra, Italy	S. Casella
Rhizobium 'hedysari' RHF	Wild type, isolated from Pisa, Italy	S. Casella
Rhizobium 'hedysari' RH19	Wild type, isolated from Sicily, Italy	S. Casella
Rhizobium 'hedysari' RH100	Wild type, isolated from the Balearic Islands	A. Toffanin
Rhizobium 'hedysari' RH44	Wild type, isolated from southern Spain	A. Toffanin
Rhizobium 'hedysari' A6	Wild type, isolated from Algeria	A. Benguedouar
Rhizobium 'hedysari' A10	Wild type, isolated from Algeria	A. Benguedouar
Rhizobium 'hedysari' RJ243	pSym-cured derivative of IS123	Ollero et al. (1989)
R.meliloti ATCC9930^T	Type strain	P. Van Berkum
R.leguminosarum bv. *trifolii* ATCC 14480^T	Type strain	P. Van Berkum
R. leguminosarum bv. *viciae* ATCC10004^T	Type strain	P. Van Berkum
R.leguminosarum bv. *phaseoli* RCR3644^T	Type strain	P. Van Berkum
R. etli CFN42^T	Type strain	P. Van Berkum
R. tropici IIB CIAT899^T	Type strain	P. Van Berkum
R. tropici IIA CFN299^T	Type strain	P. Van Berkum
R. galegae HAMBI 503		K. Lindström
pR2	R-prime bearing *R. 'hedysari' nod* genes	Ollero et al. (1993)
pRL497	IncQ, broad host range vector	Elhai and Wolk (1988)
pBPZ18	Cloned *R. 'hedysari' nod* region	This work
pBPZ14	Cloned *R. 'hedysari' nod* region	This work
pAYS14	14 kb *Hin*dIII fragment from *R. l.* bv. *trifolii* ANU843 cloned in pRL497.Source of *nodFE* probe	A. Squartini

An IS-Like Element is Present in the *nod* Gene Region of *R. 'hedysari'*

Besides the presence of some nodulation genes which have homologue counterparts in other *Rhizobium* species, a particular region of 811 bp was encountered for a total of six occurrences in the 35 kb region currently being sequenced. Two

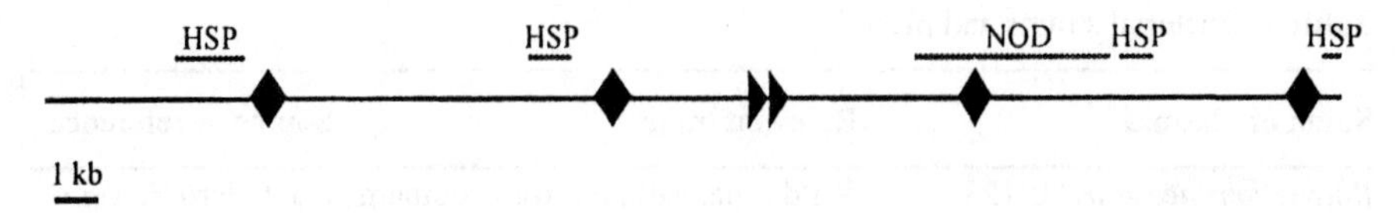

Fig. 1. Region of about 35 kb from *R. "hedysari"* IS123 sym plasmid containing common (NOD) and host-specific (HSP) *nod* genes. Diamonds (◆) indicate the newly idenified IS element ISRh1 or portions of it

of these copies contain deletions. This element presents two bordering 17-bp inverted repeats with a single mismatched base. Its sequence, shown in Fig. 2, is not significantly homologous by BLAST analysis (Altschul et al. 1990) to any of those present in the GenBank Database, while a scan to protein sequences of the SwissProt Database revealed a 55% homology to TRA1 and TRA2, two transposase enzymes from IS431MEC of *Staphylococcus aureus*. The element has been named ISRh1.

ISRh1 is Specific to *Rhizobium 'hedysari'*

A survey of the possible occurrence of such a sequence in different bacteria of the closely related members of the *Rhizobiaceae* family was carried out by hybridizing a labelled internal *Eco*RI-*Bam*HI fragment of ISRh1 to *Bam*HI-cleaved genomic DNA from the type strains of most up-to-date official species of rhizobia. A number of *Rhizobium 'hedysari'* strains from different isolation sites within the Mediterranean basin were also included. The results, shown as a computer-derived diagram of the hybridizing bands (Fig. 3), clearly indicate the absence of any hybridization signal in rhizobia other than *R. 'hedysari'*, while all strains of the latter display the element.

Hotspots for Transposition

The case of strain IS123, whose sequencing revealed the presence of ISRh1, shows six copies of the element within the 35-kb *nod* region surroundings. A comparison between the hybridizing profiles of this strain and its sym-plasmid-cured derivative RJ243 (Fig. 3, lanes l and i, respectively), show that 10 out of the 11 hybridizing bands are missing in the latter. In a different experiment, the residual band has been shown to belong to another cryptic plasmid of *R. 'hedysari'* IS123 (data not shown). Therefore a preference for plasmid-directed transposition and a bias for the symbiotic genes region are evident in this strain. The reasons for such behavior could be either (1) mechanistic, if in *cis* translocations to neighboring regions are favored, or (2) aimed at the foreign element survival in the bacterial genome. In this sense, the nodulation-host specificity zone offers peculiar advantages: (1) it is essential to *Rhizobium* lifestyle of molecular plant-microbe interactions and therefore of critical importance within its genome; (2) it defines the species identity in functional terms dictating its host plant range and it entitles

```
        10         20         30         40         50         60         70
CGGCGATGTC ACGTTGTTTG ATCAAGGTCG GAACAAGGCG CTTATGTGAG ACGTCAGGCG GTTGCGCCGG

        80         90        100        110        120        130        140
TCACGGCTTT CCACTGCGCC ATTGAGCGGA TCCGATGGAT GTGGATGGCG AGGGCTGAGT TTTTCTGGTG

       150        160        170        180        190        200        210
CGGGGGGACG AAGAGATTTC GGAGTGCCGA AAAGATCGAT ATGAACCGTT GCAAGCCGCC GACGGATCGA

       220        230        240        250        260        270        280
AATCTCTGCA TCATCCGCTC CCGTTTTCGA AGCGGCACGT GAGAATTCTC GGCTCGATTG TTCAGGCCCT

       290        300        310        320        330        340        350
TGTGCGATCG ATGTTCGACG GCGGGCATCA CCTCCCGTCT TGCAGCACCA TATGAGCGCA ATTTGTCGGT

       360        370        380        390        400        410        420
GACGATCCGC TTCGGCGTCA GGCCTTGCTT CTTCAGCAGC CTGACCAGCA ATCGCCTGGC GGCTTGGGTA

       430        440        450        460        470        480        490
TCGCGGCGGG CTTGAACGAT CTCGTCGAGA ACGTAACCGT CTTGGTCAAC GGCACGCCAG AGCCAATGTT

       500        510        520        530        540        550        560
TGCGGCCGCC GATGGAAATC ACCACCTCGT CCAGATGCCA GATATCCTTT CGCGAAGGCC TCTTTCTGCA

       570        580        590        600        610        620        630
CAACTGCCTG GCATAAGCCG CTCCGAATTT GGGACCCCAT CTCCGGATCG TCTCATGGGA GACGACGATA

       640        650        660        670        680        690        700
CCGCGCTCCA GCAGCATCTC CTCGACCATC CTCAGGCTCA AAGGGAACCG AAAATACAGC CAGACCGCAC

       710        720        730        740        750        760        770
GCGCGATAAT CTTGGGTGGG AAAGCGGTGG TTCTTGGTAG GTTACGGGCG GATTGTTCAT CCCAACCCAT

       780        790        800        810
TATCCGACAA CCGTTAAGCC GCTGACAACG TGACATCACC G
```

Fig. 2. Nucleotide sequence of the putative insertion element ISRh1. The 17-bp inverted repeats are indicated by arrowed boxes

the bearer to enter a specific ecological niche which provides trophic supply as well as shelter from competition; (3) it cannot coexist with corresponding heterologous host specificity regions, implying that interspecific gene exchange between rhizobia is followed by nodulation interference leading to the deletion of one set of host-specific genes (Ollero et al. 1993). Such phenomena may explain the absence of ISRh1 in bacterial species closely related to *R. 'hedysari'*.

Rhizobium 'hedysari' Strains can be Characterized by Their ISRh1 Hybridization Profiles

The number of copies observed in different strains of the *H. coronarium* symbiont is variable from 1 to about 20, resulting often in a complex pattern which serves a fingerprinting function, enabling the appreciation of diversity, relatedness (e.g. the conserved bands in lanes l, m, o and q), and identity (e.g. the identical profiles of strains in lanes s and t, consisting of two isolates from the same site in Algeria). Moreover, the specificity of the element suggests the use of amplification methods with primers based on ISRh1 sequence to achieve species-specific amplification and detection and strain-specific profiling in the same experiment.

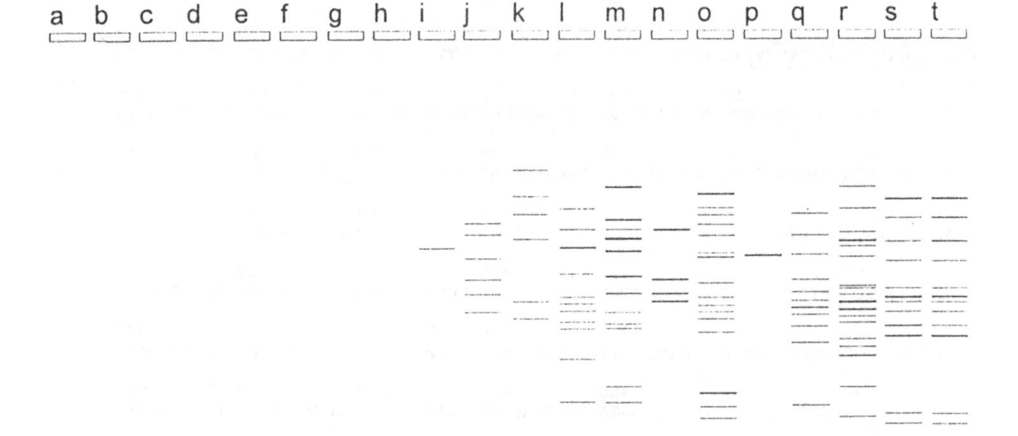

Fig. 3. Hybridization of digoxygenin-labelled *Bam*HI-*Eco*RI fragment from ISRH1 to *Bam*HI-digested total DNA from different species of *Rhizobium*: **a** *R. meliloti* ATCC9930^T; **b** *R. leguminosarum* bv. *trifolii* ATCC14480^T; **c** *R. leguminosarum* bv. *viciae* ATCC10004^T; **d** *R. leguminosarum* bv. *phaseoli* RCR3644^T; **e** *R. etli* CFN42^T; **f** *R. tropici* IIB CIAT899^T; **g** *R. tropici* IIA CFN299^T; **h** *R. galegae* HAMBI503; **i** *R. 'hedysari'* RJ243 (= strain IS123 cured of its sym plasmid); **j** *Bam*HI-digested purified plasmid pR2, a smaller derivative of sym-plasmid from strain IS123 still carrying the region of the *nod* genes; **k** lambda-*Hin*dIII molecular weight dig-labelled standard; **l** *R. 'hedysari'* IS123; **m** *R. 'hedysari'* CC1335 (835); **n** *R. 'hedysari'* HCNT1; **o** *R. 'hedysari'* RHF; **p** *R. 'hedysari'* RH19; **q** *R. 'hedysari'* RH100; **r** *R. 'hedysari'* RH44; **s** *R. 'hedysari'* A6; **t** *R. 'hedysari'* A10

Relations Between Complexity of the Hybridizing Pattern and Site of Isolation

An additional piece of information which can be derived from the results shown in Fig. 3 regards the degree of divergence of the ISRh1 hybridizing profiles. Simple cases, such as that of strain RH19 with a single band, imply a situation of ancestry with respect to increasingly complex cases. A geographical map of the mediterranean basin is sketched in Fig. 4, and the sites of isolation, from nodules of *Hedysarum coronarium*, for the different *R. 'hedysari'* strains are marked by letters corresponding to lanes in Fig. 3. The spread of a microsymbiont is guided and affected by that of the host plant within its range. It can be postulated that the spread path of the bacterial partner could be in part marked by the gradient of complexity of such profiles, created by successive transpositions. Therefore, although *Hedysarum coronarium* has later become a cultivated plant undergoing man-driven introductions, a trend in its diffusion from the yet unknown center of origin can be inferred from the status of differentiation of the transposition profiles in the rhizobial nodule isolates. Further experiments will be required to verify such hypotheses and define to which extent signature transposition profiles can be of help in tracing the paths of interactions between higher plants and their bacterial endocytobionts.

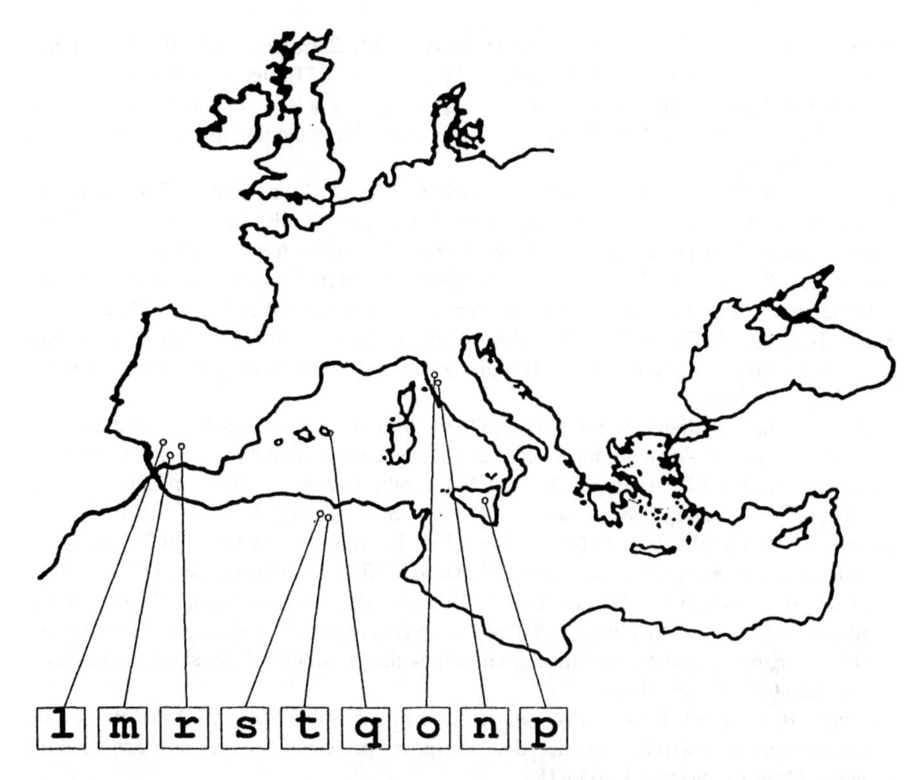

Fig. 4. Indication of the site of isolation from sulla root nodules for each wild type *R. 'hedysari'* strain. The letter code is the same used for Fig. 3

Acknowledgments: This work was supported by M.A.F., P.F. Biotecnologie Avanzate Applicate alle Piante and in part by BRIDGE program, contract BIOT-CT91-0283.

References

Altschul SF, Gish W, Myers EW, Lipman DJ (1990) Basic local alignment search tool. J Mol Biol 215:403-410

Beringer JG (1974) R-factor transfer in *Rhizobium leguminosarum*. J Gen Microbiol 84: 188-198

Cabrera E, Villa A, Ruiz-Argueso T (1979) Diversidad y caraterizaciòn de la flora nativa de rizobios que nodulan en zulla (*Hedysarum coronarium*). *In* VII Congreso Nacional de Microbiologia, Cadiz (Abstr. vol), p 294

Casella S, Gault R R, Reynolds KC, Dyson JR, Brockwell J (1984) Nodulation studies on legumes exotic to Autralia: *Hedysarum coronarium*. FEMS Microbiol Letters 22:37-45

Dazzo FB (1984) Leguminous root nodules. *In* Experimental Microbiolol. Ecology. R Burns and J Slater (eds), Blackwell Scientific Publ, Oxford, pp 431-446

Dusha I, Kovalenko S, Banfalvi Z, Kondorosi A (1987) *Rhizobium meliloti* insertion element ISRm2 and its use for identification of the *fixX* gene. J Bacteriol 169:1403-1409

Elhai J and Wolk CP (1988) A versatile class of positive selection vectors based on the nonviability of palindrome-containing plasmids that allows cloning into long polylinkers Gene 68:119-138

Espuny MR, Ollero FJ, Bellogin RA, Ruiz-Sainz JE, Perez-Silva J (1987) Transfer of the *Rhizobium leguminosarum* biovar *trifolii* symbiotic plasmid pRtr5a to a strain of *Rhizobium* sp that nodulates on *Hedysarum coronarium* J Appl Bacteriol 63:13-20

Gurfel D, Löbel R, Schiffman J (1982) Symbiotic nitrogen-fixing activity and yield potential of inoculated *Hedysarum coronarium* in Israel. Israel J of Botany 31:296-304

Kosier B, Pühler A, Simon R (1993) Monitoring the diversity of *Rhizobium meliloti* field and microcosm isolates with a novel rapid genotyping method using insertion elements. Mol Ecol 2:35-46

Lupi F, Casella S, Toffanin A, Squartini A (1988) Introduction of *Rhizobium "hedysari"* in alkaline clay-loam soil by different inoculation techniques. Arid Soil Res Rehabil 2:19-28

Ollero FJ, Espuny MR, Bellogìn RA (1989) Mobilization of symbiotic plasmid from a strain of *Rhizobium* sp (*Hedysarum coronarium*). System Appl Microbiol 11:217-222

Ollero FJ, Valverde MA, Espuny MR, Bellogìn RA (1993) In vivo formation of R-prime plasmids harbouring *nod* genes of *Rhizobium "hedysari"*. FEMS Microbiol Letters 110:269-174

Rice JD, Somasegaran P, MacGlashnan K, Bohlool BB (1994) Isolation of insertion sequence IS*RLd*TAL1145-1 from a *Rhizobium* sp (*Leucaena diversifolia*) and distribution of homologous sequences identifying cross-inoculation group relationships. Appl Environ Microbiol 60:4394-4403

Ruvkun GB, Long SR, Meade HM, van den Bos RC, Ausubel FM (1982) ISRm1: a *Rhizobium meliloti* insertion sequence that transposes preferentially into nitrogen fixation genes. J Mol Appl Genet 1:405-418

Sanger F, Nicklen S, Coulson AR (1977) DNA sequencing with chain terminating inhibitors. Proc Natl Acad Sci USA 74:5463-5467

Sarno R, Stringi L, D'Alessandro F (1978) Relazione tra il comportamento morfobiologico e produttivo e la quota di provenienza di alcune popolazioni di sulla (*Hedysarum coronarium* L) Quaderni di agronomia 9:139-168

Sambrook J, Fritsch EF, Maniatis T (1989) Molecular cloning. A laboratory manual, Second Edition, Cold Spring Harbor Laboratory, Cold Spring Harbor, NY

Soto M, Zorzano A, Olivares J, Toro N (1992) Sequence of IS*Rm4* from *Rhizobium meliloti* strain GR4. Gene 120:125-126

Squartini A, van Veen RJM, Regensburg-Tuink T, Hooykaas P JJ, Nuti MP (1988) Identification and characterization of the *nodD* gene in *Rhizobium leguminosarum* strain 1001. Mol Plant Microbe Interaction 1:145-149

Squartini A, Dazzo FB, Casella S, Nuti MP (1993) The root nodule symbiosis between *Rhizobium "hedysari"* and its drought-tolerant host *Hedysarum coronarium*. Symbiosis 15:227-238

Wheatcroft R, Watson R J (1988a) A positive strain identification method for *Rhizobium meliloti*. Appl Environ Microbiol 54:574-576

Wheatcroft R, Watson RJ (1988b) Distribution of insertion sequence IS*Rm1* in *Rhizobium meliloti* and other gram-negative bacteria. J Gen Microbiol 134:113-121

Wheatcroft R, Laberge S (1991) Identification and nucleotide sequence of *Rhizobium meliloti* insertion sequence IS*Rm3*: similarity between the putative transposase encoded by IS*Rm3* and those encoded by *Staphylococcus aureus* IS*256* and *Thiobacillus ferrooxidans* IS*T2* J Bacteriol 173:2530-2538

Effect of Drought Stress on Carbohydrate Metabolism in Nodules of *Lupinus angustifolius*

M.L Comino[1], M.R. de Felipe[2], M. Fernandez-Pascual[2], and L. Martin[1]
[1]Department of Plant Physiology, Faculty of Biology, Complutense University, 28040 Madrid, Spain.
[2]Department of Plant Physiology and Biochemistry, Centro de Ciencias Medioambientales, CSIC, Serrano 115 bis, 28006 Madrid, Spain

Key words: Carbohydrate, drought stress, *Lupinus angustifolius*, nodule, nitrogen fixation.

Summary: In this work we have investigated the effect of drought stress on nitrogenase activity (C_2H_2 reduction activity), carbohydrate metabolism, and nodule structure of *Lupinus angustifolius* plants. Nitrogenase activity was inhibited around 50%. Water deficit increased the concentration of total soluble sugars, reducing sugars and sucrose more than twice and also increased the total amylase and the acid invertase activity. The possible physiological significance of these results is discussed.

The physiological and biochemical results were correlated with structural changes observed by light microscopy. The most noticeable change was the disappearance of starch grains, after 9 days of treatment. Quantitative chemical analysis of starch in the nodules supported these observations.

Introduction

Mild to moderate drought stress has a great effect upon plant metabolism, and results in a reduction of the efficiency in key processes such as photosynthesis, respiration, nitrogen and CO_2 assimilation and lipid and protein synthesis. Dinitrogen fixation is also very sensitive to all kinds of water deficits. This can be explained both by a direct effect on the structure and physiology of nodules, but also by an indirect effect through the decrease in photosynthesis and of carbon movements in the plant.

The ability of tissues to maintain turgor pressure is an important mechanism of drought resistance. Two processes may contribute to the maintenance of turgor pressure as water potential declines: a low osmotic potential due either to a naturally high solute concentration and/or accumulation of solutes, and a high tissue elasticity (Aspinall and Paleg 1981).

The compounds involved in osmoregulation are mainly soluble sugars and amino acids. Sugars accumulate where utilization is reduced relatively to photosynthesis activity (Munns and Weir 1981).

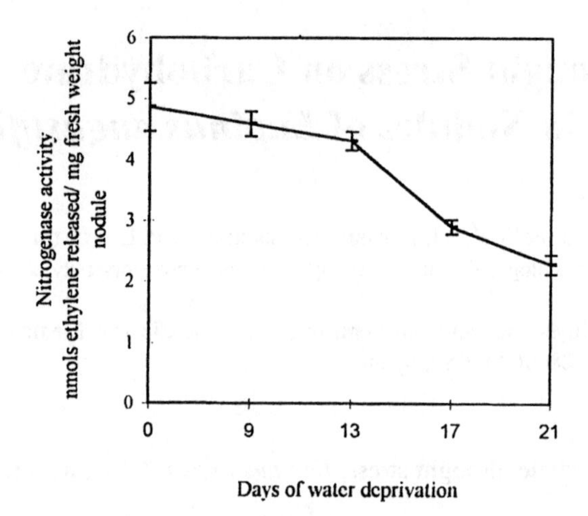

Fig. 1. Nitrogenase activity (C_2H_2 reduction) of nodules of *L. angustifolius* during the onset of drought stress. All values are the mean of seven independent samples. Vertical bars represent ± SD

Irigoyen et al. (1992) noted an increase in total soluble sugars in nodules of *Medicago sativa* plants and this may produce a significant change in the osmotic pressure.

In this work we have examined the effect of drought stress on nodules of *Lupinus angustifolius* plants. We have studied dinitrogen fixation, sugar accumulation, metabolism, and the structure of the nodules.

Material and Methods

Plant Culture. Seeds of *Lupinus angustifolius* were inoculated at the time of planting with *Bradyrhizobium* sp. (*Lupinus*) strain ISLU 16. Seeds were planted in plastic pots containing vermiculite. Upon germination all plants were watered regularly with a N-free nutrient solution and allowed to grow, in a growth chamber, during 23 days at a constant temperature of 25 °C, 80% R.H. and a 16-h photoperiod. After 23 days, plants were divided in five groups: 12 plants were watered as usual (control plants) and the other 4 groups were deprived of water for 9, 13, 17 and 21 days. All plants were recollected 44 days after sowing.

Nitrogenase Activity. Nitrogen fixation measured as acetylene reduction activity (ARA) was assayed on root systems enclosed in 100 ml tubes fitted with rubber stoppers (de Felipe et al. 1987). Ten ml of air were removed from the tubes and the same amount of acetylene was added. Gas samples were taken after 1 h and analyzed for ethylene and acetylene in a Perkin Elmer 8310 gas chromatograph equipped with a hydrogen flame ionization detector and with a column filled with Porapak R, using nitrogen as the carrier gas at a flow rate of 50 ml/min.

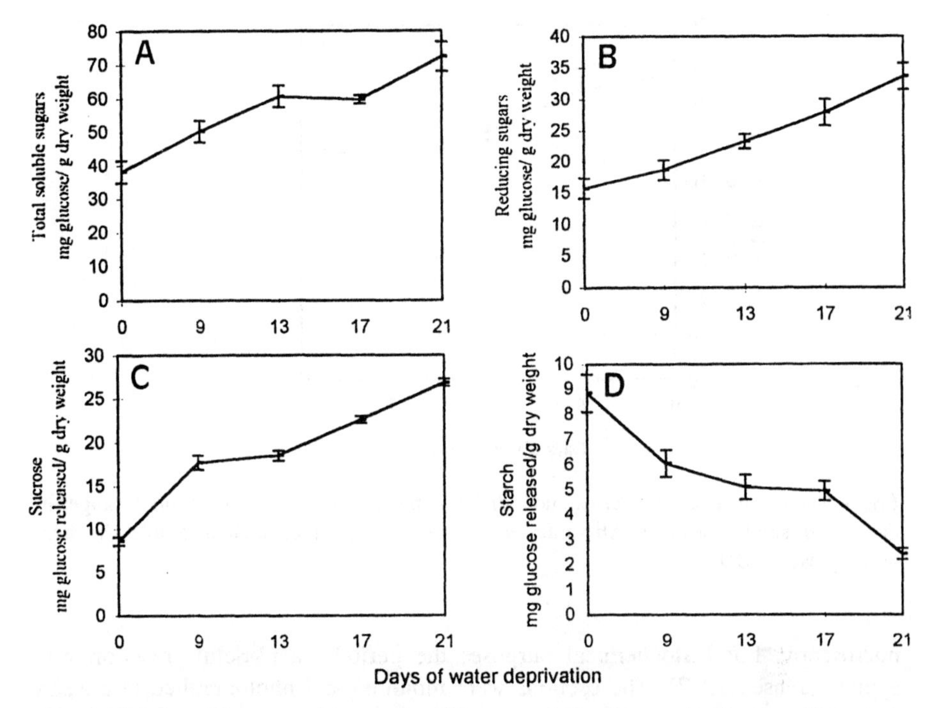

Fig. 2. Nodules of *L. angustifolius* under different periods of water deprivation: Changes in the levels of total soluble sugars **(A)**, reducing sugars **(B)**, sucrose **(C)** and starch **(D)**. All values are the mean of four independent samples. Vertical bars represent ± SD

Sugar Determinations. Sugars were quantified in 95% ethanol extracts of *Lupinus angustifolius* nodules. Total soluble sugars were analyzed with prepared anthrone according to the method described by Loewus (1952). Reducing sugars were determined according to Somogyi (1945) and Nelson (1944). The method reported by Paek (1988) was used for sucrose determination. Starch was analyzed according to Gordon et al. (1986).

Total Amylase Activity and Acid Invertase Activity. Nodules (0,5 g) were homogenized in 15 ml 50 mM acetate-Na buffer, pH 5.5. The homogenate was centrifuged at 20000 g for 20 min and the supernatant was used in the enzymatic assay. The activity of the total amylase was assayed spectrophotometrically following the method reported by Metivier and Paulilo (1980).

For the determination of acid invertase activity the method of Castrillo et al. (1992) was used. The soluble protein evaluation was performed on crude extracts as described by Bradford (1976) using bovine serum albumine as a standard.

Light Microscopy. Fresh nodules samples were taken from similar areas in all treatments. Nodules were detached from the main root. Semithin sections (1 μm) were cut in a ultramicrotome (Reichert ultracut S), and stained for light

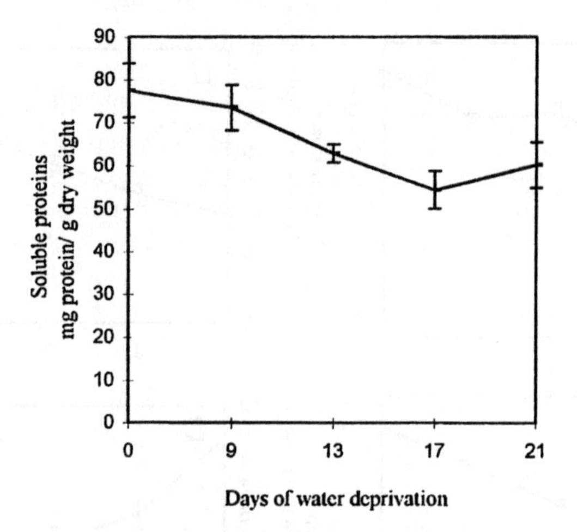

Fig. 3. Nodules of *L. angustifolius* under different periods of water deprivation: Changes in the level of soluble proteins. All values are the mean of four independent samples. Vertical bars represent ± SD

microscopy. For histochemical purposes the periodic acid-Schiff reaction was applied (Jensen 1962). The sections were mounted and photographed in a Zeiss Axiophot photomicroscope.

Results

The effect of drought stress upon acetylene reduction activity (ARA) is shown in Fig. 1. As drought developed, this function declined, reaching rates of 46% of the values of the (well-watered) control plants. Drought treatment increased the concentration of total soluble sugars (Fig. 2A) and the most significant increase was at 21 days of water deprivation. An important increase in the content of reducing sugars (more than twice the well-watered controls) was observed in nodules subjected to drought stress over 21 days (Fig. 2B). Sucrose was accumulated by progressive drought (Fig. 2C). Starch decreased with the increase of water deprivation (Fig. 2D). With water depletion the total protein levels in nodules decreased after the ninth day although a small increase was observed 21 days later (Fig. 3). Total amylase activity increased with drought stress (Fig. 4A).

The acid invertase activity increased with water deprivation progress, reaching a maximum at 17 days (Fig. 4B).

Light microscope examination of nodules showed several changes (Fig. 5). After 9 days of drought stress some regions of the cell walls became fainter stained (Fig. 5B, C, D). Otherwise, control nodules accumulated starch grains specially in the infected cells close to the inner cortex (Fig. 5A). After 9 days of treatment starch grains practically disappeared (Fig. 5B). We did not observe the

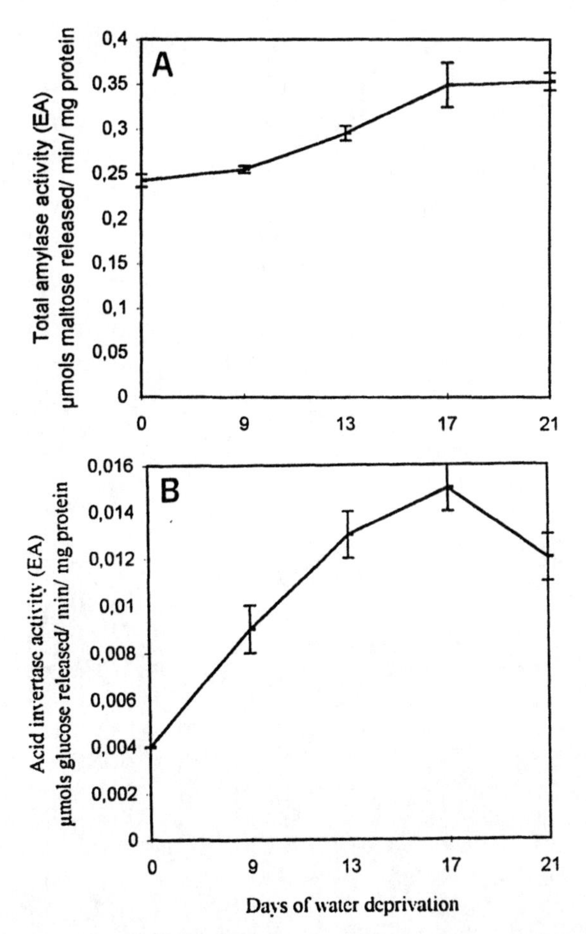

Fig. 4. Nodules of *L. angustifolius* during the onset of drought stress: **A** Total amylase activity, **B** Acid invertase activity. All values are the mean of four independent samples. Vertical bars represent ± SD

endodermis layer in control nodules (Fig. 5A) but this layer of cells appeared after 9 days of water stress (Fig. 5B).

Discussion

Water deprivation inhibited nitrogenase activity (Fig. 1). Similar responses were observed by other authors in different legumes such as *Glycine max* (Djekound and Planchon 1991), *Medicago sativa* (Becana et al. 1986) and *Vicia faba* (Guérin et al. 1990).

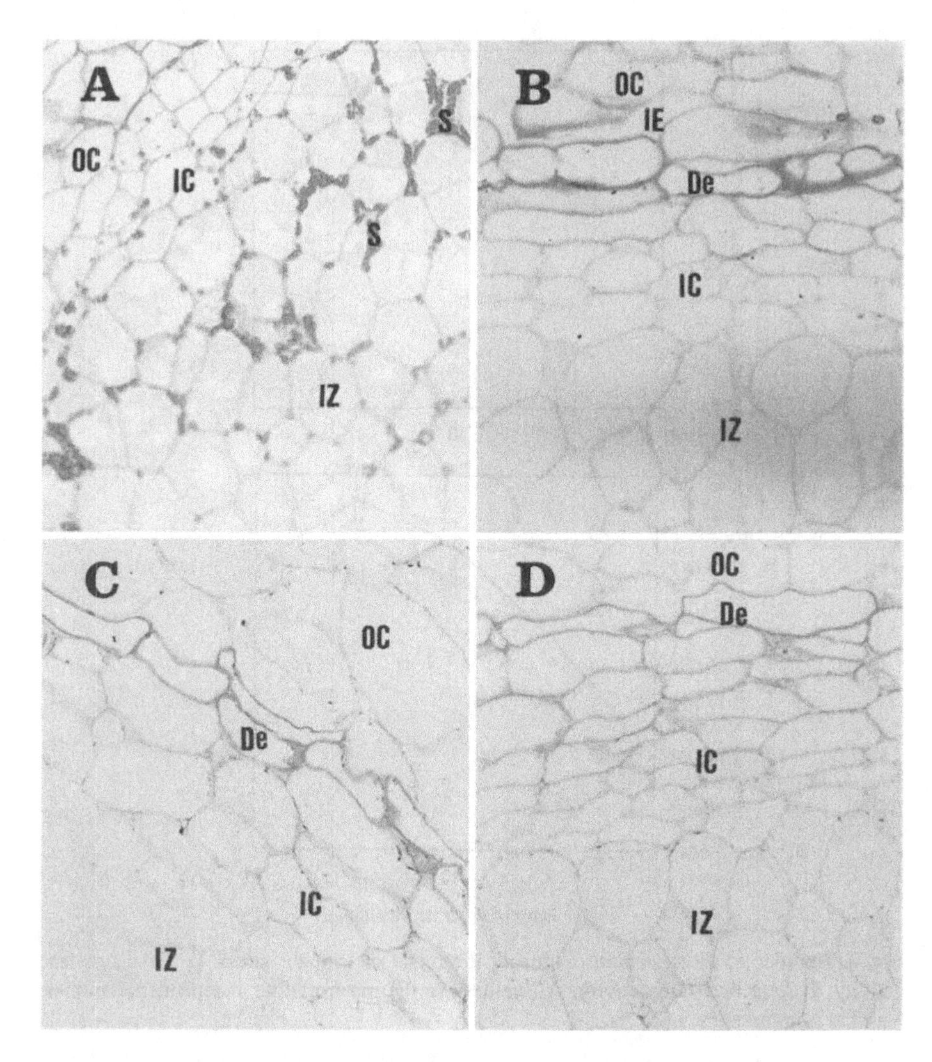

Fig. 5. Nodule sections of *L. angustifolius*, light micrographs: **A** Control nodule x400, **B** 9 days of water deprivation x400, **C** 17 days of water deprivation x400, **D** 21 days of water deprivation x400. OC Outer cortex, IC inner cortex, IZ infected zone, Ee nodule endodermis, S starch, IE intercellular space

Several factors may limit nitrogen fixation. Wery (1987) reported that this inhibition could be explained by four mechanisms: (1) a direct inhibition on *Rhizobium* nodule formation and structure of fixing nodules; (2) an indirect effect through the decrease in photosynthesis; (3) a negative effect on growth development creating a decrease in the nitrogen needs of the plant; and (4) a stimulation of mineral nitrogen assimilation. Also, other factors like restriction in oxygen diffusion, due to structural changes in the inner cortex, may limit nitrogen fixation (Díaz del Castillo et al. 1992; Guérin et al. 1991) as well.

As shown in Fig. 2, total soluble sugars increased as drought progressed. Several authors have demonstrated that under normal conditions, photosynthates translocated towards the nodules, such as sucrose, are those that are preferentially used to support nitrogen fixation. These carbohydrates are rapidly transformed in dicarboxylic acids, such as malate and succinate. Consequently bacteroids show low levels of total soluble sugars. The accumulation of soluble sugars in the nodules during drought stress may produce an important change in the osmotic pressure. This accumulation may be due to a low utilization of soluble carbohydrates in nodule metabolism during dehydration (Venkateswarlu et al. 1989). As Daie (1988) reported, under drought conditions the low carbon supply could induce the synthesis of sucrose or the remobilization of starch.

Within a great number of plant species the accumulation of reducing sugars, such as glucose, and sucrose caused a decrease of the osmotic potential. In the light of these results we have examined the concentration of reducing sugars and sucrose, as well as starch, as is shown in Fig. 2B, C, D. We observed an increase in both, reducing sugars and sucrose with the increase of drought stress. Parallel to this the starch content decreased (Fig. 2D).

In nodules soluble proteins decreased as drought progressed (Fig. 3). A similar decrease has been related by Becana et al. (1986), using nodules of *Medicago sativa* sp. These authors concluded that the decrease of soluble proteins may result from a general reduction in the levels of polyribosomes or from general decrease in the protein synthesis and from an increase in protease activity in nodules.

The increase of total amylase activity during desiccation (Fig. 4A) could be explained by the fact that drought stress induces a reduction of photosynthate supply to the nodule cells and stored starch could be used to compensate this loss. This enzyme could also be implied in the osmotic adjustment.

Otherwise an increase in the acid invertase activity was observed (Fig. 4B). Daie (1988) reported that this enzyme was involved in the supply of soluble sugars to the nodule. Acid invertase may be also involved in the maintenance of osmotic turgor during drought stress and in the formation of sucrose gradients (Hawker 1985).

With respect to light microscope examination (Fig. 5), the most important features noticed were the disappearance of starch grains after 9 days of treatment and the decrease of staining of cell walls. This could imply a partial degradation of cell wall polysaccharides. The decrease in starch grains coincided with the increase in the total amylase activity.

Drought stress has very important effects upon nodule metabolism and nodule structure. Accumulated solutes required metabolic energy and complex controls. Control of synthesized solutes probably involves the regulation of the enzymes related to the synthesis and/or breakdown of these solutes (Aspinall and Paleg 1981). The accumulation of carbohydrates could take part in a process of osmoregulation.

Acknowledgments: This study was supported by a research grant from Ramon Areces Foundation (Spain).

References

Aspinall D, Paleg LG (1981) Academic Press 1981. Sydney, Australia

Becana M, Aparicio-Tejo P, Peña J, Aguirreolea J, Sánchez-Díaz M (1986) J Exp Bot 37: 597-605

Bradford MM (1976) Analitical Biochem 72: 248-252

Castrillo M, Kruger NJ, Whatley FR (1992) Plant Sciences

Daie J (1988) CRC Critical Reviews in Plant Sciences 7: 117-137

de Felipe MR, Fernández-Pascual M, Pozuelo JM (1987) Plant and Soil 101: 99-105

Díaz del Castillo L, Hunt S, Layzell DB (1992) Physiologia Plantarum 89: 824-829

Djekound A, Planchon O (1991) Agronomy Journal 83: 316-321

Gordon AJ, Ryle GJA, Mitchell DF, Lowry KH, Powell CE (1986) Annals of Botany 58: 141-154

Guérin V, Trichant JC, Rigaud J (1990) Plant Physiol 92: 595-601

Guérin V, Plady D, Trinchant JC, Rigaud J (1991) Physiologia Plantarum 82: 306-366

Hawker JS (1985) In: Day PM, Dixon RA (eds) Sucrose. London Academic Press, pp 1-50

Irigoyen JJ, Emerich DW, Sánchez-Díaz M (1992) Physiologia Plantarum 84: 55-60

Jensen WA (1962) In: Botanical histochemistry, principles and practice. Freeman WH & Co

Loewus FA (1952) Analytical Chemistry 24: 219

Metivier J, Paulilo MT (1980) J Exp Bot 31: 1271-1282

Munns R, Weir R (1981) Aust J of Plant Physiol 8: 93-105

Nelson N (1944) J Biol Chem 153: 375-380

Paek KY, Canderlerd S, Thorpe TA (1988) Physiologia Plantarum 72: 160-166

Somogyi M (1945) J Biol Chem 195: 19-32

Turner NC, Begg JE, Rawson MM, English SD, Hearn AB (1978) Aust J Plant Physiol 5: 179-194

Venkateswarlu B, Maheswari M, Saharan N (1989) Plant and Soil 114: 69-74

Wery J (1987) In: Month L, Porceddu E (eds) Agriculture drought resistance in plants. Physiological and genetics aspects. Commision of the European Communities

Creation of Artificial Symbiosis Between *Azotobacter* and Higher Plants

É. Preininger, P. Korányi and I. Gyurján
Department of Plant Anatomy, Eötvös Loránd University, Budapest, 1088, Hungary

Key words: Strawberry, *Fragaria* x *ananassa, Azomonas insignis*, symbiosis, endocytobiosis, organogenesis

Summary: An artificial symbiosis was established between diazotrophic *Azomonas insignis* and strawberry (*Fragaria* x *ananassa*). The partnership was created by *in vitro* techniques through callus induction and organogenesis. The basis of this partnership is the bacterial dependence on the plant's metabolic activity, using maltose in the medium as a carbon and energy source which can be utilized by the plant cells only.

The presence of bacteria in the intercellular spaces of the callus tissues and regenerated plants was proven by microscopic techniques. Nitrogenase activity could also be detected in the plant tissues.

Preliminary experiments were carried out using a biolistic gun in order to ensure a higher incidence of bacterium introduction. This method may allow the incorporation of bacteria into the cells too.

Introduction

On the basis of natural symbioses, several attempts were made to establish artificial symbioses between diazotrophic prokaryotes and plants. Cyanobacteria were mixed with *Daucus* cells (Bradley 1980), but there was no successful regeneration. Gusev et al (1986) were able to regenerate tobacco shoots which contained cyanobacteria in their intercellular spaces. The root colonizing *Azospirilla* were also cultured together with plant tissues. Bacteria ensured the growth of the calli up to 18 months and nitrogenase activity was detected during this period (Vasil et al. 1979). However, regenerated plants did not contain bacteria (Berg et al. 1979). Functioning co-cultivation was established between the carrot and the adenine requiring mutant of *A. vinelandii* (Carlson and Chaleff 1974). The inoculated calli grew for a long time showing nitrogenase activity, but regeneration did not occur. From a mixed culture of carrot cells and Azotobacters, somatic embryogenesis could be induced and the regenerated plants contained bacteria in the intercellular spaces (Varga et al. 1994; Korányi et al. 1993). Acetylene reduction was also detected in potted plants.

This work aims to introduce diazotrophic *Azomonas insignis* into the strawberry (*Fragaria* x *ananassa*) which has a good in vitro system with high regeneration ability.

Materials and Methods

Maintenance of the Bacterium Partner. The CRS-HK 5 strain of *Azomonas insignis* was maintained in nitrogen free liquid and solid culture (Newton et al. 1953) at 30 °C in the dark.

Establishment of Plant - *Azomonas* Coculture via Callus Culture. Callus cultures were initiated from micropropagated strawberry shoot cultures. The youngest expanding leaves were excised from the 1-month-old shoots and the petioles removed. Leaves were then cultured together with the bacterium suspension for 16-18 h. The inoculated leaves were placed on Murashige and Skoog (1962) basal medium supplemented with 1 mg l^{-1} BA, 0.1 mg l^{-1} IBA, 0.05 mg l^{-1} NAA, 0.1 mg l^{-1} 2,4-D, 3% maltose and 0.6% agar. The pH was adjusted to 5.7.

The regeneration medium was the same as callus initiation medium.

Rooting and Growing-On. Shoots regenerated from callus were excised when they were 10 mm in length. The rooting medium was MS basal medium without any growth regulator. After 4 weeks the plantlets were transferred to pots containing nutrient-poor soil mixture (perlite:peat: compost 2:1:1).

Introduction of Bacteria into the Plant Tissues Using the Biolistic Gun. For bacterium bombardment 1.7 μm tungsten particles were used. The particle diameter had to be bigger than that of the bacteria to promote adhering of bacterial cells to the tungsten particles and to ensure introduction of bacteria into the plant tissues. The particles were suspended in ethanol at 60 mg/l concentration. For bombardments, 25 μl aliquots were taken, washed out thoroughly by several volumes of sterile water and mixed with bacteria in 165 μl sterile water. 5 μl spermidin (0.1 M) was added to the suspension, which was then put on ice for 10 min. to let the particles sediment. After removing 130 μl of supernatant, 5 μl aliquots were applied onto the plastic macroprojectiles which were accelerated by a nitrogen powered Genebooster biolistic gun at high pressure (33-35 bars). Leaves of micropropagated strawberries were placed on both callus inducing and regeneration medium for 5 days prior to bombardment. The treated cultures were left on the same medium without subculturing.

Microscopy. For light and transmission electron microscopy tissue pieces were fixed in 2% glutaraldehyde for 2 h and postfixed in 1% OsO_4 for 2 h. The buffer solution was 35 mM K-Na phosphate (pH 7.2). Samples were embedded in Durcupan ACM epoxy resin (Fluka Chemie AG). Semi-thin sections were stained with toluidine blue. Ultrathin sections were stained with uranyl and lead salts and observed with a Hitachi 7100 electron microscope.

Measurement of Nitrogen Fixing Activity. One leaf per plant was cut off to measure the nitrogenase activity using the acetylene reduction assay. Acetylene content was 30% in the vials. After a 24-h incubation, the ethylene content was detected with a Chrompack CP 9001 gas chromatograph using an Al_2O_3/KCl Plot column. Only qualitative analysis of the samples was aimed at, to compare with the control plants, rather than a quantitative measurement.

Results

Formation of Strawberry - *Azomonas* Partnership

Primary calli appeared 7-10 days after placing the bacterium inoculated leaves on callus-inducing media. Callus was produced at the petiole and at the periphery of the lamina.

Plant regeneration occurred on the same media. For our experiments organogenesis was needed as this form of regeneration provides a better possibility of bacterium introduction into the plant tissues. Leaves regenerated plants 10-14 days after callus formation and they developed vigorously and continuously without subculturing (Fig. 1). After the appearance of the first primordia, regeneration ability was continuously maintained for several months. The best leaves regenerated about 250-300 shoots during a 3-month period.

Plantlets at a minimum height of 10 mm were transferred to a rooting medium. Root primordia appeared within 1 week and in the next 3 weeks, roots developed in masses and shoots were also growing during this period.

Well-rooted strawberries were planted from in vitro conditions to pots, and were acclimatized in the greenhouse during the winter months. Plants vigorously developed new leaves while the old ones turned brown and died. This process resulted in a dynamic balance with an average leaf number of four per plant. Some of the plants produced stolons as well as flowers and fruits, all small in size. In spring, they were planted outdoors where they grew vigorously to the original size. The stolon, flower and fruit production was normal corresponding to the cultivar's features (Fig. 1).

Microscopical Examinations

The presence of bacteria was detected by light and electron microscopy. Examinations were made on callus and regenerated plant tissues. The calli contained a large number of bacteria between the cells (Figs, 2a and 3). Introduction of bacteria from the callus tissues to the regenerated plants was also shown by microscopy. *Azomonas* cells were found in the intercellular spaces of both the petiole and the lamina (Fig. 4). In the lamina, bacteria were present in the spongy parenchyma (Fig. 2b). In the petiole they were located mostly under the epidermis and in the outer cell layers of the cortex (Fig. 2c) but some of them were found among the tracheary elements.

Fig. 1. Organogenesis from bacterium-containing callus (*above*) and planted out strawberry with its fruits (*below*)

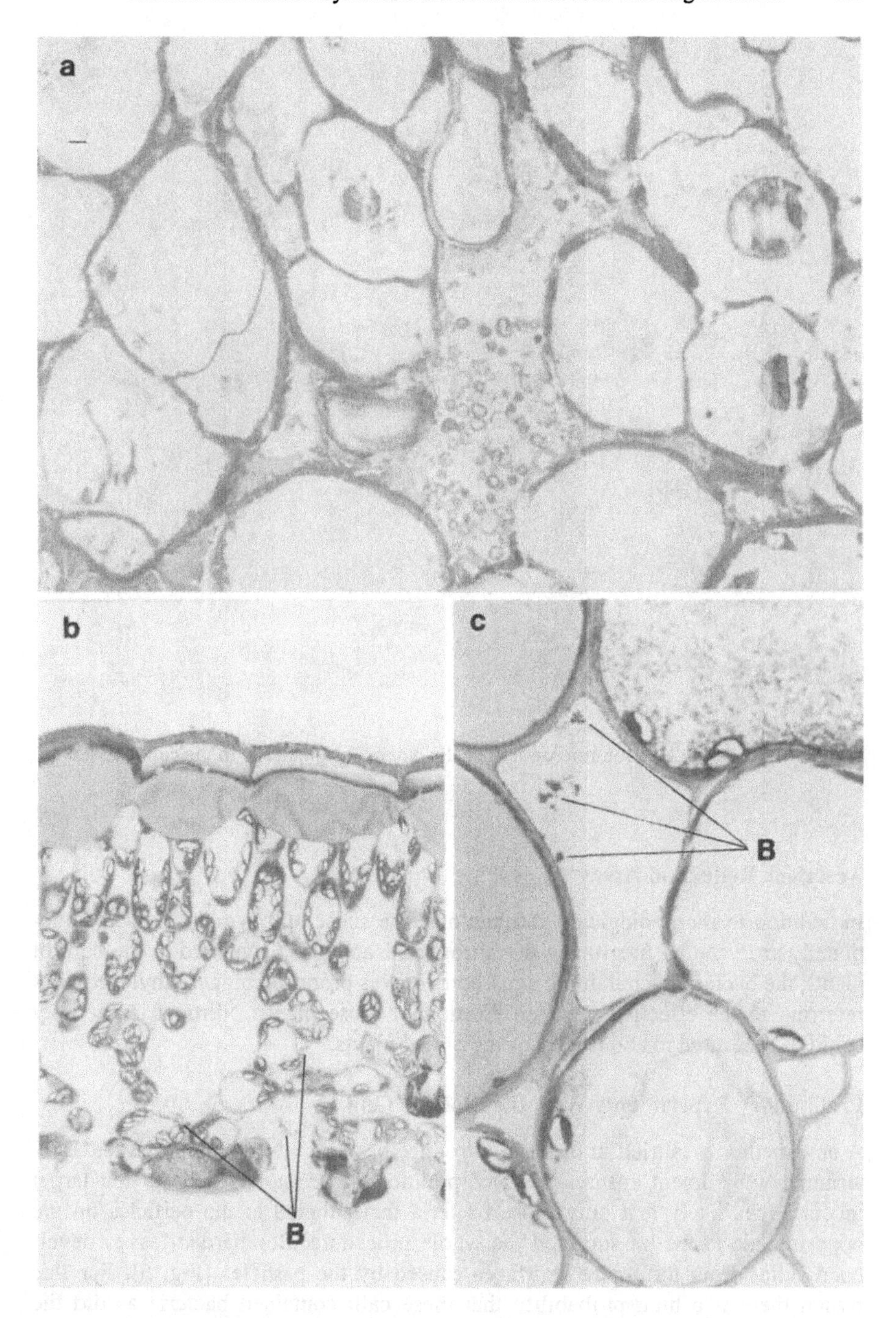

Fig. 2. *Azomonas insignis* cells in the intercellular spaces of strawberry callus (**a**), in the lamina (**b**) and in the petiole (**c**). B Bacteria

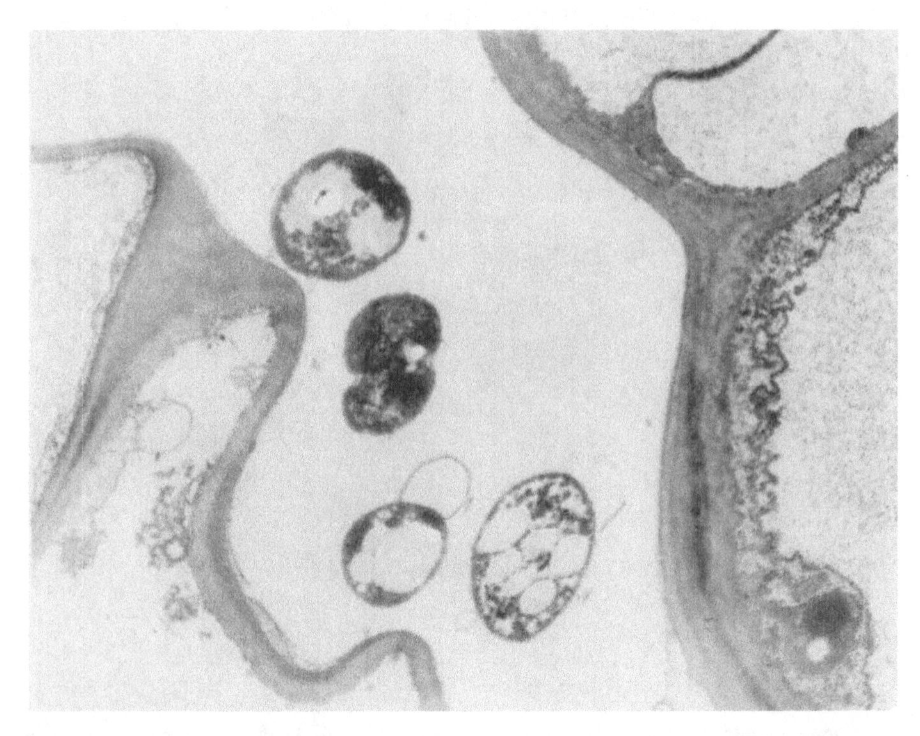

Fig. 3. Transmission electron micrograph of *Azomonas insignis* cells in the callus tissue

Acetylene Reduction Assay

In addition to morphological examinations, the successful symbioses were functionally analyzed by measuring the nitrogenase activity. Compared to the control plants, the bacterium containing strawberry leaves produced bigger ethylene peaks referring to the effective nitrogen fixation of those plants. Nitrogenase activity could be measured in about 20% of the potted plants.

Preliminary Experiments With the Biolistic Gun

A new method was tried in order to introduce bacteria into the plant tissues. Bacterium bombardment ensures the incorporation of *Azomonas* cells in the target inocule. It is likely that only those bacteria that adhered to the particles on the opposite side to the hit survived the whole procedure. Bombarded leaves developed callus from the injured surfaces caused by the particles (Fig. 5). For this reason there is a high probability that these calli contained bacteria as did the regenerated plants. With this method the random incorporation of bacteria within the cells also looks possible.

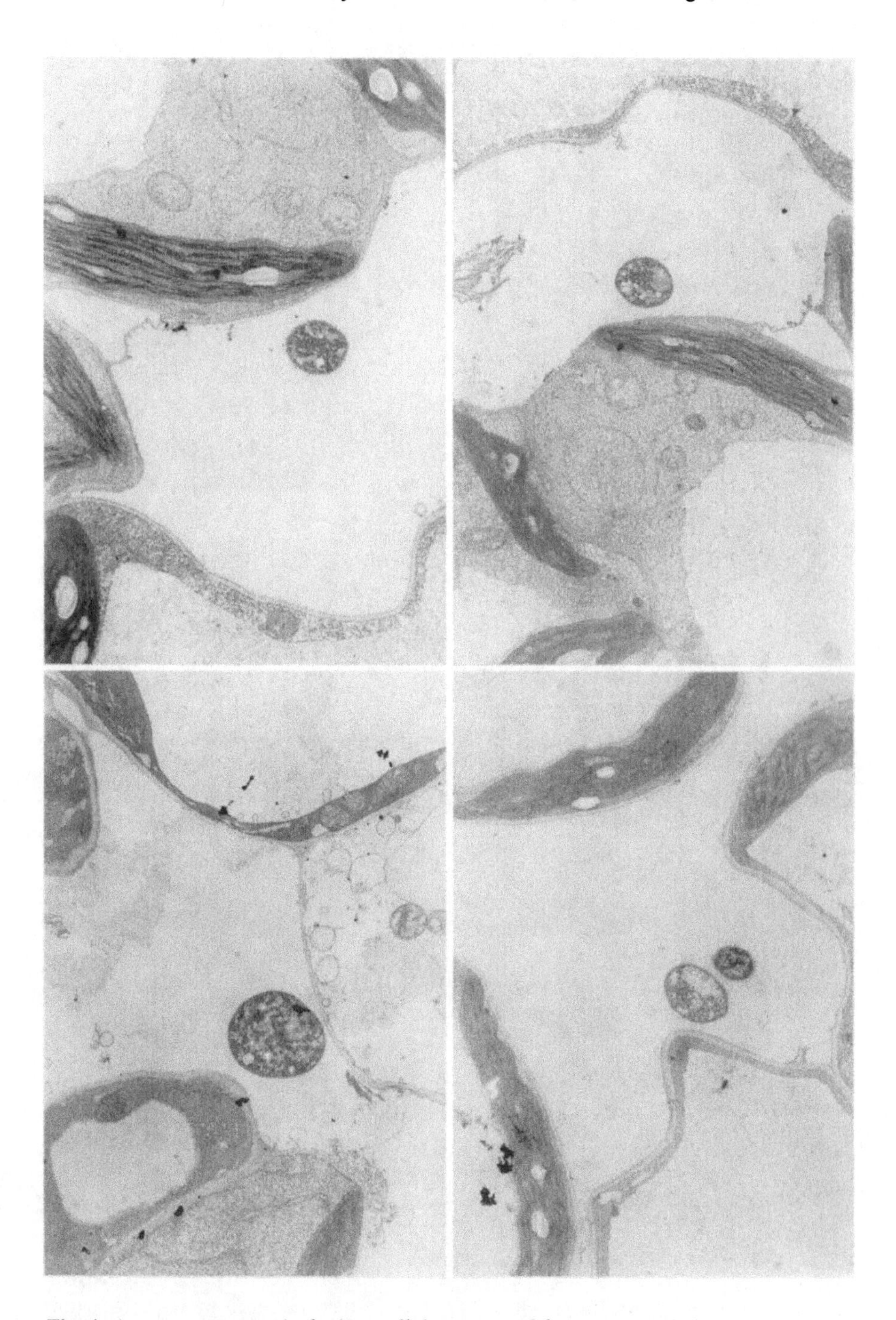

Fig. 4. *Azomonas insignis* in the intercellular spaces of the regenerated plants

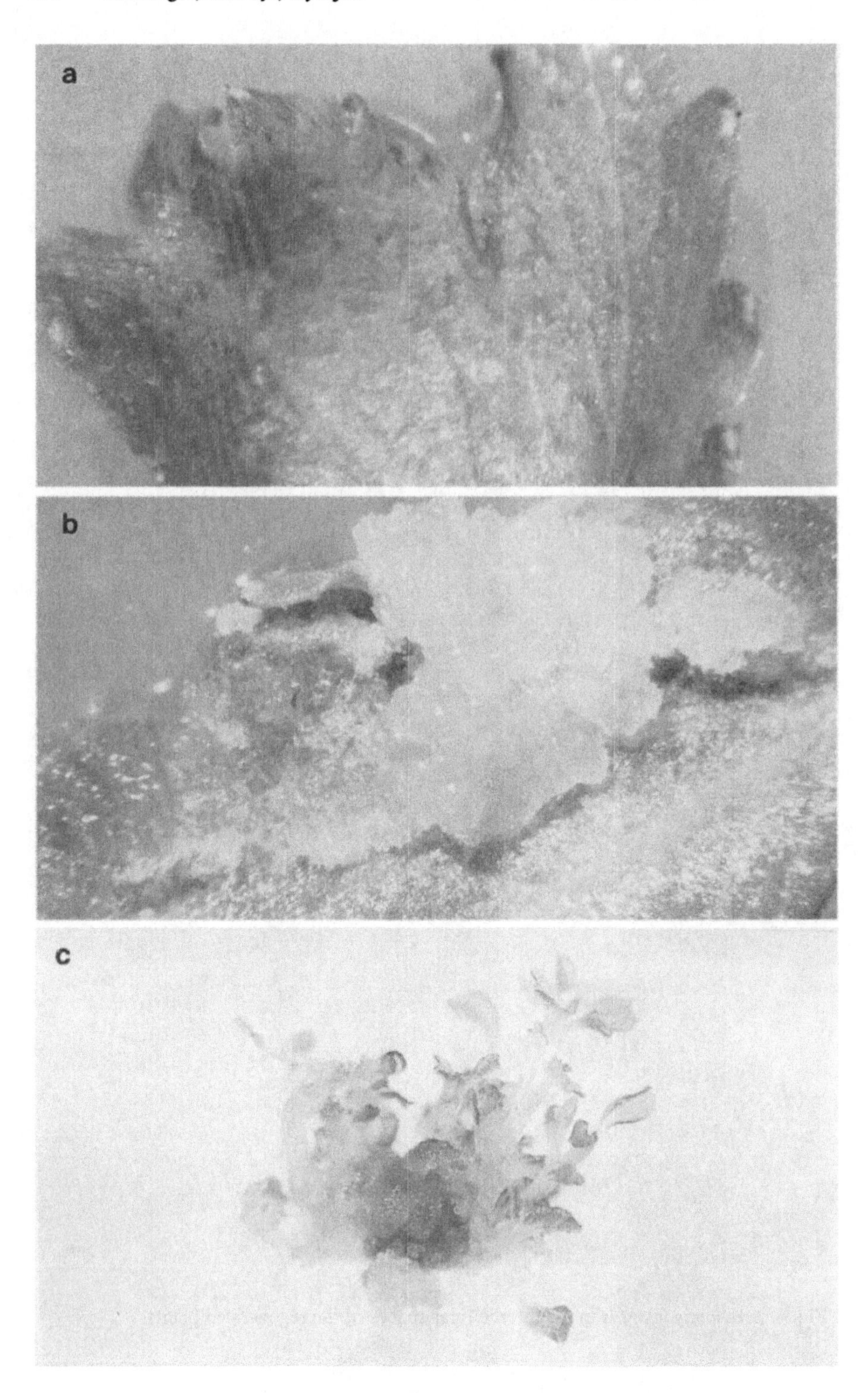
a
b
c

Discussion

A stable symbiosis was created between the diazotrophic *Azomonas insignis* and the strawberry (*Fragaria* x *ananassa*) which can be regarded as an extracellular endosymbiosis, or exocytobiosis (or endosomabiosis, the eds.). The partnership was achieved by *in vitro* tissue culture techniques started from bacterium-inoculated leaves. The bacteria gained access into the plant tissues via organogenesis. In contrast to embryogenesis, organogenesis starts from a group of cells which ensures a better possibility for bacterium incorporation. The system works with high efficiency: a high frequency of regeneration occurs producing a great number of plantlets which can be rooted, acclimatized and planted out easily without significant damage. Bacterium cells are located in the intercellular spaces of the plants which are important sites of apoplastic transport and where both of the partners excrete essential compounds for each other. *Azomonas*, like Azotobacters in general provides a reduced nitrogen source mainly as ammonia and amino acids, but also exudes vitamins and hormones (Newton et al. 1953, Karube et al. 1981, Kuhla et al. 1985). Plant cells supply the bacteria with a carbon and energy source, transforming the sugars presented in the media during the *in vitro* period and photosynthesizing later in the *ex vitro* phase.

The basis of natural symbioses is the interdependence of partners. It is important to emphasize that our media contained an inorganic nitrogen source to promote regeneration. Therefore, only one-sided bacterial dependency was ensured by the sugar source during the introduction of the bacteria into the plant tissues. Maltose present in the media can be utilized directly by the strawberry but not by *Azomonas*. Bacteria can use only those sugars as a carbon and energy source which are transformed by plant cells and excreted into the intercellular spaces. The interdependency is realized when bacterium containing regenerated plants are planted out in the soil, because they then become autotrophic organisms. As it is hoped, the system at this time turns into a real symbiosis.

The main advantage of this method is that bacterium cells infiltrate all parts of the entire plant. Since Azotobacters are aerobic organisms, in addition to the root system they can also function in the photosynthetically active leaves. Nitrogen fixation, measured by the acetylene reduction assay, could be detected in about 20% of the potted plants. This rate is caused by the haphazard introduction of bacterium cells into the plant tissues during the regeneration process. The supplementary inorganic nitrogen source added to the media for the stimulation of organogenesis did not repress the nitrogenase activity.

Using this method to establish symbiosis, the transfer of diazotrophic bacteria from generation to generation is mostly ensured in the case of vegetatively propa-

◀ **Fig. 5.** Introduction of bacteria into the leaf tissue by the biolistic gun. Injuries on the leaf surface caused by particle bombardment (**a**), callus induction from the injuries (**b**) and plant regeneration from the bacterium-containing callus (**c**)

gated plants. Preliminary experiments were carried out with a biolistic gun for the more frequent introduction of bacteria. This improved result can be achieved because the origin of the callus and regenerated plants is the site of injury on the leaves where the bacteria were bombarded. With the biolistic gun a new possibility arises: bacteria can be integrated not only in the intercellular spaces, but also directly into the cells. With the development of protoplast techniques, efforts were made to introduce bacteria within the cells. *Azotobacter vinelandii* cells were incorporated into the *Rhizopogon* fungal protoplasts (Giles and Whitehead 1977) and unicellular cell wall mutant *Chlamydomonas reinhardtii* strains were forced to take up Azotobacters (Gyurján et al. 1984). These alga-bacterium symbioses have been maintained on a nitrogen- and carbon-free medium for 7 years. Spreading out this method to the higher plants may allow the transfer of nitrogen fixing ability to even generatively propagated plants.

Some limitations can impede the method. Difficulties may arise with the tissue culture of several plants and some plants produce special secondary metabolites against the bacteria causing incompatibility between the two partners.

Acknowledgments. This work was supported by OTKA - T 6159 grant.

References

Berg RH, Vasil V, Vasil IK (1979) The biology of *Azospirillum*-sugarcane association. II.Ultrastructure. Protoplasma 101: 143-163

Carlson PS, Chaleff RS (1974) Forced association between higher plant and bacterial cells *in vitro*. Nature 252: 393-395

Karube I, Matsunaga T, Otomine Y, Suzuki S (1981) Nitrogen fixation by mobilized *Azotobacter chroococcum*. Enzyme Microb Technol 3: 309-312

Korányi P, Varga SzS, Preininger É, Gyurján I (1993) Symbiotic integration of diazotrophic bacteria with plant cells in artificial cytobioses. Endocytobiology V 465-472

Kuhla J, Dingler Ch, Oelze J (1985) Production of extracellular nitrogen-containing components by *Azotobacter vinelandii* fixing dinitrogen in oxygen-controlled continuous culture. Arch Microbiol 141: 297-302

Murashige T, Skoog F (1962) A revised medium for rapid growth and bio assay with tobacco tissue cultures. Physiol Plant 15: 473-497

Newton JW, Wilson PW, Burris RH (1953) Direct demonstration of ammonia as an intermediate in nitrogen fixation by *Azotobacter*. J Biol Chem 204: 445-453

Varga SzS, Korányi P, Preininger É, Gyurján I (1994) Artificial associations between *Daucus* and nitrogen-fixing *Azotobacter* cells in vitro. Physiol Plant 90: 786-790

Vasil V, Vasil IK, Zuberer DA, Hubbell DH (1979) The biology of *Azospirillum*-sugarcane association. I. Establishment of association in vitro. Z Pflanzenphysiol 95: 141-147

2.4 Intra- and Extracellular Interactions Between Phycobionts and Mycobionts

News on *Geosiphon pyriforme*, an Endocytobiotic Consortium of a Fungus with a Cyanobacterium

M. Kluge[1], H. Gehrig[1], D. Mollenhauer[2], R. Mollenhauer[2], E. Schnepf[3], and A. Schüßler[3]
[1]Institut für Botanik, Technische Hochschule Darmstadt, 64287 Darmstadt
[2]Forschungsinstitut Senckenberg Frankfurt, Außenst. Lochmühle, 63599 Biebergemünd
[3]Zellenlehre, Universität Heidelberg, 69120 Heidelberg

Key words: *Geosiphon pyriforme*, Glomales, *Nostoc punctiforme*, Cyanobacteria; endocyanosis.

Introduction

Geosiphon pyriforme (Kütz.) v. Wettstein is a coenocytic soil fungus and up to now the only known example of a fungus living in endocytobiotic association with a cyanobacterium, i.e. with *Nostoc punctiforme*. Due to the physiological activities of the endocytobiont the consortium as a whole is capable of C- and N-autotrophic life. Nevertheless *Geosiphon* together with its endosymbionts has to be considered as a rather primitive endocytobiotic system because the photoautotrophic partner can be experimentally separated and unified. Thus, the system may provide insights into the decisive initial steps leading in the evolution via more derived systems finally to the well-established complexes called eukaryotic cells. In spite of its potential importance for research on endocytobioses, studies on *Geosiphon* are still rare. The state of knowledge on the system available until 1993 has been reviewed recently (Kluge et al. 1994; Mollenhauer and Kluge 1994). In the present paper we report on some new results from our group obtained on the unique consortium *Geosiphon*.

Initiation and development of the partner association

A recent study by Mollenhauer et al. (1996) has extended considerably the knowledge on the process of the initial partner interaction and the following development of symbiotic association in *Geosiphon*. The cells of the cyanobacterium, *Nostoc punctiforme*, live at the first freely together with the future fungal partner in and on the soil of the same habitat. It is now clear that successful interaction of the fungus with *Nostoc* depends on the appropriate developmental stage of the cyanobacterium (for terminology of the Nostocacean life cycle (see e.g. Bilger et al. 1994; Mollenhauer et al. 1994; Dodds et al. 1995). In the soil the life cycle of

Nostoc starts from akinetes which germinate by release of motile trichomes (hormogonia). Since the hormogonia perform in dim light positive and in strong light negative phototactic movements, they gather just below the soil surface where they undergo transformation into an aseriate stage called primordium. This stage can be retransformed into hormogonia or converted into gelatinous vegetative cell masses („thalli"). It is now clear that only the primordia of *Nostoc* can interact with the future fungal partner to give rise of the symbiotic consortium.

The life cycle of the fungal partner of the association starts from spores which represent stages of perenniation. The ultrastructure of the *Geosiphon* spores has been investigated in detail by Schüßler (1995) and Schüßler et al. (1995) and provide, as it will be outlined later, useful informations for the taxonomic classification of *Geosiphon*. The spores germinate by the extrusion of seldom more than one hypha that ramifies to form an expanding mycelium of a few millimeters inside the soil. There the young hypha meet and incorporate *Nostoc* primordia. Each incorporation event leads to the formation of a pear-shaped overground bladder up to 2 mm in length formed by the fungal hypha where the incorporation took place. In this bladders the incorporated *Nostoc* cells multiply and become physiologically active.

The incorporation of *Nostoc* into the fungal hypha proceeds as follows. Upon contact of the tip of a fungal hyphe with a freshly formed *Nostoc* primordium, a portion of fungal plasma bulges out just below the apex of the hyphe. There is evidence that the plasmatic bulge remains covered with some fungal cell wall material. The bulging process is repeated several times so that finally the hyphal tip forms a raspberry-shaped mantle which encloses the contacted *Nostoc* primordium. In this stage the incorporation of the *Nostoc* into the fungal plasma takes place. Details of this process remain to be studied. Afterwards the mentioned fungal bladder develops from the raspberry-shaped structure. It is important to note that heterocytes of the *Nostoc* primordium are never mantled by the fungal cytoplasm and therefore remain outside of the hypha during the incorporation process.

There is evidence that during the first hours after incorporation the *Nostoc* filaments disintegrate and the photosynthetic pigments of the cells bleach considerably. These alterations suggest that during the initial state of the endocytotic life the incorporated cyanobacteria are exposed to a stress situation. Afterwards, during the following maturation of the *Geosiphon* bladder, the enclosed *Nostoc* cells recover gradually from the stress, i.e. the cells multiply, grow to a volume considerably larger than that of free-living cells, and arrange in filaments in which heterocytes are formed with the same frequency as in the filaments outside the bladders.

For several reasons it can be assumed that the initial reaction between the partners leading to the establishment of the symbiotic *Geosiphon* association is to large extent specific. Namely, we have observed that cells of certain strains of *Nostoc punctiforme* can be incorporated by *Geosiphon* and easily lead to the formation of functional fungal bladders, whilst with cells of other strains, although being incorporated, the formation of bladders is stopped on an early stage of development. There are, finally, *Nostoc punctiforme* strains which are not at all incorporated by the fungus.

An even more striking argument indicating a specific recognition process between the partners in *Geosiphon* is the fact that among the various developmental stages of *Nostoc* exclusively the primordia are incorporated by the fungus. There is evidence that not only the physiological quality of the primordia is different from the other stages of the *Nostoc* life cycle (Bilger et al. 1994; Mollenhauer et al. 1994) rather than also the composition of the gelatinous envelope. That is, only the envelope of the primordia, not however that of its heterocytes contains mannose (Schüßler et al. 1997). Thus, it is tempting to speculate but remains to be proven that the specific partner recognition is based on the carbohydrate composition of the cyanobacterial envelope.

Structure of the symbiotic interface between the partners in *Geosiphon*

The ultrastucture of *Geosiphon* was first investigated by Schnepf (1964). Recent studies (Schüßler 1995; Schüßler et al. 1995) revealed that inside the fungal bladder the symbiotic *Nostoc* cells are located in one tubulous compartment (the symbiosome) forming a peripherally arranged, cup-shaped structure (Fig. 1)[1]. The space between the symbiosome membrane and the wall of the *Nostoc* cells enclosed by it contains an about 30-40 nm thick layer of electron-microscopically dense and amorphous appearing material which was formerly assumed to be produced by the endosymbiont (Schnepf 1964). However, CLSM studies by means of affinity techniques with fluorescence-labeled lectins specific to GlucNAc oligomers, mannosyl/glycosyl, fucosyl, galactosyl, GalNAc and sialic acid residues revealed that the amorphous layer inside the symbiosome represents a derivative of the fungal cell wall. (Schüßler 1995; Schüßler et al. 1996). It was found that as well as the fungal cell wall the electron opaque layer of the symbiosome contains chitin. This is unequivocal proof in favor of the postulate by Schnepf (1964) that the membrane surrounding the *Nostoc* cells inside the *Geosiphon* bladder is of fungal nature. It is very interesting to note that there is striking similarity between the fungal material present in the *Geosiphon* symbiosome and the thin cell wall bordering in arbuscular mycorrhiza (AM) the symbiotic fungus from the penetrated plant cell: both are electron dense after OsO_4 fixation, amorphous in structure, and about 30-40 nm thick. In conclusion, the interface bordering in AM the host cell

---

[1] Fig. 1, slightly modified, from Mollenhauer and Kluge IJECR 10, 29-34 (1994), with permission of ATTEMPTO Verlag, Tübingen.

(i.e. the plant root cell) from the symbiont (i.e. the intracellular fungal arbuscule) seems to be equivalent to the symbiotic interface envelope of *Geosiphon*. This aspect could be very interesting in the context with the evolution of the *Geosiphon* symbiosis.

Permeability of the *Geosiphon* cell wall and uptake of nutrients

Electrophysiological experiments by Schüßler (1995) revealed that inorganic ions (Phosphate, Nitrate, Chloride) and small organic molecules (e.g. glycin, cystein) led to rapid changes in the membrane potentials suggesting that these substances are readily taken up from the outside into the bladder. On the other hand, there were none or only very slow changes of membrane potentials if hexoses (e.g. glucose) and larger amino acids were applied suggesting that such molecules could not be taken up from the environment. This different transport behavior is presumably due to the selective permeability of the bladder wall. Evidence supporting this view comes from the observation (Schüßler 1995; Schüßler et al. 1995) that in presence of solutes having molecular radii larger then 0.5 nm irreversible cytorrhysis, i.e. collapsing of the whole bladder including the cell wall, occurred, whereas in presence of small solutes cytorrhysis either did not occur or was quickly reversed. By investigating the occurrence and reversion of cytorrhysis in presence of solutes with known molecular radii as indicator for cell wall permeability it was found that the pore size of the *Geosiphon* bladder wall is about 0.5 nm, which is, compared with other cell walls, extremely low. Provided that such small pore size holds true also for the hyphal wall of *Geosiphon,* the biological advantage of the symbiotic consortium is obvious. Namely, because of the low permeability of its wall the fungus should have difficulties with saprophytic acquisition of even relative small organic molecules such as glucose, sucrose, glutamate e.c.t.. Thus, the fungus might depend largely on organic nutrients produced photoautotrophically by the endocytobionts. However, we are aware that this plausible hypothesis requires further experimental proof.

It is worth mentioning that, compared with free-living cells of the same strain, the *Nostoc* cells in the bladder show considerably higher effectivity of photosynthesis (Bilger et al., 1994) indicated by higher quantum yields in photosystem II, higher quantum flux density required to saturate the photosynthetic electron transport rates and lower susceptibility to photoinhibition. The reason for this different photosynthetic behavior is not yet known, but it is reasonable to assume that CO_2 concentration and thus the availability of the major photosynthetic substrate for the photobionts is higher inside than outside of the bladder.

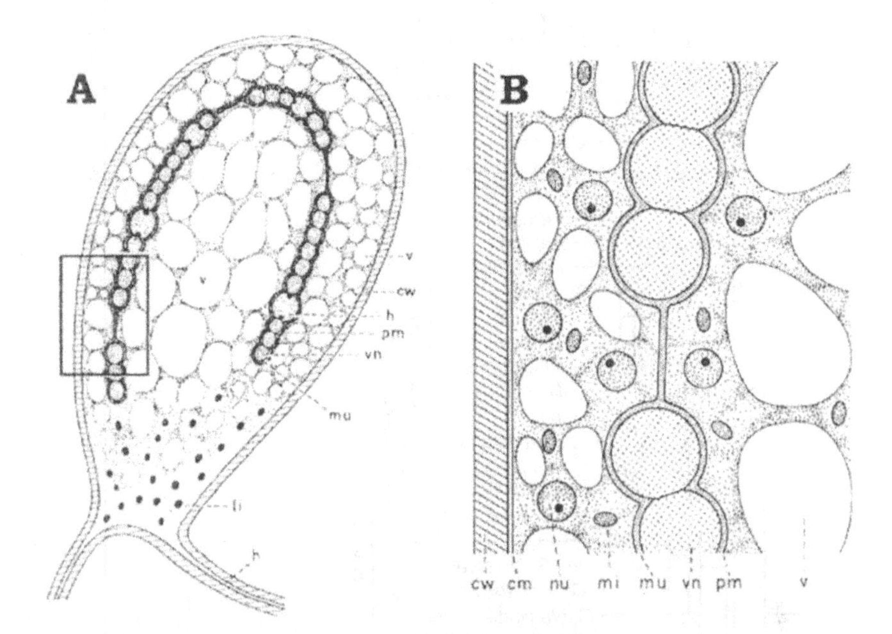

Fig. 1: Scheme of the inner structure of a *Geosiphon* bladder (according to A. Schüßler). **A** Total view. The fungal cytoplasm outside the layer of *Nostoc* cells contains many tiny vacuoles, whereas the vacuoles of the inner cytoplasm are much larger. The basal part of the bladder contains dense cytoplasm with only few tiny vacuoles, but many droplets of lipids and inclusions of glycogen. The endosymbionts are bordered against the fungal cytoplasm by a pericyanobacterial membrane. Presumably all cyanobacteria of a bladder form together with the surrounding membrane a single, tube-shaped symbiosome. **cw** cell wall of the bladder; **cm** cell membrane (plasmalemma) of the fungus; **v** vacuole; **pm** pericyanobacterial membrane; **vn** vegetative *Nostoc* cell; **h** heterocyte (above) or insert of a fungal hypha (below); **li** lipid droplet. **B** Detail of 1A in larger magnification. **nu** nucleus; **mi** mitochondrion; other abbreviations as in 1A

Taxonomic position of *Geosiphon* and its phylogenetic relation to AM fungi

Although already Knapp (1933) recognized that the fungal symbiosis partner of *Geosiphon* is a phycomycete and Mollenhauer (1992) proposed its belonging to the genus *Glomus*, until recently the exact taxonomic position of the organism was an open question. Considering the fact that *Glomus* belongs to the most important fungi forming arbuscular mycorrhiza (AM), a definite answer on this question appeared to be of great importance. Since the spores are the main characteristics in the taxonomy of AM fungi (Glomales), Schüßler et al. (1994) compared the

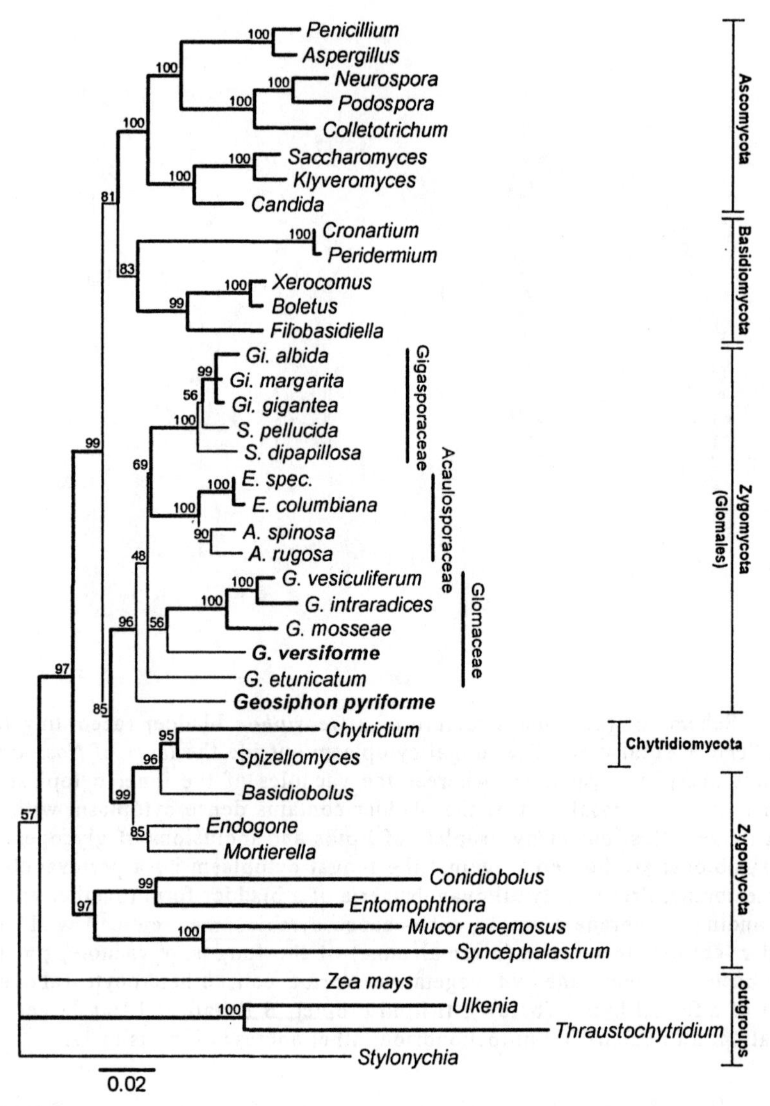

Fig. 2. Phylogenetic analyses of fungal SSU rRNA sequences. Neighbor-joining consensus tree obtained by 1000 bootstrap replicates. Bootstrap values are given as numbers above branches. Thick lines delineate the topology that is supported by 95% or more of the bootstraps. Branches with bootstrap values below 50% are reduced to polytomies (Gehrig et al. 1966, modified)

shape and ultrastructure of *Geosiphon* spores with that of various fungi belonging to the Glomales. It was found that there are striking similarities in the spore structure between *Geosiphon* and in particular *Glomus versiforme,* but also certain other species of the genus *Glomus* thereupon investigated. In contrast, there were considerable differences with members of other genera of the Glomales. Aiming to find more direct arguments for the taxonomic position of *Geosiphon* and its phylogenetic relation to other fungal classes, Gehrig et al. (1996) compared the structure of small subunit ribosomal RNA (SSU rRNA) genes of *Geosiphon pyriforme* and *Glomus versiforme* (Glomales) by means of PCR techniques. In this study two overlapping fragments (spanning a region of 1712/1715 bp) were cloned and sequenced. Using structural criteria, the sequences were aligned with SSU rRNAs of other fungi known from the literature and used for the construction of phylogentic trees. These analyses clearly showed that the order Glomales includes *Geosiphon* and forms a distinct group outside of any other group of the Zygomycetes sequenced sofar. Within the Glomales, two lineages exist, with one of them including the Gigasporaceae and Acaulosporaceae families, whereas the other representing the genus *Glomus* as defined by usual criteria. However, on the basis of the SSU rRNA data the members of this genus show considerable large phylogenetic distances suggesting that the genus is not well defined from the conventional point of view. Both *Glomus etunicatum* and *Geosiphon pyriforme* form independent, ancestral lineages within Glomales.

Altogether, the results by Gehrig et al. (1996) not only support the view of other authors (Simon et al. 1992, 1993; Walker 1992) that the genus *Glomus* probably has a polyphyletic origin rather than unequivocally show that *Geosiphon* indeed represents a fungus belonging to the order Glomales (bootstrap support 90-96%), with the question remaining open whether or not *Geosiphon* has to be placed within or outside the genus *Glomus*. A definite answer on this question depends on a better and more detailed taxonomic analysis of the genus as at the present available. In any case it is almost certain that *Geosiphon* represents a very ancestral member of the Glomales and thus is very closely related to the most important AM fungi. Experiments are now in progress in our laboratory dealing with the fascinating question whether or not *Geosiphon* itself can act as fungal partner in forming mycorrhiza with plants, simultaneously or alternatively to its endosymbiotic association with *Nostoc*.

Acknowledgments: Our work on *Geosiphon* was supported by the Deutsche Forschungsgemeinschaft (SFB 199; Teilprojekt A3).

References

Bilger W, Büdel B, Mollenhauer R, Mollenhauer D (1994) J Phycol 30: 225-230
Dodds WK, Gudder DA, Mollenhauer D (1995) J Phycol 31: 2-18
Gehrig H, Schüßler A, Kluge M (1996) J Mol Evol 43: 71-81
Kluge M, Mollenhauer D, Mollenhauer R (1994) Progr Bot 55: 130-141
Knapp E (1933) Ber Dtsch Bot Ges 51: 210-216
Mollenhauer D, Kluge M (1994) Endocytobiosis and Cell Res 10: 29-34

Mollenhauer D, Büdel B, Mollenhauer R (1994) Algol Studies 75: 189-209
Mollenhauer D, Mollenhauer R, Kluge M (1996) Protoplasma 193: 3-9
Schnepf E (1964) Arch Mikrobiol 49: 112-131
Schüßler A (1995) Dissertation, University Heidelberg
Schüßler A, Bonfante P, Schnepf E, Mollenhauer D, Kluge M (1996) Protoplasma 190: 53-67
Schüßler A, Mollenhauer D, Schnepf E, Kluge M (1994) Bot Acta 107: 36-45
Schüßler A, Schnepf E, Mollenhauer D, Kluge M (1995) Protoplasma 185: 131-139
Schüßler A, Meyer T, Gehrig H, Kluge M (1997) Eur J Phycol, in press
Simon L, Bousquet J, Lévesque RC, Lalonde M (1993) Nature 363: 67-69
Simon L, Lalonde M, Bruns TD (1992) Appl Environm Microbiol 58: 291-295
Walker C (1992) Agronomie 12: 887-857

Isoforms of Arginase in the Lichens *Evernia prunastri* and *Xanthoria parietina*: Physiological Roles and Their Implication in the Controlled Parasitism of the Mycobiont

M. C. Molina, C. Vicente, M. M. Pedrosa, and M. E. Legaz
Department of Plant Physiology, The Lichen Team, Faculty of Biology, Complutense University, 28040 Madrid, Spain

Key words: *Evernia prunastri*, *Xanthoria parietina*; arginase, isoform, lectin, mycobiont, parasitism.

Summary: *Evernia prunastri* and *Xanthoria parietina* contain several arginase isoforms. They are mainly produced by the corresponding mycobiont and are the key enzymes of putrescine biosynthesis. However, glycosylated arginases can be secreted to the intercellular spaces and they enter the phycobiont when glycosylated urease in the algal cell wall does not retain this isoform of arginase. This implies an increased production of algal putrescine that induces protoplast release after a partial hydrolysis of the cellulose component of the cell wall by putrescine-activated glucanase.

Introduction

The diamine putrescine and the polyamines spermidine and spermine are aliphatic nitrogen compounds commonly distributed in both prokaryotic and eukaryotic organisms. It is currently accepted that, in higher plants, putrescine derives from ornithine by decarboxylation but lichens share with bacteria the widespread coexistence of arginine and ornithine decarboxylations to produce putrescine (Legaz 1985).

Arginase Isoforms in *Evernia prunastri*

In the lichen *Evernia prunastri*, as well as in other lichen species, the fungal partner produces putrescine by the action of an ornithine decarboxylase whereas the algal partner produces its own diamine by the action of an arginine decarboxylase (Legaz 1985; Legaz and Vicente 1981). The level of algal putrescine is several times lower than that of fungal diamine (Escribano and Legaz 1988). However, diamine transport from the mycobiont to the phycobiont has never been observed, although the fungal partner has a specific carrier for putrescine (Escribano et al.

Table 1. Isoforms of arginase of *Evernia prunastri*

Isoform	Molecular weight (kDa)	pI[a]	K_m (mM)[b]	n_H[c]	Activated by	Inhibited by
I (Induced)	18	5.86	1.5	1.4	Atranorin Evernic acid	Usnic acid
III (Constitutive)	26	6.18	4.5	4.4	-	Atranorin Evernic acid Usnic acid
IV (Secreted)	20	5.60	4.45	1.24	Atranorin	Evernic acid Usnic acid

[a]Estimated by electrofocusing on column
[b]Direct estimation
[c]Derived from the Hill equation

1994). *E. prunastri* produces several isoforms of arginase (Legaz 1991). Three of those behave as intracellular, non-glycosylated and acidic enzymes, the molecular weight of which ranges from 16 to 20 kDa. Only one behaves as glycosylated, secreted enzyme. It is an alkaline protein that moves from the fungal cells to the intracellular spaces within the thallus.

Secreted arginase is activated by the phenol atranorin, crystallized on the lichen cortex. This implies that this isoform blends with the substrate as a hyperactive form. On the other hand, the enzyme is inhibited by evernic acid, another phenol that crystallizes on the surface of the phycobiont cell wall. This inhibition would impede the entry of active arginase into the algal cell. This could be seen as a defense mechanism since a very high amount of putrescine results in phycocide action (Birecka et al. 1981; Cheng et al. 1984; Vicente and Legaz 1983). The possibility of an interaction between an enzyme (a protein) produced by the fungus, and the surface of the algal cell could be related to a process of binding of a lectin to its ligand, and the possible entry of arginase could be explained as the reject of an incompatible partner.

Lichen Lectins and Recognition

Bubrick et al. (1981) use for the first time the category of "recognition-type protein" for a phytohaemagglutinin from the lichen *Xanthoria parietina*. This protein, and some others found for several lichen species, must fulfill the following conditions to be considered as true lectins:

1. Lichen lectins are produced by the fungal partner (Bubrick and Galun 1980). This condition is the basis of the algal selecting property of the mycobiont.

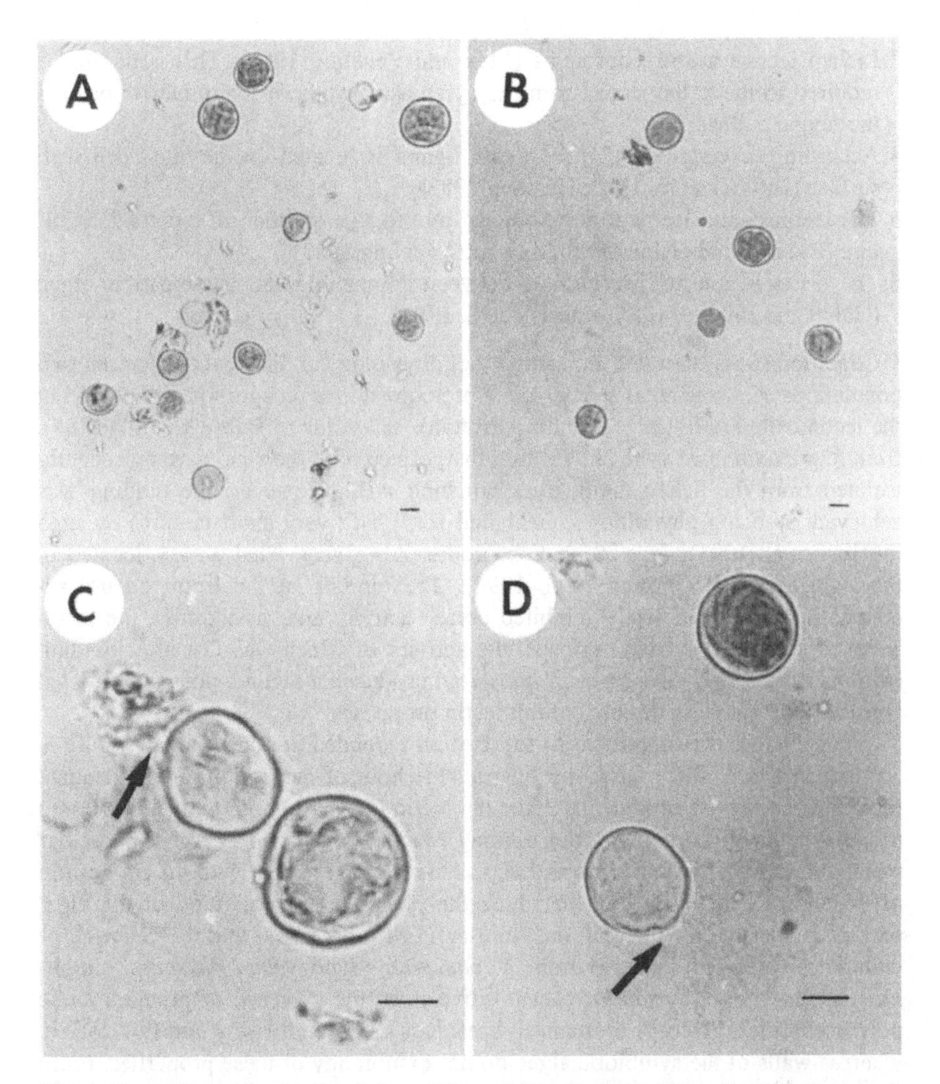

Fig. 1. Phycobionts were isolated from 6.0 g of *Xanthoria parietina* thalli, recently collected from *Robinia pseudoacacia*, according to Ascaso (1980), washed with PBS buffer, pH 7.0 (Bubrick et al. 1985) and resuspended in 6.0 ml of the same buffer. Two aliquots of 1.0 ml of this suspension were pre-incubated at 26 °C in the dark, one of them on 100 mM urea (**A** and **B**) and another in the buffer (**C** and **D**). After pre-incubation, cells were collected by centrifugation, repeatedly washed with PBS, and incubated for 3 h with 30 μg of purified SA (**A** and **C**) or ABP (**B** and **D**). Arrows indicate the point at which the cell wall is broken, through which the protoplast has been ejected. Bar = 15 μm

2. Lichen lectins are located at the fungal cell wall (Bubrick et al. 1981). This facilitates contact with the surface of a compatible partner.

3. Lichen lectins are glycoproteins (Max and Peveling 1983). This condition is required to move the lectin from the cytoplasm, where it is produced, towards the fungal cell wall.
4. According to condition (2), the lectin ligand is located on the algal cell wall surface (Bubrick et al. 1982; Hersoug 1983).
5. The lectin ligand seems to be produced by the algal partner after extended culture (Bubrick and Galun 1980; Max and Peveling 1983).
6. The lectin is able to discriminate between compatible and incompatible algae (Galun and Bubrick 1984; Kardish et al. 1991).

Condition (6) is essential, indicating that physiological differences between two populations of a sole algal species are a prerequisite for recognition. By purifying the lectin called ABP (Algal Binding Protein), according to Bubrick et al. (1985), from *X. parietina*, we were able to find that, effectively, the algal partner, recently isolated from the lichen thalli, does not bind ABP. However, the binding was achieved by using phycobionts incubated for 1 h (a very short culture) on urea (Molina et al. 1993). This incubation produced a glycosylated urease, located in the algal cell wall (Molina et al. 1994). The binding of the lectin on urease-containing algae completely inhibited urease activity and, in addition, the lectin behaved as a glycosylated arginase, the activity of which was lost after binding (Molina et al. 1993). Moreover, *X. parietina* produced a second, non-glycosylated arginase isoform that does not exhibit lectin properties.

In conclusion, it is important to say that an extended period of algal culture is not required to produce the lectin ligand. One hour of incubation on an adequate inducer is enough to produce it. Since the lectin (arginase) and its ligand (urease) are enzymatic glycoproteins, the binding reaction must depend on a particular metabolic status of both fungus and alga. This is in agreement with the suggestion of Peveling (1988) according to which the physiological condition of the algal partner is decisive for recognition. Bubrick (1988) described that the cell wall of cultured phycobionts isolated from *X. parietina, Hypogymnia physodes, Caloplaca citrina* and *C. auriantia* react positively for the presence of protein, acidic polysaccharides, glycosyl or mannosyl residues and are intensely autofluorescent, whereas walls of the symbiotic algae do not exhibit any of these properties. From our results, it is possible to discriminate between an axenic culture on a conventional, mineral medium from which the absence of nutritional inducers is patent and the algal cells are obligated to synthesize these molecules after extended culture, and the addition to the medium of an organic, specific inducer, such as urea, to rapidly produce the corresponding lectin ligand. This can explain the observation of Ott (1987), about the possible contact of germinating ascospores of *X. parietina* with coccal, free-living green algae, such as *Pleurococcus*, incompatible for true symbiosis. The "recognition" is reduced here to a lectin-ligand reaction, but other physiological conditions are required to develop a true thallus.

Lichen Lectins and Controlled Parasitism of the Mycobiont

What happens when lectin contacts an algal population that does not contain (algal cells have not induced) the lectin ligand ? According to Molina and Vicente (1993), the lectin really penetrates the algal cell when the ligand has not been induced by urea. This has also been found by using purified, fluorescein-labeled lectin (Molina and Vicente 1995). The uptake of the lectin by algal cells is followed by a rapid loss of chlorophylls (Molina and Vicente 1993).

Several new facts have been found as derived from the polymorphic nature of arginase populations in lichens (Pedrosa and Legaz 1995). Two glycosylated arginases are produced by *X. parietina* mycobiont: ABP (intracellular arginase, almost identical to that characterized by Bubrick et al. 1985), and SA (secreted arginase). Their amino acid composition is very similar but the glycosyl rest,

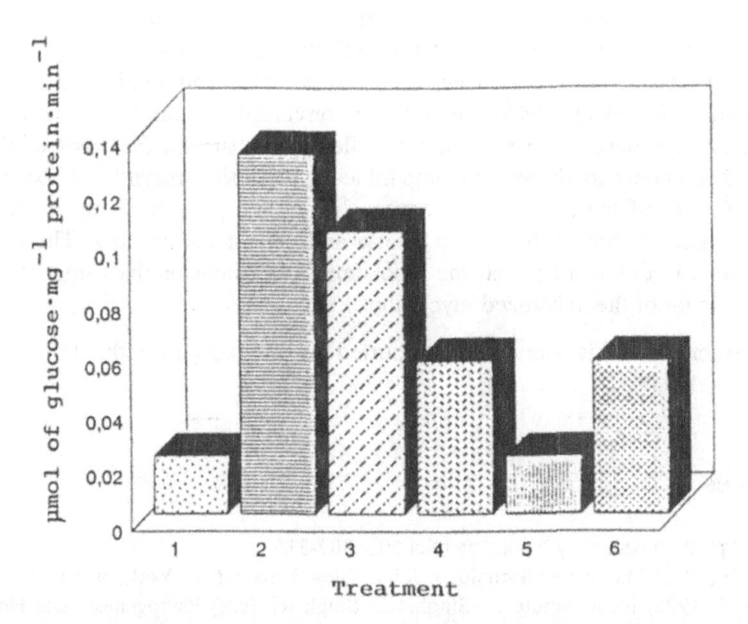

Fig. 2. Recently isolated phycobionts from *X. parietina* were incubated for 4 h in the dark, at 26 °C, and later cultured for 3 h on PBS buffer (**1**), 30 µg ABP (**2**) or 30 µg SA (**3**). Alternatively, phycobionts were cultured for 4 h on 100 mM urea and later incubated for 3 h in PBS buffer (**4**), 30 µg ABP (**5**) or 30 µg SA (**6**). After these treatments, algal cells were repeatedly washed with 10 ml of 10 mM acetate buffer, pH 3.5, and macerated with neutral alumina A-5 in 5.0 ml of the same buffer to obtain a cell-free extract. The activity of β-1,4-glucanase was measured in reaction mixtures containing 10 µg CM-cellulose, 0.5 mg protein and 10 µmol acetate, pH 3.5. Reaction was carried out at 37 °C for 5 h and stopped at 100 °C for 5 min. Reducing sugars were measured according to Somogyi (1952) and Nelson (1944) and protein according to Lowry et al. (1951). One unit of glucanase activity was 1.0 µmol of glucose produced *per* mg protein *per* min

analyzed by HPLC, is different: ABP contains glucose and N-acetyl-D-glucosamine, whereas SA contains both glucose and galactose (Molina and Vicente 1995). Both glycoproteins bind to cell wall urease but the affinity of secreted arginase for the algal ligand is higher than that shown by the intracellular enzyme. The uptake of both SA and ABP by algal cells without its ligand in their cell wall promotes the increase in the amount of algal putrescine, analyzed as free diamine, and the chloroplast is rapidly damaged (Molina and Vicente 1995).

However, the loss of chlorophylls implies another different mechanism. Light microscope observation of recently isolated phycobionts of *X. parietina* reveals that when ABP or SA are retained on the cell wall surface, algal cells remain healthy (Fig. 1A and B). However, when ABP or SA enter the algal cell and, consequently, the level of intracellular putrescine increases, the cell wall is broken and the protoplast is excluded from the cell (Fig. 1C and D). This production of protoplasts, following the increase in algal putrescine, can be related to the increase in β-1,4-glucanase activity after the corresponding treatments. The entry of ABP or SA into isolated algal cells strongly increases glucanase activity (treatments 2 and 3 in Fig. 2), whereas this activity remains at a low level when ABP or SA are retained by the induced ligand (treatments 5 and 6 in Fig. 2).

Ahmadjian (1987, 1993, 1995) seems to be convinced that the lichen association could be considered as a case of controlled parasitism on the basis of the ability of mycobionts to destroy incompatible algae during resynthesis experiments, as well as of the similarities between lichen and parasitic fungi, mainly those concerning concentric bodies, haustoria and gelatinous matrices. The new facts described in this word reveal the biochemical mechanisms that support the parasitic behavior of the lichenized mycobiont.

Acknowledgments: This work was supported by a grant from the DGICYT (Spain) No. PB93 0092.

References

Ahmadjian V (1987) Ann New York Acad Sci 503: 307-315
Ahmadjian V (1993) The Lichen Symbiosis. John Wiley & Sons, New York, pp 56-58
Ahmadjian V (1995) In: Kohmoto K, Singh US, Singh RP (eds) Pathogenesis and Host Specificity in Plant Diseases. II. Eukaryotes. Pergamon, Oxford, pp 277-288
Birecka H, Ireton KP, Bitonti AJ, McCann PP (1991) Phytochemistry 30: 105-108
Bubrick P (1988) In: Galun M (ed) Handbook of Lichenology, II. CRC Press, Boca Raton, pp 133-144
Bubrick P, Frensdorff A, Galun M (1985) Symbiosis 1: 85-95
Bubrick P, Galun M (1980) Protoplasma 104: 167-173
Bubrick P, Galun M, Ben-Yaacov M, Frensdorff A (1982) FEMS Microbiol Lett 13: 435-438
Bubrick P, Galun M, Frensdorf A (1981) Protoplasma 105: 207-211
Cheng SH, Shyr YY, Kao CH (1984) Bot Bull Acad Sin 25: 191-196
Escribano MI, Balaña-Fouce R, Legaz ME (1994) Plant Physiol Biochem 32: 55-63
Escribano MI, Legaz ME (1988) Plant Physiol 87: 519-522

Galun M, Bubrick P (1984) In: Linskens HE, Heslop-Harrison J (eds) Encyclopedia of Plant Physiology. Cellular Interactions. Springer Verlag Berlin, pp 362-401
Hersoug LG (1983) FEMS Microbiol Lett 20: 417-420
Kardish N, Silberstein L, Fleminger G, Galun M (1991) Symbiosis 11: 47-62
Legaz ME (1985) In: Vicente C, Brown DH, Legaz ME (eds) Surface Physiology of Lichens. Complutense University Press, Madrid, pp 57-72
Legaz ME (1991) Symbiosis 11: 263-277
Legaz ME, Vicente C (1981) Z Naturforsch 36c: 692-693
Lowry OH, Rosebrough NJ, Farr AL, Randall RJ (1951) J Biol Chem 193: 265-275
Marx M, Peveling E (1983) Protoplasma 114: 52-61
Molina MC, Muñiz E, Vicente C (1993) Plant Physiol Biochem 31: 131-142
Molina MC, Vicente C (1993) In: Sato S, Ishida M, Ishikawa H (eds) Endocytobiology V. Endocytobiology and Symbiosis. Tübingen University Press, Tübingen, pp 81-84
Molina MC, Vicente C (1995) Cell Adhesion Commun 3: 1-12
Molina MC, Vicente C, Muñiz E (1994) Acta Hort 381: 239-242
Nelson N (1944) J Biol Chem 153: 375-380
Ott S (1987) Bibl Lichenol 25: 81-93
Pedrosa MM, Legaz ME (1995) Electrophoresis 16: 659-669
Peveling E (1988) Naturwissenschaften 75: 77-86
Somogyi M (1952) J Biol Chem 195: 19-32
Vicente C, Legaz ME (1983) Z Pflanzenphysiol 111: 123-131

Comparison Between Recent-Isolated and Cultured Populations of Phycobionts From *Xanthoria parietina* (L.)

M.C. Molina[1], E. Stocker-Wörgötter[2], R. Zorer[2], R. Türk[2] and C. Vicente[1]
[1]Department of Plant Physiology, Faculty of Biology, Complutense University, Madrid-28040, Spain
[2] Institute of Plant Physiology, Salzburg University, Salzburg-5020, Austria

Key words: *Xanthoria parietina*, *Trebouxia*, Pseudotrebouxia, lichen, photobionts, axenic culture, natural populations, cellular division, aplanosporangium.

Summary: *Xanthoria parietina* is a lichen constituted by an Ascomycete as the fungal partner and by *Trebouxia* and *Pseudotrebouxia* as green algal phyco- or photobionts. Cellular cycle and development of the photobionts seem to be regulated by the mycobiont, although the molecular bases of these phenomena are still unknown. The phycocide characteristic of lichen phenolics as well as the capacity to enter the algae let us think that these compounds may be related to the control of photobiont growth and development. Besides, fungal isolectins with arginase activity are implicated in the complex mechanism of regulation between the mycobiont and the phycobiont. This work compares the distribution of frequencies of the cellular size of recent-isolated photobionts and cultured photobionts under aposymbiotic conditions during two months. The observed high yield of aplanosporangia in culture, in contrast to the results obtained under natural conditions, is an evidence for strict control of the mycobiont upon the phycobiont.

Introduction

Several ultrastructural studies of lichens have demonstrated that the photobiont population divides much more quickly at the edges of the thallus than at other sites, where it remains constant (Hill 1985; Honegger 1991). Some authors consider that photobiont's cellular division, and so the size of the population, is regulated by the mycobiont. Honegger (1987) proposed that phenolic compounds could act as controlling molecules of this process (Honegger 1991). This hypothesis is supported by the phycocide nature of some of these phenolics (Avalos and Vicente 1987) and their capacity to regulate some enzymatic activities of the photobiont (Pedrosa and Legaz 1991; Molina et al. 1994).

In addition, some isolectins with arginase activity have been described as molecules related to complex control and microselection mechanisms of phycobiont growth induced by the mycobiont (Molina et al. 1993; Molina and Vicente 1993, 1995).

Another possibility, involves the movement of genetic material through intercellular spaces. Ahmadjian *et al.* (1991) found that some DNA fragments can be seen as plasmids which act as factors for algal genetic transformation. These plasmid-like DNA fragments presumably originate from the fungal partner.

The present work attempts to establish the difference between the distribution frequency of recently isolated phycobionts and axenically cultured phycobionts for two months. We also show a microscope probe concerning the presence of *Trebouxia* and *Pseudotrebouxia* on axenic culture, in which differences can be observed in the mode of cellular division between both photobionts (Ahmadjian 1982).

Material and Methods

Plant Material. *Xanthoria parietina* (L.) Th Fr., growing on *Robinia pseudoacacia* L. was collected in Pedraza (Segovia). Thalli were air-dried and stored at 4°C in the dark, no longer than 2 weeks. Voucher specimen deposited in MCBA-40621.

Photobiont Fresh Isolation. Photobiont cells were isolated from three thalli from *Xanthoria parietina* weighing approximately 0.3 g (Ascaso 1980). After mechanical disruption of the thalli in distilled water, cell suspensions were centrifuged in a sucrose-KJ gradient. Algal cells, which remained in the interphase, were collected by centrifugation and repeatedly washed with distilled water. The algal cells isolated from each thallus were mixed to eliminate interthalline variability. After this, fresh algal cells were observed under the light microscope (Reichert-Jumg-Polivar). The area of 107 algal cells was estimated by using a size-shape program and a PS/2 IBM 30 computer equipped with a digitizer table.

Photobiont Culture. Well developed thalli were selected under a binocular microscope and later washed in a vial for half an hour with phosphate buffered saline solution, pH 7.4 containing Tween 80 (Bubrick and Galun, 1986). During this initial cleaning, the washing solution was changed three times. Finally, the podetia (without apothecia) were cut into pieces of about 5 mm length. These pieces were broken between two slides to select algal cells according to the micropipette method (Ahmadjian 1973, Stocker-Wörgötter and Türk 1989). The droppers with photobionts were transferred to sterilized solid Basal Bold's Medium (Deason and Bold 1960) including soil extract (Esser 1976). Different subcultures (all containing both genera, *Trebouxia* sp. and *Pseudotrebouxia* sp.) were prepared to get a culture without fungal contamination. After 2 months, 3 of these clean subcultures were mixed in liquid medium to eliminate interculture variability. Then, 120 algal cells were observed as described above.

Culture Conditions. Tissues were incubated in a culture chamber at 20±2°C under a 14:10 h light-dark regime. Fluorescent lamps (Philips 65 W/RW) with a photon flux density of 60-100 $\mu E\ m^{-2}\ s^{-1}$ were used (Stocker-Wörgötter and Türk 1993). After 2 months the algal cells were observed as above.

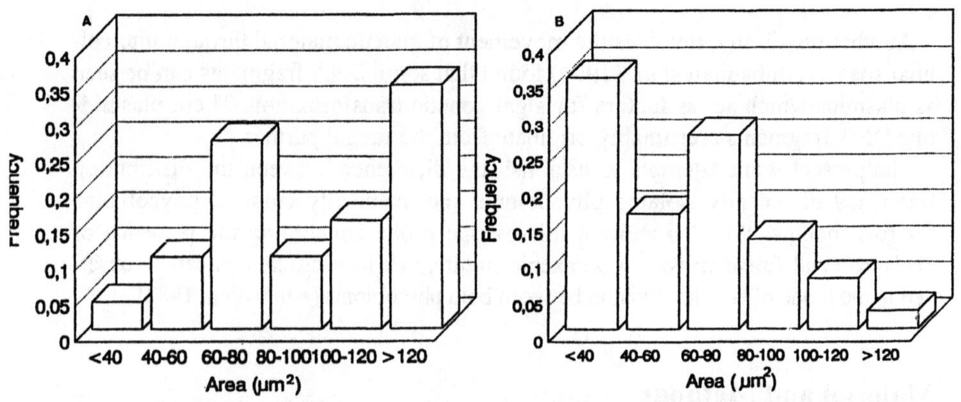

Fig. 1. Photobionts (*Trebouxia* and *Pseudotrebouxia*) of *X. parietina* thalli. Frequency distributions of cellular area within mixed algal populations. **A** Algal cells of a fresh isolation; **B** algal cells of 2 month old axenic cultures. Deviations from the mean value (three replicas) were not greater than 10%

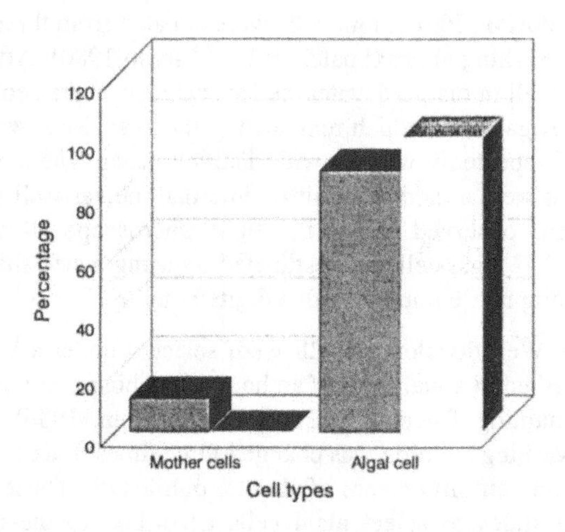

Fig. 2. Photobionts (*Trebouxia* and *Pseudotrebouxia*) of *X. parietina* thalli. Percentage aplanosporangia (mother cells) of freshly isolated (dark) and of cultured (light) photobionts. Deviations from the mean value (three replicas) were not greater than 10%

Results

Figure 1 shows the results obtained after estimation of frequency distribution in cell size. A high heterogeneity in cell size was observed, both in the photobiont population recently isolated, and in the culture population. The main difference between both populations was the frequency of cells smaller than 40 μm^2 and greater than 120 μm^2. Smaller algae (<40 μm^2) were the most frequent cells in axenic culture, decreasing in frequency when cell size is increased (Fig. 1B) while a photobiont

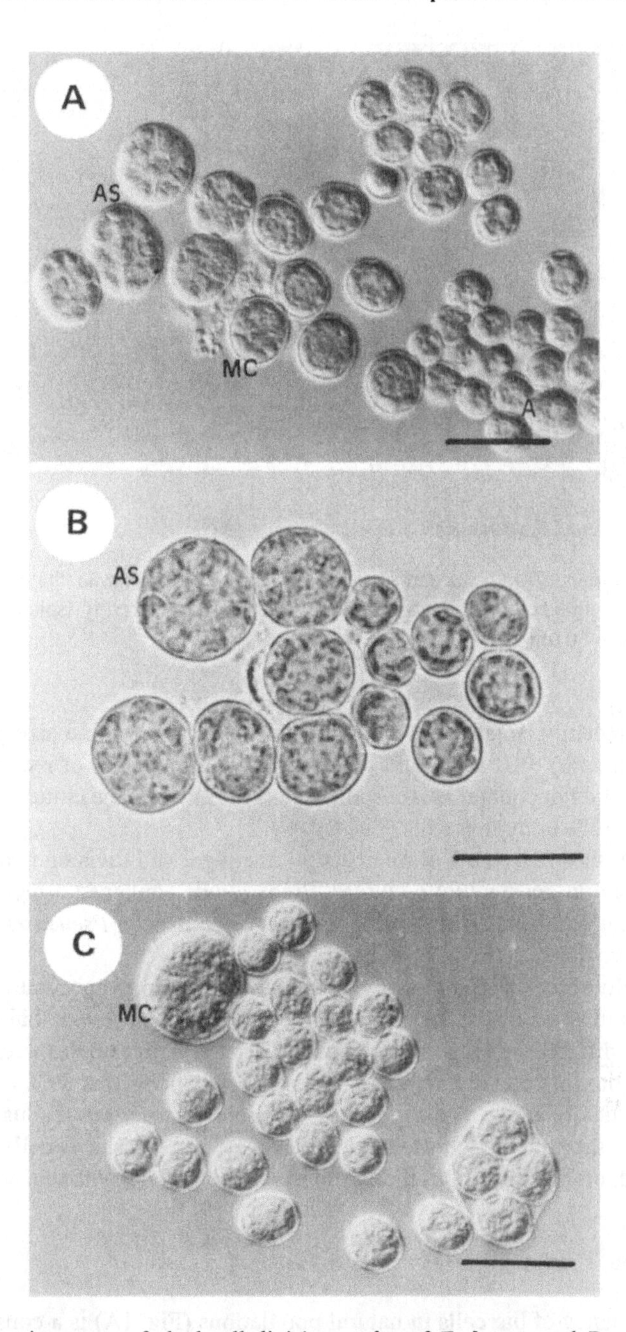

Fig. 3. Light microscopy of algal cell division cycles of *Trebouxia* and *Pseudotrebouxia* (aposymbiotic culture of mixed algal isolates from *X. parietina*). **A** *Trebouxia* sp.: a high number of aplanosporangia is observed. Full arrow shows not vegetative division. **B** and **C** *Pseudotrebouxia* sp.: open arrow shows vegetative division (**B**) or a tetrad as a result of vegetative division (**C**). A Aplanospore; AS Aplanosporangium; MC mature cell. Bar = 15 μm

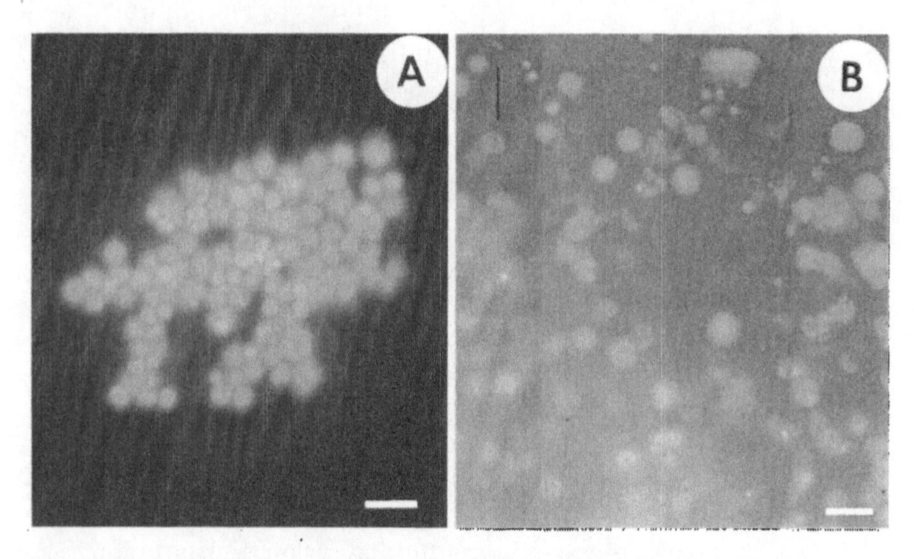

Fig. 4. Photobionts (*Trebouxia* and *Pseudotrebouxia*) of *X. parietina* thalli. Fluorescence microscopy of algal cells from aposymbiotic culture (**A**) and freshly isolated from lichen thalli (**B**). Bar = 50 μm

population recently isolated has a high frequency of cell with a size greater than 120 μm^2 (Fig. 1A). It is important to note that the population of recently isolated photobionts did not contain aplanosporangia (Fig. 2) while the cultured population consisted of 12% of mother cells (Fig. 2) .

Figure 3A shows, by optical microscopy, the algal cell division cycle from genus *Trebouxia* in aposymbiotic culture. Preparations contained a high amount of aplanosporangia. It was also possible to observe cells from *Pseudotrebouxia* during vegetative division (Fig. 3B and C).

Finally, fluorescence microscopy of photobionts from aposymbiotic culture (Fig. 4A) and photobionts recently isolated from *X. parietina* thalli (Fig. 4B) shows clear differences in the number of red cells in both samples, red cells being the most photosynthetically active ones. Figure 4A shows that 99% of cells were photosynthetically active while phycobionts isolated from the thallus have more variability in the kind of fluorescence. In this way, Fig. 4B shows cells fluorescing red, dull red, orange and gray, in order of descending photosynthetic activity.

Discussion

A high frequency of big cells in natural populations (Fig. 1A) is a consequence of photobionts lacking cellular division, since aplanosporangia have not been developed (Fig. 2). The photobionts recently isolated contain large mature cells, but they do not begin cell cycle and cellular division. Probably this is a result of regulation of the photobiont population by the mycobiont through the production of secondary compounds like phenolic acids (Honneger 1987; Avalos and Vicente 1987) or lectins with enzymatic activity (Molina et al. 1993; Molina and Vicente 1993; 1994). Hence, Slocum et al. (1980), observed that, independent from environmental conditions, only 1 from 184 algal cells analyzed in *Parmelia caperata* thalli was an aplanosporangium.

Nevertheless, algal cells in culture show a high percentage of small cells (Fig. 1B) or autospores (Fig. 3A) as a consequence of vegetative mitotic division, very frequently in *Pseudotrebouxia* sp. (Fig. 3B and C), and non-vegetative mitotic division charasteristic in *Trebouxia* sp. (Fig. 3A). Hence, Ahmadjian (1982) describes both types of division in *X. parietina* axenic culture, because *Trebouxia italiana*, *Pseudotrebouxia decolorans* and *P. aggregata* have been described as members of the photobiont population of *X. parietina* in aposymbiotic conditions. The high amount of autospores and aplanosporangia could be explained as a consequence of the aposymbiotic culture, i.e., unmodified growth of the photobiont population in absence of a mycobiont controling and limiting the population size. On the other hand, photosynthetic activity from cultured algae is very high and independent of cell size because about 99% of algal cells show red fluorescence (Fig. 4A). However, populations of recently isolated photobionts show a wide fluorescence range from red (photosynthetically active cells) to gray (degenerative cells) according to Holopainen and Kauppi (1989) as shown in Fig. 4B. These variations of photosynthetic activity could be explained as a consequence of the mycobiont control and of low cellular turnover.

In conclusion, this chapter shows how the mycobiont could modify and control the frequency of distribution of photobiont population size and its photosynthetic activity.

Ackowledgements: We are very grateful to Miss Segovia and Dr. Reyes for English revision and those belonging to Türk's laboratory for their collaboration. This work was supported by a grant from the Comunidad Autónoma de Madrid (B.O.C.M. , March 16th, 1993).

References

Ahmadjian V (1973). In: Ahmadjian V, Hale ME (eds) The lichens. New York, Academic Press, pp 565-579

Ahmadjian V (1982). In: Round FJ, Chapman DJ (eds) Progress in Phycological Research, Elsevier Biomedial Press, Amsterdam, pp 179-233

Ahmadjian V Brink JJ, Shehaata A I (1991). In: Molecular biology of lichens. Search for plasmid DNA and the question of gene movement between bionts. Proc Intern Symp Lichenol, Nippon Paint Co, Osaka, pp 2-21

Ascaso C (1980). Ann Bot 45: 483

Avalos A, Vicente C (1987). Plant Cell Reports 6: 74-76

Bubrick P, Galun M (1986). Lichenologist 18: 47-49

Deason DR, Bold HC (1960). Univ Texas Publ 6022: 1-70

Hill DJ (1985). In: Brown DH (ed) Lichen Physiology and Cell Biology. Plenum Press, London, pp 303-317
Holopainen T, Kauppi M (1989). Lichenologist 21: 119-134
Honegger R (1987). Bibl Lichenol 25: 59-71
Honegger R (1991). Ann Rev Physiol Plant Mol Biol 42: 553-578
Molina MC, Vicente C (1993). In: Sato S, Ishida M, Ishikawa H (eds) ENDOCYTOBIOLOGY V, Tübingen University Press, Tübingen, pp 69-74
Molina MC, Vicente C (1995). Cell Adh Com 3: 1-12
Molina MC, Muñiz E, Vicente C (1993). Plant Physiol Biochem 31: 131-142
Molina MC, Vicente C, Muñiz, E (1994). Acta Horticulturae 381: 417-420
Pedrosa MC, Legaz E (1991). Symbiosis 11: 345-357
Slocum RD, Ahmadjian V, Hildredth KC (1980). Lichenologist 12: 173-187
Stocker-Wörgötter E, Türk R (1989). Nova Hedwigia 48: 207:228
Stocker-Wörgötter E, Türk R (1993). Crypt Bot 3: 283-289

Presence and Identification of Polyamines and their Conjugation to Phenolics in Some Epiphytic Lichens

J.L. Mateos and M.E. Legaz
Department of Plant Physiology, The Lichen Team, Faculty of Biology, Complutense University of Madrid, 28040 Madrid, Spain

Key words: *Quercus pyrenaica,* epiphytic lichens, phenolics, polyamines, xylem exudate.

Summary: *Evernia prunastri, Ramalina calicaris, Ramalina farinacea* and *Parmelia sulcata* thalli contain the diamine putrescine, and the polyamines spermidine and spermine as free molecules as well as conjugated ones. The xylem sap from oak branches supporting these epiphytic lichen species do contain atranorin, as a lichen phenolic and spermidine. Some lichen phenolics are found to be conjugated to acid insoluble polyamines.

Introduction

Epiphytic lichens play a secondary role as pathological organisms against the phorophytes on which they grow, and it became a common practice to destroy corticolous lichens growing on fruit trees with fungicides (Hale 1983). In addition, it has been observed that oak branches which support a heavy population of epiphytic lichens appear to be absolutely defoliated (Estevez et al. 1982). There may be an extensive penetration of the rhizines through the cork, cortex and cambium as far as the living wood (Ascaso et al. 1980) and these hyphae readily tear off areas of both xylem and phloem, and disperse either individually or in small groups toward the vessels (Ascaso 1985). As a consequence of this, small shrubs and trees densely covered with lichens can clearly be stunted and damaged (Hale 1983). Asahina and Kurokawa (1983) described the inhibition of tea-shoot development by these epiphytes. Brown and Mikola (1974) noticed a decrease in the growth rate of *Pinus sylvestris*, probably caused by an inhibition of mycorrizal association produced by terricolous lichens.

Medullary hyphae can penetrate into host tissues, through both mechanical and enzymic actions (Brodo 1973) (Yagüe et al. 1984), to reach xylem vessels (Ascaso et al. 1980). Some authors have suggested that this physical penetration can be acompanied by an injection of lichen phenolics into the tissues of the phytophore, with a special regard to usnic and sekikaic acids synthesized by *Ramalina tayloriana* (Ozenda and Clauzade 1970). This fact can be explained somehow by means of lichens injection of metabolic inhibitors that, after the acropetal translocation via xylem, reach the leaves and induce abscision. In a field trial to evaluate this hypothesis, previous reports show that evernic acid, a phenolic compound naturally synthesized by *Evernia prunastri* thallus, can be injected into the host xylem (Avalos et al. 1986) where it is acropetally transported. Additionally, evernic acid has been shown to inhibit both respiration and appearance of foliar buds, and also retards leaf formation (Legaz et al. 1988).

On the other hand, the diamine putrescine (PUT) and the polyamines, spermidine (SPD) and spermine (SPM) occur ubiquitously in plant tissues and physiological effects of these metabolites in plants have been recently reviewed (Smith 1985, Evans and Malmberg, Galston and Kaur-Shawney 1990). Processes involved include cell division, embryogenesis, flower induction and development, fruit growth and senescence, and response to stress.

The presence and abundance of polyamines (PAs) in xylem and phloem exudates indicates that they might be translocated in plants (Friedman et al. 1986). This long-distance translocation further supports the hypothesis that PAs do play a regulatory role in plant growth and response to stress.

PAs can be produced in nature as cationic free molecular bases, but are frequent as conjugates, then bound to phenolics, other low-molecular weight compounds or even macromolecules, in both plant and animal kingdoms (Escribano and Legaz 1988, Guggisberg and Hesse 1983, Martin-Tanguy 1985, Tiburcio et al. 1990, Twaardowski et al. 1982). In *Evernia prunastri* thallus, PUT, SPD and SPM are bound to evernic acid (Mateos and Legaz 1994).

With all of these premises, in this paper we attempt to examine the occurrence of PAs and lichen phenolics in vascular exudates of *Quercus pyrenaica* as well as its possible conjugation in some lichen epiphytes, as evidence about the correlation between lichen phenol accumulation and defoliating action.

Material and Methods

Young shrubs of *Quercus pyrenaica* Lam. growing in Valsaín (Segovia, Spain), with or without a lichen population mainly composed by *Evernia prunastri* (L.) Ach., *Ramalina calicaris* (L.) Fr., *Ramalina farinacea* (L.) Ach. and *Parmelia sulcata* Tayl. were used throughout this work. Portions of about 5 cm length of both apical and basal ends of branches were cut and placed with the morphological upper end at the basal position in centrifuge tubes and spun at 3000 g for 30 min at 2 °C to extract the xylem exudate (modified from Tiburcio et al. 1985). Mean exudation amount was about 2 ng g^{-1} . This dry residue was immediately mixed with cold $HClO_4$ to a final concentration of 5% (v/v), and stored at -20 °C until required.

Samples of 1.0 g of air-dried thalli were floated on 15 ml of 20 mM MES buffer, pH 6.8, for 2 h at 26 °C in the dark. Extraction and dansylation of PAs as well as post-derivatization cleanup and HPLC analysis were carried out according to that described by Escribano and Legaz(1988). Bound PAs, both soluble (SH) and insoluble (PH) acid fractions were analyzed after hydrolysis with 12 N HCl for 18 h at room temperature (modified from Escribano and Legaz 1988).

Lichen phenolics were removed from the cortex by washing off samples with 50 ml pure acetone (HPLC grade) for 5 min at room temperature in the dark, for avoiding phenolic photooxidation. Acetonic phase was dried in running air and the dry residue was chromatographed. Phenol composition contained in xylem exudates were analyzed as well, under the same conditions.

In parallel to the analysis of those PAs accumulated in both SH and PH fractions, we evaluated qualitatively which phenolic compound was recovered as a conjugate. The analytical procedure was a modification of Pedrosa and Legaz (1993). Thus, an aliquot (1.0 ml) of each PA hydrolysate was dried *in vacuum* and resuspended in an equal volume of distilled water. Phenols were removed from the aqueous suspension with 5.0 ml diethylether:ethylacetate (65:35, v/v) by vigorous shaking for 30 min at 40 °C. Organic phases were filtered through Whatman No. 2 filters, and then collected. This procedure was repeated four times. Aqueous phases were again subject to separation with 4 x 5.0 ml chloroform:acetonitrile (60:40, v/v). The organic phases were collected and filtered under the same conditions as above. Ether and chloroformic phases were mixed in the same tube and evaporated. Dry powders were kept at -20 °C until chromatographed.

Phenolic dry residues from both PH and SH fractions were redissolved in 1.0 ml acetonitrile:acetic acid:bidistilled water (70:1:29, v/v) to be chromatographed. Separation of these compounds was performed on a Nucleosil 5 C8 column (Scharlau, 125 mm x 4 mm, i.d.) using as mobile phase acetonitrile:acetic acid:bidistilled water (70:1:29, v/v) at a flow rate of 0.7 ml min^{-1}. The program of detector was: wavelength, 270 nm for up to 2.8 min, and 280 nm from 2.81 min to 10 min. The liquid chromatograph was a Varian 5000 fitted together with a Vista CDS 401 computer.

Results

E. prunastri, R. calicaris, R. farinacea and *P. sulcata* contained PUT, SPD and SPM after incubation on MES buffer for 2 h (Table I). The amount of free SPD was higher than the other PAs, with the exception of *R. calicaris* where SPM was mostly accumulated. PA conjugation to macromolecules (acid insoluble fraction, PH) showed a similar behaviour. However, PH-PA titers were lower than those found as free molecules (Table 1). The different PA levels observed for each collected epiphyte could be explained in terms of the variable metabolic interconversion patterns among them. No PA concentration was found in the SH fraction under the incubation conditions described above.

PAs were also studied in the xylem fluid. HPLC study confirms the occurrence of SPD in xylem exudates (Table 2), whereas levels of PUT and SPM were abso-

Table 1. Accumulation patterns of PAs in diffrent lichen epiphytes (μg polyamine g^{-1} dry thallus ± SD)

Lichen sample	S-PUT	S-SPD	S-SPM	PH-PUT	PH-SPD	PH-SPM
E. prunastri	0.37 ± 0.0	6.27 ± 0.01	1.75 ± 0.09	0.41 ± 0.01	1.94 ± 0.0	0.38 ± 0.0
R. calicaris	0.34 ± 0.0	2.07 ± 0.01	1.03 ± 0.04	0.15 ± 0.0	1.36 ± 0.01	0.31 ± 0.004
R. farinacea	0.60 ± 0.01	2.54 ± 0.01	1.52 ± 0.01	0.32 ± 0.0	1.35 ± 0.0	0.17 ± 0.015
P. sulcata	0.74 ± 0.0	2.55 ± 0.03	1.03 ± 0.04	0.31 ± 0.02	1.06 ± 0.04	0.15 ± 0.0

S-PA: free soluble polyamine; PH-PA: acid-insoluble bound polyamine; SD: standard deviation as a mean of four replicates.

Table 2. Polyamines from xylem exudates (ng polyamine g^{-1} dry xylem ± SD)

Sample[a]	PUT	SPD	SPM
I	b.d.l.[b]	6.1 ± 0.01	b.d.l.
II	b.d.l.	13.2 ± 0.0	b.d.l.
III	b.d.l.	55.0 ± 0.0	b.d.l.

[a]I: xylem sap collected from leafless and lichenless branches; II: xylem sap collected from branches with leaves but not lichens; III: xylem sap collected from defoliated branches invaded by lichens.
[b]b.d.l.: below detection limit.

lutely undetectable. The lowest concentration of this triamine was obtained when leafless and lichenless branch samples were collected. Therefore, a notable increase in SPD titer is observed when using branches which support a high density lichen population. These appeared largely defoliated when collected.

Analysis by HPLC of oak xylem extracts from the youngest branches shows no significant peaks (Fig.1A). A similar result was observed for those phenolics partitioned from a xylem exudate from several folious branches (Fig.1B), which did not have epiphytic lichens growing on them. Nevertheless, a sharp peak with a retention time of 3.62 min was identified as atranorin, by spiking. This compound was only detected in vascular exudates of branches with a large population of epiphytes (Fig. 1C). Here, it is important to note that there was a compound most likely identified as the aglycone everninic acid. However, no esterase activity which could hydrolyse evernic acid to everninic acid, has been found previously in the xylem sap or the leaf tissues of *Q. pyrenaica* [10].

Phenolics from the different lichen samples as well as their conjugates to PAs are summarized in Table 3. Evernic, usnic and salazinic acids were found to be the only phenolic compounds recovered from the PH fraction in *E. prunastri*, *R. calicaris*, and *P. sulcata*, respectively. No phenolics were detected in the acid soluble fraction (SH).

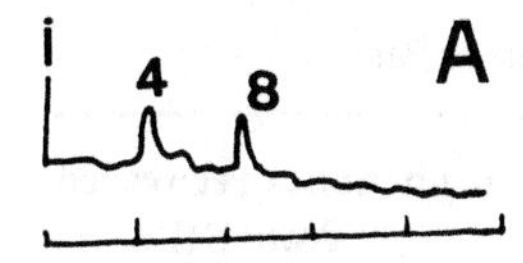

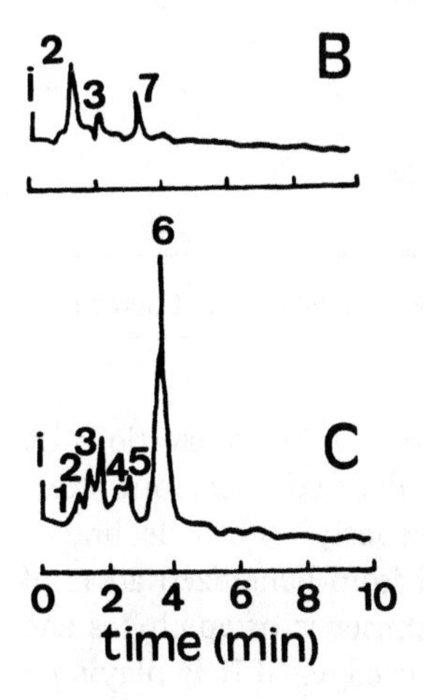

Fig. 1. HPLC traces of the xylem exudate obtained from the apical ends of branches of *Quercus pyrenaica* with no lichens or leaves (**A**); with leaves but no lichens (**B**); and defoliated and fully invaded by epiphytic lichens (**C**). i Injection; 3 evernínic acid; 6 atranorin; 1, 2, 4, 5, 7 and 8 unidentified

Discussion

The results showed that SPD was the only PA present in vascular exudates of the three different branches studied (Table 2) which has a significant amount in exudates of xylem sap collected from defoliated branches. The presence of SPD in the bleeding exudate of cut stumps proceeding from those non-invaded areas, which represent the ascending xylem sap (Bollard 1960, McClure and Israel 1979, Pate 1980) implies that PAs can be synthesized in the root system and exported to the shoot. The PA transport was also indicated by the data of Friedman et al. (1986), who detected specially SPD in stem exudates of various plant species. The abundance of inorganic and organic nitrogenous compounds, including amino acids, in the xylem sap is well known (Miflin and Lea 1977, Hanson and Cohen 1985). Thus, SPD like other nitrogenous metabolites can be found in the vascular system of *Q. pyrenaica*.

It has been shown that leaf abscision in senescent wheat plants also results in a temporary accumulation of both PUT and SPD (Peeters et al. 1993). In plants, PA content increased dramatically in response to different stress factors, such as water stress (Foster and Walters 1991), fungal infection (Foster and Walters 1992) and ion deficiency (Reggiani et al. 1993). This increase likewise leads to PUT accumulation, but this does not extend to whole species as it is shown in the present

Table 3. Phenolic production by epiphytic thalli and its conjugation to Pas

Lichen Samples	free thalline phenolic composition	phenolics recovered from PH	phenolics recovered from SH
Evernia prunastri	evernic acid atranorin	evernic acid	n.d.
Ramalina calicaris	sekikaic acid usnic acid	usnic acid	n.d.
Ramalina farinacea	protocetraric acid usnic acid	n.d.	n.d.
Parmelia sulcata	salacinic acid atranorin	salacinic acid	

PH: acid-insoluble bound phenolics; SH: acid-soluble bound phenolics; n.d.: not detected.

report where PUT and SPM levels were zero. This agrees with observations by Tiburcio et al. (Tiburcio et al. 1986) for tobacco plants subjected to osmotic stress where an increased amount of SPD occurred simultaneously with a decline in PUT titers. Our data on the high SPD levels recovered from lichenized areas of the oak tree seem to suggest a correlation between the presence of epiphytes and the defoliating effects suffered by the host. To our knowledge, SPD is playing a role itself in defoliation, since the biosynthesis of the triamine is enhanced during leaf abscision. Indeed, abscisic acid (ABA) functions as a hormonal accelerator to promote abscisionof leaves, and it has been recently demonstrated (Aurisano et al. 1993) that it induces a stress-like PA pattern in wheat seedlings. Nevertheless, besides this hypothesis, our results point out that the increase in SPD amount is not properly due to an induced *de novo* synthesis of the PA which takes place during abscision, but to an injection of SPD from epiphytic lichens into the xylem vessels. In fact, this possibility should not be discarded since concentrations of soluble (S) as well as acid insoluble (PH) SPD are quantitatively more significant in all lichen samples analyzed (Table 1). However, the fact of obtaining a higher amount of SPD in xylem sap samples collected from defoliated branches invaded by lichens than in those lichenless supports that the origin of the polyamine could be, at least, partially due to its injection from the lichen (Table 2).

Otherwise, the hypothesis about injection of some lichen metabolites into the phytophore on which they develop can be further supported by the results shown in Fig. 1. The entry of atranorin into the xylem sap exists, since this phenol was only detected in the apical zones of defoliated branches. Defoliation of trees having lichens is a very well-known effect (Estevez et al. 1982), but it still remains unclear whether it is a consequence of accelerated senescence or an inhibition of new leaf formation. On this basis, it is reasonable to consider atranorin as an uncoupler of photolysis, being translocated via xylem to the leaves to stop biomass production and to induce abscision (Ozenda and Clauzade 1970). This action can be corroborated by the observation that lichen phenols produce irreversible damage in several membrane systems (Kinraide and Ahmadjian 1970, Vicente 1985).

Until recently, research on plant PAs was generally approached on the pools of free compounds. Nonetheless, it is now apparent in many cases that covalently bound forms of diamines and PAs can account for a significant portion of the metabolic pools. PUT, SPD and SPM are bound to phenolics in some plants. The binding is achieved through an amide linkage to one or two molecules of coumaric, caffeic and/or caffeic ferulic acids, forming hydroxycinnamoyl acid amides (Martin-Tanguy 1985). The possibility that PAs can be conjugated to phenols is countenanced by our results (Table 3). Surprisingly, this qualitative analysis reveals that the recovery of phenols was only produced from a PH fraction. These low molecular weight compounds should be found at the SH fraction. However, they seemingly appeared at PH where PAs are supposed to be bound to macromolecules such as proteins or nucleic acids (Appelbaum et al. 1988; Serafini-Fracassini et al. 1992). This fact has been meaningfully explained, in the case of *E. prunastri*, since evernic acid is a constituent of a PA complex integrated by proteins as well as genomic DNA (Mateos and Legaz, submitted). However, no previous reports show the occurrence of PA-conjugated metabolites after hydrolyzing the binding between them, as was carried out in this study.

Concerning the metabolic significance of these conjugates, it has been suggested that they may act as a storage reservoir of PAs. In lichens, the accumulation of these PA-to-phenolic conjugates may be considered as an homeostatic mechanism that prevents excessive fluctuations of cellular pH (Legaz et al. 1993).

In our study, the presence and identification of PAs and their possible binding to phenolics play an important role in the selection of the phenol to be injected into the xylem and contribute to defoliation of the phytophore. Atranorin, the only phenolic compound detected in vascular exudates (Fig. 1), was not found as a conjugate at any studied epiphyte (Table 3), from which it is hypothesized that the injection to the xylem and its later acropetal translocation only occurs when the phenolic is found as a free molecule. This is in correspondence to that argued for PA conjugates whose free bases may be released sometime during physiological events and also as a means of PA transport (Martin-Tanguy 1985) and, particularly to free PAs insofar as their polycationic nature is involved in their physiology (Peeters et al. 1993).

Acknowledgments. This work was supported by a grant from the Dirección General de Investigación Científica y Tecnológica (Spain). No. PB93 0092. We are grateful to Prof. Vicente for his help and constructive discussion.

References

Appelbaum A, Canellakis ZN, Applewhite PB, Kaur-Shawney R, Galston AW (1988). Plant Physiol 88: 996-998

Asahina Y, Kurokawa S (1983). Miscel Rep Res Inst Nat Resources 20: 80-85

Ascaso C (1985). In: Vicente C, Brown DH, Legaz ME (eds) Surface Physiology of Lichens. Complutense University Press, Madrid, pp 89-104
Ascaso C, González C, Vicente C (1980). Cryptog Bryol Lichénol 1: 43-49
Aurisano N, Bertani A, Mattana M, Reggiani R (1993). Physiol Plant 89: 687-692
Avalos A, Legaz ME, Vicente C (1986). Biochem Syst Ecol 14: 381-384
Bollard EG (1960). Ann Rev Plant Physiol 11: 141-166
Brodo IM (1973). In: Ahmadjian V, Hale ME (eds) The Lichens. Academic Press, New York, pp 401-443
Brown MT, Mikola P (1974). Acta Forest Fenn 141: 1-6
Escribano MI, Legaz ME (1988). Plant Physiol 87: 519-522
Estévez MP, Orús MI, Vicente C (1982). In: Vicente C, Municio AM (eds) Estudios sobre Biología. Complutense University Press, Madrid, pp 117-132.
Evans PT, Malmberg RL (1989). Ann Rev Plant Physiol 40: 235-269
Foster SA, Walters DR (1992). J Plant Physiol 140: 134-136
Foster SA, Walters DR (1991). Physiol Plant 82: 185-190
Friedman R, Levin N, Altman A (1986). Plant Physiol 82 1154-1157
Galston AW, Kaur-Shawney R (1990). Plant Physiol 94: 406-410
Guggisberg A, Hesse M (1983). In: Brossi A (ed) The Alkaloids: chemistry and pharmacology. Academic Press, New York, pp 85-188
Hale ME (1983). In: The Biology of Lichens. Edward Arnold, London, pp 146-148
Hanson SD, Cohen ID (1985). Plant Physiol 78: 734-738
Kinraide WTB, Ahmadjian V (1970). Lichenologist 4: 234-239
Legaz ME, Escribano MI, Vicente C (1993). In: Sato S, Ishida M, Ishikawa H (eds) ENDOCYTOBIOLOGY V. Tübingen University Press, Tübingen, pp 59-68
Legaz ME, Pérez-Urria E, Avalos A, Vicente C (1988). Biochem Syst Ecol 3: 253-259
Martin-Tanguy J (1985). J Plant Growth Regul 3: 381-399
Mateos JL, Legaz ME (1994). In: Abstracts of the 20th International Symposium on Chromatography, Bornemouth, UK, p Th-34
McClure PR, Israel DW (1979). Plant Physiol 64: 411-416
Miflin BJ, Lea PJ (1977). Ann Rev Plant Physiol 28: 299-329
Ozenda P and Clauzade G (1970). In: Les Lichens. Mason et Cie, Paris, 125-149
Pate JS (1980). Plant Physiol 31: 313-340
Pedrosa MM and Legaz ME (1993). In: Sato S, Ishida M, Ishikawa H (eds) ENDOCYTOBIOLOGY V. Tübingen University Press, Tübingen, 75-80
Peeters KMU, Geons JM, Van Laere AJ (1993). J Exptl Bot 268: 1709-1715
Reggiani R, Aurisano N, Mattana M, Bertani A (1993). J Plant Physiol 141: 136-140
Serafini-Fracassini D, Torrigiani P, Branca C (1992). Physiol Plant 60: 351-357
Smith TA (1985). Ann Rev Plant Physiol 36: 117-143
Tiburcio AF, Kaur-Shawney R, Galston AW (1990). In: Davies D (ed) The biochemistry of Plants. Academic Press, San Diego, 16: 283-325
Tiburcio AF, Kaur-Shawney R, Galston AW (1986). Plant Physiol 82: 375-378
Tiburcio AF, Kaur-Shawney R, Inhersoll RB, Galston AW (1985). Plant Physiol 78:323-326
Twaardowski T, Puliskowska J, Wiewiórowski M (1982). Bull Acad Pol Sci 29: 129-140
Vicente C (1985). In: Vicente C, Brown DH, Legaz ME (eds) Surface Physiology of Lichens. Complutense University Press, Madrid, 11-32
Yagüe E, Orús MI, Estévez MP (1984). Planta 160: 212-218

2.5 Vertebrate Evolution and Medical Significans

Intestinal Methanogens and Vertebrate Evolution: Symbiotic Archaea are Key Organisms in the Differentiation of the Digestive Tract

J.H.P. Hackstein[1] and P. Langer[2]
[1]Department of Microbiology and Evolutionary Biology, Faculty of Science, Catholic University of Nijmegen, 6525 ED Nijmegen, The Netherlands
[2]Institut für Anatomie und Zellbiologie der Justus-Liebig-Universität Giessen, Aulweg 123, 35385 Giessen, Germany

Introduction

Intestinal methanogenic bacteria play an important role in the complex biota of vertebrate digestive tracts (Miller and Wolin 1986). They are the terminal consumers of hydrogen that is generated during the anaerobic decomposition of biopolymers. Since studies on terrestrial arthropods (Hackstein and Stumm 1994) revealed strong evolutionary constraints on the association between methanogens and invertebrates, we screened more than 250 vertebrate species for methane emissions in order to rule out the conditions for the presence of intestinal methanogenic bacteria.

Results and Discussion

Our measurements revealed that there is no simple correlation between particular feeding habits and the competence for methanogens: both carnivorous and herbivorous species can house methanogens. Moreover, among non-methanogenic species, both types of feeders are found. Nevertheless, methane status generally is shared by most species belonging to a particular higher taxon (Hackstein and van Alen 1996). In the chiropteran, insectivoran, and carnivoran lineages, not a single methane producer is found, whereas all ruminants and old-world monkeys and apes produce methane (Hackstein et al. 1995). Also, among ant-eating animals there is a strong correlation between methane emission and taxonomic position: the giant ant-eater and the tamandua belong to the edentata, and members of this taxon generally are methane producers. The aardvark, too, is a methane producer; it belongs to the tubilidentata that are closely related to the methanogenic taxa of the proboscidea, sirenia and hyracoidea. In contrast, the carnivore-related ant-eating pangolins do not emit methane. Only rodent and South-American primate taxa exhibit a certain variability with respect to the trait methane production (Hackstein et al. 1995; Hackstein and van Alen 1996).

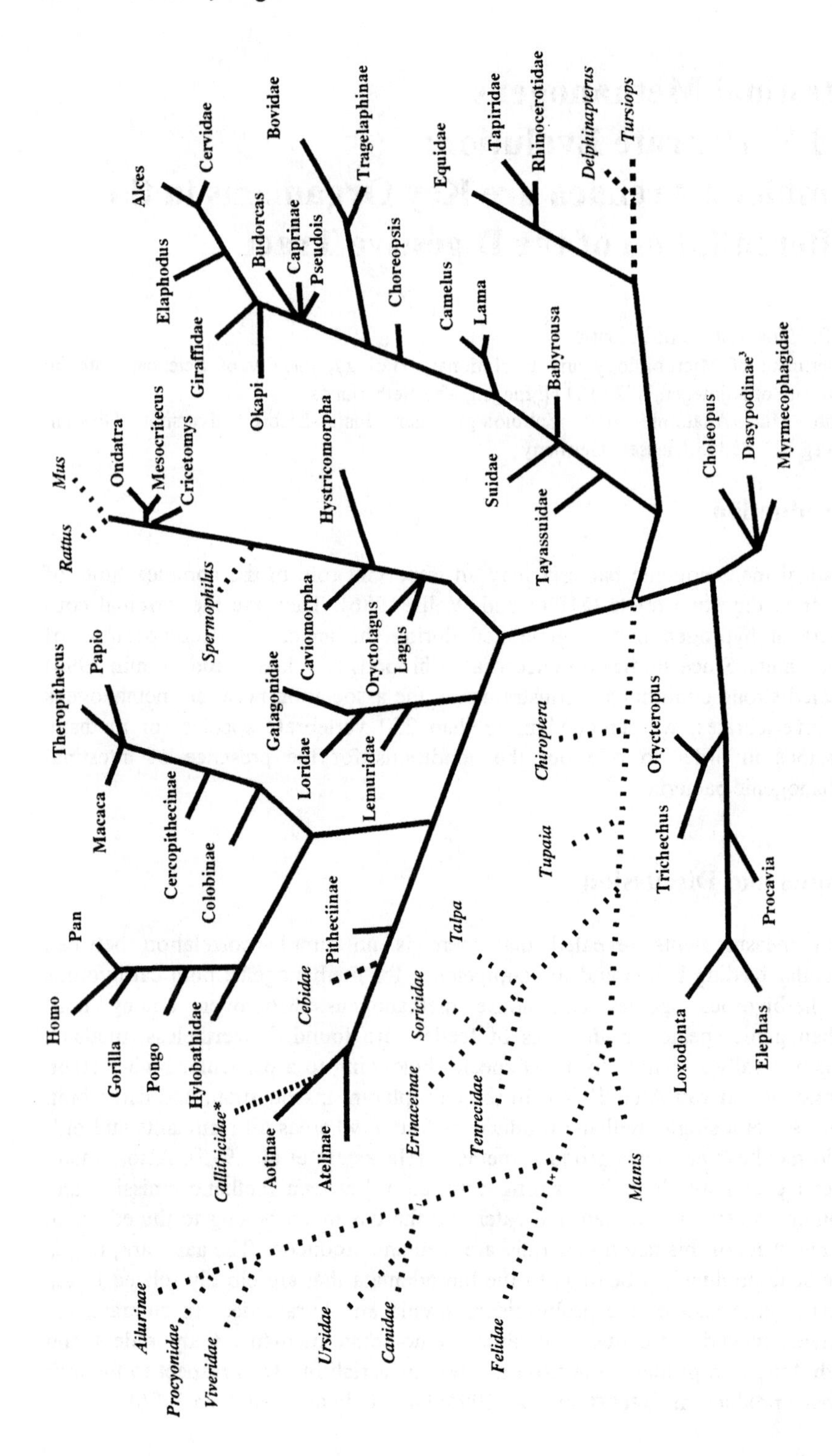
Homo
Gorilla
Pan
Pongo
Hylobathidae
Macaca
Theropithecus
Papio
Cercopithecinae
Colobinae
Callitricidae*
Aotinae
Atelinae
Cebidae
Pitheciinae
Lemuridae
Loridae
Galagonidae
Oryctolagus
Lagus
Caviomorpha
Hystricomorpha
Spermophilus
Rattus
Mus
Ondatra
Mesocritecus
Cricetomys
Giraffidae
Okapi
Elaphodus
Alces
Cervidae
Bovidae
Budorcas
Caprinae
Pseudois
Tragelaphinae
Choreopsis
Camelus
Lama
Equidae
Tapiridae
Rhinocerotidae
Delphinapterus
Tursiops
Babyrousa
Suidae
Tayassuidae
Choleopus
Dasypodinae[3]
Myrmecophagidae
Orycteropus
Trichechus
Procavia
Loxodonta
Elephas
Chiroptera
Tupaia
Talpa
Soricidae
Erinaceinae
Tenrecidae
Manis
Ailurinae
Procyonidae
Viverridae
Ursidae
Canidae
Felidae

Vertebrate Evolution

Integrating the trait "methane emission" into evolutionary trees (Fig. 1) reveals that the ability to emit methane is a primitive-shared (plesiomorphic) character among amniotes. Furthermore, the loss of this character obeys Dollo's law: the loss of competence for methanogens is a definitive event. There are no atavistic trends that restore the loss of competence. Consequently, the competence for intestinal methanogens must be controlled genetically. Studies on human family trees support this assumption, since the competence for methanogens segregates like a single autosomal dominant Mendelian character (Hackstein et al. 1995). Thus, one has to conclude that taxonomic position is crucial for competence for methanogenic bacteria.

Another condition for the presence of methanogenic bacteria seemed to be morphological adaptations of the intestinal tract, i.e. rumina or caeca (Langer and Snipes 1991). However, methanogens are also found in the intestinal tracts of certain primitive hindgut fermenters such as the hystricomorphs (Hackstein and van Alen 1996). Therefore, at least elaborated foregut fermenting devices are not a prerequisite for housing methanogens. On the contrary, the complete absence of such structures from non-methanogenic taxa such as chiropters, insectivores, and carnivores suggests that the presence of methanogens has been crucial for the evolution of fermenting devices. A methanogenic (fermenting), but not ruminating foregut evolved in the highly specialized phytophagic langurs (Primates), whereas in the strictly herbivorous but non-methanogenic pandas (Carnivora) the evolution of a significant intestinal differentiation did not occur (Hackstein and van Alen 1996).

There is a high but not absolute correlation between methanogenesis and the presence of morphological differentiations of the colon/caecum that can host fermenting microbiota (Hackstein and van Alen 1996). Colon-caecum fermenting species are characteristic for the more ancient mammalian taxa. They flourished at the beginning of the eocene, and their radiation preceded the advent of the ruminants (Langer 1994). Since competence for intestinal methanogens is a plesiomorphic character among amniotes, it is likely that the presence of methanogens in the digestive tracts of primordial hindgut fermenting mammals was a necessary condition for and not a consequence of the evolution of differentiated hindgut. Thus, the absence of methane emissions among mammals with elaborated hindgut structures seems to be a secondary event. The situation among rodents supports this assumption: caviomorphs and hystricomorphs are methane producers whereas most murids do not emit methane (Fig. 2). Consequently, the presence of methanogens seems to be cause and not consequence for the evolution of fermenting differentiations of the intestinal tract.

◀ **Fig. 1.** Phylogenetic tree of mammals based on the publication of Miyamoto and Goodman (1986). Bold, roman characters and solid lines indicate methane producers, dotted lines and italics non-producers. (Modified after Hackstein and van Alen 1996)

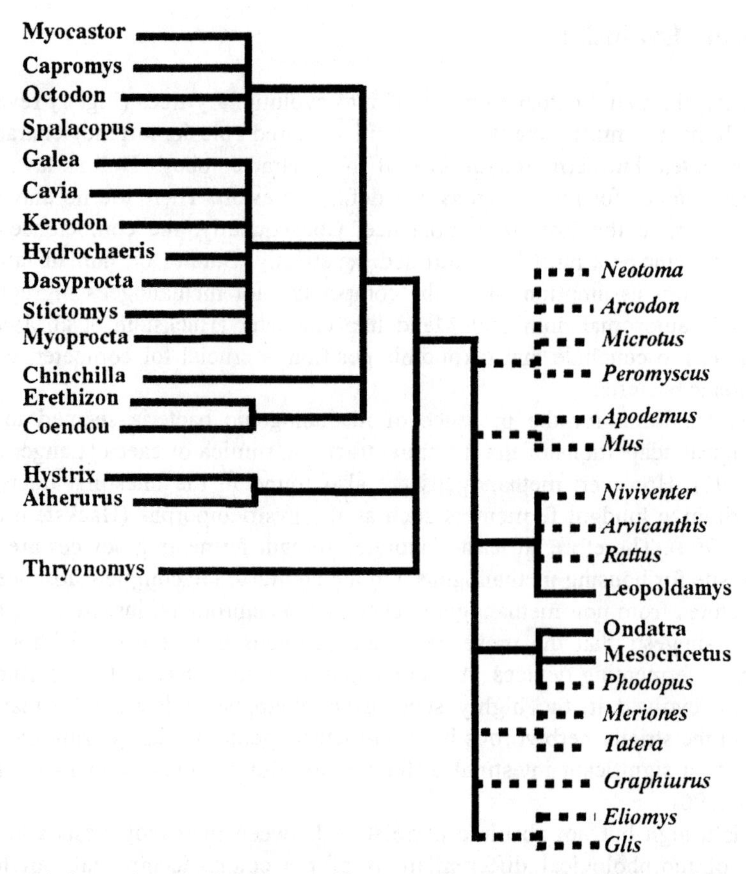

Fig. 2. Phylogenetic tree of rodents. Bold, roman characters and solid lines indicate methane producers; dotted lines and italics, non-producers. (Modified after Catzeflis et al. 1995)

The distribution of producers and non-producers also suggests an additional correlation with the size of the animals, generation times, and the weaning period (Langer and Hackstein 1997a,b). The comparative analysis of methanogenesis and weaning periods of 78 different species of mammals reveals that methanogenic species, regardless of their intestinal differentiations, have weaning periods exceeding 30 days, or 65% of the duration of the lactation period, respectively (Fig. 3). Animals with shorter weaning periods do not emit methane, and it is questionable whether they are still capable of degrading biopolymers alloenzymatically. These are generally small animals that have short generation times, and among certain small rodents (e.g. the Cricetinae) the evolution of non-fermenting forestomachs seems to be have compensated the loss of "methanogenic" fermentations. Thus, these forestomachs are functionally not equivalent to the rumina and other fermenting forestomachs of larger mammals.

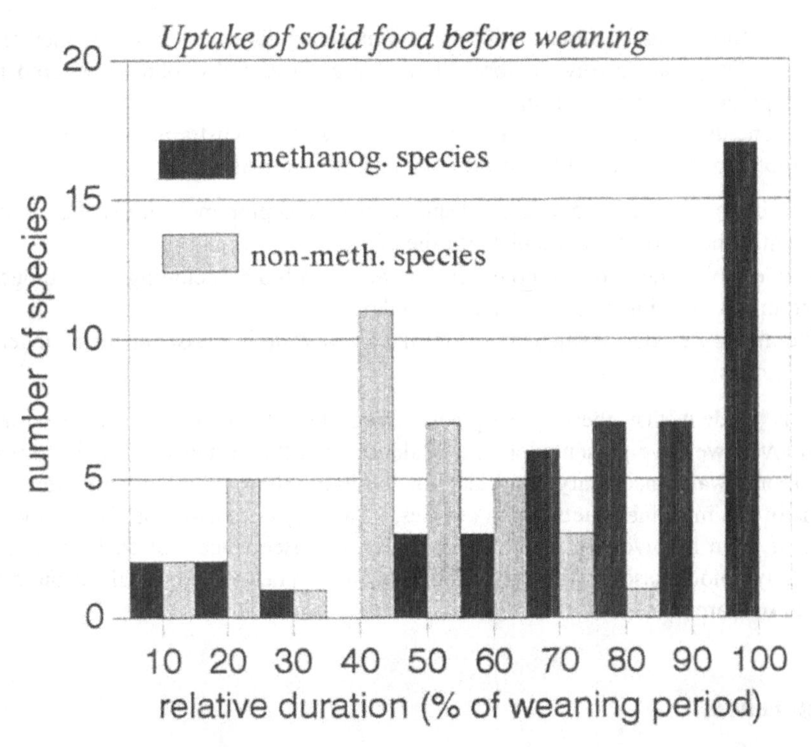

Fig. 3. Correlation between the relative duration of the weaning period of mammals and their methane status. With a few exceptions, methanogenic species have a weaning period exceeding 60% of the duration of the lactation period

Obviously, the symbiosis of methanogenic bacteria is crucial for the alloenzymatic digestion of biopolymers, and the establishment of such symbiotic associations apparently has a number of requirements that constrained the evolution of mammals. The weaning period, i.e. the part of the lactation period when the young suckle milk and take up solids at the same time, is characterized by substantial changes in diets and digestive enzymes. In species with an alloenzymatic digestion, the weaning period must allow the colonization of the intestinal tract by suitable microorganisms. Figure 3 suggests that the colonization of the intestinal tract by complex, methanogenic microbiota that can digest biopolymers is a rather time-consuming process that exceeds 30 days. There are only a few exceptions to this rule; they can be understood if the ecological and nutritional aspects are taken into account.

It is evident that constraints such as descent, size, and weaning period, were of considerable importance for the evolution of mammals. Whereas large animals (i.e. larger than 1 kg) with long generation times could explore low protein diets by the evolution of more sophisticated adaptations for the alloenzymatic digestion of their food, small animals with short generation times had to quit methanogenesis. They had to rely on food of higher quality, and evolution favored the development of non-fermenting devices.

In conclusion, there are several prerequisites that constrain the presence of symbiotic methanogens in the intestinal tracts of mammals:

1. The hosts require a genetic predisposition (a receptor molecule?) that is frequently shared by most members of the taxon.
2. The establishment of complex intestinal microbiota including methanogens requires a weaning period exceeding 30 days.
3. Preliminary evidence suggests a minimal size (or body mass) for methanogenic mammals

It is evident that these prerequisites restricted the evolution of mammals. Moreover, we have presented ample evidence, that the presence of methanogenic symbionts was a necessary condition for the evolution of fermenting differentiations of the intestinal tracts. Such devices, however, were essential for the evolution of large herbivorous mammals and - as a consequence - of their predators. Thus, symbiotic associations with methanogenic archea were crucial for the evolution of mammals.

References

Catzeflis FM, Hänni C, Sourrouille P, Douzery E (1995) Re: Molecular systematics of hystricognath rodents: The contribution of sciurognath mitochondrial 12S rRNA sequences. Molecular Phylogenetics and Evolution 4: 357-360

Hackstein JHP, Stumm CK (1994) Methane production in terrestrial arthropods. Proc Natl Acad Sci USA 91: 5441-5445

Hackstein JHP, van Alen TA, op den Camp H, Smits A, Mariman E (1995) Intestinal methanogenesis in primates - a genetic and evolutionary approach. Dtsch tierärztl Wschr 102/4: 152-154

Hackstein JHP, van Alen TA (1996) Faecal methanogens and vertebrate evolution. Evolution 50: 559-572

Langer P (1994) Food and digestion of Cenozoic mammals. In: Chivers DJ, Langer P (eds) The digestive system in mammals. Cambridge University Press, Cambridge, UK, pp 9-23

Langer P, Hackstein JHP (1997a) An evolutionary view of the weaning process in mammals. submitted

Langer P, Hackstein JHP (1997b) Digestive tract and reproduction in rodentia. submitted

Langer P, Snipes RL (1991) Adaptations of gut structure to function in herbivores. In: Tsuda T, Sasaki Y, Kawashima R (eds) Physiological aspects of digestion and metabolism in ruminants. Academic Press, San Diego, 1991, pp 349-384

Miller TL, Wolin MJ (1986) Methanogens in human and animal intestinal tracts. Syst Appl Microbiol 7: 223-229

Miyamoto MM, Goodman M (1986) Biomolecular systematics of eutherian mammals: phylogenetic patterns and classification. Syst Zool 35: 230-240

Pleomorphic Bacterial Intracytoplasmic Bodies: Basic Biology and Medical Significance

G.J. Domingue, Sr.
Tulane University School of Medicine and Graduate School, Department of Urology, SL-42, and Department of Microbiology and Immunology, 1430 Tulane Avenue, New Orleans, Louisiana 70112, USA

Introduction and Rationale

There are numerous diseases of humans exhibiting symptomatology suggestive of a bacterial etiology, yet routine bacteriologic methods fail to demonstrate the presence of organisms. These diagnostic failures often lead to frustrating clinical situations and difficulty in management of the patient's disease. The inability to substantiate by bacteriologic culture that there is an infectious disease often leads to inappropriate use of antimicrobials in these patients and may contribute to the prolonged persistence of dormant bacterial forms in vitro. We reported (Green et al. 1974a) that intracytoplasmic bodies derived from relatively stable L-forms of streptococci were demonstrated in human embryonic kidney tissue culture cells infected with streptococcal L-forms. These intracytoplasmic bodies (0.05-0.15 μ) were non cultivable; however, there was continual persistence of these small bodies in the tissue culture cells despite disintegration of the L-form. Twenty-five days later, there was conversion of the tiny bodies to the parent *Streptococcus faecalis* (wall-containing bacterium) from which the relatively stable L-form was derived. During the period of persistence of small bodies, no L-form or parent bacteria could be cultured in vitro. While persistence of the small bodies appeared to render no visible pathology, reversion to the wall-containing streptococci resulted in death of the tissue culture cells. Unlike the relatively stable (nonreverting) L-form, a stable streptococcal L-form derived from a different strain of *Streptococcus faecalis* could be cultured 73 days after initial infection, indicating that the organisms had survived and multiplied in vivo. At no time during the course of this study did the stable L-form revert to its parent bacterial form. Throughout all of the cell infection experiments, the stable L-form, when cultured, grew out more readily and in greater numbers than did the relatively stable, reverting L-form strain.

Concomitant with these tissue culture experiments, we studied the reproductive cycle of the relatively stable streptococcal L-form in vitro (Green et al. 1974b). Based on electron microscopic observations, we proposed that the small, dense, nonvesiculated L-forms observed in culture constituted the central (core) element of the cycle. It was seen to divide and bud rapidly. In addition, the dense forms appeared to be capable of growth and development within vesicles of mature mother forms. When these forms were released from the vesicles into the surrounding liquid medium, further growth occurred, resulting in the development of immature and ultimately mature mother forms. Under conditions unfavorable for L-form growth, these dense bodies developed first into transitional forms and then into cell wall-containing bacteria. We hypothesize that these dense forms might therefore be considered as undifferentiated "stem cells" with the capacity to develop along several different routes, depending upon the stimulus received. The foregoing hypothesis brings up several interesting considerations. One such consideration involves the origin of the undifferentiated dense forms or "stem cells." The concept of a large vesiculated mother form as the more mature form is basic and is corroborated by Maxted (1968), who observed development of dense particles over a 24-h period by light microscopy, and by Bibel (Bibel and Lawson 1972), who demonstrated by electron microscopy the appearance of such bodies in significant numbers in the late-exponential phase of growth.

From our observations, it appears logical that the vesiculated "mother" form is capable of fostering growth of the smaller elementary bodies into undifferentiated dense forms which are then extruded. There is good evidence that elementary bodies of at least 0.24 μ are in fact capable of growing into these undifferentiated dense forms (Coussons and Cole 1968). Bibel (Bibel and Lawson 1972), Fass (1973), and Weibull et al. (1965) assumed that smaller elementary bodies within vesicles did not play a major role in replication, because their morphologic data suggested that they might be deficient in or devoid of deoxyribonucleic acid and therefore should be considered nonviable. Morphologic conclusions concerning nucleic acid content have limitations. It would be useful to corroborate morphologic observations with extensive nucleic acid determinations. Specifically, it would be important to utilize ultracentrifugal methods (below and above 78000 g) to separate intracytoplasmic bodies according to size and study their biochemical content by modern nucleic acid-based analyses. More importantly, it would be worthwhile to determine whether bodies of different sizes generate colonies in culture (L-forms or cell wall-containing bacteria). It seems possible that these particles might be capable of maturation within the vesicle of the mother form or even after rupture of the vesicle, as long as they remain in communication with (attached to) the mother form. Could these bodies be synonymous with some of the smaller elementary corpuscles described by Dienes et al. (Dienes 1968, Dienes and Bullivant 1968, Dienes and Madoff 1968, Dienes et al. 1968) as short chains and fragments of filaments in close association with large forms? Certain similarities are obvious when one considers the apparent pinching-off of these bodies from long, irregularly constricted filaments. Another consideration involves reversion. Certain morphological similarities are apparent between the reverting L-forms shown by Schonfeld and DeBruijn (1973) and those observed in our studies. In our research, intermediate forms demonstrated the formation of cell wall in the region of the dense body; it remains to be clarified, however, whether the forming ordinary bacterium separates completely from the L-form or whether the remaining portions of the L-form are ultimately incorporated in the wall-containing bacterium. We favor the latter interpretation because we have not observed transformation of autonomous dense bodies into bacterial forms as hypothesized by Schonfeld and DeBruijn (1973).

In regard to that point in the developmental cycle of the L-form when reversion to the parent bacterial form is triggered, it may be assumed from morphological data that the accumulation of electron-opaque material and the formation of mesosome-like structures are synonymous with aging and death, as well as prerequisites for reversion. Therefore, reversion, as well as aging, with subsequent death of the L-form, may be the effect of a common cause, namely, the depletion of available nutrients and the accumulation of toxic products in the growth medium. Reversion of the L-form may aptly illustrate one of the immutable laws of nature: an organism, when faced with unfavorable environmental conditions, must adapt or die. This may raise the question of how a compromised L-form mobilizes the energy necessary for reversion to the wall-containing bacterial phase. It must be emphasized that our hypothesis is based on morphological studies. Further biochemical and physiological experiments such as those of King and Gooder (1970), as well as corollary data involving quantitative growth curves similar to those of Wyrick and Rogers (1973), would help to support our argument, since this would permit a relationship of culture growth phase and age to morphology. Elucidation of the question whether some elementary bodies mature into "stem cells" would aid in correlation of our hypothesized developmental cycle for the streptococcal L-form in pure culture with that of persisting bacterial bodies in human cells. We further hypothesized that these small dense bodies observed in pure cultures of streptococcal L-forms could be the persisting agents of bacterial infection in patients with infectious diseases that fail to yield bacteria on culture of clinical specimens. Since making those initial observations, we have been successful in proving that dormant bacteria (often in the form of 0.22 μ filterable bodies) exist in infected human tissues and body fluids. We have demonstrated clinical relevance for several diseases of the urinary tract and have clearly established that cell wall-defective bacteria can be induced within a suitable host; that they can survive and persist in a latent state within the host; and that they can induce pathologic responses compatible with disease.

Clinical Relevance

The clinical examples that follow demonstrate the persistence of dormant bacterial forms in infectious diseases of the urinary tract. When efforts are made to employ specialized laboratory methods including modern, molecular techniques, a definitive bacteriologic diagnosis can be made that is clinically useful. If appropriate antimicrobial therapy is instituted, the patient's dormant infection can be eradicated.

Dormant Bacteria as a Cause of Idiopathic Hematuria

Often the clinical course in idiopathic hematuria is highly suspicious for urinary tract infection. However, negative urine cultures and the absence of ordinary bacteria and white blood cells in the urine sediment as well as failure of therapeutic approaches should lead to extensive urological and laboratory investigation. A classic example of the significance of dormant bacteria in idiopathic hematuria is illustrated by the following case (Domingue et al. 1993): A 22-year-old woman had recurrent respiratory and genitourinary infections requiring antimicrobial therapy for almost a year before becoming pregnant with a right fallopian ectopic pregnancy. Two weeks following surgery she had low grade fever, severe headaches, and gross hematuria. No edema or hypertension was noted and the urine was free of red blood cell casts. She was treated with cephalexin (a cephalosporin) for probable urinary tract infection. Because she failed to respond, she was referred to a urology specialist about a month after the onset of hematuria. During the next 6 weeks a number of diagnostic studies were performed and all were unremarkable except that cystoscopy revealed bloody urine coming only from the right ureter. Persistent microscopic hematuria and intermittent gross hematuria were noted; however, no bacteria, white blood cells, or casts were seen. The patient received antimicrobials periodically but continued to have low grade fever and severe headaches. Two months after the onset of hematuria, a series of urine cultures were submitted during a 3 week period for special L-form as well as routine cultures. The L-form cultures were consistently positive for *Streptococcus agalactiae* and another Gram-positive coccus, tentatively identified as *Staphylococcus haemolyticus*, but later reclassified as *Enterococcus faecalis* when it was definitely determined that the catalase test was truly negative for the isolate and that it had all other characteristics of *Enterococcus*. These two organisms had in common only sensitivity to nitrofurantoin. Routine urine cultures were negative after 24-48 h incubation, although a few yielded small numbers of the same organisms if cultured for a prolonged period (> 72 h). Electron microscopy of urine collected directly from the right ureter revealed cell wall-aberrant bacteria in various states of reversion to wall-containing bacteria. The patient was started on a 6-week course of nitrofurantoin 10 weeks after the onset of hematuria. Within 4 days the hematuria resolved and it has not recurred at 3-year follow-up. Two weeks after nitrofurantoin was discontinued, the urine yielded no growth on routine and L-form cultures. Clinical status improved gradually during the next few months. During the subsequent years the patient remained well, delivered a normal baby and has not had a recurrence of hematuria.

We concluded that the patient had a dormant bacterial infection due to persisting bacterial variants. The pathology that can be produced by dormant bacteria cannot be appreciated unless clinical laboratories provide the methodology for their detection. Cell wall-defective bacteria can be so dysmorphic as to resemble debris when viewed by bright light microscopy. Oil immersion phase microscopy of a wet preparation of the urine sediment allows their recognition; such a specimen stained with the fluorescent dye, acridine orange, and viewed with ultraviolet light microscopy will reveal the content of nucleic acids. Urine sediment, heat fixed on a slide and Gram-stained, is a routine technique for identifying bacteria; fragile cell wall deficient/defective bacteria are totally disrupted by such a method. The detection of aberrant bacterial bodies in blood is best accomplished by separating the red blood cells from serum, washing them with an isotonic electrolyte solution, then quickly lysing with hypotonic saline; the lysate should then be immediately cultured on enriched media and under conditions suitable for growth of osmotically fragile forms. This technique tends to capture the microorganisms adherent to the red blood cell membrane, free of the growth inhibiting substances in serum.

The etiology of the illness of our patient is not rare or unique. Many idiopathic collagen disorders could have a related pathogenesis. Confirmation of this dynamic is only possible if it is considered and clinical laboratories are designed to recognize it. Many examples exist of persistent infections that produce only episodic illness. Some possible mechanisms for this activation are the transient immune dysfunction that can be brought about by severe physical or emotional stress. In addition, the introduction of a co-infecting symbiotic organism can adversely alter the relationship of host and microbe. Normal microbiota is a statistical concept that derives from the immune competence of most of the population. It is unwise to dismiss the pathogenic capacities of any microbe in a patient with a mysterious illness.

Genesis of the Urinary Oval Fat Body by Intracellular Lipid-Absorbing Dormant Bacterial Bodies in Nephrotic Syndrome

Since 1975, the urine sediments of numerous children with nephrotic syndrome have been serially examined by oil immersion phase microscopy during periods of relapse and remission (Woody et al. 1983). Most of these children were steroid-responsive and therefore had no biopsy. One patient whose disease was steroid-resistant had a renal biopsy showing focal and segmental hyalinosis and sclerosis. Except for persistent depression of serum IgG and IgA, all clinical laboratory disturbances of the steroid responsive patients returned to normal during periods of remission. None of the patients' urine showed conventional microorganisms, pyuria or hematuria. During periods of relapse, the urine sediment of nephrotic patients was consistently seen to contain large numbers of aberrant bacterial forms when viewed by oil immersion phase microscopy. These tended to aggregate in what appeared to be a mucinous matrix; during periods of remission, few to no organisms were seen. The organisms were usually 1-2 μ in diameter; most had a dense spherical core somewhat separated from an enclosing plasma membrane. Occasionally, a small dense body about 0.2 μ in diameter lay on this membrane. When stained with acridine orange and viewed by ultraviolet light, this membrane-associated body displayed the intense yellow-green fluorescence characteristic of DNA; bacterial variants seen with renal epithelial cells also had a yellow-green fluorescence when stained with acridine orange. When urine sediment was sealed under a coverslip and observed for a period of 1-2 days, this membrane-associated body appeared to be the origin of a filamentous transformation of the organism. When hyperlipidemia accompanied relapse, many organisms were seen to contain lipid that partially or completely obscured their cores.

In the early phase of relapse, large, frequently multinucleate, renal epithelial cells that contained various numbers of organisms were often present in the urine sediment. When the intracellular organisms were few in number, they tended to lie near the cell nucleus, but in larger cells the cytoplasm was usually crowded with bacterial variants. Like those organisms free in urine, many were obscured by absorbed lipid. Cells with large populations of organisms were seen to accumulate astonishing amounts of lipid. If the urine sediment was observed under a sealed coverslip for 1-2 days, this lipid was sometimes seen to escape from the cell and lie around it in large droplets; the organisms filling the cell were then revealed.

Electron microscopy further defined the structure of the bacterial variants, showing a cytoplasmic membrane but no cell wall. Some larger bacterial variants contained more than one body of diverse size and density. Vacuoles containing an elementary body were occasionally seen within some of the organisms. Swelling and autolysis appeared to cause some distortion of the urinary forms; an undistorted organism, found in renal biopsy, gave a sharper view of the distribution of trilaminar membranes. Our observations allow us to conclude that the lipid is being absorbed by intracellular bacterial variants rather than, as is the current concept, by intrinsic cellular mechanisms. The presence of these bacterial variants has been shown by acridine orange stains for nucleic acids and by electron microscopy.

Buried Bacterial Variants in Urinary Casts and Renal Epithelial Cells

Urinary casts are composed of a matrix of mucoprotein secreted into the renal tubules. Cells, cellular debris, and plasma proteins that are filtered through injured glomeruli or shed into the tubules become entrapped in this matrix. In chronic renal disease, increased morphologic and functional heterogeneity of nephron units occurs; therefore, casts which reflect the pathology of the nephron from which they emerge can vary markedly in appearance. Granules in casts were once thought to be by-products of degenerate cells. More recently, evidence based on immunofluorescent tagging suggests that they are aggregates of serum proteins. Reconciliation of this evidence with the concept that granules in casts can contain revertible bacterial variants is allowed if the organism is viewed as serving as a nidus for the aggregation. The size, character, and uneven distribution of granules in casts is better explained by this view than by the view that granules result from a general precipitation of contained serum proteins. Since we have demonstrated that blood which is sterile by conventional culture methods probably always contains some form of bacterial variants, their presence in glomerular filtrate, and therefore in casts, is predictable.

Several investigators have reported the presence of cryptic bacterial forms when blood and other body fluids are cultured in ordinary media for prolonged periods.

Lysis of erythrocytes, with release of adsorbed microorganisms, is possibly responsible for such growth. We have shown that positive cultures can be obtained more frequently, and within a relatively short period, by using a rich variant medium and lysing the blood with hypotonic solutions beforehand. These results are not altered when the lysate is filtered through a 0.22 μ filter. This suggests that organisms adsorbed on the erythrocyte are in an elemental form and that some of them have the essential biochemical machinery to revert to the parent organism if nutrients and cultural conditions are optimal.

During the 2.5 years before the death of a 6.5-year-old female with renal Fanconi's syndrome, ten blood cultures were made (Domingue et al. 1979). There was no growth in routine culture media, but when media designed for the growth of bacterial variants were used, large spheroidal forms, measuring 3-6 μ in diameter developed in all the cultures within a week. Six of the positive variant cultures reverted to classic Gram-positive coccal forms in less than a month. When sealed under a coverslip and observed for several days by phase microscopy, some of the large spheroidal bodies could be seen to extrude mycelial-like filaments. Inspection of granules in the urinary casts by phase microscopy showed them to be from 2-8 μ in diameter and to contain a core. Occasionally, a 10-12 μ laminated granule could be seen. In urine specimens under sealed coverslips, granules in some of the casts became transformed, within 2-3 days, into mycelial-like filaments, while granules in other casts extruded large segmenting filaments that became classic streptococcal-like forms.

The presence of cystic bodies within many of the renal pelvic epithelial cells overlying the medullary pyramids was an important finding. These cysts contained bodies measuring 2-8 μ in diameter, and the cysts enlarged and distorted the cells that they occupied. Many of these bodies were attached to the cyst membrane. With Brown and Brenn's tissue stain, the core of these intracystic bodies had the same differential staining as the granules in the casts. Involved cells showed variations of staining with Jones' methenamine silver stain. Some cells were filled with tiny bodies, about 1 μ in diameter, that stained intensely with silver, while the cystic structures within other cells contained large, pink-staining bodies, some with an inner silver-staining lamination. By electron microscopy, cysts with a unit membrane were seen within some renal pelvic epithelial cells. The plasmalemma of the cysts, by a process of budding, seemed to be the source of 3-80 nm intracystic bodies with extraneous material attached to their outer surfaces. An amorphous, dense substance, located in the central portion of the cysts, contained many filamentous structures. Indirect immunofluorescent studies on frozen sections showed fluorescence of casts with serum derived from guinea pigs immunized with bacterial variants. No fluorescence of casts resulted from application of serum from nonimmunized guinea pigs or from immunized guinea pig antiserum preabsorbed with cultured variants. The variants cultured from the patient's blood fluoresced strongly with immunized guinea pig antiserum. This serologic evidence confirms the antigenic relationship between the bacterial variants demonstrated in casts and those isolated from the patient's blood. Granules in the urinary casts from this patient with renal Fanconi's syndrome were seen to evolve into streptococcal-like forms. The close morphologic and histochemical resemblance of these granules to bodies seen within the epithelial cells of the renal collecting system and in hepatocytes strongly suggests that they are identical and are, in fact, bacterial variant forms, whose parent organism is a Gram-positive coccus. The patient's lysed blood, when cultured, produced

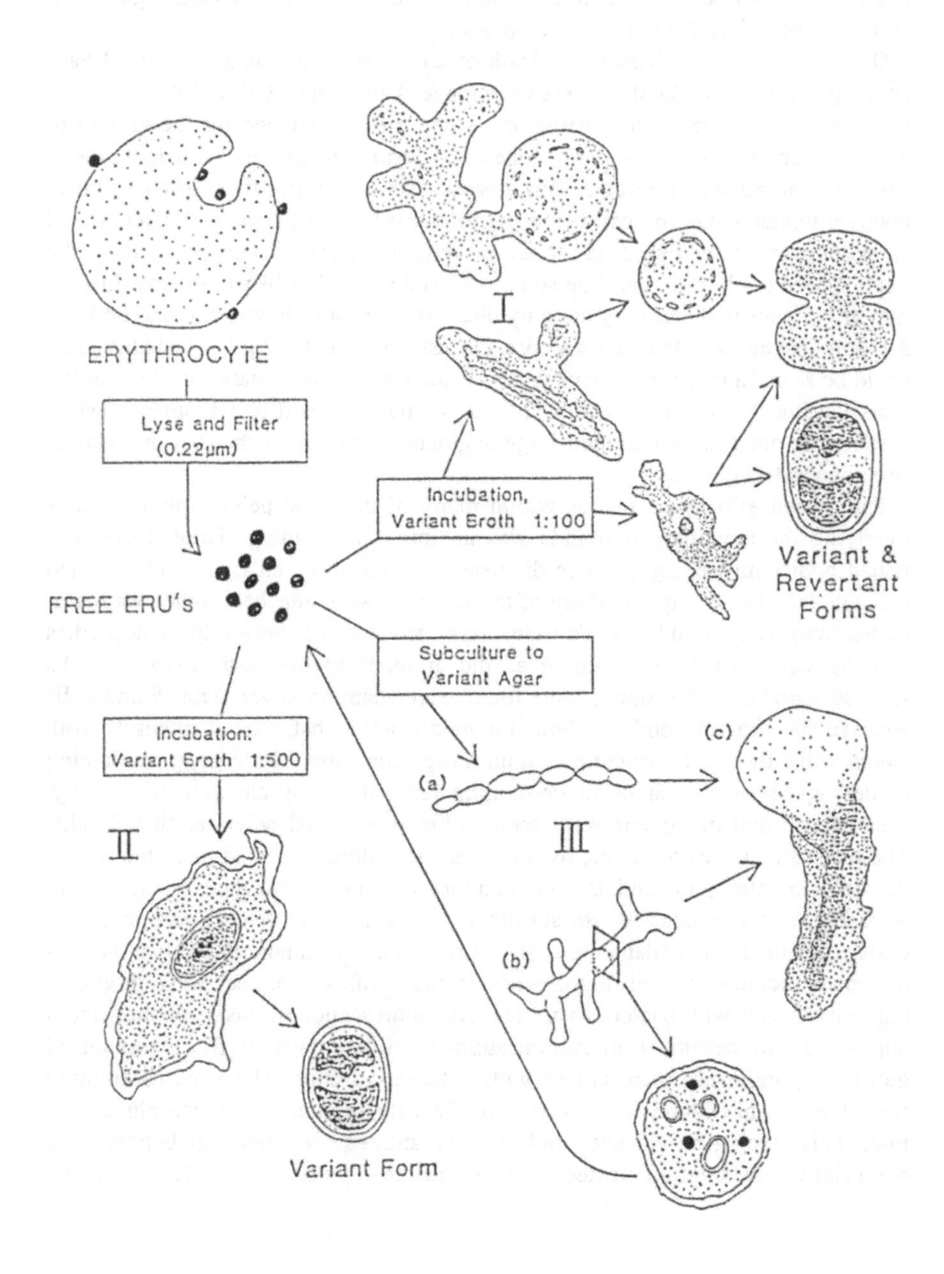

identical forms, which also reverted to Gram-positive cocci. Therefore, an equation seems fulfilled between blood, tissue, and urinary forms. These findings are comparable with our earlier work described with human embryonic kidney fibroblasts infected with cell wall deficient enterococci, demonstrating large vesiculated variants containing dense bodies within the fibroblasts (Green et al. 1974a). These data support our belief that the cystic inclusions in our patient's renal epithelial cells are bacterial variants. The absence of tissue reaction to these cysts is also consistent with the tissue culture model.

This chronically ill child appears to have amplified a process of cryptic parasitization with bacterial variants that probably occurs in many nephropathies. In such patients, release of antigens and toxins (Hryniewicz 1976, Smith 1971, Horne et al. 1977, Rodwell et al. 1972) by bacterial variants could lead to mesangial and basement membrane deposits, as well as to occlusive reactions in the renal microcirculation.

◀ **Fig. 1.** Proposed interrelationships and hypothesized developmental cycle of isolated filterable forms
We hypothesize, from our cultural and morphological data, that elementary reproductive units (ERU) adherent to erythrocytes are freed upon lysis, and pass through a 0.22 μ filter. Depending upon the conditions of culture, three morphologically distinct groups of forms are observed.
Lysate cultured in variant broth at 1:100 dilution yields a population of large spheroid and elongate forms (group I). Also observed are small round bodies, thought to represent ERUs, entrapped within an amorphous flocculent material, probably lysate debris. It is hypothesized that finely granular, L-form spheroids lacking cell walls arise from, or in association with, these large forms; such L-forms may also arise directly from ERUs. The granular spheroids then give rise, through a complex series of intermediate forms, to revertant bacterial organisms with thin, but distinct, cell walls.
Incubation of lysate in variant broth at 1:500 dilution yields variant bacterial forms with thin cell walls in association with flocculent material (group II). Those forms described in group I are rarely, if ever, observed in 1:500 cultures. In contrast also to reversion in group I cultures, which may take many months, the appearance of revertants within group II cultures is rapid, and occurs frequently after as little as 24-48 h of incubation.
Subculture of group II forms onto variant agar yields a third morphologically distinct group of forms (group III). These include: elongate cocci (a); branching forms (b); and intermediate forms (c) (here greatly enlarged). Whereas coccal forms exhibit a thin to absent cell wall, the branching forms and the branching portion of intermediate forms exhibit a thick, though electron lucent cell wall, or coat. Within branching forms from older cultures there is observed a variety of dense bodies and vesicles (see enlargement (b)) which are hypothesized as possessing reproductive potential as ERUs and capable of reinitiating the cycle under appropriate culture conditions

Conclusions

Our fine structural studies have documented the growth on variant bacterial media of morphologically heterogenous bacterial forms derived from blood lysates, urine, and renal tissue. We have shown in several experimental studies that there are immunologic and bacteriologic interrelationships between these structures. Additionally, the morphologic associations and patterns of growth permit the formulation of an hypothesis concerning the life cycle of forms in this system (Fig. 1). It is clear that the erythrocyte is a key element in this cycle: the incubation of blood plasma in enriched broth usually yields no growth. Our observation of dense bodies within invaginations of the erythrocyte cell surface and upon the plasmalemma suggests that these may represent the minimal reproductive units which pass the 0.22 μ filter used in the preparation of the filtered lysate for culture. Most ranged in size from 0.17 to 0.20 μ in diameter, and would thus pass into the lysate. Once within a culture medium of variant broth, these particles are surrounded by a thick cell coat or capsule. Some of this material may be of lysate origin (proteins perhaps including hemoglobin), as suggested by Nelson (Nelson 1958a,b), or may at least in part be synthesized by the developing organism. Once free of the coating material, these forms are recognizable as bacterial variants of diverse morphology. Many of the variants observed were not dissimilar to those described in studies of L-forms of bacteria (Green et al. 1974b). The routine presence of both variants and organisms of classical bacterial morphology, together with intermediate forms within reverting cultures in clinical specimens, strongly suggests the origin of these classical bacteria from these variant forms. The subculture of the variants arising within the variant broth to variant agar has provided an opportunity to document a cultural and morphologic lability among coccal and filamentous forms. Again, intermediate forms between these two morphologic types strongly suggest their origin from a common source, and their interconvertability. Finally, the production of both vesicles and small dense granules within short filamentous forms suggests that these may represent, upon their release from the parent cell, a population of elementary reproductive units capable of reinitiating the developmental cycle (Fig. l), or of associating with erythrocytes in blood.

It is acknowledged that we have created an artificial situation for rapid release of these organisms by exposing whole cells to osmotic lysis which allows for subsequent growth in vitro of the bacterial variants in an enriched medium. Nevertheless, we believe that an analogous situation does exist in vivo, as exemplified by the clinical findings previously described in this manuscript. It is theorized that while the organisms are cell-associated, there is probably no immunologic reaction to the organisms. Obviously, various host environmental factors could affect the cellular multiplication of these forms. Overloading of host cells with organisms which slowly replicate would very likely lead, with time, to lysis of whole cells and release of the elementary bacterial forms. If external conditions (after whole cells lysis) are conducive to maintaining the viability of these elementary bacterial particles, they conceivably could mobilize the necessary energy needed to partially or completely convert to ordinary wall-containing bacteria, or remain in a stabilized cell wall deficient state. All of these events could be of pathologic consequence to the host. In reverting partially or completely, overt signs of disease may appear. During the transformation from a minimal reproductive unit to a transitional or fully reverted classical form, there may be synthesis of toxins, changes in antigenic structure and formation of physiologic by-products that could be pathologic to the host tissues. The variant forms per se might be equally as detrimental to host tissues, but the effect may be more subtle and hence not as readily recognizable histopathologically. More importantly, the immunologic event(s) triggered by: release of bacterial antigens (following lysis of parasitized whole cells), microbially altered host-cell antigens, or newly created bacterial-host cell antigens could precipitate a variety of immunopathologic conditions including autoimmune-like disease.

It is hoped that additional studies, including immunologic and modern nucleic acid technologies for characterization of the bacterial forms described in our studies, will contribute to the substantiation or modification of our proposed developmental cycle constructed on the basis of morphologic and cultural evidence. It is predicted that a better understanding of the dynamics of the host-pathogen interaction in bacterial persistence would most likely lead to further knowledge of the relationship of these forms to disease.

Acknowledgments. I am grateful to Drs. Mary T. Green, Paul F. Heidger, Jr., and Hannah B. Woody for their collaboration and contributions to various phases of the research described. Sincere appreciation is expressed to Kamma C. Pontoppidan and Liset G. Human for excellent technical assistance. The studies reported in this chapter were supported by The Hume Research Fund, Department of Urology, Tulane University School of Medicine, New Orleans, Louisiana, USA.

References

Bibel DJ, Lawson JW (1972) Scanning electron microscopy of L-phase streptococci. II. Growth in broth and upon Millipore filters. Can J Microbial 18: 1179-1184

Coussons RT, Cole RM (1968) The size and replicative capacities of small bodies of group A *Streptococcus* L-forms. pp 327-331. In: R Caravano (ed) Current research on group A *Streptococcus*. Excerpta Medica Foundation, New York

Dienes L (1968) Morphology and reproductive processes of bacteria with defective cell walls. p 74-93 In: LB Guze (ed) Microbial protoplasts, spheroplasts and L-forms. The Williams and Wilkins Co., Baltimore

Dienes L, Bullivant S (1968) Morphology and reproductive processes of the L-forms of bacteria. II. Comparative study of L-forms and mycoplasma with the electron microscope. J Bacteriol 95: 672-687

Dienes L, Madoff S (1968) Morphology of the L-forms of group A *Streptococcus* and other bacterial species. p 332-334 In: R Caravano (ed) Current research on group A *Streptococcus*. Excerpta Medica Foundation, New York

Dienes L, Madoff S, Bullivant S (1968) Study of L-forms as seen in thin sections with the electron microscope. p 342-345 In: R Caravano (ed) Current research on group A *Streptococcus*. Excerpta Medica Foundation, New York

Domingue GJ, Woody WB (1997) Bacterial persistence and expression of disease. Clinical Microbiology Reviews 10: 320-344

Domingue GJ, Thomas R, Walters F, Seirano A, Heidger PM Jr (1993) Cell wall-deficient bacteria as a cause of idiopathic hematuria. J Urol 150: 483-485

Domingue GJ, Woody HB, Farris BK, Schlegel JU (1979) Bacterial variants in urinary casts and renal epithelial cells. Arch Int Med 139: 1355-1360

Fass RJ (1973) Morphology and ultrastructure of staphylococcal L colonies: light, scanning, and transmission electron microscopy. J Bacteriol 113: 1049-1053

Green MT, Heidger PM Jr, Domingue G (1974a) Demonstration of the phenomena of microbial persistence and reversion with bacterial L-forms in human embryonic kidney cells. Inf and Immun 10: 889-914

Green MT, Heidger PM Jr, Domingue G (1974b) Proposed reproductive cycle for a relatively stable L-phase variant of *Streptococcus faecalis*. Inf and Immun 10: 915-927

Horne D, Hakenbeck R, Tomasz A (1977) Secretion of lipids induced by inhibition of peptidoglycan synthesis in streptococci. J Bacteriol 132: 704-171

Hryniewicz W (1976) Streptococcal L-forms and some of their properties. In: Roux J (ed) Spheroplasts, Protoplasts and L-forms of Bacteria. Paris, INSERM pp 39-55

King JR, Gooder H (1970) Reversion to the streptococcal state of enterococcal protoplasts, spheroplasts, and L-forms. J Bacteriol 103: 692-696

Lynn RJ, Haller GJ (1968) Bacterial L-forms as immunogenic agents. In: Guze LB (ed) Microbial Protoplasts, Spheroplasts and L-forms. Baltimore, Williams & Wilkins Co., pp 270-278

Maxted WR (1968) Observations on the induction, growth and nature of streptococcal L-forms. P.320-326 In: R Caravano (ed) Current research on group A *Streptococcus*. Excerpta Medica Foundation, New York

Nelson EL (1958a) The development in-vitro of particles from cytoplasm. I. Observations on particle development in bacteriologic media. J Exp Med 107: 755-768

Nelson EL (1958b) The development in-vitro of particles from cytoplasm. II. Particles from hemoglobin and deoxyribonucleic acid. J Exp Med 107: 769-782

Rodwell AW, Peterson JE, Rodwell ES (1972) Macromolecular synthesis and growth of mycoplasmas. In: Pirie NW (ed) Pathogenic Mycoplasmas. Amsterdam, Associated Scientific Publishers, pp 123-139

Schonfeld JK, DeBruijn WC (1973) Ultrastructure of the intermediate stages in the reverting L-phase organisms of *Staphylococcus aureus* and *Streptococcus faecalis*. J Gen Microbiol 77: 261-271

Smith PF (1971) The biology of mycoplasmas. New York, Academic Press Inc. pp 16-34

Weibull C, Mohri T, Afzelius BA (1965) The morphology and fine structure of small particles present in cultures of a Proteus L-form. J Ultrastruct Res 12: 81-91

Woody HB, Walker PD, Domingue GJ (1983) Genesis of the urinary oval fat body by intracellular lipid-absorbing cell wall-deficient bacteria. In: GJ Domingue (ed) Cell Wall-Deficient Bacteria: Basic Principles and Clinical Significance. Addeson Wesley, Reading, Massachusetts

Wyrick PB, Rogers HJ (1973) Isolation and characterization of cell wall-defective variants of Bacillus subtilis and Bacillus licheniformis. J Bacterio 116: 456-465

List of Authors and Participants (bold-typed)

Photograph of many of the participants and a Puzzle: Who is Who?

In the spotlight are at the front (centre, from left to right):

M. Sugiura (Miescher-Ishida-Prize Winner), K.W. Jeon (ISE President), M. Ishida (ISE Honorary Member)

Subject Index

D

P

Q

R

S

MIX
Papier aus verantwortungsvollen Quellen
Paper from responsible sources
FSC® C105338

Printed by Books on Demand, Germany